Basic Engineering Thermodynamics

A Venkatesh
Professor Emeritus, Mechanical Engineering Department
East West Institute of Technology,
Bangalore, India
(Formerly Professor of Mechanical Engineering, IIT Madras)

Universities Press

Universities Press (India) Private Limited

Registered Office
3-6-747/1/A & 3-6-754/1, Himayatnagar,
Hyderabad 500 029 (A.P), India
e-mail: *www.universitiespress.com*

Distributed by
Orient Blackswan Private Limited

Registered Office
3-6-752, Himayatnagar, Hyderabad 500 029 (A.P), India

Other Offices
Bangalore / Bhopal / Bhubaneshwar / Chennai
Ernakulam / Guwahati / Hyderabad / Jaipur / Kolkata
Lucknow / Mumbai / New Delhi / Patna

First published 2007
Reprint 2009, 2011, 2013

ISBN: 978 81 7371 587 7

Set in Times New Roman 11 Points by
Sigma publishing Services, Chennai 600 033

Printed at
Charita Impressions
Hyderabad 500 020

Published by
Universities Press (India) Private Limited
3-6-747/1/A & 3-6-754/1, Himayatnagar,
Hyderabad 500 029 (A.P), India

Foreword

Professor A Venkatesh has taught both Basic Thermodynamics and Applied Thermodynamics courses for over three decades at IIT Madras, and has been very popular with students. His research interests lie in solar energy utilization and refrigeration systems, both of which have thermodynamics as their base.

It has been observed that students encounter thermodynamics in three stages: during their first brush with the subject, they do not really understand it; in the second stage, they understand bits and pieces of it and in the third stage, they know they do not completely understand it but by then are used to it and it does not bother them anymore!

There are reasons for this discomfort. Thermodynamics involves very little mathematics: you can get by with algebra and calculus. It employs logic and verbal arguments. There are more thought experiments than actual ones. It deals essentially with ideal situations, systems and processes: Thermotopia. Historically, it is interesting that thermodynamic theory came after the invention of devices and artefacts: the steam engine and the second law of thermodynamics are all about our inability to achieve something, and have been referred to as laws of impotence.

There are some who believe that we have reached a stage of saturation in developments based on thermodynamics; however, time and again, it opens up new vistas. For example, irreversible thermodynamics, second law analysis, availability and energy analysis, pinch analysis for heat exchangers, information theory, thermodynamics of life processes, chaos theory, catastrophe theory, thermodynamics of rate processes, etc.

In this book, Professor Venkatesh has provided a thorough, comprehensive and traditional treatment of basic thermodynamics, which is of direct interest and relevance to students of Mechanical, Chemical and Aeronautical Engineering. He has included a large number of illustrative solved problems as well as a number of problems to be solved, and objective questions.

I am sure that students and teachers will find this book to be a one-stop source of teaching-learning material suitable for academic requirements.

Professor R Natarajan

Former Chairman, All India Council for Technical Education
Former Director, Indian Institute of Technology Madras
Vice President, Indian National Academy of Engineering

This book is dedicated to
my wife, Janaki Venkatesh,
and to my two children, Malini and Bimal,
without whose love, affection and support
I would not be what I am today.

Preface

After teaching the subject for over three decades, first as Professor of Mechanical Engineering at IIT Madras, then continuing, since my retirement in 2001, at a couple of other institutes around Bangalore, I have discovered that there are two kinds of books on thermodynamics in the market.

On the one hand there are several good books on the subject written by reputed authors. Invariably, these books are voluminous, containing more information than required by beginners who wish to learn just the basics. On the other, there are books that oversimplify the subject, under the pretext that the information contained in them is sufficient to score well in examinations.

The second category of books does a disservice to the student as it does not develop in the student an intimate understanding of the subject which would benefit him or her. The fact that thermodynamics is an abstract subject makes it more difficult for students to understand its concepts by reading these books because very brief explanations are provided in them. What students need are elaborate explanations and a large number of exercises and illustrative problems to make them understand all difficult concepts.

The fact that there is a need for a book that bridges the gap between advanced but voluminous books and simple but inadequate books is what prompted me to write this book. This is reflected in the adequate detail that has been given while keeping in mind the fact that the reader needs this book to provide a basic and introductory course of thermodynamics.

Teaching basic thermodynamics to first-year students as well as advanced concepts to other undergraduate and postgraduate students has given me a good understanding of how the subject matter should be presented so that even a beginner will find the material useful and beneficial. I have used my experience to select only those topics that, in essence, form the concepts of basic thermodynamics. I have attempted to explain these difficult concepts in such a manner that it will be easy for the student even when reading it for the first time. Examples and illustrative problems have been provided exhaustively in each chapter to solidify the concepts.

Chapters 1–8 focus primarily on the basics of the subject such as the concepts of system, control volume, properties of working substances in general, the zeroth law of thermodynamics, heat and work interactions, the first law, the second law and its corollaries and entropy. In Chapters 9, 10 and 11 the concepts explained thus far have been applied to analyse problems involving particular types of working substances such as ideal gases, ideal gas mixtures, steam and real gases, none of which undergo changes in chemical composition or aggregation during a given process. On the other hand, Chapter 12 gives a flavour of combustion processes during which the composition and chemical aggregation of the working substance do change.

In several books on the subject, evaluation of the properties of the fluids precedes the basics of the subject and the analysis of processes; however, like some other authors, I have followed the opposite. Although both methods have their own advantages I have immense faith in teaching the subject in the latter manner.

During the course of my career, I have been involved in the preparation of numerous examination papers on the subject for various universities and competitive examinations around the country. The experience gained from this endeavour has made me prepare numerous objective type questions which have been given at the end of the book to help students who are preparing for various competitive examinations.

In preparing this book, I have referred to several good books written on the subject. I have listed these books at the end. I wish to acknowledge my indebtedness to these books which have been a source of inspiration to me.

Many people have, directly or indirectly, helped me produce this book. I acknowledge all their help and wish to thank them profusely. It also gives me pleasure to acknowledge the excellent support I received from Mr Madhu Reddy and other members of Universities Press at every stage, starting with the scrutiny of the manuscript to the printing of the book. In particular, I am grateful to the team comprising Ms Shubashree Desikan, Ms Bibi Xavier, copyeditor, Mr Henrick Rose, Mr Charles and Mr A Baskar of the production department for their efforts in bringing out the book in time. The book would not have turned out the way it did without their untiring efforts and I would like to thank each person.

I sincerely believe that this book, *Basic Engineering Thermodynamics*, will be of immense help to students and teachers of all engineering institutions and that the book will enjoy wide acceptance.

Constructive criticism and suggestion for improvement from the users of this book are welcome.

Contents

FUNDAMENTAL CONCEPTS AND DEFINITIONS 1

1.1 Introduction

Thermodynamics is a science that deals with all aspects of energy conversion, energy exchange and energy saving. It is also concerned with the basics of the study of materials, chemical reactions and plasma. Because of its wide application in many fields, thermodynamics has long been identified as an essential part of engineering curricula in technical institutions the world over. Basic thermodynamics is characterized by a few fundamental definitions and concepts and because of its fundamental nature, it is considered an essential prerequisite for the study of all thermal engineering subjects. In applied thermodynamics, the basics of thermodynamics are applied to the design and analysis of various engineering devices involving energy interactions. Study of thermodynamics, therefore, is of profound importance and an inextricable part of engineering education.

This book deals with the necessary fundamental aspects of thermodynamics.

1.1.1 Thermodynamics

Definition: Thermodynamics is the science of the interactions among systems and the consequent changes experienced by the interacting systems. Thermodynamic interactions occur by exchange of *Heat* and/or *Work* which are, as it will be made clear in due course, energies in transition. Under the influence of heat and/or work the interacting systems undergo changes in their *properties*. Thus, thermodynamics is the science of the study of heat and work interactions and their effects on the *properties* of the interacting *systems*. In short, it can be said that thermodynamics is the science of energy interactions.

1.1.2 Scope of Thermodynamics

Like any other science, thermodynamics also derives its strength from experimental observations. These observations are formulated into the zeroth, first, second and third laws of thermodynamics. In the chapters that follows, these laws are not only explained in detail but are also applied to analyse several happenings in energy transfer and conversion devices of practical importance. A study of the properties of various kinds of substances used in energy conversion devices forms an important part of a course in thermodynamics.

Thermodynamics is concerned only with the states of systems before and after a certain interaction and is insensitive to intermediate ones. Also, thermodynamics has no intrinsic timescale. Suppose an interaction occurs between two systems, the changes in properties of the two systems at the end of the interaction depends only on the magnitudes and directions of the interaction and not on their rates of occurrence. For example, if one unit of heat (or work) is transferred to a system the resulting changes in properties are proportional to this unit of heat (or work) and not to its rate of being transferred. Irrespective of whether one unit of heat (or work) is transferred in one second, one minute or one hour, the changes in properties experienced by the interacting systems remain the same.

1.1.3 Microscopic and Macroscopic Thermodynamics

As discussed earlier, thermodynamics deals with behaviour of matter under external influences. Matter comprises myriads of particles which are its atoms or molecules. These particles are in random motion and collide continuously. The condition of matter at any instant is thus governed by that of the constituent particles. The behaviour of matter can thus be analysed through a study of the conditions of its atoms or molecules. This approach is adapted in microscopic thermodynamics. Microscopic treatment is cumbersome and tedious as it has to take into account the conditions of innumerable particles of the system. However, in statistical thermodynamics this problem is less severe, as it considers a statistical group of particles behaving identically. In macroscopic thermodynamics, which is the subject matter of this book, the study is confined to gross or time-averaged behaviour of a large number of particles which are observable or measurable at the macroscopic level. Macroscopic thermodynamics can only be applied to a continuum. A continuum is a system whose size is more than the mean free path of the molecules. In other words, a system can be treated as a continuum if it were to contain a large number of particles to make the statistical averages meaningful. It is generally agreeable to consider a system as a continuum if its size were to be greater than either 10 times the mean free path or 10 times the mean molecular path.

1.1.4 Engineering Thermodynamics

Engineering thermodynamics is the science that deals with design and analysis of practical energy conversion devices such as power plants that develop shaft power from heat, refrigerators, heat pumps, air conditioners, heat exchangers, nozzles, diffusers etc. It also studies the properties of various substances used in these devices.

1.2 System and Control Volume

1.2.1 System

A system is an identifiable collection of matter on which attention is focused during thermodynamic analysis of a problem. It is recognized as the material enclosed in a boundary called the *system boundary* which can either be real or imaginary. Everything outside the system boundary, with which the system interacts is called the surroundings. During any event, the system boundary can change in size, shape and orientation with respect to a stationary observer but material cannot cross the system boundary in either or both directions. Heat and/or work interactions between a system and its surroundings can occur across the system boundary.

1.2.2 Control Volume

A control volume is a zone in space on which attention is focused during the analysis of a problem. It is identified as the volume inside a real boundary called the *control surface.* Everything outside the control surface with which the control volume interacts is called the *surroundings.* During a certain happening the size, shape and orientation of the control surface cannot change. But, material can cross the control surface in either or both directions. Heat and/or work interactions between a control volume and its surroundings can occur across the control surface.

In some books on the subject a system is also referred to as a 'closed system'. This is probably a misnomer as the definition 'system' already emphasizes that a system is one which is indeed closed with respect to material transfer across its boundary. Similarly, control volume is also referred to as an 'open system'. When the definition of a system clearly describes it as closed with regard to material transfer, it cannot be prefixed with 'open' to give any meaningful alternative.

Schematic representations of system, system boundary, control volume, control surface and surroundings are given in Figures 1.1*a* and *b*.

A few common examples of *system* are (i) a lead acid battery, (ii) the weight on a pressure cooker, (iii) a refrigerator when its door is closed, (iv) a coil spring in an automobile, (v) an electric bulb, (vi) the piston in a reciprocating engine, (vii) air in a football in play, (viii) a satellite orbiting earth, (ix) a gas enclosed in a piston–cylinder device during compression and expansion processes, (x) steam undergoing a cycle in a steam power plant, (xi) a refrigerant in either a refrigerator or a heat pump and (xii) air undergoing a cycle in a closed-cycle gas turbine power plant.

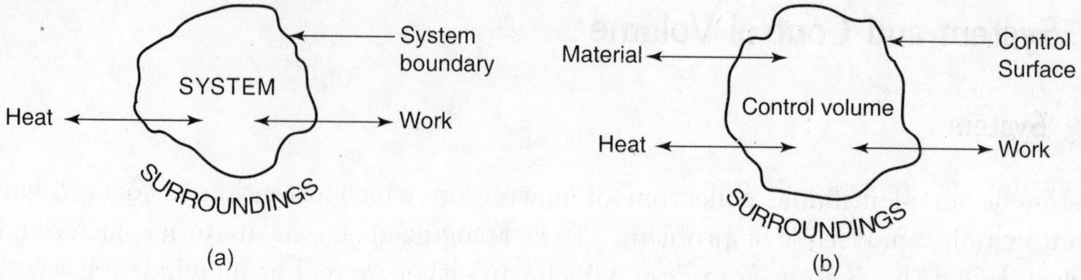

Fig. 1.1*a* System; *b* control volume.

(***Note*:** Before any thermodynamic analysis is taken up it is essential that either a system or a control volume is first identified and the analysis confined strictly to either of them.)

A few examples of *control volume* are (i) turbines, (ii) compressors, (iii) heat exchangers, (iv) nozzles, (v) diffusers, (vi) open-cycle gas turbine power plant, (vii) internal combustion engine, (viii) a window air conditioner, (ix) a pressure cooker, (x) a domestic water pump lifting water from a sump to an overhead tank, (xi) a ceiling fan, (xii) a boiler in a steam power plant .

1.3 Dimensions and Units

Dimensions are physical quantities. Mass [M], length [L], time [τ] and temperature [T] are the fundamental dimensions. The symbols used for these quantities, given within brackets, denote the corresponding dimension. The dimensions listed above are not exhaustive but are only representative.

Units are arbitrarily selected magnitudes of the dimensions. For example, millimetre (mm), centimetre (cm), metre (m), kilometre (km) and so on are units of length. Similarly, milligram (mg), gram (g), kilogram (kg) and so on are units of mass; second (s), minute (min), hour (h) are units of time and kelvin (K) is the unit of temperature. The SI system recognizes kg, m, s and K as the fundamental units of mass, length, time and temperature, respectively.

1.3.1 Primary Dimensions

A small group of dimensions with the help of which the dimensions of all other physical quantities can be specified are called *primary dimensions*. In the SI system of units the primary group of dimensions consist of mass, length, time and temperature.

1.3.2 Derived Dimensions

There are a number of other dimensions which can be derived in terms of these primary dimensions. These are called *derived dimensions*. A few examples of derived dimension are force [F], velocity [L/τ], acceleration [L/τ^2], density [M/L^3], pressure [F/L^2], work [FL], power [FL/τ] and specific heat $\left[\frac{FL}{MT}\right]$. To obtain consistent units for all derived dimensions it is essential to insert into the dimensional equations the respective fundamental units of [M],[L], [τ] and [T].

1.3.3 Force

The relation among [F], [M], [L] and [τ] is very important. The four are related by Newton's law:

$$F = ma,$$

where m is the mass and a is the acceleration and the corresponding dimensional relation is [F] = [ML/τ^2].

The unit of force is obtained in terms of the fundamental units selected for the primary dimensions, [M], [L] and [τ]. Therefore, as [F] = [ML/τ^2], the unit of force is kg m/s^2. In SI system this unit is called a newton (N). Thus, 1 N = 1 kg m/s^2. A force of 1 newton when acting on a body of 1 kg mass makes it move with an acceleration of one m/s^2. Given in Table 1.1 are a few physical quantities, their dimensions and fundamental units.

Example 1.1

Find the velocity of a body, in m/s, which travels a distance of 80 km in 2 hours.

Table 1.1 Fundamental dimensions, a few derived dimensions and their basic units

Quantity	Dimension	Basic unit
Mass	[M]	kg
Length	[L]	m
Time	[τ]	s
Temperature	[T]	K
Force	[F] = [ML/τ^2]	kg m/s^2 = N (Newton)
Velocity	[L/τ]	m/s
Acceleration	[L/τ^2]	m/s^2
Density	[M/L^3]	kg/m^3
Pressure	[F/L^2]	N/m^2 = Pa (Pascal)
Work	[FL]	Nm = J (Joule)
Power	[FL/τ]	Nm/s = W (Watt)
Specific Heat	$\left[\frac{FL}{MT}\right]$	J/kg K

▶ **Data:**

Distance travelled = 80 km = 80 × 1000 = 8 $\times 10^4$ m, Time of travel = 2h = 2 × 3600 = 7200 s (here, 1000 is the factor that converts km to m and 3600 is the factor that converts h to s).

▶ **Solution:**

When fundamental units are used for the primary dimensions the unit of the derived dimension, velocity, will automatically be in its fundamental unit of m/s^2:

$$\text{Velocity} = \frac{\text{distance travelled}}{\text{time}} = \frac{8 \times 10^4}{7200} = 11.1 \text{ m/s}^2.$$

Example 1.2

Find the force, in kN, acting on a body of mass 120 kg moving with an acceleration of 10 m/s^2.

▶ **Data:**

$m = 120$ kg, $a = 10$ m/s^2.

▶ **Solution:**

$F = ma = 120 \times 10 = 1200$ kg m/s^2 = 1200 N

$$1 \text{ kN} = 1000 \text{ N}$$

Therefore, $$F = \frac{1200}{1000} = 1.2 \text{ kN}.$$

Example 1.3

Find the weight of a body of mass 60 kg at a place where the acceleration due to gravity is 9.8 m/s^2.

▶ **Data:**

$m = 60$ kg and $g = 9.8$ m/s^2.

▶ **Solution:**

Weight, W = force with which the body is attracted towards the centre of the earth.
Therefore, $W = mg = 60 \times 9.8 = 588$ kg m/s^2 = 588 N.

1.4 Thermodynamic Properties

Thermodynamic properties are observable, measurable or calculable characteristics of a system. Some of the commonly encountered properties, with their symbols within parentheses, are pressure (p), volume (V), temperature (t or T), density (ρ), energy (E), enthalpy (H), entropy (S), and specific heats (c_p, c_v). Such properties are classified into two categories, *intensive* and *extensive*.

1.4.1 Intensive Properties

Intensive properties are those that are independent of the mass of a system. A few examples are pressure, temperature and density.

1.4.2 Extensive Properties

Extensive properties are those that are dependent on the mass of a system. Some examples are mass, volume, energy, enthalpy and entropy.

1.4.3 Specific Properties

Specific properties are extensive properties per unit mass. Their values are obtained when the corresponding extensive properties are divided by the mass of the system. Examples of these are specific volume (v), specific energy (e), specific enthalpy (h) and specific entropy (s). While extensive properties are denoted by capital letters, the corresponding specific properties are denoted by the same letters in lower case. In general, if 'Z' is any extensive property of a system of mass 'm', the corresponding specific property 'z' is given by, $z = \dfrac{Z}{m}$. Specific properties can also be considered as intensive properties.

1.4.4 Pressure

For a student of engineering, understanding the concept of pressure is very important and hence an attempt is made here to explain this property in detail. The pressure exerted by a fluid at rest at a given point is the same in all directions and is defined as the normal component of the force exerted per unit area. Thus, the unit of pressure is N/m^2, which is also called a pascal (Pa). The pressure of a fluid is measured by a pressure gauge, the most common among which is either a manometer or a Bourdon pressure gauge. Generally, pressure gauges measure the pressure of a system to which they are connected with respect to the local atmospheric pressure (p_{atm}). In other words, they read the extent to which pressure is either greater than or less than the atmospheric pressure.

1.4.4.1 Absolute Pressure:

The pressure of a system measured with respect to zero datum (absolute vacuum) is called the absolute pressure (p_{abs}).

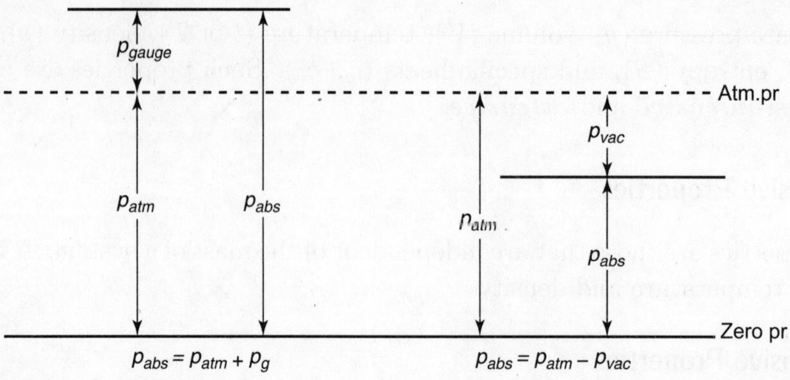

Fig. 1.2 Relations among various pressures.

1.4.4.2 Gauge Pressure:

When the absolute pressure of a system is greater than the atmospheric pressure the difference between the absolute pressure and the atmospheric pressure is called the gauge pressure (p_g) of the system. This is because a gauge connected to the system would read this pressure. Thus,

$$p_g = p_{abs} - p_{atm} \tag{1.1}$$

1.4.4.3 Vacuum Pressure

When the absolute pressure of a system is less than the atmospheric pressure the difference between the atmospheric pressure and the absolute pressure is called the vacuum pressure (p_{vac}). This is because a vacuum gauge connected to the system would read this pressure. The relation among these pressures are, hence:

$$p_{vac} = p_{atm} - p_{abs} \tag{1.2}$$

The above-mentioned relations among various pressures are illustrated in Figure 1.2.

1.4.4.4 Manometer and a Barometer

When a U-tube manometer is used for pressure measurement, one of its limbs is connected to the system whose pressure is to be measured, the other limb is exposed to the atmosphere. When reading gauge pressure, the manometric fluid rises in the column exposed to the atmosphere and when reading vacuum pressure, it descends in the open limb. In either case, the manometer reading, p_g or p_{vac}, is given as the vertical distance between the menisci in the two limbs. If h is this reading, measured in m, the gauge reading, in N/m^2, is given by

$$p_g \quad \text{or} \quad p_{vac} = \rho g h$$

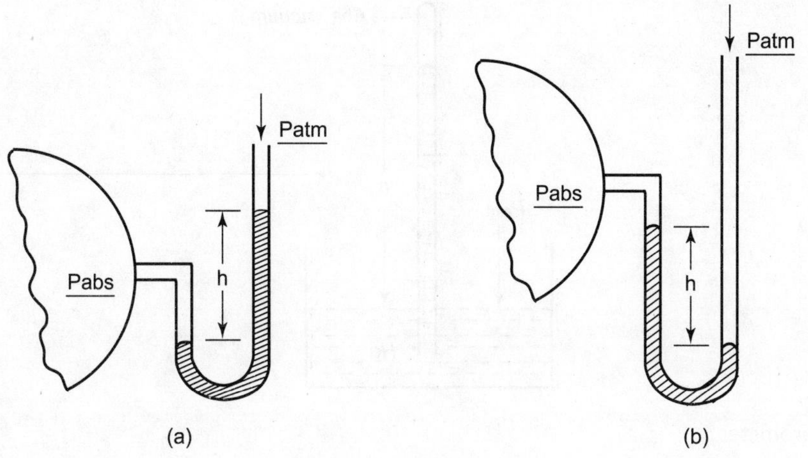

Fig. 1.3 U-tube manometer *a* $P_{abs} > P_{atm}$; *b* $P_{abs} < P_{atm}$.

where ρ is the density of the manometric fluid and g is the acceleration due to gravity. This fact is explained in the following way:

Shown in Figure 1.3*a* is a manometer connected to a system whose pressure is greater than the atmospheric pressure. The manometer reads $h-$ the vertical difference in height between the two menisci. The difference between p_{abs} and p_{atm} is equal to the weight of the manometric fluid in the column, h, divided by the area of cross-section, A, of the manometric tube. Therefore,

$$p_{abs} - p_{atm} = p_g = \frac{\text{weight of the fluid in column '}h\text{'}}{\text{area of the manometric tube}}$$

$$= \text{mass of the fluid in column } h \times \frac{\text{acceleration due to gravity}}{\text{area}}$$

$$= \frac{(Ah\rho) \times g}{A},$$

that is, $p_{abs} - p_{atm} = p_g = \rho g h$ (1.3)

Shown in Figure 1.3*b* is a manometer connected to a system whose pressure is less than the atmospheric pressure. In this case also

$$p_{atm} - p_{abs} = p_{vac} = \rho g h \quad (1.4)$$

p_g or p_{vac} will be obtained in N/m^2 when ρ, the density of the manometric fluid, is in kg/m^3, g, the acceleration due to gravity is in m/s^2 and h the gauge reading, is in m.

An U-tube manometer with one limb exposed to the atmosphere and the other exposed to absolute vacuum reads the atmospheric pressure. Such a manometer is called a barometer. A straight tube can work as a barometer and it is not necessary to use an U-tube.

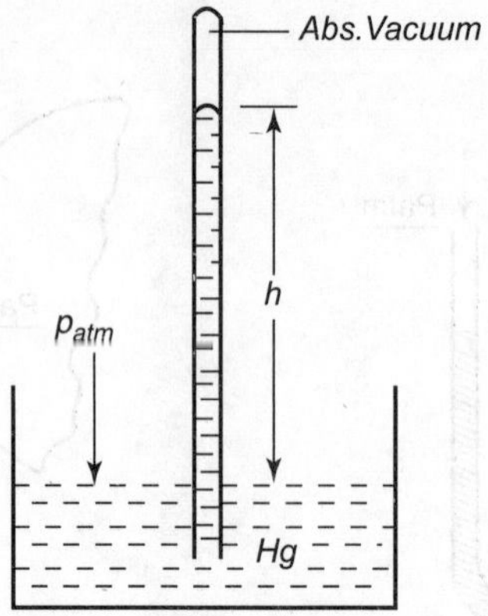

Fig. 1.4 A barometer.

A long glass tube is filled with mercury to the brim. It is then turned upside down keeping the open end temporarily closed, such that mercury does not spill out. The temporarily closed end of the tube is then dipped in a pool of mercury and then opened holding the tube vertically. Some mercury in the tube will slide into the pool and the height of the mercury in the vertical tube stabilizes. This is shown in Figure 1.4. A small length of the tube in the top portion, above the mercury meniscus, which appears empty is under absolute vacuum. The height between the surface of the mercury in the pool and the top meniscus of the mercury in the tube is a measure of atmospheric pressure. This height h, is converted to pressure units through the relation $p_{atm} = \rho g h$. The arrangement shown in Figure 1.4 is similar to a U-tube in which one end is connected to the atmosphere and the other to absolute vacuum.

1.4.4.5 Practical Units of Pressure

A pascal (1 N/m^2) is too small an unit for practical measurements in engineering practice. Hence, several other, bigger units are used. They are listed in Table 1.2 along with their abbreviations and their equivalents in N/m^2.

Example 1.4

A turbine is supplied with steam at a pressure of 12 bar gauge. After expansion, the steam leaves the turbine at a pressure of 75 mm Hg vacuum to enter the condenser. Find the steam pressures, in kPa abs, at the inlet and exit of the turbine. Assume the following: density of mercury = 13.6 g/ml, atmospheric pressure = 760 mm Hg and g = 9.805 m/s^2.

Table 1.2 Various units of pressure

No.	Unit of pressure	Abbreviation	Equivalent N/m^2
1	pascal	Pa	1.0
2	kilopascal	kPa	10^3
3	bar	Bar	10^5
4	atmosphere	Atm	1.01325×10^5
5	megapascal	MPa	10^6

(***Note:*** While mentioning the pressure of a system, besides its magnitude in any unit, it must also be clearly specified whether it is the absolute pressure, gauge pressure or vacuum pressure. In the absence of any such specification the given pressure is assumed to be the absolute pressure of the system.)

▶ **Data:**

At the inlet to the turbine, $p_g = 12 \text{ bar} = 12 \times 10^5 \text{ N/m}^2$. (Note: 1 bar $= 10^5$ N/m^2).

At the exit, $p_{vac} = 75 \text{ mm Hg} = \dfrac{75}{1000} = 0.075$ m Hg (Note: 1000 mm = 1 m).

Atmospheric pressure $= 760 \text{ mm Hg} = \dfrac{760}{1000} = 0.76$ m Hg.

$$\rho_{\text{Hg}} = 13.6 \text{ g/ml} = 13.6 \times \frac{10^{-3}}{10^{-6}} = 13.6 \times 10^3 \text{ kg/m}^3, \quad \text{g} = 9.805 \text{ m/s}^2.$$

▶ **Solution:**

Exit pressure in N/m^2 is given by

$$p_{vac} = \rho g h = (13.6 \times 10^3) \times 9.805 \times 0.075 = 10{,}001 \text{ N/m}^2$$

Also, $$p_{atm} = \rho g h = (13.6 \times 10^3) \times 9.805 \times 0.76 = 101{,}344 \text{ N/m}^2$$

At the inlet, $$p_{abs} = p_{atm} + p_g = 101{,}344 + 12 \times 10^5 = 1{,}301{,}344 \text{ N/m}^2$$

$$= \frac{1{,}301{,}344}{1000} = 1301.344 \text{ kPa}$$

At the exit, $$p_{abs} = p_{atm} - p_{vac} = 101{,}344 - 10{,}001 = 91{,}343 \text{ N/m}^2 = 91.343 \text{ kPa}.$$

1.4.5 Specific Volume

The specific volume, v, of a system is its volume per unit mass and is defined, for a continuum, as follows:

$$v = \lim_{\text{d}V \to \text{d}V^*} \left(\frac{\text{d}V}{\text{d}m} \right)$$

where dV^* is the limiting minimum volume upto which the system can be treated as a continuum. The fundamental unit of specific volume is m^3/kg.

1.4.6 Density

The density of a system is its mass per unit volume. Hence, it is also equal to the reciprocal of its specific volume. Thus, $\rho = \left(\frac{1}{v}\right)$, and its fundamental unit is kg/m^3.

1.4.7. Temperature

The temperature, t or T, is the property that distinguishes between hot and cold bodies. It is measured using either the celsius scale or the thermodynamic temperature scale. The unit of temperature in the former scale is °C (degree celsius) and in the latter scale it is K (kelvin). More details on t and T are available in chapters 3 and 7, respectively. The relation between the two is

$$T = t + 273.15$$

Example 1.5

A cylinder of volume 0.04 m^3 contains air. Given that the mass of air is 50 g, find its specific volume and density.

► **Data:**

$V = 0.04 \text{ m}^3$, $m = 50 \text{ g} = \frac{50}{1000} = 0.05$ kg.

► **Solution:**

$$\text{Specific volume,} \quad v = \frac{V}{m} = \frac{0.04}{0.05} = 0.8 \text{ m}^3/\text{kg}.$$

$$\text{Density,} \quad \rho = \frac{1}{v} = \frac{1}{0.8} = 1.25 \text{ kg/m}^3.$$

Example 1.6

Water boils at 100°C at a pressure of 1 Atm. What is the equivalent of this temperature in the kelvin scale?

► **Data:**

$t = 100°\text{C}$, $T = ?$

► **Solution:**

$T = t + 273.15 = 100 + 273.15 = 373.15$ K.

Example 1.7

A refrigerator cabinet is maintained at a temperature of 263.15 K. What is the corresponding temperature in degree celsius?

▶ **Data:**

$T = 263.15$ K, $t = ?$

▶ **Solution:**

$T = t + 273.15$, therefore, $263.15 = t + 273.15$, or $t = -10°\text{C}$.

1.5 State, Path and Process

1.5.1 Thermodynamic State

The condition of a system at any given instant is called its state. The state of a system is nothing but the totality of its properties. In other words, it is the complete description of a system in terms of its properties.

1.5.2 State Diagram

This is a diagram on cartesian coordinates, with any two independent properties being represented on the x and y axes. This diagram represents the state of a system at different instants.

1.5.3 State Point

This is a point on a state diagram showing the condition of a system at a given instant, in terms of the two independent properties selected for the x and y axes.

1.5.4 Path and Process

A path is a line joining successive state points on a state diagram, during a change of state. A process is that which causes the system to change its state. Shown in Figure 1.5 is a sample state diagram.

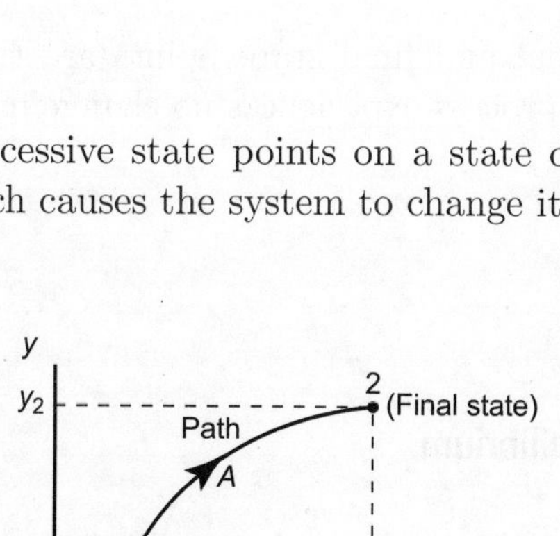

Fig. 1.5 A sample state diagram.

1.5.5 Isolated System

An isolated system is one which has no interactions at all with its surroundings. The properties of an isolated system in equilibrium never change.

1.5.6 Adiabatic System

An adiabatic system is one which has no heat interaction with its surroundings. A system whose boundary is covered by an insulating material tends to an adiabatic system.

1.5.7 Quasi-Static Process

A quasi-static process is one that takes place so slowly that the system passes successively through equilibrium states. In other words, it is a process in which the system and the surroundings are in equilibrium all through. This is one of the essential requirements for the path of the process to be shown distinctly on a state diagram.

1.5.8 Non-Cyclic Process

A system is said to have executed a non-cyclic process if its initial and final state points are not the same.

1.5.9 Cyclic Process

In a cyclic process, the initial and final state points are the same. In other words, a system subjected to a cyclic process experiences no change in any property on completing the process.

1.6 Equilibrium

1.6.1 Thermodynamic Equilibrium

A system is said to be in thermodynamic equilibrium if no change in any of its properties occurs when it is isolated from its surroundings. In such a case, there would be no unbalanced forces and no temperature gradients within the system, and no chemical reaction or transfer of material from one part of the system to another. Therefore, there can be no spontaneous change in any macroscopic property of a system in equilibrium. Thus, the condition of thermodynamic equilibrium calls for mechanical, thermal and chemical equilibrium, simultaneously. Thermodynamics mainly studies the behaviour of systems that are in equilibrium or passing through equilibrium states during a process.

1.6.2 Mechanical Equilibrium

A system is said to be in mechanical equilibrium when there are no unbalanced forces within the system and that the system is in force-balance with its surroundings. If an unbalanced force exists, either the system alone or the system and the surroundings, both, undergo changes until mechanical equilibrium is reached.

1.6.3 Thermal Equilibrium

A system is said to be in thermal equilibrium if the temperature within the system is uniform and is equal to the temperature of its surroundings. In other words, a system is said to be in thermal equilibrium if no change in any property occurs when it is put in contact with its surroundings through a *diathermic wall.* A diathermic wall is one that allows only heat to be transferred across it but not work, when brought between a system and its surroundings.

1.6.4 Chemical Equilibrium

A system is said to be in chemical equilibrium if it experiences no change in either its chemical composition or aggregation with time.

1.7 Illustrative Problems

1.1 The acceleration due to gravity, g, is given as a function of elevation above the sea-level, H, by $g = (980.6 - 3.088 \times 10^{-6} H)$, where g is in cm/s^2 and H is in cm. If a plane weighs 4500 N at sea-level what is its weight at an altitude of 5000 m? What is its percentage difference from the sea-level weight ?

▶ **Data:**

$g = 980.6 - 3.088 \times 10^{-6} H$, $W_{sea\ level} = 4500$ N

▶ **Solution:**

At sea level, as $H = 0$, hence, $g = 980.6 \text{ cm/s}^2 = 9.806 \text{ m/s}^2$.

$$W_{sea\ level} = 4500 \text{ N} = m \times g = m \times 9.806$$

$$\text{Therefore,} \quad m = \frac{4500}{9.806} = 458.9 \text{ kg}$$

$$g_{5000} = [980.6 - 3.088 \times 10^{-6} \times (5000 \times 100)] \times 10^{-2} = 9.7906 \text{ m/s}^2.$$

$$W_{5000} = m \times g_{5000} = 458.9 \times 9.7906 = 4492.9 \text{ N}.$$

$$\% \text{ difference in weight} = \left[\frac{(4500 - 4492.9)}{4500}\right] \times 100 = 0.157\%$$

1.2 Consider a helium-filled balloon of diameter 10 m. The balloon is supposed to carry two people weighing 70 kg each. The mass of the balloon fabric, the ropes and the cage is 20% of the mass of the people. The density of helium is $\frac{1}{7}$ times the density of air. Determine the acceleration of the balloon when it is just released from the ground assuming the density of air to be 1.16 kg/m^3.

▶ **Data:**

Diameter of the balloon, $D_b = 10$ m, mass of the people, $m_p = 2 \times 70 = 140$ kg, mass of the balloon fabric, ropes and cage, $m_f = 0.2 \times 140 = 28$ kg, $\rho_{\text{He}} = \left(\frac{1}{7}\right)\rho_{air}$ and $\rho_{air} = 1.16 \text{ kg/m}^3$.

▶ **Solution:**

$$\text{Upward force of buoyancy} = \rho_{air} \times V_{balloon} \times g = \rho_{air} \times \frac{\pi}{6} D_b^3 \times g$$

$$= 1.16 \times \frac{\pi}{6} \times 10^3 \times 9.81 = 5958.3 \text{ N}$$

Downward gravitational force on helium, people and the balloon fabric =

$$m_{net} \times g = (m_{\text{He}} + m_p + m_f) \times g = (\rho_{\text{He}} \times V_{balloon} + m_p + m_f) \times g$$

$$= \left[\left\{\left(\frac{1}{7}\right) \times 1.16 \times \left(\frac{\pi}{6}\right) 10^3\right\} + 140 + 28\right] \times 9.81$$

$$= 254.77 \times 9.81 = 2499.3 \text{ N}$$

Let a be the acceleration of the balloon at take off. Applying force-balance,

$$m_{net}\, a = \text{upward forces} - \text{downward forces}$$

$$\text{Therefore,} \quad 254.77a = 5958.3 - 2499.3, \quad \text{or} \quad a = 13.6 \text{ m/s}^2.$$

1.3 A spring scale is used to measure weight. It is proposed to employ it for determining the mass of a sample of moon-rock on the surface of the moon. The spring was originally calibrated for the earth's gravitational acceleration of 9.81 m/s^2. The scale reads 5.5 kg on the moon and the moon's gravitational acceleration is 1.4 m/s^2. Determine the mass of the sample. What would have been the reading on a beam balance on the moon?

▶ Data:

$g_{earth} = 9.81$ m/s^2, $g_{moon} = 1.4$ m/s^2, spring balance reading on the moon = 5.5 kg (i.e., spring balance reading = $5.5 \times g$ at the place of calibration).

▶ Solution:

A spring balance reads weight (i.e., force). Hence, its reading for a given mass depends upon the value of g. The spring balance is calibrated on earth where $g_{earth} = 9.81$ m/s^2. This means when the spring balance reads 5.5 kg, for any g, the reading corresponds only to a weight of $5.5 \times 9.81 = 53.95$ N. Therefore, the weight of the rock sample on the moon = 53.95 N. Thus,

$$m_{rock} \times g_{moon} = 53.95 \text{ N}, \quad \text{or} \quad m_{rock} = \frac{53.95}{g_{moon}} = \frac{53.95}{1.4} = 38.54 \text{ kg}.$$

A beam balance always reads mass and not weight as the same g acts on both pans of the balance, thereby nullifying the effect of g. Therefore, irrespective of the value of g, a beam balance reads 38.54 kg both on the earth and on the moon.

1.4 The lower half of a 12-m-high cylindrical container open to atmosphere at the top is filled with water (density = 1000 kg/m^3) and the upper half with an oil of specific gravity 0.85. Determine the pressure difference between the top and the bottom of the cylinder in mm Hg. Assume that the density of mercury is 13,600 kg/m^3.

▶ Data:

The arrangement is as in Figure 1.6, $\rho_{water} = 1000$ kg/m^3, ρ_{oil} = specific gravity × $\rho_{water} = 0.85 \times 1000 = 850$ kg/m^3, $h_{oil} = h_{water} = 6$ m, $\rho_{Hg} = 13{,}600$ kg/m^3.

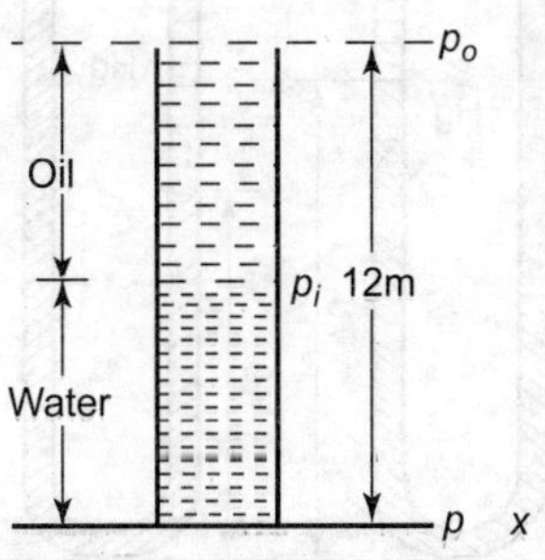

Fig. 1.6 Figure for Illustrative Problem 1.4.

▶ Solution:

Let p, p_o and p_i be the pressures at the bottom of the cylinder, top of the cylinder and at the interface of oil and water in the cylinder, respectively.

It is evident, from Figure 1.6, that,

$$\begin{aligned}(p - p_o) &= (p - p_i) + (p_i - p_o) \\ &= (\rho_{water} \times h_{water})g + (\rho_{oil} \times h_{oil})g \\ &= (1000 \times 6) \times 9.81 + (850 \times 6) \times 9.81 \\ &= 108{,}891 \text{ N/m}^2.\end{aligned}$$

And, $\quad = (\rho_{\text{Hg}} \times h_{\text{Hg}})g = 13{,}600 \times h_{\text{Hg}} \times 9.81.$

$$\text{Therefore,} \quad h_{\text{Hg}} = \frac{108{,}891}{(13{,}600 \times 9.81)} = 0.816 \text{ m} = 816 \text{ mm}.$$

1.5 In Figure 1.7, bulbs A and B contain water (density = 1.0 g/ml) at different pressures. These bulbs are connected to two mercury (density = 13.6 g/ml) manometers. The open ends of these two manometers are connected by an inverted U-tube manometer containing oil (density = 0.873 g/ml). Calculate $(p_A - p_B)$, in kPa.

▶ Data:

$\rho_{wat} = 1000 \text{ kg/m}^3$, $\rho_{\text{Hg}} = 13{,}600 \text{ kg/m}^3$ and $\rho_{oil} = 873 \text{ kg/m}^3$. The other data are as in Figure 1.8 given below. Figure 1.8 is the same as Figure 1.7 except for the additional points a, b, c, d, e, f and g marked on it. These points are required for the solution.

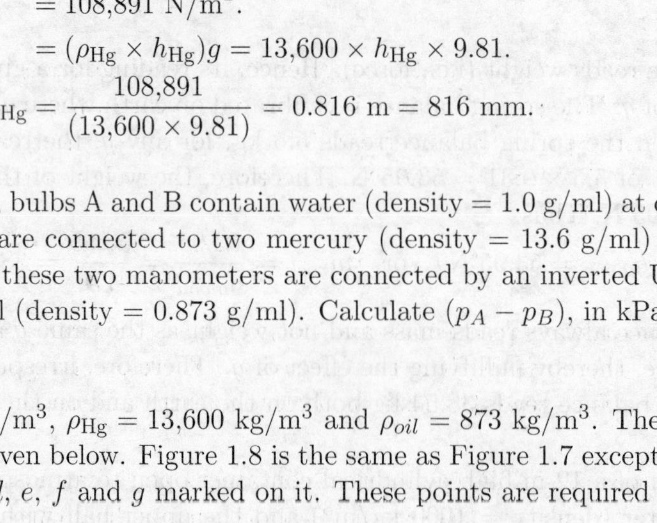

Fig. 1.7 Figure for Illustrative Problem 1.5.

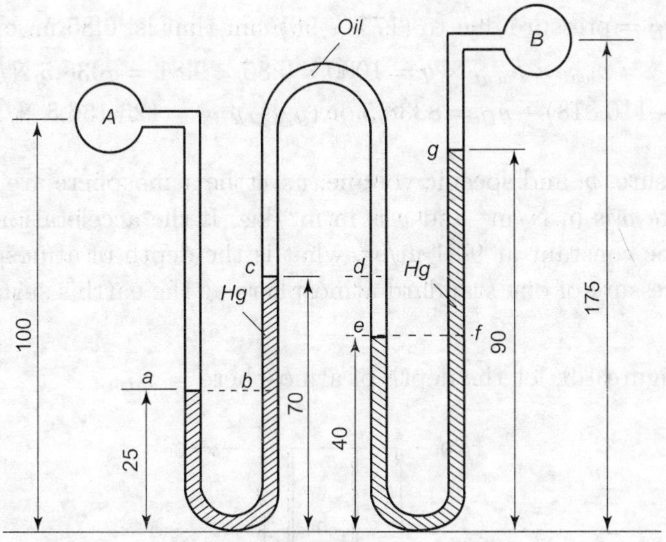

Fig. 1.8 Figure for solving Illustrative Problem 1.5.

▶ **Solution:**

$p_a = p_A +$ pressure due to (100 – 25) cm, i.e., 0.75 m, of water column

$= p_A + \rho_{wat} \times h_{wat} \times g = p_A + 1000 \times 0.75 \times 9.81 = p_A + 7357.5 \text{ N/m}^2.$

The hydrostatic principle states that the pressure at all points at the same height of a stationary column of a single fluid is the same. Thus, $p_a = p_b$, $p_c = p_d$ and $p_e = p_f$.

Therefore, $p_b = p_a = p_A + 7357.5 \text{ N/m}^2.$

$p_b - p_c =$ pressure due to (70 – 25) cm, that is, 0.45 m of Hg column

$= \rho_{\text{Hg}} \times h_{\text{Hg}} \times g = 136{,}00 \times 0.45 \times 9.81 = 60{,}037 \text{ N/m}^2.$

Therefore, $p_c = (p_A + 7357.5) - 60{,}037 = p_A - 52{,}679.5 \text{ N/m}^2$

Similarly, $p_d = p_c = p_A - 52{,}679.5$

and, $p_e = p_d +$ pressure due to (70 – 40) cm, that is, 0.30 m, of oil column

$p_e = p_d + \rho_{oil} \times h_{oil} \times g$

$p_e = (p_A - 52{,}679.5) + 873 \times 0.3 \times 9.81 = (p_A - 50{,}110) \text{ N/m}^2.$

Also, $p_f = p_e = (p_A - 50{,}110)$

and, $p_f - p_g =$ pressure due to (90 – 40) cm, that is, 0.50 m of Hg column

$= \rho_{\text{Hg}} \times h_{\text{Hg}} \times g = 13{,}600 \times 0.50 \times 9.81 = 66{,}708 \text{ N/m}^2.$

Therefore, $p_g = p_f - 66{,}708 = p_A - 50{,}110 - 66{,}708 = p_A - 116{,}818 \text{ N/m}^2.$

$$p_g - p_B = \text{pressure due to } (175 - 90) \text{ cm, that is, } 0.85 \text{ m, of water column}$$
$$= \rho_{wat} \times h_{wat} \times g = 1000 \times 0.85 \times 9.81 = 8338.5 \text{ N/m}^2.$$

Hence, $(p_A - 116{,}818) - p_B = 8338.5$ or $(p_A - p_B) = 125{,}156.5 \text{ N/m}^2 \approx 125.16$ kPa.

1.6 Assume pressure, p, and specific volume, v, of the atmosphere are related by $pv^{1.4} = 80,340$, where p is in N/m^2 and v is in m^3/kg. If the acceleration due to gravity is assumed to be constant at 9.81 m/s^2, what is the depth of atmosphere necessary to produce a pressure of one standard atmosphere at the earth's surface?

▶ **Solution:**

As shown in Figure 1.9, let the depth of atmosphere = z_{atm}.

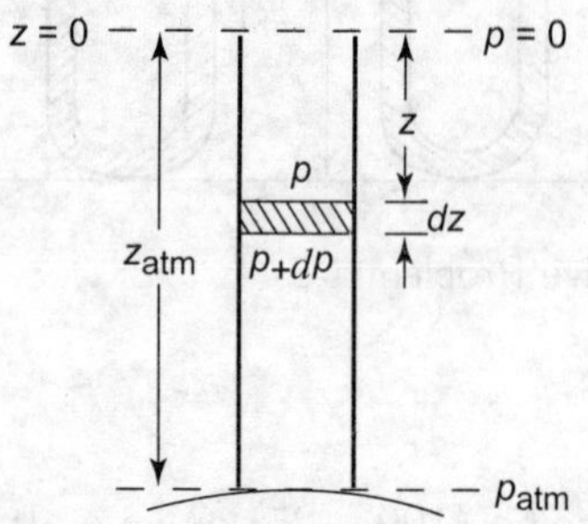

Fig. 1.9 Figure for Illustrative Problem 1.6.

Considering a small element of atmospheric column of thickness dz at a depth of z from the extremity of the atmosphere (where $p = 0$), dp the pressure difference across the element is given by

$$\mathrm{d}p = \rho g \mathrm{d}z \tag{i}$$

From the data $$pv^{1.4} = 80{,}340$$

hence, $$v = \left(\frac{80{,}340}{p}\right)^{\frac{1}{1.4}}$$

Now, $$\rho = \left(\frac{1}{v}\right)$$

and therefore, $$\rho = \left(\frac{p}{80{,}340}\right)^{0.7143}$$

Substituting for ρ in (i),

$$\mathrm{d}p = \left(\frac{p}{80{,}340}\right)^{0.7143} \times g\mathrm{d}z$$

or $$\left(\frac{\mathrm{d}p}{p^{0.7143}}\right) = (80{,}340^{-0.7143}) \times 9.81 \text{ d}z$$

Therefore, $$3.077 \times 10^{-3}\mathrm{d}z = p^{-0.7143} \text{ d}p \tag{ii}$$

It can be seen from Figure 1.9 that the boundary conditions are:

$$p = 0, \quad \text{at} \quad z = 0,$$

and

$$p = p_{atm} \quad \text{at} \quad z = z_{atm},$$

and

$$p_{atm} = 1.01325 \times 10^5 \ \text{N/m}^2.$$

Integrating (ii) between the above boundary conditions,

$$3.077 \times 10^{-3} \int_{z=0}^{z=z_{atm}} dz = \int_{p=0}^{p=p_{atm}} p^{-0.7143} dp$$

$$3.077 \times 10^{-3} z_{atm} = \frac{p_{atm}^{(1-0.7143)}}{(1-0.7143)}$$

$$= \frac{(1.01325 \times 10^5)^{0.2857}}{0.2857}$$

Therefore, $z_{atm} = 30{,}626 \text{ m} = 30.626 \text{ km}.$

Problems

1.1 A force of 8 N acts on a mass of 0.1 kg. Find the acceleration of the mass in cm/s^2.

1.2 During takeoff in a space ship, a 70-kg astronaut is subjected to an acceleration that is five times the pull of the earth's standard gravity ($= 9.81 \text{ m/s}^2$). If the takeoff is vertical, what force does he exerts on the seat?

1.3 If the weight of a body on the moon $\left(g_{moon} = \dfrac{g_{earth}}{6}\right)$ is 1.935 N, find its mass and weight on the surface of the earth.

1.4 A pressure cooker cooks faster than an ordinary cooker because higher pressure and temperature are maintained inside. A small weight (petcock) with a suitable pin sits in a cylindrical opening of cross-sectional area 5 mm^2 on the sealed lid of the cooker. The steam can escape from the cooker only through this opening when the weight is raised. Determine the mass of the petcock for an operating pressure of 100 kPa gauge. Assume atmospheric pressure to be at its standard value.

1.5 An astronaut on the moon's surface ($g = 1.8 \text{ m/s}^2$) collects a rock sample that weighs 4.5 kg on the spring balance he had carried with him that was calibrated on the surface of the earth ($g = 9.81 \text{ m/s}^2$). The rock sample was brought to the earth and was weighed first with the same spring balance and then with a beam balance. What would the two readings be, in magnitude and dimension.

1.6 An artificial satellite is reported to have encircled the earth at a speed of 26,500 km/h at a height of 917 km above the earth's surface. Taking the mean diameter of the earth to be 12,700 km and assuming the orbit to be circular, calculate the value of the acceleration due to earth's gravity at this height. If the weight of the satellite is reported to be 950 N at sea-level, what would its weight be at the operational altitude ?

1.7 If the atmospheric pressure is 1.013 bar and a Bourdon pressure gauge connected to a system reads 2.5×10^5 N/m^2, express the gauge pressure and the absolute pressure of the system in kPa.

1.8 A turbine is supplied with steam at a pressure of 15 bar gauge. After expansion in the turbine the steam passes to a condenser which is maintained at a vacuum of 710 mm Hg by means of pumps. The barometric pressure is 775 mm Hg. Express the inlet and exhaust steam pressures in N/m^2. Assume density of mercury to be 13.6 g/ml.

1.9 Convert the following readings of pressure to N/m^2 absolute, assuming the barometer reads 760 mm Hg. (a) 40 cm Hg vac., (b) 190 cm Hg gauge and (c) 100 cm water gauge. Take the densities of water and mercury to be 1.0 and 13.6 g/ml, respectively.

1.10 In Figure 1.10, given that the pressure in bulb A is 65 kPa vacuum, find the value of H in cm. Assume density of Hg is 13,600 kg/m^3.

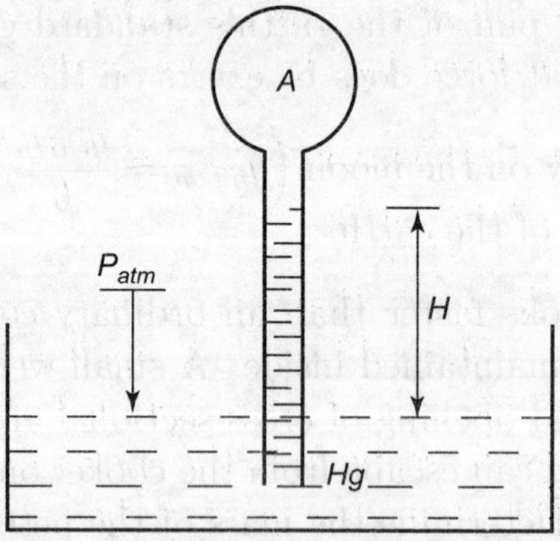

Fig. 1.10 Figure for Problem 1.10.

1.11 A vacuum gauge connected to a tank reads 30 kPa at a location where the barometer reads 755 mm Hg. Determine the absolute pressure in the tank assuming density of Hg to be 13,590 kg/m^3.

1.12 A manufacturer of internal combustion engines, at Chennai, is to supply a supercharged gasoline engine to a customer in Udhagamandalam. The engine specifications calls for a guaranteed absolute pressure of 120 kPa at which the petrol–air mixture enters the cylinder. What reading should the customer get on a mercury (density = 13.6 g/ml) manometer, in cm of Hg, used to measure the inlet pressure and how much will it differ, in percentage, from that obtained during the manufacturer's test at Chennai. Assume density of Hg to be constant at 13.6 g/ml and use the following data:

Place	Altitude (m)	Atm. pressure (kPa)	Accn. due to gravity (m/s^2)
Chennai	5	100	9.81
Udhagamandalam	1760	85	9.79

1.13 A person recorded the barometric readings at the top and bottom of a building and found them to be 741 and 761 mm Hg, respectively. Assuming the average densities of air and mercury to be 1.2 kg/m^3 and 13.6 g/ml, respectively, estimate the height of the building.

1.14 The basic barometer can be used as an altimeter in an aeroplane. When an aeroplane is trying to land, the ground control reports a barometric reading of 750 mm Hg while the pilot's reading is 660 mm Hg. Estimate the altitude of the aeroplane, assuming the average density of air to be 1.1 kg/m^3.

1.15 A 170-cm-tall skin-diver descends to the bottom of a salt lake of depth 24 m and stands vertically on the floor of the lake. The density of lake water is 1056 kg/m^3. If $g = 9.81$ m/s^2 and atmospheric pressure is 1 bar, calculate the absolute pressure exerted on the head of the diver in kPa.

1.16 A vertical cylinder containing argon gas is fitted with a frictionless leak-proof piston of mass 50 kg and a cross-sectional area of 0.025 m^2. The atmospheric pressure is 95 kPa and the local acceleration due to gravity is 9.79 m/s^2. What is the argon pressure inside the cylinder ?

1.17 An inclined tube manometer is used to measure the gas pressure in a pipe as shown in Figure 1.11. The manometric fluid is mercury. Determine the gas pressure in the pipe if the atmospheric pressure is 1.01×10^5 Pa and that the density of mercury is 13,600 kg/m^3.

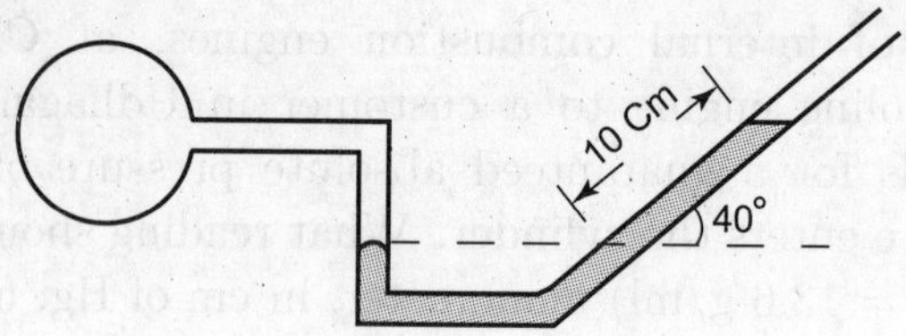

Fig. 1.11 Figure for Problem 1.17.

1.18 A mercury manometer is used to measure the pressure of steam flowing through a pipe. Figure 1.12 shows the readings at a certain instant when there is 5 cm of condensed water in the left limb of the manometer. Given that the barometric pressure is 100 kPa and that the densities of water and mercury are, respectively. 10^3 and 13.6×10^3 kg/m^2, find the absolute pressure of steam in bar.

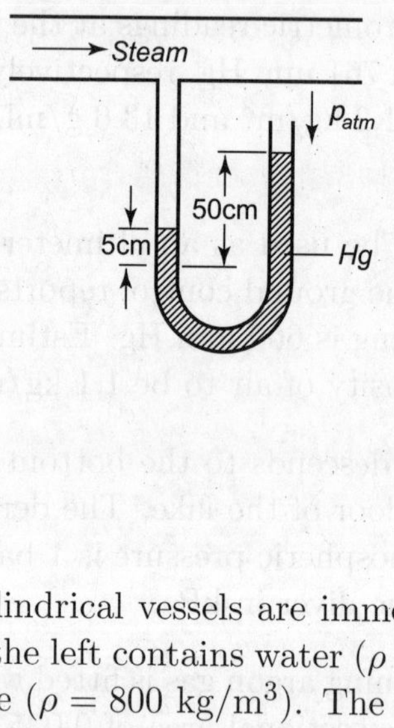

Fig. 1.12 Figure for Problem 1.18.

1.19 Two inverted identical cylindrical vessels are immersed in two tanks, as shown in Figure 1.13. The tank on the left contains water ($\rho = 1000$ kg/m^3) while the one on the right contains kerosene ($\rho = 800$ kg/m^3). The two tanks are filled to the brim and the free surfaces of the liquids are exposed to the atmosphere at 0.1 MPa. The two vessels are connected by means of small pipes and valves to a vertical cylinder fitted with a frictionless piston of negligible mass. Initially, when the valves are closed, while the air above the liquid surfaces in the two vessels are at atmospheric pressure, the Bourdon pressure gauge connected to the piston–cylinder mechanism to read the pressure of air in the cylinder, reads zero. The valves are then opened and a set of small weights are added to the top of the piston successively so that the piston descends slowly but steadily through a certain distance. Find the depths through which the liquid surfaces in each vessels descends in the same period, when the Bourdon gauge reads 40 kPa.

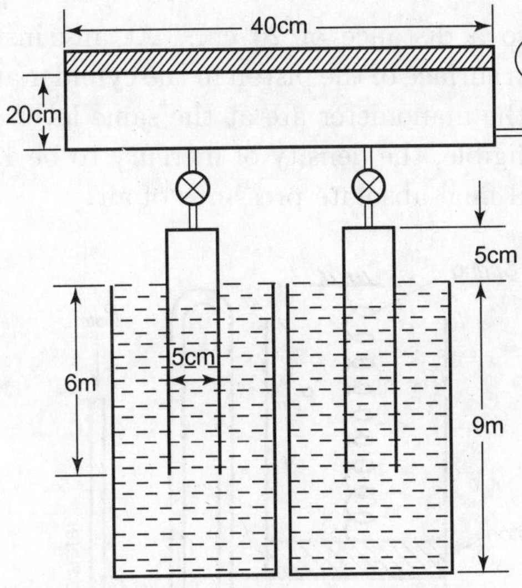

Fig. 1.13 Figure for Problem 1.19.

1.20 A 50-kg frictionless piston fits inside a vertical pipe of diameter 20 cm and is pulled upward to a height of 6.5 m. The pipe's lower end is immersed in water and the atmospheric pressure is 100 kPa. The gravitational acceleration is 9.78 m/s^2. The piston just rests on the surface of water. (a) Determine the force required to hold the piston at 6.5 m mark and (b) determine the pressure of water on the underside of the piston.

1.21 Figure 1.14 shows a channel of width b, normal to the plane of the paper, height z and length L. The pressure on the free surface of the liquid in the channel is p_0. Derive an expression for the force F exerted by the liquid on the piston in terms of p_0, z, b, g and the density of liquid.

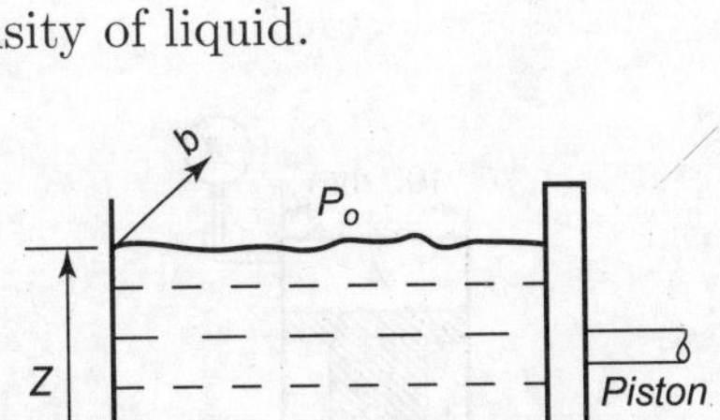

Fig. 1.14 Figure for Problem 1.21.

1.22 A cylinder of diameter 25.25 cm in which a piston is retained by a spring contains 0.01 m^3 of air at a temperature of 25°C. A mercury manometer connected to the cylinder initially reads as shown in Figure 1.15. Air is heated steadily until the piston rises slowly to a distance of 20 cm. At any instant during the process described, the bottom surface of the piston in the cylinder and the mercury meniscus in the open limb of the manometer are at the same level. Assuming the weight of the piston to be negligible, the density of mercury to be 13.6 g/ml and p_{atm} to be 1 bar, find initial and final absolute pressures of air.

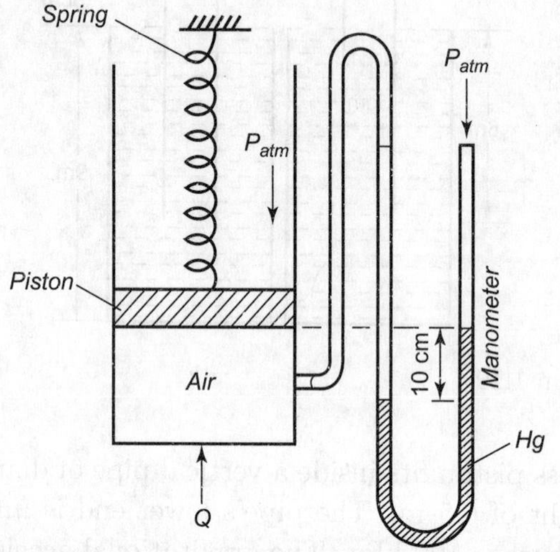

Fig. 1.15 Figure for Problem 1.22.

1.23 A gas is contained in two cylinders A and B, connected by a leak-proof piston of two different diameters as shown in Figure 1.16. The mass of the piston is 10 kg and the gas inside cylinder A is at 250 kPa. Calculate the gas pressure in cylinder B.

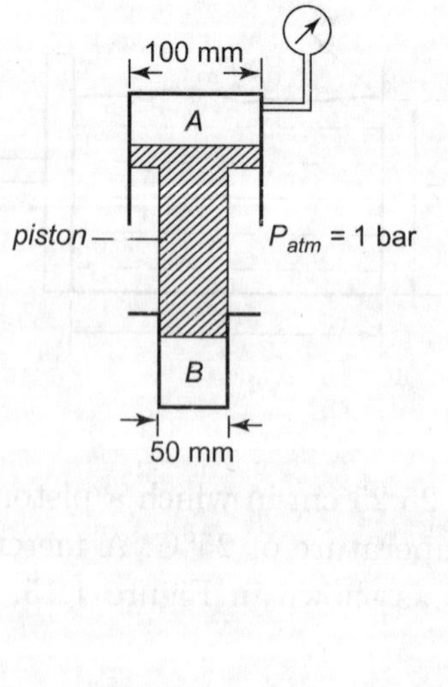

Fig. 1.16 Figure for Problem 1.23.

1.24 A gas is confined in a vertical, frictionless piston–cylinder device. The piston has a mass of 4 kg and a cross-sectional area of 35 cm^2. A compressed spring above the piston exerts a force of 60 N on the piston. Also the top surface of the piston is exposed to the atmosphere at 95 kPa. Determine the pressure inside the cylinder.

1.25 A 20-cm-diameter vertical pipe, open at both ends, is immersed in a lake such that the lower end is in water. A 50-kg leak-proof, frictionless piston slides in this pipe. Initially, the piston is held such that its bottom surface just touches the water in the pipe. The piston is then pulled up through a distance of 6 m, during which process the lower face of the piston is always in contact with the water surface. The barometric pressure is 100 kPa and $g = 9.8$ m/s^2. Determine (a) the force required to hold the piston at the 6 m mark and (b) the pressure of water underneath the piston.

WORK

2

2.1 Introduction

As explained earlier, thermodynamics is the science of the interactions between *heat* and *work* and the changes in the *properties* of the interacting *systems.* Hence before venturing upon a study of thermodynamic analyses of processes it is essential that one knows thoroughly the meanings of the words heat, work, properties and systems.The last two namely, *propserties* and *systems*, have been dealt with in detail in Chapter 1. This chapter deals with *work* in detail. Heat is discussed in Chapter 3.

2.2 Mechanics Definition of Work

Work is said to be done when the point of application of a force moves through a certain distance. The magnitude of work done is equal to the product of the force and the distance through which the point of application of the force moves, in the direction of force. Thus, the unit of work is newton metre (Nm). In Chapter 4 on the first law of a thermodynamics it is shown that, in the SI system, 1 Nm is also equal to 1 joule (J), an unit of heat transfer. Hence Nm and J can be used interchangeably.

Example 2.1

Figure 2.1 shows a block resting on a plane inclined at θ to the horizontal. The block is acted upon by a horizontal force F. Under the action of the force, the block moves a distance l, from position 1 to position 2. Find the work done by the force, in kJ, given that $F = 180$ N, $l = 12$ m and $\theta = 30°$.

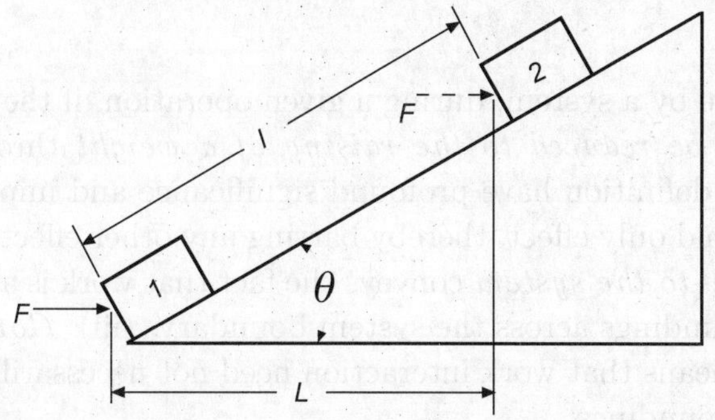

Fig. 2.1 Block moving up an inclined plane.

► **Solution:**

Work, $W \neq F \times l$, where l is the distance moved up the plane, but $W = F \times L$, where L is the distance travelled by the point of application of force *in the direction of force.*

It can be seen from Figure 2.1 that $L = l \cos \theta$

$$\text{Therefore,} \quad W = F \times (l \cos \theta) = 180 \times (12 \times \cos 30°)$$
$$= 180 \times (12 \times 0.866) = 1870.6 \text{ Nm} = 1870.6 \text{ J} = 1.8706 \text{ kJ}.$$

2.2.1 Power

Power is defined as the rate of doing work. As the unit of work is joule (J), the unit of power has to be J/s. 1 J/s is called a watt (W). The practical unit of power is either a kilowatt (kW), which is equal to 10^3 W, or a megawatt (MW), which is equal to 10^6 W.

2.3 Thermodynamic Definition of Work

2.3.1 Necessity

The mechanics definition of work has certain limitations. To identify and also to evaluate the magnitude of work, it is necessary, according to the definition, to physically observe the point of application of the force and be able to measure the distance the point moves in the direction of force. In many situations, it becomes difficult to either identify the point of application of force or to measure the distance it moves. Flow of current through a resistor is one such example. This process involves electrical work interaction but it is not possible, in the macroscopic level, to either find the point of application of force or the distance it moves. This necessitates defining work in thermodynamics in a more universal manner, which not only encompasses the mechanics definition stated earlier but also includes all other possible modes of work interactions.

2.3.2 Definition

Work is said to be done by a system, during a given operation, if the *sole effect, external to the system, could be reduced to the raising of a weight* through a distance. The words in italics in the definition have profound significance and importance. (i) The *sole effect* means the one and only effect, thereby barring any other effect besides the raising of a weight. (ii) *External to the system* conveys the fact that work is an interaction between a system and its surroundings across the system boundary. (iii) *Could be reduced to the raising of a weight* means that work interaction need not necessarily result in the raising of a weight in the surroundings.

Suppose a system interacts with its surroundings and the consequent change in the surroundings does not happen to be the raising of a weight as the only effect. In such a case, the interaction may or may not be work. To assess whether it is work or not a specific exercise may be needed to find out whether the *same interaction* could have resulted in the raising of a weight as the sole effect in a suitably imagined fictitious surroundings. If the answer to this proposition is 'yes', the interaction is work and if not the interaction may not be work.

The magnitude of work interaction is calculated as the product of the weight raised (mg) and the distance it moves up vertically (against the gravitational force) in the surroundings.

2.3.3 Sign Convention

It is now clear that a system cannot do work on itself as it is an interaction across the system boundary. There must be a donor and a receptor for the work interaction to occur. The donor does work on the receptor. It is customary to give a positive sign for the work interaction of the donor and a negative sign for the work interaction of the receptor. Thus, work done by a system (work out) is considered to be positive and work done on the system (work in) is considered negative. Figure 2.2 schematically shows the sign convention adapted for work.

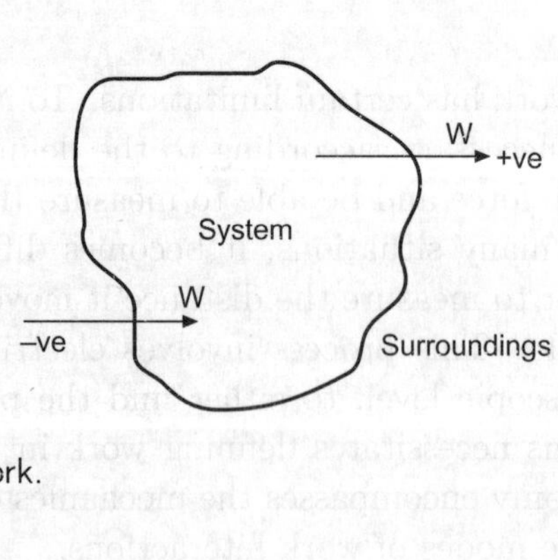

Fig. 2.2 Sign convention for work.

In any given process, the magnitudes of work interaction for both the system and the surroundings are the same but for their direction. Hence,

$$W_{System} + W_{Surroundings} = 0. \tag{2.1}$$

Example 2.2

An agent slowly raises a body of mass, 2 kg, through a vertical distance of 5 m at a place where $g = 9.8\ \text{m/s}^2$. Find the magnitude and direction of work for the systems agent and body.

► Solution:

Consider the agent as the system. As shown in Figure 2.3*a*, the agent is raising the body against the gravitational force. To overcome the gravitational pull on the body, the agent has to do work on the body. Therefore, according to the mechanics definition of work,

$$\begin{aligned} W_{agent} &= \text{resisting force} \times \text{distance moved} \\ &= \text{weight of the body} \times \text{distance} \\ &= (mg) \times L = 2 \times 9.8 \times 5 = 98\ \text{Nm} \end{aligned}$$

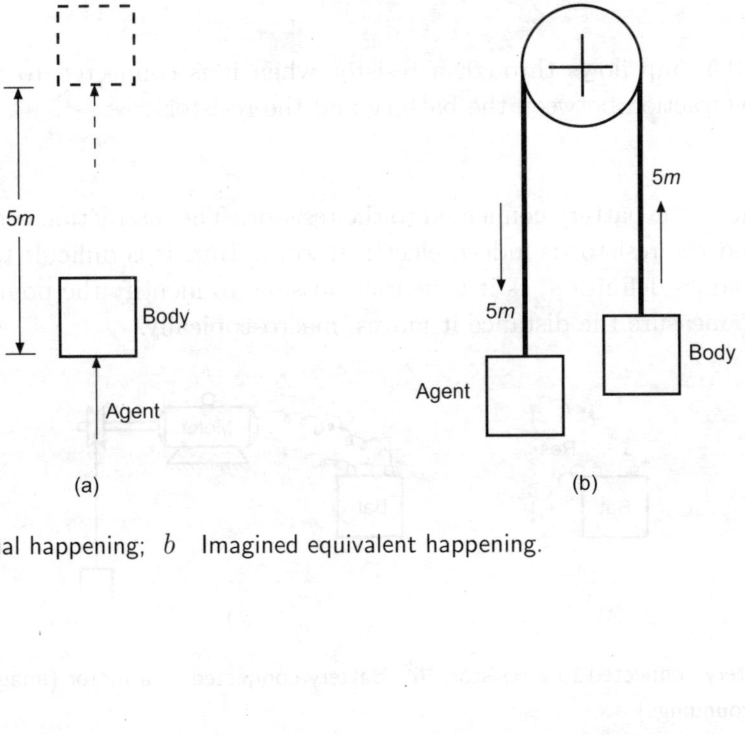

Fig. 2.3*a* Actual happening; *b* Imagined equivalent happening.

The same result can be obtained by using the thermodynamic definition of work. Shown in Figure 2.3*b* is the same interaction of the agent and body, but in a different, suitably imagined, surroundings. Here, the agent and the body are imagined to be suspended over a frictionless pulley by a light inextensible cable. As the pulley rotates slowly, the body is lifted upwards. For the body to be lifted slowly, in the limiting case, mass of the body must be equal to the mass of the agent. In both cases, *a* and *b*, the agent is subjected to the same interaction, namely, the raising of the same weight through the same distance. Considering the agent as the system, the only effect external to it is the raising of the body through a distance of 5 m. As the sole effect external to the agent is the raising of the weights, according to thermodynamic definition, work is done by the agent.

$$W_{agent} = \text{weights raised} \times \text{distance} = (mg)h = 2 \times 9.8 \times 5 = 98 \text{ Nm}$$

It is advisable to present the answer in either of the two following formats:

(i) $W_{agent} = +98$ Nm or (ii) 98 Nm of work is done by the agent.

(***Note***: It is wrong to state +98 Nm of work is done by the agent. Either use '+' or 'by' and not both.)

As $W_{agent} + W_{body} = 0$, $W_{body} = -W_{agent} = -98$ Nm.

Therefore, (i) $W_{body} = -98$ Nm, or, (ii) 98 Nm of work is done on the body.

Example 2.3

A current of 0.5 amp flows through a resistor when it is connected to a 12 V battery. Identify the interaction between the battery and the resistor.

▶ Solution:

Figure 2.4*a* shows the battery connected to the resistor. The interaction occurring between the battery and the resistor is indeed electrical work. But, it is difficult to infer this just from the mechanics definition, as it is neither possible to identify the point of application of force nor to measure the distance it moves, macroscopically.

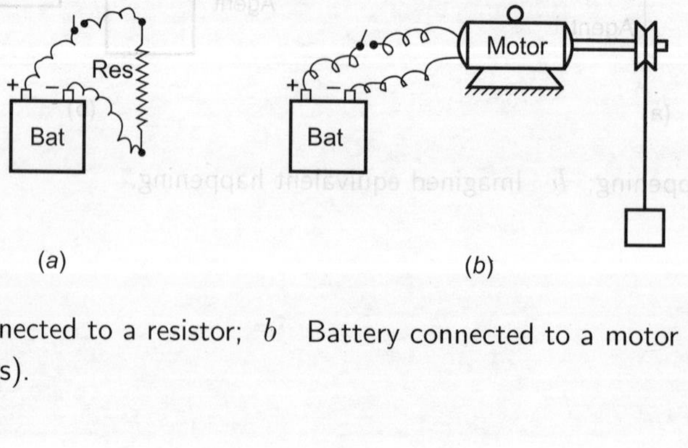

Fig. 2.4*a* Battery connected to a resistor; *b* Battery connected to a motor (imaginary surroundings).

Let it now be imagined, as shown in Figure 2.4*b*, that the same battery supplies the same current against an equivalent resistance offered by the windings of a motor. In such a scenario, the interactions the battery has with its surroundings in both cases, *a* and *b*, are the same as it supplies the same current against the same resistance.

The motor shaft is fixed with a pulley. An inextensible, light, cable wound around the pulley carries a weight at its end. As the battery supplies current to the motor, the pulley on the motor shaft rotates, thereby, lifting the weight connected to the cable. Assuming that the friction in the motor bearing is negligible and that the windings are of a thicker gauge, so that there is no heating of the motor windings, the only effect external to the battery is the raising of a weight. Hence, the interaction between the battery and the motor (and thus the original resistor) is work. The work is positive for the battery, in conformity with the definition.

2.4 Displacement Work

Displacement work is the mode of work involved when a system boundary sweeps a volume while interacting with its surroundings. The volume swept may be against the surroundings or against the system itself, depending upon whether the work is done *by* the system or *on* the system. Before explaining the methodology adapted in evaluating displacement work, the reader is cautioned that displacement work can be evaluated only for a *quasi-static process.*

2.4.1 Quasi-Static Process

A quasi-static process is one which occurs so slowly that, at any instant during the process, the uniform system pressure on the moving part of the system boundary is equal to the external resisting pressure. In other words, the system is in mechanical equilibrium with its surroundings all through the process. Thus, the path of a quasi-static process on any state diagram is the line joining the successive equilibrium states from the initial to the final state. A quasi-static process is also called a *resisted process* or a *reversible process.*

2.4.2 Evaluation of Displacement Work in Quasi-Static Processes

Suppose the boundary of a system of uniform pressure sweeps a volume slowly against the resisting pressure of the surroundings. Let the pressure at an instant, acting on the moving boundary of area, A, be p and let the boundary move through a small distance $\mathrm{d}L$. The differential work interaction in this period, δW_d, is given by

$$\delta W_d = Force \times \text{normal distance}$$
$$= (pA)\mathrm{d}L = p(A\mathrm{d}L) = p\mathrm{d}V$$

where $\mathrm{d}V$ is the volume swept by the boundary. This is further explained with the help of a simple example. Consider a horizontal piston–cylinder device shown in Figure 2.5.

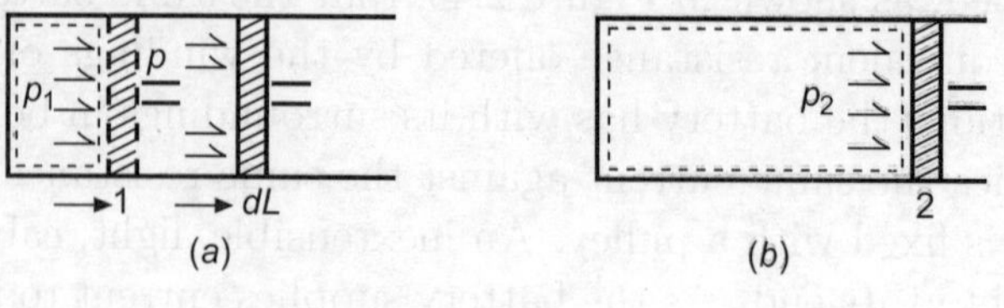

Fig. 2.5 Piston–cylinder device to explain displacement work.

Initially, when the piston is at the extreme left position, let the cylinder contain a certain mass of a fluid which is considered as the system. Let the system be now subjected to a process which makes the piston move slowly from position 1, shown in Figure 2.5*a*, to position 2, shown in Figure 2.5*b*. Let the pressure of the system be p at any one instant during the movement of the piston between the two extreme positions. This pressure is also equal to the resisting pressure on the piston. Consider a small interval of time from this instant during which the piston travels a negligible distance $\mathrm{d}L$. Let the area of cross-section of the cylinder be A. The force acting on the piston is given by

$$F = pA$$

The displacement work done by the fluid as the piston moves a distance $\mathrm{d}L$, according to mechanics definition of work, is given by

$$(\delta W_d) = F \times \mathrm{d}L = (pA)\mathrm{d}L = p(A\mathrm{d}L) = p\mathrm{d}V$$

where $\mathrm{d}V$ is the volume swept by the system boundary (which is equal to the piston displacement volume in the small period considered here).

The net work done by the fluid as its boundary moves from 1 to 2 is given by

$$W_d = \int_1^2 \delta W_d = \int_1^2 p\mathrm{d}v \tag{2.2}$$

Because of this relation, displacement work is also called $p\mathrm{d}V$ work.

Shown in Figure 2.6, in cartesian coordinates, is a typical path 1–2 of a quasi-static process of the type happening during expansion of a system. The displacement work $W_d = \int p\mathrm{d}v$. As p is represented on the y axis and V on the x axis, $\int p\mathrm{d}v = \int y\mathrm{d}x =$ area bound by the path 1–2 with the x axis. This area is shown hatched on the p–V diagram. It can hence be generalized that the area bound by the path of a quasi-static process with the x axis on a p–V plane is always equal to the displacement work.

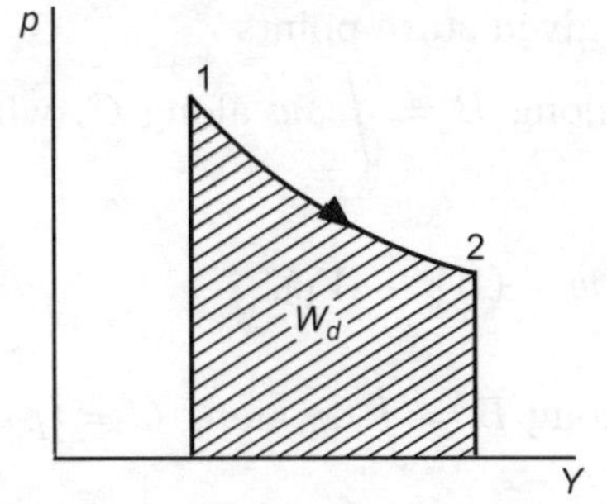

Fig. 2.6 Typical path 1–2 and displacement work in a quasi-static process.

2.4.3 Path and Point Functions

Figure 2.7 illustrates three quasi-static processes A, B and C undergone by a system. These processes occur between the same two state points, 1 and 2, of a given system. The area below each path must be equal to the corresponding displacement work. All three areas have different values. Accordingly, the displacement work for each path is different. Therefore, between two given state points of a system, the displacement work is not unique as it depends upon the path. It is for this reason that work is said to be a *path function.*

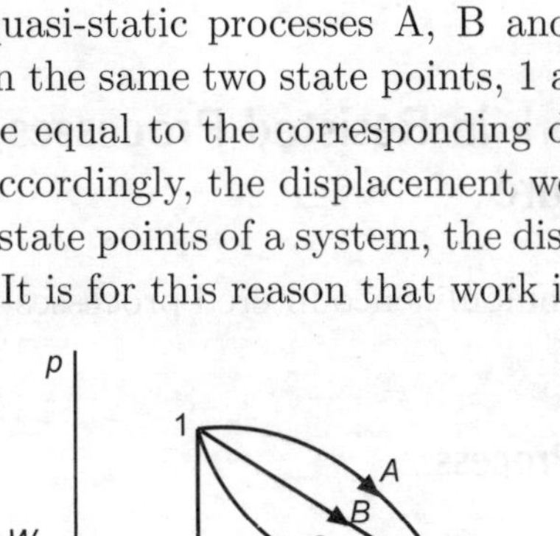

Fig. 2.7 Displacement work in a quasi-static process.

It is to be observed that for each path in Figure 2.7 the initial and final state points are the same. In other words, the initial and final thermodynamic properties are the same for all three processes. Therefore, between two state points, the change in any property is independent of the path. It is for this reason that properties of a system are called *point functions.*

Hence, it is mathematically true that while δ W is an *inexact differential* or *imperfect differential*, dp, dV etc are *exact differentials* or *perfect differentials.* The inexact differential or path function is symbolized in this book by δ. For example, it is δW and not dW. Similarly, it is dp, dV, dT etc and not δp, δV or δT. It is shown later that heat is also a path function for the same reason. Hence, differential heat transfer, which is also an inexact differential, is symbolized by δQ and not by dQ.

In summary, between two given state points

- $\int \delta w$ along $A \neq \int \delta w$ along $B \neq \int \delta w$ along C, where A, B, C refer to different paths.
- For any given path, $\int \delta w \neq (W_2 - W_1)$.
- $\int \mathrm{d}p$ along $A = \int \mathrm{d}p$ along $B = \int \mathrm{d}p$ along $C = (p_2 - p_1)$.
- $\int \mathrm{d}V$ along $A = \int \mathrm{d}V$ along $B = \int \mathrm{d}V$ along $C = (V_2 - V_1)$.

Being a path function, work at a point has no meaning, That is, W_2 and W_1 have no meaning. Generally, $\int \delta w$ along a given path is represented as ${}_1W_2$ or W_{1-2}.

2.5 Displacement Work in Resisted Processes Undergone by Systems of Uniform Pressure

Discussed below are a few commonly encountered processes in engineering practice involving displacement work.

2.5.1 Constant Pressure Process

A process is said to be a constant pressure process if the pressure on all parts of the moving system boundary remains constant as the system moves from the initial to the final state. A simple example of a constant pressure process is the one undergone by a fluid in a vertical frictionless piston–cylinder device, as shown in Figure 2.8*a*.

The system pressure or the resisting pressure is equal to the combined weight of the piston and the mass placed on it per unit cross-sectional area of the cylinder, plus the

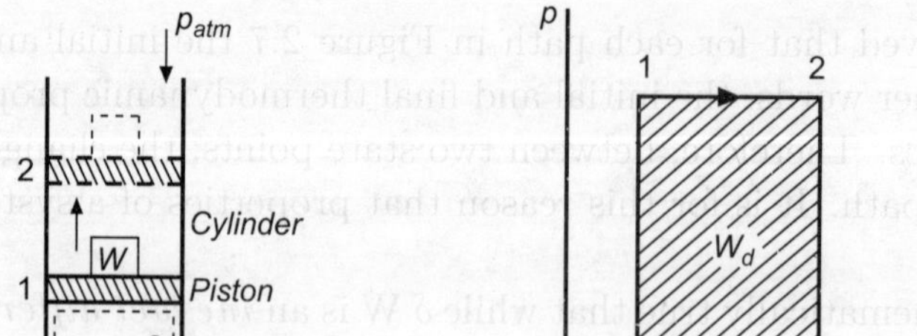

Fig. 2.8*a* Constant pressure process; *b* $p-V$ diagram.

atmospheric pressure. Supposing the piston moves from position 1 to position 2 during a certain process undergone by the fluid, say heating, as $p =$ constant, the displacement work W_d is given by:

$$W_d = \int_1^2 p\mathrm{d}V = p\int_1^2 \mathrm{d}V = p(V_2 - V_1) \tag{2.3}$$

If $V_2 > V_1$, it is an expansion process (or a heating process) and W_d for the fluid is positive , and if $V_2 < V_1$ it is a compression process (or a cooling process) and W_d for the fluid is negative. The path 1–2 shown on the $p-V$ diagram in Figure 2.8*b* is for an expansion process. The area below this path, shown hatched, is equal to the displacement work.

2.5.2 Constant Volume Process

A system is said to undergo a constant volume process if, during the process, the system boundary does not sweep any volume. An example of such a process is that undergone by a system enclosed in a rigid vessel or in a piston–cylinder device in which the piston is held stationary, as shown in Figure 2.9*a*. In such a case,

$$W_d = \int_1^2 p\mathrm{d}V = 0, \text{ as } \mathrm{d}V = 0 \tag{2.4}$$

The path of the process is shown by 1–2 in Figure 2.9*b*. It is to be noted that the area of the path bound by the x axis is zero.

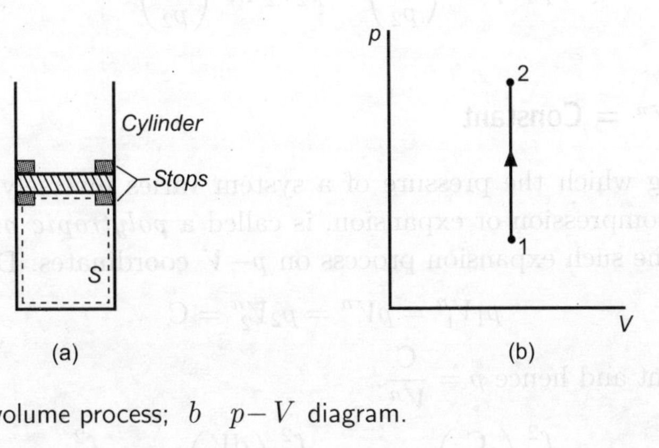

Fig. 2.9*a* Constant volume process; *b* $p-V$ diagram.

2.5.3 Process $pV =$ Constant

This process during which the pressure of a system varies inversely as its volume is called a *hyperbolic process.* Shown in Figure 2.10 is the path of one such expansion process.

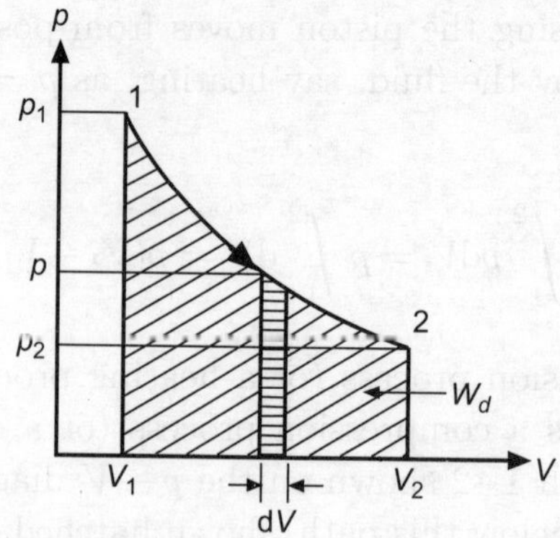

Fig. 2.10 p– V diagram for a hyperbolic process.

During this process,

$$p_1V_1 = pV = p_2V_2 = \text{C}$$

where C is a constant and hence $p = \dfrac{\text{C}}{V}$.

$$W_d = \int_1^2 p\text{d}V = \int_1^2 \left(\frac{\text{C}}{V}\right)\text{d}V = \text{C}\int_1^2 \left(\frac{\text{d}V}{V}\right) = \text{C}\ln\left(\frac{V_2}{V_1}\right)$$

Therefore,

$$W_d = p_1V_1 \ln\left(\frac{V_2}{V_1}\right) = p_2V_2 \ln\left(\frac{V_2}{V_1}\right) \tag{2.5}$$

Also, as

$$\left(\frac{V_2}{V_1}\right) = \frac{p_1}{p_2}$$

$$W_\text{d} = p_1V_1 \ln\left(\frac{p_1}{p_2}\right) = p_2V_2 \ln\left(\frac{p_1}{p_2}\right) \tag{2.6}$$

2.5.4 Process pV^n = Constant

This process, during which the pressure of a system varies inversely with V^n, where n is the exponent of compression or expansion, is called a *polytropic process.* Figure 2.11 shows the path of one such expansion process on p– V coordinates. During this process,

$$p_1V_1^n = pV^n = p_2V_2^n = \text{C}$$

where C is a constant and hence $p = \dfrac{\text{C}}{V^n}$.

$$\begin{aligned}W_\text{d} &= \int_1^2 p\text{d}V = \int_1^2 \left(\frac{\text{C}}{V^n}\right)\text{d}V = \text{C}\int_1^2 \left(\frac{\text{d}V}{V^n}\right) = \text{C}\int_1^2 V^{-n}\text{d}V \\ &= (\text{C}V_2^{1-n} - \text{C}V_1^{1-n})/(1-n) = (p_2V_2^nV_2^{1-n} - \text{p}_1V_1^nV_1^{1-n})/(1-n) \\ &= (p_2V_2 - p_1V_1)/(1-n) \\ &= (p_1V_1 - p_2V_2)/(n-1)\end{aligned} \tag{2.7}$$

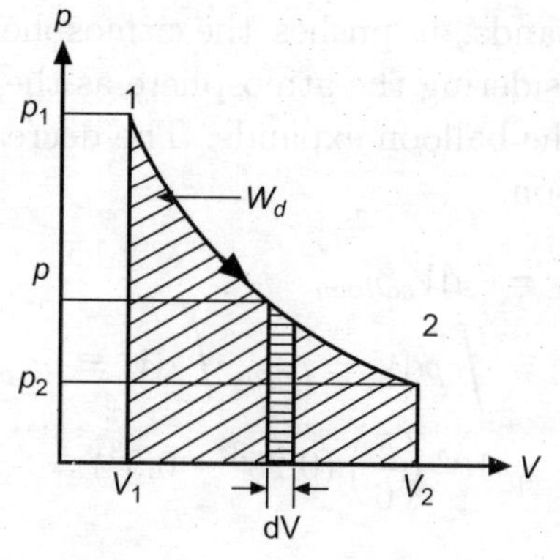

Fig. 2.11 $p-V$ diagram for a polytropic process.

Example 2.4

(a) A spherical balloon has a diameter of 25 cm and contains air at a pressure of 1.5 bar. The diameter of the balloon increases to 30 cm in a certain process during which the pressure in the balloon is observed to be proportional to the diameter. Calculate the magnitude and direction of work for air inside the balloon. (b) Given that the atmospheric pressure is 1 bar find the magnitude and direction of work for the atmosphere and the balloon.

► **Data:**

$p_1 = 1.5 \times 10^5$ N/m^2, $d_1 = 0.25$ m, $d_2 = 0.3$ m, $p_{atm} = 10^5$ N/m^2.

► **Solution:**

(a) As p is proportional to d, $p =$ kd, where k is the *proportionality constant.*

Therefore, $$p_1 = \text{k}d_1, \text{ or k} = \frac{p_1}{d_1} = \frac{1.5 \times 10^5}{0.25} = 6 \times 10^5 \text{ N/m}^3.$$

$$p_2 = \text{k}d_2 = 6 \times 10^5 \times 0.3 = 1.8 \times 10^5 \text{ N/m}^3.$$

As the balloon is spherical, $$V = \frac{\pi d^3}{6}, \text{ d}V = \frac{3\pi d^2 \text{d}d}{6} = \left(\frac{\pi d^2}{2}\right) \text{d}d$$

$$W_{air} = \int_1^2 p\text{d}V = \int_1^2 \text{k}d\left(\frac{\pi d^2}{2}\right)\text{d}d = \int_1^2 \left(\frac{\text{k}\pi}{2}\right) d^3 \text{d}d$$

$$= \frac{(\text{k}\pi/2)[d_2^4 - d_1^4]}{4}$$

$$= (6 \times 10^5) \times \left(\frac{\pi}{8}\right)\left[0.3^4 - 0.25^4\right] = 988 \text{ Nm}$$

(W_{air} is positive as air does work on the balloon fabric to increase the latter's diameter).

(b) While the balloon expands, it pushes the atmosphere at a constant pressure of $p_{atm} = 10^5$ N/m^2. Considering the atmosphere as the system, its volume decreases during the process as the balloon expands. The decrease is equal to the increase in the volume of the balloon.

Therefore, $\mathrm{d}V_{atmosphere} = -\mathrm{d}V_{balloon}$

$$W_{atm} = \int p\mathrm{d}V = p_{atm}\int \mathrm{d}V = p_{atm}[-(V_2 - V_1)]$$
$$= 10^5\left(\frac{\pi}{6}\right)(0.25^3 - 0.3^3) = -596 \text{ Nm}$$

The work done by the air in the balloon is 988 Nm. Work done on the atmosphere is 596 Nm.

As $$W_{air} + W_{balloon} + W_{atm} = 0$$
$$W_{balloon} = -(W_{air} + W_{atm}) = -(988 - 596) = -392 \text{ Nm.}$$

(***Note***: This work goes into stretching the balloon fabric. If the balloon fabric were to be inelastic and light then this work would have been zero, and the work done by the air would have been equal to the work done on the atmosphere.)

Example 2.5

In a quasi-static process, a certain pure substance is expanded from 10 bar, 0.1 m^3 to 2 bar, 0.35 m^3. Find the work interaction for the pure substance.

▶ **Data:**

$p_1 = 10$ bar, $V_1 = 0.1$ m^3, $p_2 = 2$ bar, $V_2 = 0.35$ m^3.

▶ **Solution:**

Let the expansion process follow the path pV^n = constant.

As $$p_1V_1^n = p_2V_2^n,\ (V_2/V_1)^n = p_1/p_2$$

or, $$n = \frac{\ln(p_1/p_2)}{\ln(V_2/V_1)}$$
$$= \frac{\ln(10/2)}{\ln(0.35/0.1)} = 1.2847$$

Now, $$W_d = (p_1V_1 - p_2V_2)/(n-1) \tag{2.8}$$
$$= \frac{(10 \times 10^5 \times 0.1 - 2 \times 10^5 \times 0.35)}{(1.2847 - 1)}$$
$$= 1.05374 \times 10^5 \text{ J}$$
$$= 105.37 \text{ kJ}$$

Example 2.6

(a) A certain mass of air is compressed from 1 bar, 0.1 m^3 to 6 bar in a piston–cylinder device according to $pV^{1.4}$ = constant. Find the work of compression for air. (b) Had the compression been carried out hyperbolically between the same initial state and the same final pressure as in (a), what would be the work done on air ?

► **Data:**

(a) $p_1 = 1.0$ bar, $V_1 = 0.1$ m^3, $p_2 = 6$ bar, and $p_1V_1^{1.4} = p_2V_2^{1.4}$

► **Solution:**

$$\begin{aligned} V_2 &= (p_1/p_2)^{1/1.4} V_1 \\ &= (1/6)^{1/1.4} \times 0.1 = 0.0278 \text{ m}^3. \\ W_d &= (p_1V_1 - p_2V_2)/(n-1) \\ &= \frac{10^5(1 \times 0.1 - 6 \times 0.0278)}{(1.4-1)} \\ &= -16{,}700 \text{ J} = -16.7 \text{ kJ} \quad \text{(i.e., 16.7 kJ of work is done on air)} \end{aligned}$$

► **Data:**

(b) $p_1 = 1.0$ bar, $V_1 = 0.1$ m^3, $p_2 = 6$ bar, and $p_1V_1 = p_2V_2$

► **Solution:**

$$\begin{aligned} V_2 &= \left(\frac{p_1}{p_2}\right) V_1 = \left(\frac{1}{6}\right) \times 0.1 = 0.0167 \text{ m}^3. \\ W_d &= p_1V_1 \ln\left(\frac{V_2}{V_1}\right) \\ &= 1 \times 10^5 \times 0.1 \times \ln\left(\frac{0.0167}{0.1}\right) = -17{,}897 \text{ J} = -17.9 \text{ kJ} \end{aligned}$$

2.5.5 Piston Displacement Work

From the engineer's point of view, the work interaction in a piston–cylinder device, such as in an internal combustion engine, is of utmost importance. To explain displacement work in Section 2.4.2, a piston–cylinder mechanism was considered which housed a fixed mass of a fluid. But this is not what really happens in a piston–cylinder device of an internal combustion engine. The mass of fluid in the cylinder does not remain fixed for it to be considered a system.

There is a steady flow of fuel–air mixture (in the case of spark ignition engines) or air (in the case of compression ignition engines) into the cylinder during the suction stroke (a stroke is the distance the piston travels between its two extreme positions during the reciprocating motion of the piston, and, each stroke corresponds to half a rotation of the crank shaft of the engine) and a steady flow of combustion products out of the cylinder during the exhaust stroke. During one machine cycle, besides suction and discharge strokes there could also be separate compression and expansion strokes. The net work output of the piston, called *indicated work*, is the algebraic sum of the work interactions the piston will have with the working substance in communication with it during all the strokes corresponding to a machine cycle.

This work is equal to $\int p\mathrm{d}V$ over the machine cycle. Pressure acting on the piston during each stroke changes with the position of the piston in the cylinder. Hence, to evaluate $\int p\mathrm{d}V$ it is necessary to know the pressure acting on the piston at each of its positions during the cycle.

2.5.5.1 Engine Indicator

An engine indicator is a device which measures the pressure against the piston for each of its positions during a machine cycle, when mounted on an engine being tested for its performance. The indicator marks a closed curve on the indicator card of the device. The height of the diagram is proportional to the pressure and the length is proportional to the stroke of the engine. The area of the diagram, thus, is proportional to the net work on the piston in a machine cycle.

2.5.5.2 Mean Effective Pressure

Mean effective pressure (abbreviated usually either as *mep* or p_m) is an imaginary constant pressure difference between the forward and return motions of the piston, over the stroke length, which would develop the same indicated work as in the actual case of varying pressure.

The volume swept by the piston in a stroke is given by LA, where A, the cross-sectional area of the cylinder $=\left(\frac{\pi}{4}\right)D^2$, where, D is the bore of the cylinder and L is its stroke length. Thus, indicated work per machine cycle $= p_m LA$.

If the area of the indicator diagram $= a$ and its length $= l$, the mean pressure difference is then proportional to a/l.

Thus, $$p_m = (a/l)\mathrm{S} \tag{2.9}$$

where S, the proportionality constant, called the spring constant of the indicator, is defined as the displacement of the pencil point of the indicator for unit pressure on the indicator piston. S is given in $\mathrm{N/m^2 \cdot m}$, or simply in $\mathrm{N/m^3}$.

2.5.5.3 Indicated Power

Indicated power, P_{ind}, is the rate at which work is done on the piston in an internal combustion engine. Thus,

$$P_{ind} = \frac{(xy)p_m \text{LAN}}{60{,}000} \text{ kW}, \tag{2.10}$$

where, x = number of cylinders; $y = 1$ or 2, depending upon whether the engine is single-acting or double-acting (and, in the absence of any specific information y is taken to be one); p_m, the mep, in N/m^2 is either equal to area of the $p-V$ diagram divided by the stroke volume or equal to $\frac{aS}{l}$ and N is the number of engine cycles per minute (number of working strokes per minute). If n is the speed of the engine (crank-shaft speed) in rev/min, for a two-stroke engine $N = n$, and for a four-stroke engine $N = \frac{n}{2}$. (For details the reader may refer to some standard books on internal combustion engines.)

Example 2.7

A three-cylinder four-stroke internal combustion engine running at 500 rev/min has a bore of 7.5 cm and a stroke of 10 cm. A test on this engine gave an indicator diagram of 6 cm^2 area and a length of 5 cm. The spring constant of the indicator spring is 2 bar/mm. Find the mep, in kPa, and the indicated power in kW.

► **Data:**

$x = 3$, $y = 1$, four-stroke engine, $n = 500$ rev/min and therefore $N = \frac{n}{2} = 250$ cycles/min, $D = 0.075$ m, $L = 0.1$ m, $a = (6 \times 10^{-4})$ m^2, $l = 0.05$ m, $S = 2$ bar/mm $= (2 \times 10^5)/10^3 = 2 \times 10^8$ N/m^3.

► **Solution:**

$$p_m = \frac{a}{l}S = \left[\frac{(6 \times 10^{-4})}{0.05}\right] \times 2 \times 10^8 = 2.4 \times 10^6 \text{ N/m}^2 = 2400 \text{ kPa}.$$

$$\text{A} = \left(\frac{\pi}{4}\right)\text{D}^2 = \left(\frac{\pi}{4}\right) \times 0.075^2 = 4.418 \times 10^{-3} \text{ m}^2$$

$$P_{ind} = (xy)\left(\frac{p_m LAN}{60{,}000}\right)$$

$$= \frac{3 \times 1 \times (2.4 \times 10^6) \times 0.1 \times (4.418 \times 10^{-3}) \times 250}{60{,}000} = 13.25 \text{ kW}$$

2.6 Displacement Work in Unresisted Processes

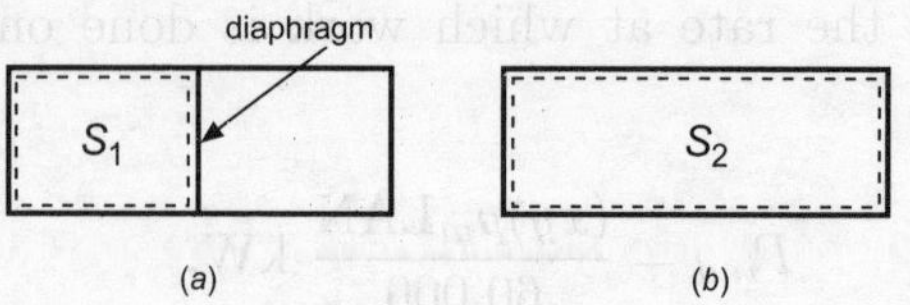

Fig. 2.12 Unresisted process.

An unresisted process is not a quasi-static process. It can be shown that displacement work in an unresisted process is equal to zero.

Consider, as shown in Figure 2.12*a*, a cylindrical container divided into two halves by a diaphragm. Let the right-hand side portion be evacuated and the left-hand side portion be filled with a gas to a certain pressure.

Let the gas, enclosed initially in the boundary S_1 be considered the system. Suppose the diaphragm is suddenly ruptured by some means, the gas in the left part of the vessel suddenly moves towards right to fill the cylinder completely and the final system boundary will be S_2, as shown in Figure 2.12*b*. Although the system boundary sweeps a volume in this process, it does so against zero pressure. Hence, the process is unresisted. The displacement work in such a process is given by $W_d = \int p\mathrm{d}v = 0$, as the resisting pressure $p = 0$.

Alternatively, the initial system boundary can also be considered to be the total internal volume of the cylinder, comprising the left and right portions. This system boundary still contains the same quantity of gas in the left portion, as the right portion has no material. When the diaphragm is ruptured and the gas expands rapidly to occupy the entire cylinder, the imagined system boundary does not change. Thus the system does not sweep any volume. As $\mathrm{d}V = 0$, there is no displacement work.

2.7 Other Modes of Work

2.7.1 Shaft Work

The importance of shaft work in engineering lies in the fact that most power plants deliver their output through rotating shafts. Shaft work, W_s, is evaluated as follows:

$$W_s = \int \delta W_s = \int T\mathrm{d}\theta = T\theta \tag{2.11}$$

where T is the steady torque on the shaft and θ is the angle of rotation of the shaft under the influence of T. The unit of W_s is Nm.

Expressing θ in terms of n, the speed of the shaft in rev/min, and the duration in minutes,

$$W_s = T(2\pi n\tau) = 2\pi nT\tau \tag{2.12}$$

Shaft power output p_s is calculated from:

$$P_s = \frac{2\pi nT}{60{,}000} \text{ kW} \tag{2.13}$$

Example 2.8

A four-stroke, six-cylinder petrol engine with a bore of 10 cm and a stroke of 15 cm develops a power of 24.87 kW in the engine crank-shaft. The torque in the crank shaft is 474.9 Nm. An engine indicator mounted on the engine gives an indicator diagram of length 5 cm and area 9.37 cm^2. The spring constant is 0.5 bar/mm. Find the mechanical efficiency of the engine.

▶ Data:

$N = n/2$, where n = engine speed, $x = 6$, $y = 1$ (assumed), $A = \frac{\pi d^2}{4} = \frac{\pi(0.1)^2}{4}$ m^2, $L = 0.15$ m, $P_{shaft} = 24.87$ kW, $T = 474.9$ Nm, $l = 0.05$ m, $a = 9.37 \times 10^{-4}$ cm^2, $S = 0.5 \times 10^5/10^{-3}$ N/m^3 $= 0.5 \times 10^8$ N/m^3.

▶ Solution:

$$p_m = \frac{aS}{l} = 9.37 \times 10^{-4} \times 0.5 \times 10^8/0.05 = 937{,}000 \text{ N/m}^2.$$

$$P_{ind} = (xy)p_m \frac{LAN}{60{,}000} \text{ kW} \tag{2.14}$$

$$= (6 \times 1) \times 937{,}000 \times 0.15 \times \left[\frac{\pi(0.1)^2}{4}\right] \times \frac{n}{2 \times 60{,}000}$$

$$= 0.055194 \text{ nkW}$$

$$P_{shaft} = \frac{2\pi nT}{60{,}000} \tag{2.15}$$

$$= 2\pi n \times 474.9/60{,}000 = 0.04973\ n \text{ kW}$$

$$\text{Mechanical Efficiency} = \frac{P_{shaft}}{P_{ind}} = 0.04973\ n/0.055194\ n = 0.9 = 90\%$$

2.7.2 Electrical Work

When a steady current of I amp flows out of a system at a point A and reenters it at a point B, the electrical potential of A being V volts higher than B, the system is said to do positive electrical work at a rate of P_e given by

$$P_e - IV \tag{2.16}$$

The unit for P_e is J/s or W.

Therefore, electrical work in a duration of τ minutes is given by

$$W_e = IV \times 60\tau \tag{2.17}$$

The unit for W_e is Nm or J

Example 2.9

(a) A fluid contained in a horizontal cylinder with a frictionless leak-proof piston is continuously agitated by a stirrer passing through the cylinder cover. The cylinder diameter is 400 mm. During a stirring process lasting 10 min the piston moves slowly through a distance of 490 mm against the atmospheric pressure of 1 bar. The net work done on the fluid during the process is 1965 Nm. If the speed of the electric motor driving the stirrer is 750 rev/min, compute the torque in the driving shaft and shaft power output of the motor. (b) The electric motor in (a) is supplied with a current of 0.9 amp from a 24 V battery. Compute the net work for the battery and the motor.

► **Data:**

$d = 0.4$ m, $\tau = 10$ min, $L = 0.49$ m, $p_{atm} = 1$ bar, $W_{net} = -1965$ Nm, $n = 750$ rev/min, $I = 0.9$ A and $V = 24$.

► **Solution:**

The schematic of the arrangement is shown in Figure 2.13.

(a) Considering the fluid in the cylinder as the system, stirrer work (W_{str}) is done on the fluid and in turn the fluid does work on the piston (W_d). If W_{net} is the net work for the fluid, then

$$W_{net} = W_{str} + W_d$$

Here, $$W_{net} = -1965 \text{ Nm and } W_d = \int p\mathrm{d}V$$

$$\therefore \quad W_d = p_{atm} \times \text{diplacement volume} = p_{atm} \times \left(\frac{\pi d^2}{4}\right) \times L$$

$$= 1 \times 10^5 \times \left(\frac{\pi \times 0.4^2}{4}\right) \times 0.49 = 6157.5 \text{ Nm}$$

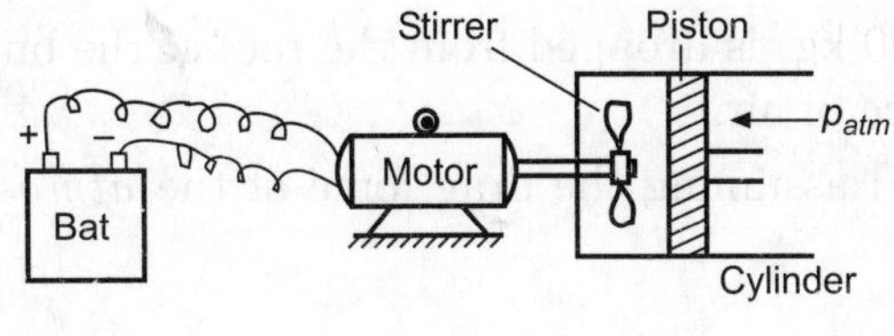

Fig. 2.13 Schematic for solving Example 2.9.

Therefore, $\quad -1965 = W_{str} + 6157.5$

or $\quad W_{str} = -8122.5$ Nm (negative sign indicates that the work is done on the fluid)

$$P_{shaft} = \frac{W_{str}}{\text{duration in seconds}} = \frac{8122.5}{10 \times 60} = 13.54\text{W}$$

$$= \frac{2\pi nT}{60} = 2\pi \times 750 \times \frac{T}{60}$$

Therefore, $\quad T = 0.172$ Nm

(b) Considering the battery as the system, the work done by it is given by:

$$W_{bat} = IV\tau = 0.9 \times 24 \times 10 \times 60 = 12{,}960 \text{ Nm}$$

The motor as a system has two work interactions to take into account, namely the electrical work from the battery and the shaft work delivered by it. Therefore,

$$W_{motor} = W_{ele} + W_s$$

But, $\quad W_{ele} = -W_{bat}$ and $W_s = -W_{str}$

Thus, $\quad W_{motor} = -12{,}960 + 8122.5 = -4837.5$ Nm

Illustrative Problems

2.1 Evaluate the work done in the following processes. The systems to be considered are in italics:

(a) Bags of building materials, each of 50 kg mass, have to be transported from the ground to the roof of a building, which is 4 m above the ground level. This is done by hooking *a bag* to one end of *a rope*, which passes over *a pulley* , and *a person* pulls the other end of the rope slowly. Neglect air resistance and the friction in the bearings of the pulley. Also consider the rope to be light and inextensible. Assume $g = 9.807$ ms^2.

(b) *A person* slowly lowers *a bag* (of mass 50 kg) from the roof of the building to the ground, using the same pulley and rope in problem (a).

(c) *A bag* (of mass 50 kg) is dropped from the roof of the building as in problem (a). Neglect drag force of air.

(d) Redo problem (c) assuming the drag force of the *atmosphere* on the bag to be 5 N.

▶ **Solution:**

(a) The arrangement is shown in Figure 2.14.

(i) Consider the pulley as the system. It rotates against the friction in its bearing. As the friction is given to be zero, the pulley has to do no work for its movement (unresisted process). Thus, $W_{pulley} = 0$.

(ii) Consider the rope as the system. Work is to be done on the rope to lift it. As the rope is supposed to be light, this work is negligible. Also, as the rope is inextensible, no work is done in stretching it. Thus, $W_{rope} = 0$.

(iii) Considering the person pulling the rope as the system, it may be seen that the only effect external to the person is the raising of the bag. Therefore, work is done by the person.

$$\begin{aligned} W_{person} &= \text{weights raised in the surroundings} \times \text{distance} \\ &= \text{mass of bag } \times g \times \text{distance through which it moves up} \\ &= 50 \times 9.807 \times 4 = 1961.4 \text{ Nm} = 1.9614 \text{ kJ} \end{aligned}$$

The surroundings for the person are the pulley, the rope and the bag. Therefore,

$$W_{person} + W_{pulley} + W_{rope} + W_{bag} = 0$$

that is $\quad 1.9614 + 0 + 0 + W_{bag} = 0$, or $W_{bag} = -1.9614\,\text{kJ}$.

(Work is done by the person on the bag.)

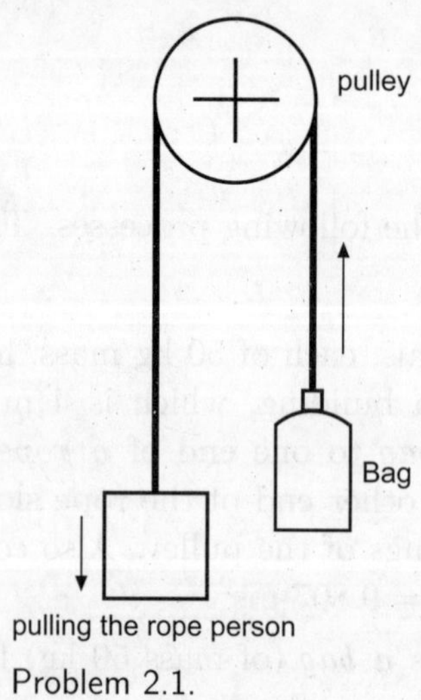

Fig. 2.14 Figure to solve Illustrative Problem 2.1.

(b) Figure 2.14 can be considered here as in part (a). The bags are being lowered slowly. This does not mean that the person is pushing the bag down. Slowly, the bags are going down against the resistance offered by the person. Considering the bag as the system, as there is no change in pulley and the rope, the only effect external to it is equal to the rise in resistance offered by the person, which may be assumed to be equal to an equivalent weight attached to the other end of the rope. For slow movement, the resistance offered is equal to the weight of the bags. Thus,

$$W_{bag} = \text{resistance offered by the person} \times \text{distance imagined to be moved up}$$

$$= (50g) \times 4 = (50 \times 9.807) \times 4 = 1.9614 \text{ kJ}$$

As $\quad W_{bag} + W_{pulley} + W_{rope} + W_{person} = 0$

$$1.9614 + 0 + 0 + W_{person} = 0, \quad \text{or} \quad W_{person} = -1.9614 \text{ kJ}$$

(Work is done by the bag on the person.)

(c) As the bag drops down, the only resistance it may have to overcome is air-resistance (drag force of air). As the drag force of air is given to be negligible, the bag has no surroundings to work against. Its fall is thus equal to an unresisted process. Hence, $W_{bag} = 0$.

(d) When air-resistance is considered, the bag moves down against this resistance. Therefore, there is work interaction between the bag and the atmosphere. Imagine that the bag and the 5 N weight, which is the atmospheric resistance, are connected to either ends of a light, inextensible cable passing over a frictionless pulley. The bag moves down against the atmospheric resistance that goes up by an equal distance. Therefore, the only effect external to the bag is the raising of 5 N.

W_{bag} = resisting weight lifted up × distance = 5 × 4 = 20 Nm. As $W_{bag} + W_{atm} = 20 + W_{atm} = 0$, or $W_{atm} = -20$ Nm

(Work is done by the bag on the atmosphere.)

2.2 A $p-V$ diagram of an engine is as shown in Figure 2.15. Determine the mep.

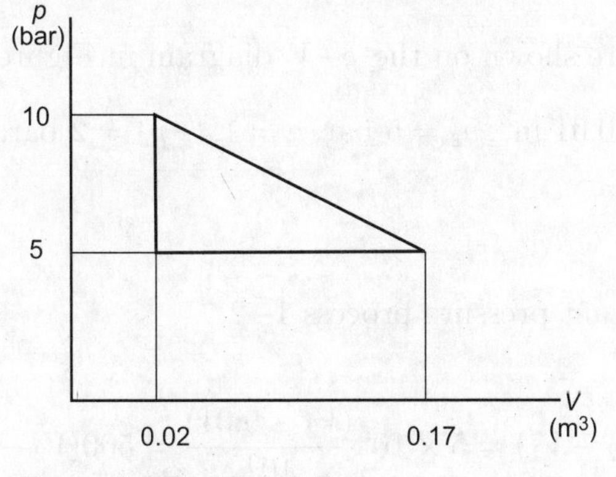

Fig. 2.15 $p-V$ diagram for Illustrative Problem 2.2.

► **Solution:**

$$\begin{aligned} p_m &= \text{average difference in pressure over the stroke length} \\ &= \text{area of the } p\text{–}V \text{ diagram/stroke volume} \\ &= \text{area of the triangle/stroke volume} \\ &= 0.5 \times [(10-5) \times 10^5 \times (0.17-0.02)]/(0.17-0.02) \\ &= 2.5 \times 10^5 \text{ N/m}^2 = 2.5 \text{ bar} \end{aligned}$$

2.3 A gaseous system undergoes three quasi-static processes in sequence. The gas initially at 5 bar, 0.01 m^3 is expanded at constant pressure. It is then further expanded according to the relation $pV^{1.4}$ = constant to 2 bar, 0.025 m^3. The gas is then returned to its initial state during which process pV= constant. Calculate the work interaction in each of the three processes and the net work for the system.

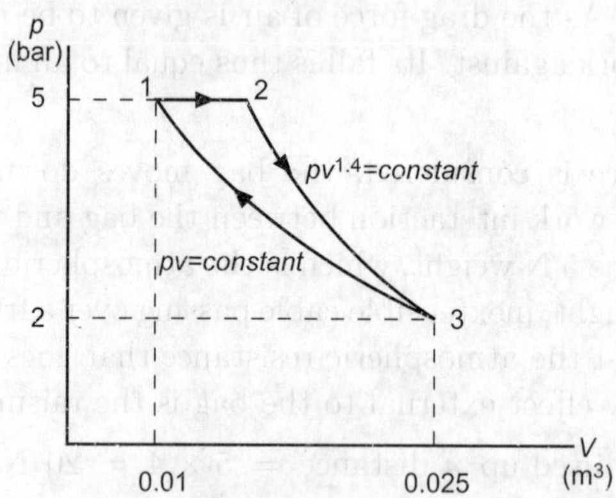

Fig. 2.16 p– V diagram for Illustrative Problem 2.3.

► **Data:**

The three processes are shown on the p– V diagram in Figure 2.16.

$$p_1 = 5 \text{ bar}, V_1 = 0.01 \text{ m}^3, p_2 = 5 \text{ bar}, n = 1.4, \ p_3 = 2 \text{ bar}, V_3 = 0.025 \text{ m}^3.$$

► **Solution:**

Considering the constant pressure process 1–2,

$$W_{1-2} = p(V_2 - V_1) = 5 \times 10^5 \frac{(V_2 - 0.01)}{10^3} = 500(V_2 - 0.01) \text{ kJ} \quad \text{(i)}$$

As process 2–3 follows the path $pV^{1.4}$ = constant,

$$p_2 V_2^{1.4} = p_3 V_3^{1.4}$$

or $$V_2 = V_3\left(\frac{p_3}{p_2}\right)^{1/1.4} = 0.025 \times \left(\frac{2}{5}\right)^{1/1.4} = 0.013\ \text{m}^3.$$

Substituting for V_2 in (i),

$$W_{1-2} = 500 \times (0.013 - 0.01) = 1.5\ \text{kJ}$$

Considering the process 2–3 that follows the path $pV^{1.4}$ = constant,

$$W_{2-3} = \frac{(p_2V_2 - p_3V_3)}{(1.4-1)} = \frac{(5 \times 0.013 - 2 \times 0.025) \times 10^5}{0.4}$$
$$= 3750\ \text{J} = 3.75\ \text{kJ}.$$

Similarly, for the process 3–1, during which pV = constant,

$$W_{3-1} = p_3V_3 \ln\,(V_1/V_3) = 2 \times 10^5 \times 0.025 \ln\,(0.01/0.025)$$
$$= -4581\ \text{J} = -4.581\ \text{kJ}.$$
$$W_{net} = W_{1-2} + W_{2-3} + W_{3-1} = 1.5 + 3.75 - 4.58 = 0.67\ \text{kJ}.$$

2.4 Consider the system consisting of air in a frictionless vertical piston–cylinder mechanism, shown in Figure 2.17. The cylinder diameter is 20 cm. Initially, as shown in Figure 2.17, when the spring just touches the top surface of the piston but does not exert any force on it, the pressure and volume of the system are found to be 90 kPa and 0.1 m^3. The mass of the piston is 16 kg and the atmospheric pressure is 100 kPa. In a certain process, during which the force exerted by the spring on the piston is proportional to the square of the displacement of the piston from its initial position, the piston rises slowly but steadily. When the piston is half-way between the two stops (when the volume of air is 0.15 m^3) the system pressure is found to be 200 kPa. Finally, the piston rests against the top stops and the system pressure is 550 kPa. (a) Show the process undergone by the system on a p–V diagram. (b) Calculate the net work interaction for (i) the air, (ii) the piston, (iii) the spring and (iv) the atmosphere.

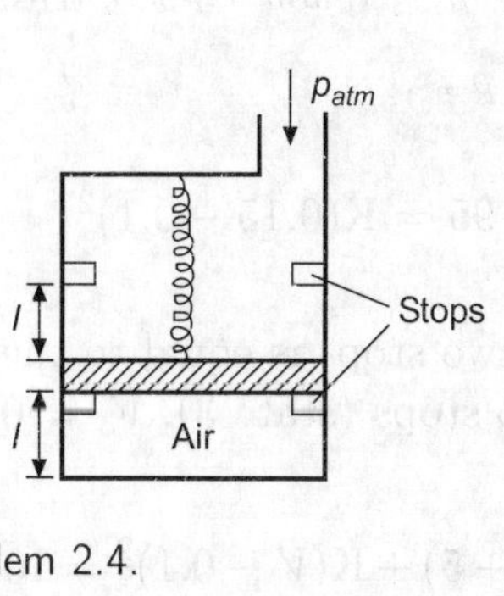

Fig. 2.17 Figure for Illustrative Problem 2.4.

▶ **Solution:**

(a) Initially (state 1) the piston rests on the stops. The pressure and volume of air are $p_1 = 90$ kPa and $V_1 = 0.1$ m^3. The piston can rise from this position only when the system pressure p_2 reaches the resisting pressure. At this instant (state 2), while $V_2 = V_1$, $p_2 = p_{atm} + p_{spr} + p_{piston}$. As per data, at this instant the spring will just touch the piston without exerting any pressure. Thus $p_{spr} = 0$.

Therefore, $p_2 = p_{atm} + p_{piston} = 100 +$ weight of piston/area

$$= 100 + \left\{ \frac{(16 \times 9.81)}{[(\frac{\pi}{4}) \times (0.2)^2]} \right\} \times 10^{-3}$$

$$= 100 + 5 = 105 \text{ kPa} \qquad \text{(i)}$$

From state 2, the piston starts rising until it just touches the upper stops. During this process the pressure of air inside keeps increasing due to the varying force exerted by the spring. It is given that the spring force is proportional to the square of the piston displacement, say L. But, p_{spr} is proportional to spring force as $p_{spr} = F_{spr}/A$, where A the area of cross-section of the cylinder is constant.

Therefore, $p_{spr} = \text{K}L^2$, where K is a constant (ii)

Let V be the volume of air when the piston is displaced up by a certain distance L from the bottom stops.

$$\text{The volume swept by the piston} = (V - 0.1) = \left(\frac{\pi}{4}\right) d^2 L$$

As $\left(\frac{\pi}{4}\right) d^2 L = (V - 0.1)$, L is proportional to $(V - 0.1)$,

or L^2 is proportional to $(V - 0.1)^2$

Substituting for L^2 in (ii), $p_{spr} = \text{K}(V - 0.1)^2$. (iii)

It is given that when $V = 0.15$ m^3, $p_{air} = 200$ kPa where p_{air} is given by

$$p_{air} = (p_{atm} + p_{pis}) + p_{spr}$$

Therefore, $200 = 100 + 5 + p_{spr}$,

Or, $p_{spr} = 95$ kPa.

Substituting for p_{spr} in (iii), $95 = \text{K}(0.15 - 0.1)^2$

Or, K = 38,000 kPa/m^6.

As the volume between the two stops is equal to the initial volume, the volume of air when the piston touches the top stops (state 3), $V_3 = 0.2$ m^3. The pressure at state 3 is given by

$$p_3 = (p_{atm} + p_{pis}) + p_{spr} = (100 + 5) + \text{K}(V - 0.1)^2 = 105 + 38{,}000(0.2 - 0.1)^2 = 485 \text{ kPa.}$$

The final pressure p_4 is given to be 550 kPa. Hence, between states 3 and 4 the pressure changes from 485 to 550 kPa at a constant volume of 0.2 m^3. The pressure and volumes at the four states, 1, 2, 3 and 4 are represented on a $p-V$ plane in the diagram shown in Figure 2.18.

(b) $W = \int_1^4 p dv$ = area bound by the path 1-2-3-4 with the x axis.

$$= \int_1^2 p dV + \int_2^3 p dV + \int_3^4 p dV$$

$$= 0 + \int_2^3 p dV + 0, \text{ as 1–2 and 3–4 are constant-volume processes.}$$

$$= \int_2^3 [p_{atm} + p_{pis} + p_{spr}] dV$$

$$= \int_2^3 p_{atm}\, dV + \int_2^3 p_{pis}\, dV + \int_2^3 p_{spr}\, dV$$

$$= \int_2^3 100\, dV + \int_2^3 5\, dV + \int_2^3 38{,}000(V - 0.1)^2\, dV$$

$$= 100(V_3 - V_2) + 5(V_3 - V_2) + \frac{38{,}000[(V_3 - 0.1)^3 - (V_2 - 0.1)^3]}{3}$$

$$= 100(0.2 - 0.1) + 5(0.2 - 0.1) + \frac{38{,}000[(0.2 - 0.1)^3 - (0.1 - 0.1)^3]}{3}$$

$$= 10 + 0.5 + 12.67 = 23.17 \text{ kJ}.$$

(ii) Work done on the atmosphere = 10 kJ and therefore $W_{atm} = -10$ kJ
Work done on the piston = 0.5 kJ and therefore $W_{pis} = -0.5$ kJ
Work done on the spring =12.67 kJ and therefore $W_{spr} = -12.67$ kJ

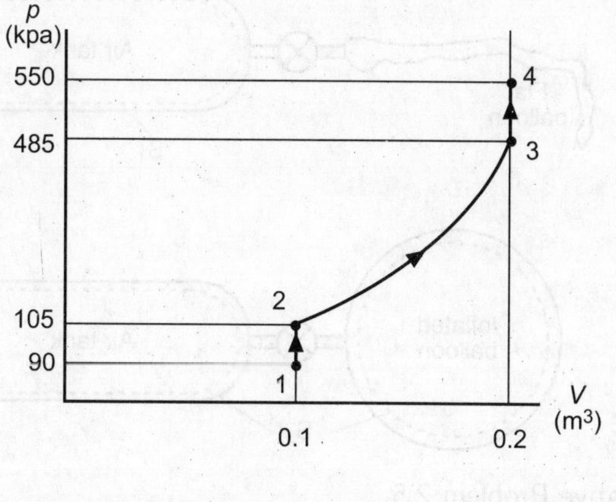

Fig. 2.18 $p-V$ diagram.

2.5 A balloon made from a light, inelastic material which is initially flat is inflated by slowly filling it with air from a tank of compressed air. The final volume of the balloon is 5 m^3. The barometer reads 76 cm Hg. Considering the air in the tank, balloon and the connecting pipe as the system, determine the magnitude and direction of work. Assume the density of Hg to be 13.6 g/ml.

▶ **Solution:**

In Figure 2.19, while (a) shows the initial system boundary S_1, (b) shows the final system boundary S_2.

$$W = W_{air\ in\ tank} + W_{air\ in\ balloon}$$

$W_{air\ in\ tank} = 0$, as air inside the tank does not sweep any volume.

Therefore, $W = W_{air\ in\ balloon} = \int p dV$, where p is the resisting pressure for the expansion of air. This is equal to the atmospheric pressure as the balloon fabric offers no resistance to either lift it as it gets filled by air (because the balloon fabric is light) or stretch it (because the fabric is inelastic).

$$\begin{aligned} W &= p_{atm}(V_2 - V_1), \text{ where } (V_2 - V_1) = \text{volume of the balloon} \\ &= (\rho_{\text{Hg}}\ g\ h_{\text{Hg}}) \times \text{balloon volume} \\ &= (13.6 \times 10^3 \times 9.81 \times 0.76) \times 5 = 506{,}981 \text{ Nm} \approx 507 \text{ kJ} \end{aligned}$$

2.6 Determine the work done by the atmosphere as air slowly enters an evacuated rigid bottle of volume 0.1 m^3. Assume that the atmospheric pressure is 100 kPa.

▶ **Solution:**

Broken lines in Figure 2.20*a* and *b* show, respectively, the initial and final system boundaries, S_1 and S_2.

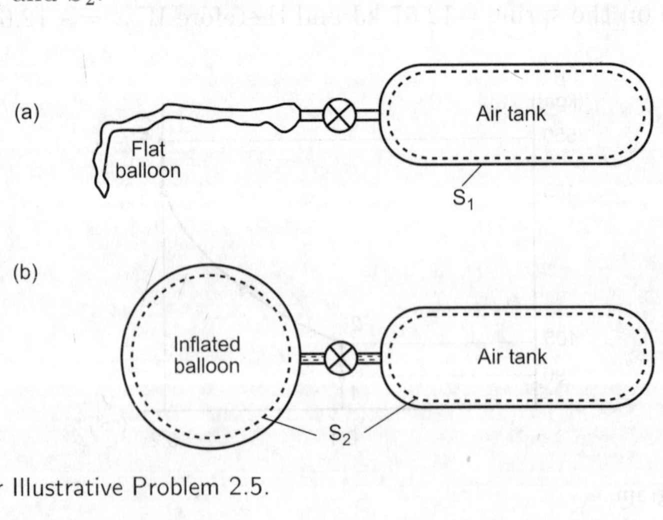

Fig. 2.19 Figure for Illustrative Problem 2.5.

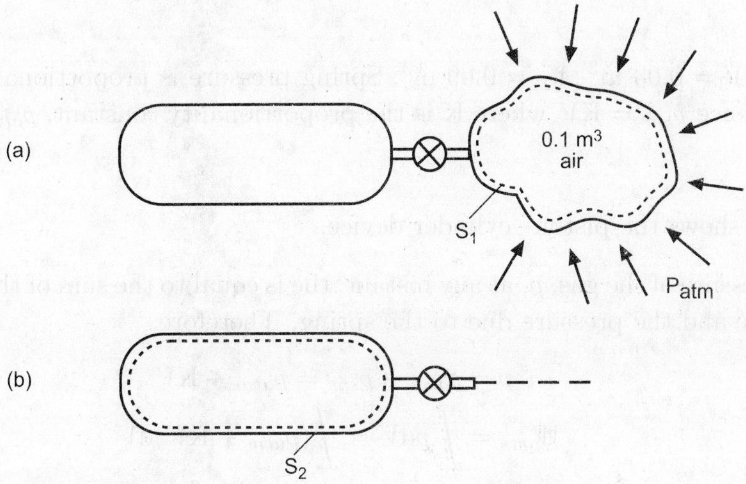

Fig. 2.20 Initial and final system boundaries.

This is a resisted process, as air enters the bottle slowly. A volume of 0.1 m^3 of air from the atmosphere which would ultimately enter the bottle could be imagined to be the system. Therefore,

$$W = \int_{free\ air\ boundary} p\mathrm{d}V + \int_{air\ in\ bottle} p\mathrm{d}V$$

The second term is zero as the air in the bottle does not sweep any volume as the bottle is given to be rigid. Therefore,

$$W = \int_{free\ air\ boundary} p\mathrm{d}V = p_{atm}(V_2 - V_1)$$

Here the initial volume of the free air boundary is $= 0.1\ \mathrm{m}^3$ and the final volume is zero.

Hence, $W = 100 \times (0 - 0.1) = -10$ kJ (Work is done by the rest of the atmosphere on the air that enters the bottle finally).

2.7 In a frictionless, horizontal piston–cylinder mechanism, the piston is forced against the gas by a spring which exerts a force directly proportional to the volume of the gas. In addition to the spring force, the atmospheric pressure of 1 bar acts upon the outer side of the piston. (a) Considering the gas as the system, calculate the work when from an initial state of 0.03 m^3, 2 bar the gas volume increases to 0.09 m^3. Plot the state changes on the $p-V$ plane. (b) Considering the spring as the system find the work interaction. (c) Account for the difference in the magnitudes of work in (a) and (b). (d) What conclusion can be drawn from a study of the $p-V$ diagram?

▶ **Data:**

$p_1 = 2$ bar, $V_1 = 0.03$ m^3, $V_2 = 0.09$ m^3, Spring pressure is proportional to the spring force and hence $p_{spr} = \mathrm{K}V$ where K is the proportionality constant, $p_{atm} = 1$ bar.

▶ **Solution:**

Figure 2.21 shows the piston–cylinder device.

(a) The pressure of the gas, p, at any instant, the is equal to the sum of the atmospheric pressure and the pressure due to the spring. Therefore,

$$p = p_{atm} + p_{spr} = p_{atm} + \mathrm{K}V$$

$$W_{gas} = \int \mathrm{pd}V = \int (p_{atm} + \mathrm{K}V)\mathrm{d}V$$

$$= p_{atm}(V_2 - V_1) + \mathrm{K}\frac{(V_2^2 - V_1^2)}{2} \tag{i}$$

Initially, $p = (p_{atm} + p_{spr})$

and therefore, $p_{spr} = p - p_{atm} = 2 - 1 = 1 \text{ bar} = 10^5 \text{ N/m}^2$

Also, initially, $p_{spr} = \mathrm{K}V_1 = \mathrm{K} \times 0.03$

or $$K = \frac{p_{spr}}{V_1} = \left(\frac{10^5}{0.03}\right) \text{ N/m}^2 \cdot \text{m}^3$$

Substituting for K in (i),

$$W_{gas} = 10^5 \times (0.09 - 0.03) + \left(\frac{10^5}{0.03}\right) \times \frac{(0.09^2 - 0.03^2)}{2}\text{J} \tag{ii}$$

$$= 6 + 12 = 18 \text{ kJ}.$$

As per the data, pressure of the gas varies linearly with volume according to $p = (p_{atm} + KV)$.

Hence, $p_2 = 10^5 + \left(\frac{10^5}{0.03}\right) \times 0.09 = 4 \times 10^5 \text{ N/m}^2 = 4$ bar. With these results the p– V diagram can now be drawn as in Figure 2.22.

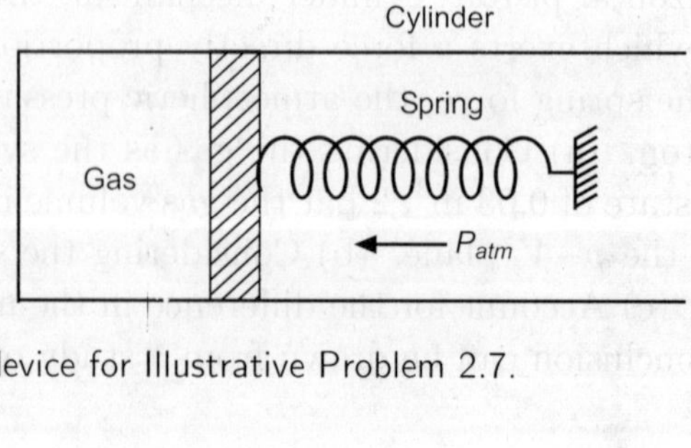

Fig. 2.21 Piston–cylinder device for Illustrative Problem 2.7.

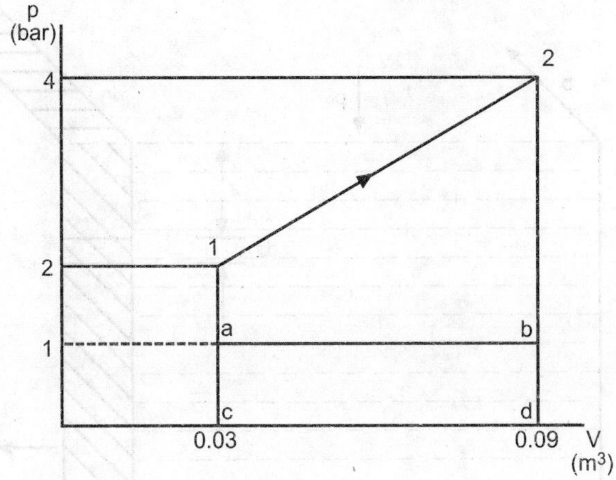

Fig. 2.22 $p-V$ diagram.

(b) In (ii) the first term on the right-hand side = work done by the gas on the atmosphere and the second term = work done by the gas on the spring. They are positive with respect to the gas and hence, $W_{spr} = -12$ kJ.

(c) Magnitude of work in (a) = 18 kJ and the magnitude of work in (b) = 12 kJ. Therefore, the difference between the two = 6 kJ. This must be equal to work done on the atmosphere. Therefore, $W_{atm} = -6$ kJ. This can also be confirmed from the fact that

$$W_{sys} + W_{surrounding} = 0$$

That is, $$W_{gas} + (W_{atm} + W_{spr}) = 0$$

Therefore, $18 + (W_{atm} - 12) = 0$ or $W_{atm} = -6$ kJ

(d) A study of the $p-V$ diagram reveals the same result.

Area below 1–2 = area 1–2–d–c–1 = 18 kJ = magnitude of W_{gas}.

Area below a–b = area a–b–d–c–a = 6 kJ = magnitude of W_{atm}.

Area 1–2–b–a–1 = (area 1–2–d–c–1) – (area a–b–d–c–a) = 12 kJ = magnitude of W_{spr}.

2.8 Consider Problem 1.21. Calculate the work done by the liquid on the piston when the length L is increased slightly, but slowly, from L_1 to L_2. Express your answers in terms of p_o, b, Z_1, L_1, L_2 and g where z_1 corresponds to L_1. It is given that the volume of water in the channel remains constant.

► **Solution:**

Referring to Figure 2.23, at any point on the piston at a depth of Z, the pressure p is given by $p = p_0 + \rho g Z$.

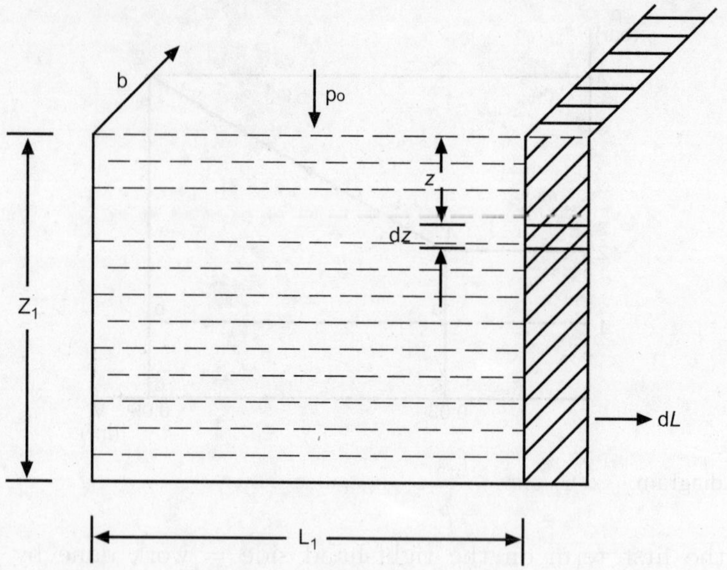

Fig. 2.23 Schematic diagram for Illustrative Problem 2.8.

The force, F, exerted by water on a small element of piston of area $(b \times \mathrm{d}Z)$, at a depth Z, is given by,

$$F = p \times \text{area} = (\rho_0 + \rho g Z)(b \times \mathrm{d}Z)$$

As the piston moves out a distance of d L in the direction of F, slowly, work done by the water on the element of area b dZ is given by,

$$\delta W = F\mathrm{d}L = (p_o + \rho g Z)(b \times \mathrm{d}Z)\mathrm{d}L$$

As the piston moves through a small distance of $L_2 - L_1$, slowly, the depth of water in the tank decreases from Z_1 to Z_2. Therefore, Z may be assumed to vary from zero to $(Z_1 + Z_2)/2$. Therefore,

$$W = \int \delta W = \int_{L_1}^{L_2} \int_0^{(Z_1+Z_2)/2} (p_o + \rho g Z)(b \times \mathrm{d}Z)\mathrm{d}L$$

$$= p_o b \left[\frac{(Z_1 + Z_2)}{2}\right](L_2 - L_1) + \rho g b \left[\frac{(Z_1 + Z_2)}{2}\right]^2 (L_2 - L_1)$$

As the volume of water remains the same in the channel, $bZ_1L_1 = bZ_2L_2$, and therefore, $Z_2 = Z_1L_1/L_2$. Substituting for Z_2 in the above equation for W,

$$W = b(L_2 - L_1)\left(\frac{Z_1}{2}\right)\left(1 + \frac{L_1}{L_2}\right)\left[p_o + \rho g Z_1 \left(\frac{1 + L_1/L_2}{4}\right)\right]$$

2.9 A sealed vessel of negligible mass having the shape of a rectangular prism with a base area of 150 cm^2 and a height of 15 cm floats on the surface of a large lake. Calculate the work required to move the vessel vertically to the bottom of the lake. Assume the following: Density of water = 1000 kg/m^3, Depth of lake is = 2 m and atmospheric pressure is 1 bar.

▶ **Data:**

A, the bottom surface area of the vessel $= 150 \times 10^{-4} = 0.015$ m^2, $L_1 = 15/100 = 0.15$ m, $\rho = 1000$ kg/m^3, $L_2 = 2$ m.

▶ **Solution:**

Consider an instant when the bottom surface of the vessel is at a depth of $L (L < L_1)$ in the tank, as shown in Figure 2.24.

At the instant shown in the figure the resisting force for the downward movement of the vessel is given by

$$F = A \times [\text{(pressure on the bottom surface of the vessel)} - \text{(pressure on the top surface of the vessel)}]$$

$$= A[(p_{atm} + \rho g L) - p_{atm}] = \rho A g L$$

Work required to move down the vessel through a small distance $\mathrm{d}L$ is given by

$$W = F\mathrm{d}L = \rho A g L \mathrm{d}L$$

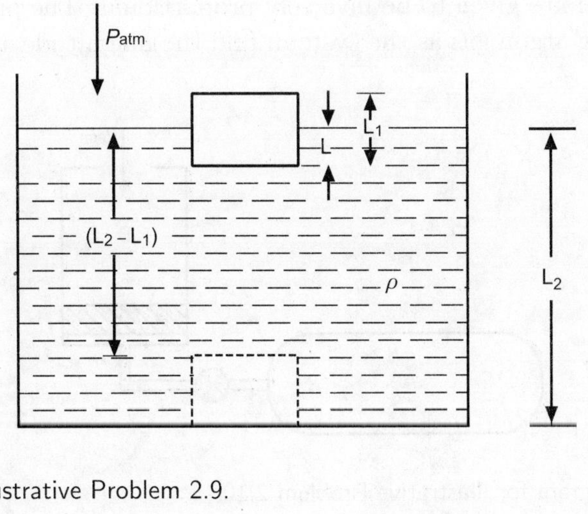

Fig. 2.24 Figure for Illustrative Problem 2.9.

Work required to move the vessel until the top surface of the vessel becomes flush with the water level in the tank is given by

$$W_1 = \int \delta W = \int_{L=0}^{L=L_1} \rho AgL dL = \rho AgL_1^2/2$$

$$= \frac{1000 \times 0.015 \times 9.81 \times 0.15^2}{2} = 1.655 \text{ J}$$

During further movement of the vessel to the bottom of the tank, at any instant, the resisting force is constant and is given by

$$F = A \times \text{pressure due to a column of water } L_1 = \rho AgL_1$$

The vessel moves down a distance of $(L_2 - L_1)$ against this constant force F. The work during this process is given by

$$W_2 = F(L_2 - L_1) = \rho AgL_1(L_2 - L_1)$$

$$= 1000 \times 0.015 \times 9.81 \times 0.15 \times (2 - 0.15) = 40.83 \text{ J}$$

The net work required to move the vessel from top to bottom of the tank is given by,

$$W = W_1 + W_2 = 1.655 + 40.83 = 42.485 \text{ J}$$

2.10 As shown in Figure 2.25, a tank 'A' of volume 400 litres is connected to a large vertical cylinder B through a valve. Cylinder B is fitted with a leak-proof piston of negligible mass. Initially, the piston is at the bottom of the cylinder. Friction between the piston and the cylinder can be neglected. The top surface of the piston is open to the atmosphere (p_{atm}= 101.325 kPa). Tank A contains argon gas at 250 kPa and 30°C. The valve connecting tank A and cylinder B is opened. Argon gas enters the cylinder to raise the piston slowly. During this process, the pressure and volume of argon are given to be inversely proportional. The piston comes to a stop finally. Treating argon gas as the system, find the magnitude and direction of work.

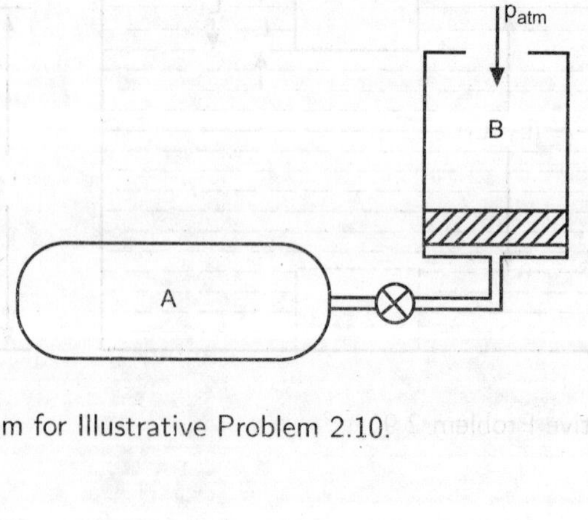

Fig. 2.25 Schematic diagram for Illustrative Problem 2.10.

▶ **Data:**

$p_1 = 250$ kPa, $t_1 = 30°$ C, $V_1 = 400 \times 10^{-3}$ m^3, mass of the piston is negligible, $p_{atm} = 101.325$ kPa. During the process p is inversely proportional to V and therefore $p_1V_1 = p_2V_2$.

▶ **Solution:**

When the valve is opened, argon gas enters the cylinder to raise the piston slowly up against the atmospheric pressure. The piston stops rising when the pressure of argon equals the atmospheric pressure.

Hence, $p_2 = p_{atm} = 101.325$ kPa.

Because $p_1V_1 = p_2V_2$,

$$V_2 = \frac{p_1V_1}{p_2} = \frac{250 \times 400 \times 10^{-3}}{101.325} = 0.987 \text{ m}^3$$

But $V_2 = V_1 +$ piston displacement volume

Therefore, piston displacement volume $= V_2 - V_1 = 0.987 - 400 \times 10^{-3} = 0.587$ m^3.

$$W_{Ar} = \int_A p\text{d}V + \int_B p\text{d}V$$

The first term on the right-hand side is zero, as there is no change in volume in A. Hence,

$$W_{Ar} = W_{atm} = \int_B p\text{d}V$$

As the mass of the piston is negligible and as the piston moves without friction, work done by the argon on the piston is equal to the work done by the piston on the atmosphere. Therefore,

$$\begin{aligned} W_{Ar} &= \int p_{atm}\text{d}V = p_{atm}\int \text{d}V \\ &= p_{atm} \times \text{piston displacement volume} \\ &= 101.325 \times 0.587 = 59.478 \text{ kJ.} \end{aligned}$$

Problems

2.1 Indicate whether the work in the following cases is positive, negative or zero. The systems to be considered are in italics:

(a) *A coil spring* stands on end on a table. *A book* placed on it compresses the spring.

(b) *A body* is lowered by *a crane* through a certain vertical distance.

(c) *A paddle* wheel run by a motor stirs *a liquid* in a vessel.

(d) When the plunger of a tyre pump is pushed down it forces *air* into the tyre.

2.2 Evaluate the magnitude and direction of work interaction in the following processes, for the systems indicated in italics:

(a) *An agent* slowly raises *a body* of mass 2 kg through a vertical distance of 3 m at a place where g = 9.78 m/s^2.

(b) *A body* of mass 15 kg falls freely through a vertical distance of 10 m at a place where $g = 9.7$ m/s^2. Neglect the drag force offered by *the atmosphere.*

(c) Redo problem (b), assuming the drag force to be 5 N.

(d) *The electric motor* of a crane is supplied with a current of 14 amp from a 200 V *DC Supply* for a period of 15 s while the crane lifts a *machine* of mass 2000 kg through a vertical distance of 2 m. The weight of the inextensible *cable* and the resistance of the *pulleys* over which the cable passes are negligible. Assume $g = 9.81$ m/s^2.

2.3 A four-cylinder, single-acting four-stroke gas engine of 150-mm bore develops an indicated power of 18 kW while running at 215 rev/min. Calculate the area of the indicator diagram that would be obtained. Assume that the engine stroke is 10 times the length of the indicator diagram and that the spring constant of the indicator is 250×10^5 N/m^3.

2.4 A single-acting, single-cylinder water pump has a cylinder of 20 cm diameter and a piston stroke of 30 cm. The machine cycle consists of two strokes of piston, namely, the suction stroke at a constant pressure of 110 kPa and a delivery stroke at a constant pressure of 400 kPa. The pump is driven at a speed of 60 rev/min. Find the rate, in kW, at which the piston does work on water.

2.5 A single-acting, four-stroke, three-cylinder petrol engine of bore, 10 cm, stroke, 15 cm, develops a torque of 475 Nm at the output shaft running at 300 rev/min. An engine indicator with a spring constant of 0.5 bar/mm is mounted on the engine. The indicator diagram length and area are, respectively, 5 cm and 18.75 cm^2. Find the mechanical efficiency (ratio of shaft power output to the indicated power) of the engine.

2.6 A series of indicator diagrams for a diesel engine cylinder yields the following average results: area of the indicator diagram = 1.5 cm^2; length of the diagram = 5 cm; spring constant of the indicator = 16×10^6 N/m^3; the engine speed = 300 rev/min; cylinder bore = 25 cm and the engine stroke = 45 cm. The engine takes two shaft revolutions to complete each machine cycle. Find the power developed by the engine.

2.7 A four-stroke, six-cylinder petrol engine of bore, 10 cm, and a stroke of 15 cm, develops a torque of 800 Nm at the output shaft. An engine indicator mounted on the engine gives a diagram of length 5 cm. The spring constant of the engine indicator is 0.5 bar/mm. Find the area of the indicator diagram, in cm^2, if the ratio of shaft power output to the indicated power is 0.8.

2.8 A single-cylinder, double-acting, two-stroke oil engine has a cylinder diameter of 200 mm and a stroke of 250 mm. When the engine is run at a speed of 300 rev/min, the mep is found to be 9 bar. Evaluate the work done per cycle and the indicated power.

2.9 Figure 2.26 shows the $p-V$ diagram of an engine cycle. What is the mep in N/m^2?

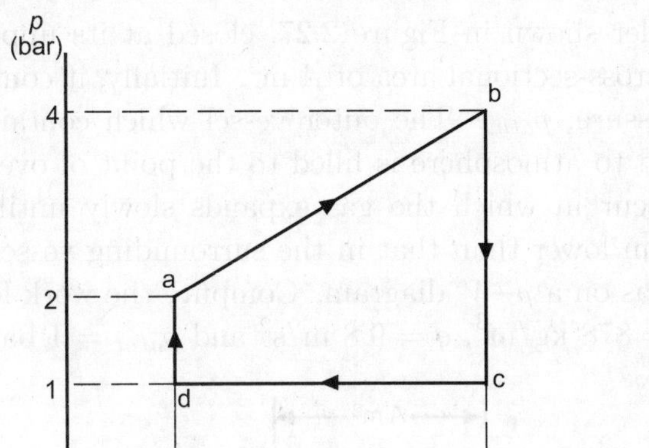

Fig. 2.26 Diagram for Problem 2.9.

2.10 A system consisting of a quantity of gas exerts a steady pressure of 1.5 bar (gauge) on a 25-cm diameter piston as the piston moves slowly a distance of 15 cm. How much work is done by the system? Assume atmospheric pressure to be 1 bar.

2.11 Consider a piston–cylinder machine. The external normal force in N, applied to the piston, is given by: $F = (-7+15L)\times 10^3$ where L in m is the distance from the closed end of the cylinder to the piston. How much work is done on the piston when it moves from the position $L = 1$ m to $L = 1.5$ m. Sketch the $p-V$ diagram for the process undergone by the fluid in the machine and show the work graphically.

2.12 The relation between the pressure and volume of a gaseous system during an expansion process is $p = 2V + 320$, where p is in N/m^2 and V is in m^3. Find the magnitude and direction of work for the system as it expands from an initial volume of 0.2 m^3 to a final volume of 0.4 m^3.

2.13 A spherical balloon has a diameter of 25 cm and contains air at a pressure of 1.5 bar.

(a) In a certain process, the diameter of the balloon increases to 30 cm during which the pressure is proportional to the diameter. Calculate the work done by the air inside the balloon.

(b) Given that the balloon expands against the atmospheric pressure of 1 bar, calculate the work done by the balloon on the atmosphere.

(c) Account for the difference in works in (a) and (b).

2.14 A rigid cylinder is evacuated and closed with a valve. When the valve is partly opened, air from the atmosphere enters the bottle slowly. Given that the atmospheric pressure is 1.013×10^5 N/m^2, calculate the work interaction for 1 m^3 of air that enters the bottle.

2.15 The inverted cylinder shown in Figure 2.27, closed at its upper end and open at the bottom, has a cross-sectional area of A m^2. Initially, it contains V_1 m^3 of a gas at atmospheric pressure, p_{atm}. The outer vessel which contains a liquid with its free surface exposed to atmosphere is filled to the point of overflowing. A process is now made to occur in which the gas expands slowly until the liquid level in the cylinder is 20 cm lower than that in the surrounding vessel. Show the process undergone by the gas on a $p-V$ diagram. Compute the work for the gas assuming $A = 0.01$ m^2, $\rho_{liq} = 878$ kg/m^3, $g = 9.8$ m/s^2 and $p_{atm} = 1$ bar.

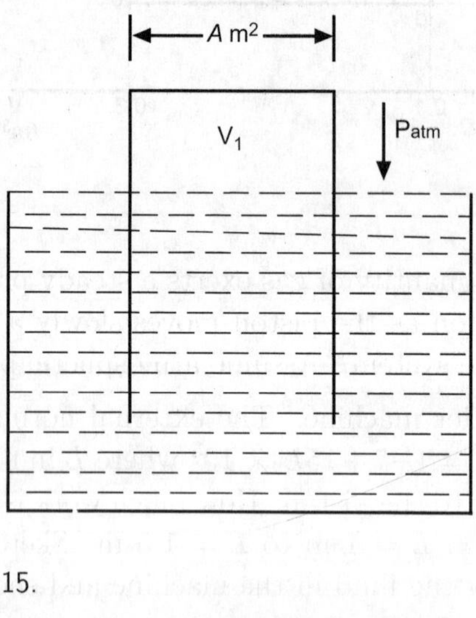

Fig. 2.27 Figure for Problem 2.15.

2.16 As shown in Figure 2.28, a horizontal cylinder of cross-sectional area 150 cm^2 is fitted with a freely floating, frictionless, leak-proof piston, there being no piston-rod. The cylinder, which initially contains 0.0045 m^3 of air at atmospheric pressure of 1 bar, is connected to a light, inextensible balloon; the balloon is initially flat. The other end of the cylinder is connected via a valve to a rigid vessel containing nitrogen at high pressure. The balloon is slowly inflated by admitting nitrogen to the other end of the cylinder during which the piston moves through a distance of 250 mm. When fully inflated, the volume of the balloon is 0.003 m^3. Any change in the air pressure, p, is related to the corresponding volume, V, by pV = constant. (a) Sketch the trace that would be obtained on an indicator connected to the air side of the cylinder and (b) evaluate the work for each of the following systems: atmosphere, balloon, air, piston and nitrogen.

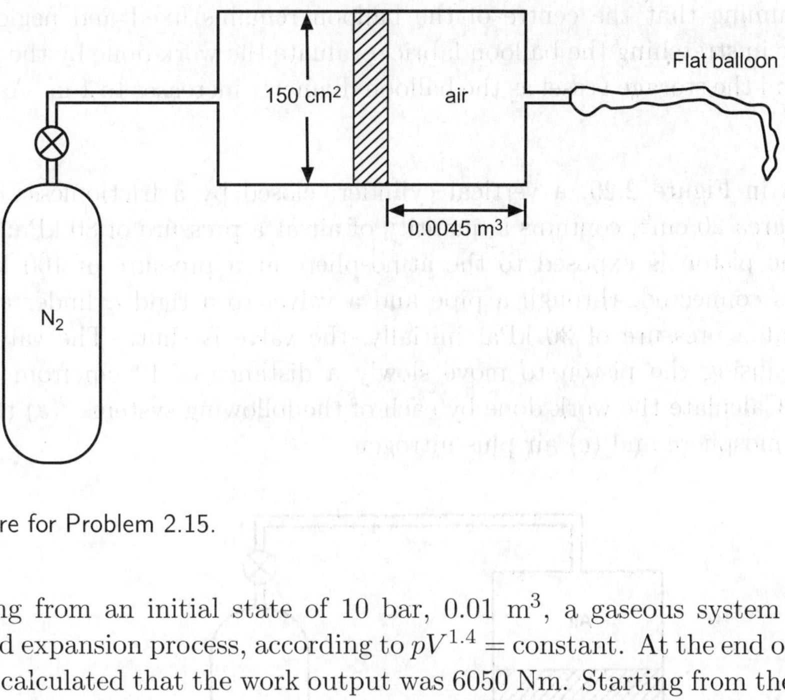

g. 2.28 Figure for Problem 2.15.

.17 Starting from an initial state of 10 bar, 0.01 m^3, a gaseous system undergoes a resisted expansion process, according to $pV^{1.4}$ = constant. At the end of the process, it was calculated that the work output was 6050 Nm. Starting from the same initial state, had the system undergone a resisted expansion process, according to pV = constant, to the same final volume as in the previous case, what would be the work output? Sketch the two processes on a p– V diagram .

.18 A piston–cylinder device contains initially 1 kg of a fluid at 1 bar, 0.8 m^3. The fluid is compressed to 5 bar according to $pV^{1.4}$ = constant. At the end of this process, the fluid is subjected successively to a constant volume process and a constant pressure process that brings the fluid back to its initial state. (a) Show the cycle executed by the fluid on a p– V diagram. Calculate the work interaction in each process and thus the network for the cycle. (b) Find the mep for the cycle.

.19 A gaseous system undergoes a cycle with the following three quasi-static processes in sequence. From state 1 (p_1 = 1 bar, V_1 = 0.3 m^3) the system undergoes a constant pressure compression process to state 2 (V_2 = 0.2 m^3). Between states 2 and 3 (p_3 = 4 bar) the system undergoes a compression process according to pV^n =

constant. Between states 3 and 1 the gas expands according to pV = constant. Sketch the cycle on a $p-V$ plane and find the magnitude and direction of work in each of the processes and the net work for the system.

2.20 An open tank is filled to the brim with a liquid of density 1120 kg/m^3. A flexible spherical balloon 0.5 m in diameter is immersed in the liquid with its centre 2.4 m below the free surface of the liquid. Gas from a storage vessel is used to inflate the balloon thereby causing the tank to overflow. The atmospheric pressure is 1 bar. Assuming that the centre of the balloon remains fixed and neglecting the work done in stretching the balloon fabric, evaluate the work done by the gas in the balloon and the storage vessel as the balloon diameter increases to 1 m. Assume g = 9.81 m/s^2.

2.21 As shown in Figure 2.29, a vertical cylinder, closed by a frictionless, leak-proof piston of area 20 cm^2, contains a quantity of air at a pressure of 80 kPa; the lower face of the piston is exposed to the atmosphere at a pressure of 100 kPa. The cylinder is connected, through a pipe and a valve, to a rigid cylinder containing nitrogen at a pressure of 30 kPa; initially, the valve is shut. The valve is then opened, causing the piston to move slowly a distance of 12 cm from its initial position. Calculate the work done by each of the following systems: (a) the piston, (b) the atmosphere and (c) air plus nitrogen.

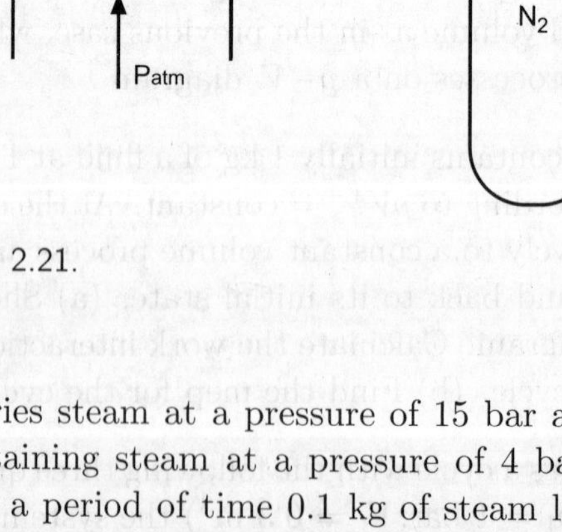

Fig. 2.29 Figure for Problem 2.21.

2.22 A steam main carries steam at a pressure of 15 bar and a temperature of 250°C. A rigid vessel containing steam at a pressure of 4 bar is connected, via a valve, to the main. Over a period of time 0.1 kg of steam leaks from the main into the vessel, thereby raising the pressure of the contents of the vessel to 5 bar. For the system comprising the steam finally in the vessel find the magnitude and sign of work interaction. (Take the specific volume of steam at 15 bar, 250°C to be $0.152\ m^3/kg$).

2.23 Using a water pump, a sealed overhead tank of volume 5 m^3 containing initially air at 1 bar is to be filled with water from a large lake. During the process of filling, as water enters the tank at its bottom, the air in the tank cannot escape and at any instant the air pressure is inversely proportional to the volume it occupies. When 80% of the tank is filled with water the average height of water in the tank is 20 m above the surface of the lake. The electric motor driving the pump draws 2.5 amp from the mains at 220 V. Given that the net work for the pump–motor combination is 2×10^5 Nm, find the time required, in min, for the given filling process to complete. Assume density of water to be 1000 kg/m^3.

2.24 In a frictionless piston–cylinder mechanism, the piston is forced against the gas by a spring which exerts a force proportional to the square of the volume of the gas. In addition to the spring force the atmospheric pressure of 1 bar acts upon the outer side of the piston. (a) Considering the gas as the system find the work interaction when from an initial state of 1 m^3, 2 bar the gas volume increases to 4 m^3. Plot the state changes on a $p-V$ plane. (b) Considering the spring as the system find the work interaction for the same process. (c) Account for the difference between the magnitudes of work in (a) and in (b).

2.25 Figure 2.30 shows a piston-cylinder mechanism employed to pump a corrosive fluid. The 250 mm diameter cylinder is closed by a light flexible diaphragm attached to the 150 mm diameter piston. The stroke of the piston is 220 mm and the length of the cylinder is 110 mm. The shape of the diaphragm in its extreme positions is the frustum of a cone. During the delivery stroke, an agent pushes the thin piston rod slowly inwards against the constant fluid pressure of $1.8 \times 10^5\ N/m^2$ gauge. The atmospheric pressure acts on the outer side of the diaphragm and piston. Evaluate the work done by the agent .

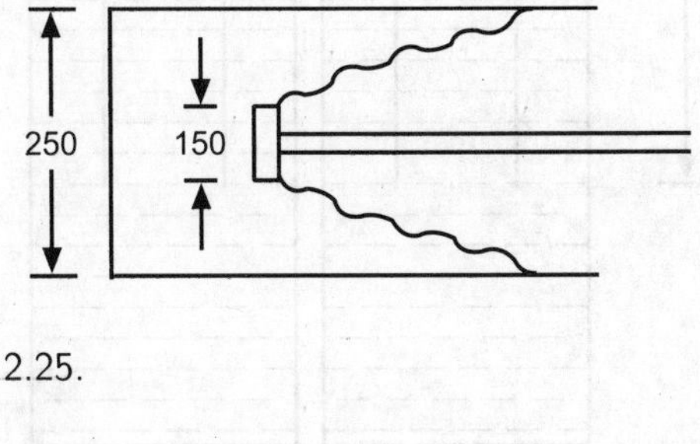

Fig. 2.30 Figure for problem 2.25.

2.26 Two inverted cylindrical vessels are immersed in two identical tanks as shown in Figure 2.31 and the vessels are kept fixed in position. The two tanks are filled to the brim, one with water (density =1000 kg/m^3) and the other with kerosene (density = 800 kg/m^3). The free surfaces of the liquids are exposed to atmosphe at 1 bar. Initially the pressure of air above the liquid in each of the two vesse is atmospheric. The two vessels are connected to a frictionless piston-cylind device containing nitrogen via two valves. The weight of the piston is negligib and initially the Bourdon pressure gauge reads zero (Figure 2.31 shows the initi configuration). After opening the valves a set of small weights are added, one aft another, on the piston such that the piston descends slowly. Finally, the pistc comes to a stop 4 cm from the bottom of the cylinder and the Bourdon pressu gauge now reads 0.4 bar. Considering air and nitrogen together as the syste determines the net work interaction. Sketch the process on a $p-V$ plane. Assun that the force on the piston varies linearly with piston displacement and negle the volume of the connecting pipes.

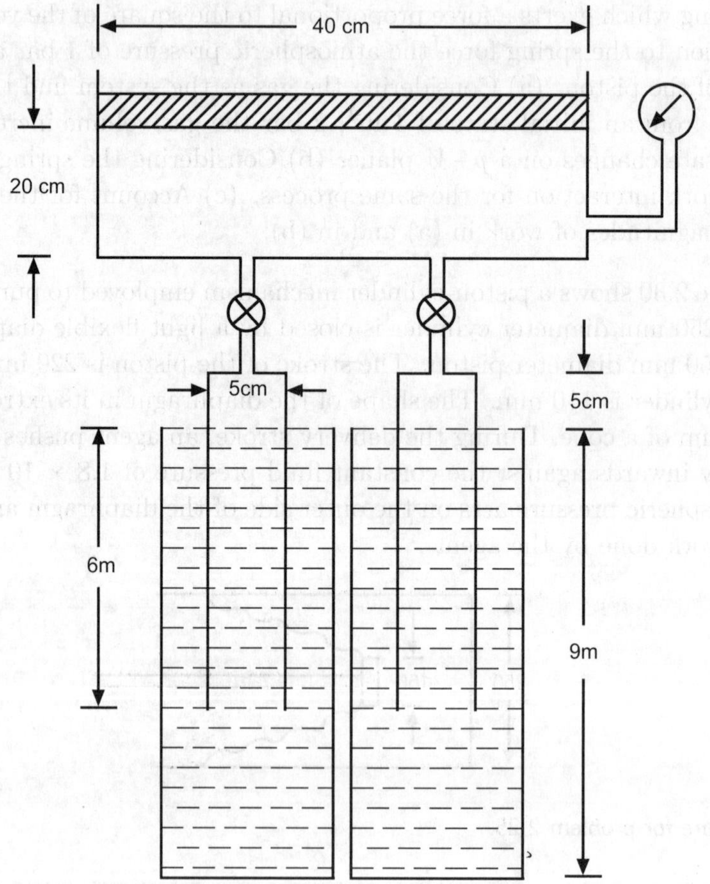

Fig. 2.31 Figure for Problem 2.26.

2.27 A fluid contained in a horizontal cylinder fitted with a frictionless leak-proof piston is continuously agitated by means of a stirrer passing through the cylinder cover. The cylinder diameter is 450 mm. During a stirring process occupying 10 min the piston moves slowly outwards to a distance of 220 mm against the atmospheric pressure of 1 bar. The net work done on the fluid during the process is 2000 Nm. (a) Given that the speed of the electric motor driving the stirrer is 600 rev/min, estimate the torque in the driving shaft and the shaft power output of the motor. (b) If the electric motor is supplied with a current of 0.40 amp from a 24 V accumulator, evaluate the net work for (i) the accumulator and (ii) the motor.

2.28 An electric resistance wired to the surroundings, is placed within the cylinder of a piston–cylinder machine. A current of 5 amp is established through the resistor with a drop of 100 V across it. At what speed must the piston move with a constant force of 90,000 N to maintain the rate of displacement work equal to electrical work?

2.29 The torque in an engine shaft, T (in Nm), is related to the speed of the shaft, ω (in rev/min), by $T = 195\cos(\pi\omega/2000)$. Determine the power output of the shaft when it is running at 475 rev/min.

2.30 A propeller of a ship is driven by a steam turbine through an 8:1 reduction gear. The mean-resisting torque imposed by the water on the propeller is 775×10^3 Nm and the shaft power delivered by the turbine to the reduction gear is 14 755 kW; the turbine speed is 1440 rev/min. Evaluate (a) the torque developed by the turbine, (b) the power delivered to the propeller shaft and (c) the net rate of work for the reduction gear.

2.31 In an internal combustion engine, the torque T in Nm, is found to vary with the shaft speed, n in rev/min, according to $T = 395\sin[\pi/2)(n/3000)]$. Calculate the power delivered by the engine at 1000 rev/min.

2.32 An electric device is working at a constant voltage of 120 V. The current drawn by the device varies with time according to $I = 9.57e^{-t}/60$ where I is in amp and t is in seconds. Determine the power delivered to the device at $t = 2$ s.

TEMPERATURE AND HEAT 3

3.1 Introduction

The most important concepts in thermodynamics are heat, work, systems and their properties. Except *heat* all other concepts have been dealt with in the earlier chapters. This chapter examines *heat*. The concept of *heat* is inextricably linked to *temperature* and therefore, before discussing heat, the concept of temperature is first explained.

3.2 Temperature

The concept of temperature originates from the way our senses react to hot and cold bodies. Among similar bodies, those that feel hotter to the touch are supposed to be at a higher temperature than others that feel colder or less hot. Experiments have shown that certain measurable characteristics of materials change with a change in temperature. Hence, measuring such changes is an indirect estimation of the change in temperature. In fact, thermometers, which are devices used to measure temperature, operate on this principle. Listed below are some changes that occur in materials as the temperature increases, that is, as the material moves from colder to hotter conditions. Also, against each, the nomenclature of the type of thermometer that works in response to the said change are also given in italics: (i) the length of a metal rod increases—*metallic thermometer*, (ii) the volume of a given mass of liquid increases—*mercury in glass thermometer*, (iii) the resistance of a metallic resistor increases—*resistance thermometer*, (iv) the resistance of a semiconductor decreases—*thermister*, (v) the volume of a gas at constant pressure increases—*constant pressure gas thermometer*, (vi) the pressure of a gas at constant volume increases—*constant volume gas thermometer*, (vii) when two dissimilar metals are connected to make a junction and two such junctions are connected in series there exists an emf in the circuit when the two junctions are exposed to different temperatures—*thermocouple*.

3.2.1 Temperature Equilibrium

It is a matter of common experience that when a hot body is brought in contact with a cold body, with the advent of time, the former feels less hot and the latter less cold. During this period, the properties of both bodies undergo changes. Eventually, both feel alike to our touch. The two bodies are then said to have the same temperature or to be in *temperature equilibrium* or *thermal equilibrium.* From this instant onwards, the properties of both bodies in contact cease to change. This fact may be rewritten as follows: *Two systems are said to be in thermal equilibrium if the properties of either system do not change when they are brought in communication with each other through a diathermic wall.* A *diathermic wall* is one that permits only heat interaction between the communicating systems and not work interaction. On the contrary, an adiabatic wall is one that may permit work interaction but not heat interaction.

3.2.2 Zeroth Law of Thermodynamics

The zeroth law of thermodynamics states that two systems are in thermal equilibrium if they are separately in thermal equilibrium with a third system.

In other words, if system S_1 is in thermal equilibrium with S_3 and S_2 is in thermal equilibrium with S_3, then S_1 and S_2 shall be in thermal equilibrium. This fact is further emphasized with the help of arrangements in Figures 3.1*a* and *b*. In Figure 3.1*a* two systems S_1 and S_2, separated by an adiabatic wall, are simultaneously in communication with system S_3 through a diathermic wall. The whole assembly is enclosed in an adiabatic enclosure. After sometime S_1 and S_2 reach thermal equilibrium with S_3. Once thermal equilibrium is reached, there will be no further changes in the properties of S_1 and S_2.

At this instant, if the adiabatic and the diathermic walls are swapped, as shown in Figure 3.1*b*, such that S_1 and S_2 are communicated through the diathermic wall, no change in properties of S_1 or S_2 will occur, thereby proving the zeroth law of thermodynamics.

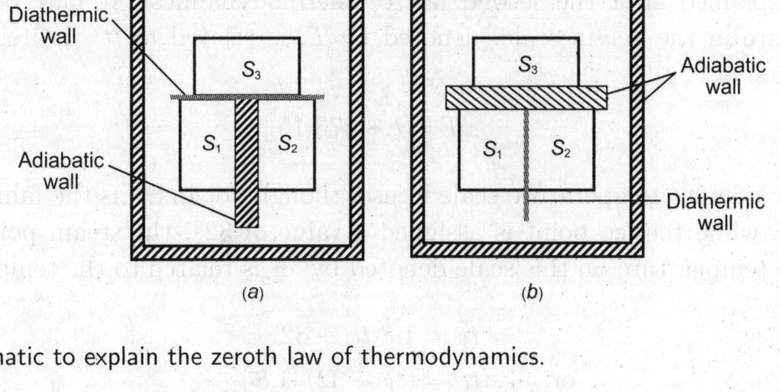

Fig. 3.1 Schematic to explain the zeroth law of thermodynamics.

3.2.3 Significance of the Zeroth Law

It can be said that temperature of a system is a property that determines whether or not a system is in thermal equilibrium with other systems. The significance of this law lies in the fact that it forms the basis for temperature measurement. The zeroth law is not a mathematical proposition, but represents the essence of empirical information on temperature and thermal equilibrium.

Let S_2 be a quantity of melting ice at standard atmospheric pressure, S_3 a thermometer and S_1 a system whose temperature is to be measured. The temperature of S_2 *(ice point)* is universally assigned a fixed value of zero degree (0°C) on the celsius scale of temperature. This is one of the System International fixed points. Let the thermometer S_3 be brought in communication with S_2. The mercury level in the capillary of the thermometer keeps dropping. Once equilibrium between S_2 and S_3 is reached, there is no further change in the level of mercury. Let this level be marked as 0°C, in accordance with the value assigned for the fixed point. Now, let S_1 be brought in communication with S_3 the thermometer. If the mercury level were to stay put at the zero mark, then S_1 and S_3, are in thermal equilibrium. According to the zeroth law, S_1 must then be in temperature equilibrium with S_2. As S_2 is assigned the value of 0°C, S_1 must also be at 0°C.

Similarly, considering boiling water at standard atmospheric pressure as S_2, which is given a fixed value of 100°C *(steam point)*, the above exercise can be repeated again to get the 100°C mark on the thermometer and use it to measure the temperature of S_1.

To measure temperatures between the *ice point* and the *steam point* the distance between the two marks are divided into 100 equal parts, and each graduation is marked as 1°C on the celsius scale.

3.2.4 Unit of Temperature

The unit of temperature in SI is the *degree celsius*, abbreviated as °C. One degree is equal to, as explained above, one hundredth of the difference between the steam point and the ice point. The other unit of temperature in SI is the kelvin, abbreviated as K. This unit will be explained after the second law of thermodynamics. It may be noted that the temperature in the kelvin scale, denoted by T, is related to 't' the temperature in the celsius scale by

$$T = t + 273.15 \tag{3.1}$$

The other popular temperature scale in use, though not in SI, is the fahrenheit scale. On this scale, while the ice point is assigned a value of 32°, the steam point is given 212°. Thus, the temperature on this scale denoted by °F is related to the temperature in °C by

$$t_F = 1.8\, t_C + 32 \tag{3.2}$$

$$\text{or,} \quad t_C = (t_F - 32)/1.8 \tag{3.3}$$

3.2.5 International Fixed Points

These are reproducible temperatures, usually phase change temperatures of certain substances. They are measured using primary gas thermometers and are assigned with certain fixed values which are universally accepted. Their primary purpose is to serve as standards or reference temperatures in calibrating thermometers in various ranges. Table 3.1 lists the international fixed points and the values assigned to them in degrees celsius.

Table 3.1 International fixed points in °C

1.	Normal boiling point of oxygen (oxygen point)	=	– 182.97
2.	Triple point of water	=	0.01
3.	Normal boiling point of water (steam point)	=	100.00
4.	Normal boiling point of sulphur (sulphur point)	=	444.60
5.	Normal boiling point of antimony (antimony point)	=	630.50
6.	Normal boiling point of silver (silver point)	=	960.80
7.	Normal boiling point of gold (gold point)	=	1063.00

Example 3.1

(a) The average temperature of the human body is 98.6°F. What is the value of this temperature in °C ? (b) The annual average ambient temperature in peninsular India is said to be 27°C. What is the corresponding temperature on °F ?

▶ Solution:

(a) $t_C = \dfrac{(t_F - 32)}{1.8}$

Therefore, $$t_C = \frac{(98.6 - 32)}{1.8} = 37°\text{C}$$

(b) $$t_F = 1.8\, t_C + 32$$

Therefore, $$t_F = 1.8 \times 27 + 32 = 80.6°\text{F}$$

Example 3.2

If the temperature of a system is 27°C, find its temperature in kelvin.

▶ Solution:

$T = t + 273.15 = 27 + 273.15 = 300.15 \text{ K} \approx 300 \text{ K}$ (and not 300° K)

Example 3.3

If the temperature of a system is 250 K, find its temperature in °C.

► **Solution:**

$t = T - 273.15 = 250 - 273.15 = -23.15°\text{C} \approx -23°\text{C}.$

3.3 Measurement of Temperature—Thermometers

3.3.1 Constant Volume Gas Thermometer

A gas thermometer is a primary thermometer in which the working fluid is a permanent gas. There are two types of gas thermometers, namely (i) constant volume gas thermometer and (ii) constant pressure gas thermometer. Shown in Figure 3.2 is a constant volume gas thermometer.

It consists of a tiny gas bulb 'B' connected to a U-tube mercury manometer through a capillary. The bulb contains a mass of permanent gas. The bottom of the manometer is connected to a mercury reservoir through a flexible tube. There is an indicating point marked at the top of the manometer limb connected to the bulb. The open limb and the mercury reservoir are exposed to atmosphere. The bulb is exposed to a system whose temperature is to be measured. When the bulb attains thermal equilibrium with the system the mercury reservoir is moved up or down until the mercury meniscus in the left limb coincides with the indicating mark. This procedure is a must in all measurements to ensure that, during each measurement, the volume of the gas considered is the same. The volume of the gas in the thermometer now equals the bulb volume plus the volume of the capillary and the left limb upto the indicating mark. The difference in height between the mercury menisci in the two limbs, 'h', is a measure of the gauge pressure of the gas in the bulb. The absolute pressure, p, of the gas is given by $p = p_o + \rho gh$. This pressure is proportional to the temperature of the gas in the sensing element. The value of the temperature obtained depends upon the nature of the gas. The value of this temperature, represented by θ, is suggested to be obtained from the relation,

$$\theta = \frac{273.16\, p}{p_{tp}} \tag{3.4}$$

in accordance with the basis of using a single fixed point to obtain the kelvin scale. In the above-mentioned relation, p_{tp} is the pressure read by the same constant volume thermometer, when its sensing bulb is exposed to triple point of water. The single fixed point used is the triple point of water (refer to Chapter 10) at which point, all the three phases of water are in equilibrium. The triple point temperature is 0.01°C which corresponds to a temperature of 273.16 K. The temperature θ can be converted to kelvin through the ideal gas temperature scale.

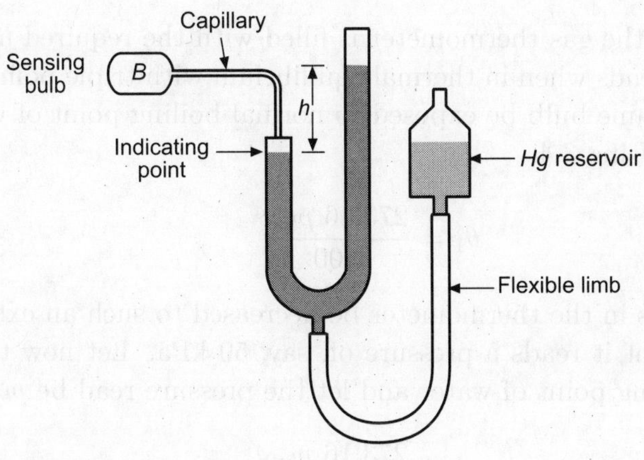

Fig. 3.2 A constant volume gas thermometer.

3.3.2 Ideal Gas Temperature Scale

Assuming that the permanent gas in the bulb of a constant volume gas thermometer behaves as an ideal gas (refer to Chapter 9), the pressure p, the volume V and the temperature T on ideal gas temperature scale are related by

$$pV = m\mathrm{R}T$$

In this relation, m the mass of the gas and R the particular gas constant and V the volume of the gas are constants and hence p is proportional to T. Most real gases do not strictly follow the above relation. Using θ to indicate the empirical real gas temperature, the ratio of two pressures read by the gas thermometers exposed to two different temperatures is given by

$$\frac{p_2}{p_1} = \frac{\theta_2}{\theta_1}$$

Applying this relation to the gas in the bulb of a constant volume gas thermometer initially exposed to triple point of water which is assigned a temperature of 273.16 K, and then to an unknown pressure p,

$$\frac{p}{p_{tp}} = \frac{\theta}{273.16}$$

that is, $$\theta = \frac{273.16\, p}{p_{tp}} \quad \text{as in (3.4)}$$

Now let us consider finding the ideal gas temperature corresponding to θ. As an example let us find the ideal gas temperature at steam point.

The sensing bulb of the gas thermometer is filled with the required mass of a real gas so that the pressure it reads when in thermal equilibrium with triple point of water is, say, 100 kPa. Let now the same bulb be exposed to normal boiling point of water and let the pressure read be p_{bp1}. Using 3.4,

$$\theta_1 = \frac{273.16\ p_{bp1}}{100}$$

Now, let the mass of gas in the thermometer be decreased to such an extent that when it is exposed to triple point it reads a pressure of, say, 50 kPa. Let now the same bulb be exposed to normal boiling point of water and let the pressure read be p_{bp2}. Using 3.4,

$$\theta_2 = \frac{273.16\ p_{bp2}}{50}$$

Let such experiments be conducted reducing the initial pressure gradually to a very low value.

Plot the values of θ with p_{tp} and extrapolate the resulting curve to $p_{tp} = 0$. It will be found that θ tends to 373.15 as p_{tp} tends to 0. And, as p tends to zero the gas tends to behave as an ideal gas (Chapter 9).

Similar sets of experiments can be carried out using different gases, such as H_2, N_2, air and O_2. In each case the result obtained shows that although the values of θ for a given p_{tp} may vary from gas to gas, depending upon its nature, the value of θ corresponding to steam point tends to be 373.15 as p_{tp} tends to zero in each case. This fact is shown in the plot of θ vs p_{tp} in Figure 3.3.

The ideal gas temperature is hence defined by

$$T = 273.16 \lim_{p_{tp} \to 0} \frac{p}{p_{tp}}$$

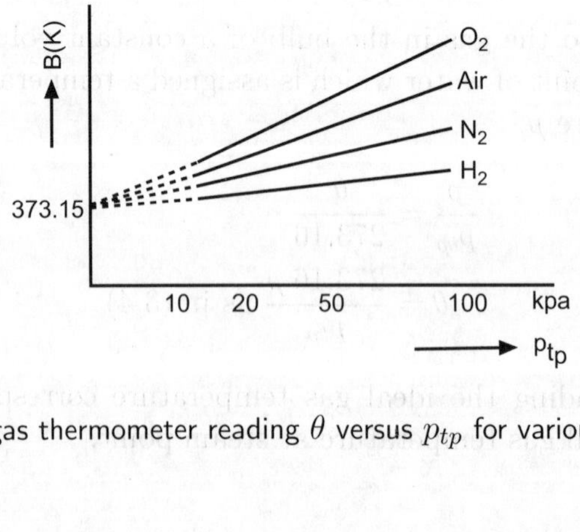

Fig. 3.3 Constant volume gas thermometer reading θ versus p_{tp} for various gases.

3.3.3 Constant Pressure Gas Thermometer

In this thermometer, the pressure of the gas in the sensing element is held constant and the volume that the gas now occupies is an indication of the temperature to which the bulb is exposed to. The constant volume gas thermometer, shown in Figure 3.2, can also be used as a constant pressure thermometer. In such a case, before reading the volume of gas each time, the reservoir is so moved that the difference in height 'h' between the two menisci is kept the same for all measurements.

3.3.4 Mercury-in-Glass Thermometer

This consists of a tiny sensing element (a small glass bulb) filled with a certain quantity of mercury. This mercury is in communication with an evacuated capillary. The thermometric scale is marked on the glass stem surrounding the capillary. When the sensing element is exposed to a system whose temperature is to be measured the volume of mercury in the sensing bulb changes and accordingly the mercury level in the capillary keeps changing and reaches a steady value when it attains thermal equilibrium. The temperature of the system is read from the scale.

Example 3.4

The readings t_A and t_B of two celsius thermometers A and B agree well at ice point (0°C) and the steam point (100°C). At other points between these two, the temperatures are related by $t_A = p + qt_B + rt_B{}^2$, where p, q and r are constants. When the two thermometers are immersed in a well-stirred oil bath, while A reads 51, B reads 50. (a) Determine what A reads when B reads 30. (b) Discuss the question of which thermometer is correct.

Solution:

$$t_A = p + qt_B + rt_B{}^2 \qquad \text{(i)}$$

From the data, when $\quad t_A = 0, t_B = 0$

Substituting for t_A and t_B in (i)

$$p = 0 \qquad \text{(ii)}$$

Also, when $t_A = 100, t_B = 100$, therefore, from (i),

$$100 = 0 + q100 + r100^2$$

or, $$q + 100r = 1 \qquad \text{(iii)}$$

Further, when $t_A = 51, t_B = 50$

and hence, $51 = q50 + r50^2$

or, $$q + 50r = \frac{51}{50} \quad \text{(iv)}$$

Solving for q and r between (iii) and (iv)

$$q = 1.04 \text{ and } r = -0.0004 \quad \text{(v)}$$

Substituting for p, q and r from (ii) and (v) in (i),

$$t_A = 1.04t_B - 0.0004t_B{}^2$$

(a) When $t_B = 30$, $t_A = 1.04 \times 30 - 0.0004 \times 30^2$

Hence, $t_B = 30.84°\text{C}$

(b) Other than at ice point, and steam point, the two thermometers, A and B, read different values. Hence, if A is assumed to be correct, B becomes faulty and vice versa. Both cannot be correct. Therefore, either one of the thermometers is correct or both are faulty.

3.3.5 Resistance Thermometer

The sensing element in this type of thermometer happens to be a fine platinum resistance wire enclosed in a suitable glass tube or a metallic tube. Generally, the resistance of the wire is selected to be around 100 ohm at 0°C. The resistance of the wire, R_t, at any other temperature t°C, in a restricted calibrated range, is assumed to be given by

$$R_t = R_0(1 + at + bt^2)$$

where, R_0 is the resistance at 0°C and a and b are calibration constants. The measure of the change in resistance is a measure of the temperature. The change in resistance is normally measured by connecting the sensing resistance to one limb of the Wheatstone Bridge, as shown in Figure 3.4. Here, R_1 and R_2 are fixed resistances, R_3 a variable resistance and R_4, the resistance of the sensing element of the resistance thermometer. The bridge is connected to a constant voltage source E and a galvanometer G. The bridge is initially balanced by varying R_3 such that there is no current flow through the galvanometer (identified by null deflection in the galvanometer). Then, $\dfrac{R_1}{R_2} = \dfrac{R_3}{R_4}$. The value of R_4 keeps changing with the temperature which is to be measured. Whenever R_4 changes, there is an imbalance in the bridge, as indicated by the deflection in the galvanometer. This deflection is a measure of the temperature. Alternatively, the bridge can also be brought to balance by either increasing or decreasing R_3 to the required extent, to restore the bridge to balance. The resistance cut in or cut out is an indication of the change in resistance of R_4, which can be calibrated against temperature.

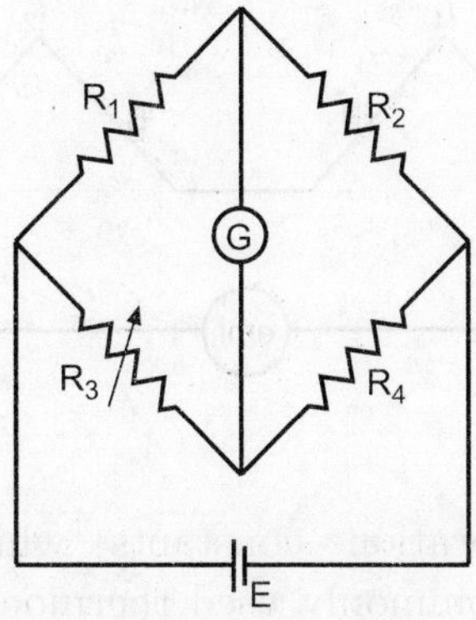

Fig. 3.4 The Wheatstone Bridge, which is used for measuring change in resistance.

Example 3.5

The resistance of the windings in a certain motor is found to be 80 ohms at a room temperature of 25°C. When operating at full load under steady state conditions, the motor is switched off and the resistance of the windings, measured immediately, was found to be 95 ohms. The resistance of the material of the windings at a temperature of t°C is given by $R_t = R_0\ (1 + 0.00395\ t)$, where R_0 is a constant (= resistance at 0°C). Find the temperature attained by the coil during full load.

▶ **Solution:**

$$R_t = R_0(1 + 0.00395t)$$

At $$t = 25,\ R_t = 80$$

Substituting these values for t and R_t in the above equation,

$$80 = R_0(1 + 0.00395 \times 25),\ \text{or},\ R_0 = 72.81 \text{ ohm}.$$

Substituting the values for R_t and R_0 for full load conditions,

$$95 = 72.81(1 + 0.00395t) \text{ or } t = 77.16°\text{C}.$$

3.3.6 Thermocouple

Figure 3.5 shows the schematic of a thermocouple used for measuring temperatures. Here, A and B are two dissimilar metals. The two junctions of these metals which are connected in series are exposed to temperatures t_1 and t_2. Whenever $t_1 \neq t_2$ there is an emf in the circuit. This emf, ε, in the circuit, which is proportional to $(t_1 - t_2)$ is measured either using a millivoltmeter or a potentiometer. Generally, t_2 is selected to be 0°C by immersing the so-called reference junction in melting ice. Then $(t_1 - t_2) = t_1$ and the 'emf' measured is an indication of temperature t_1. The emf, ε, in the range of calibration, is supposed to vary with temperature 't' according to

$$\varepsilon = a + bt + ct^2 + dt^3$$

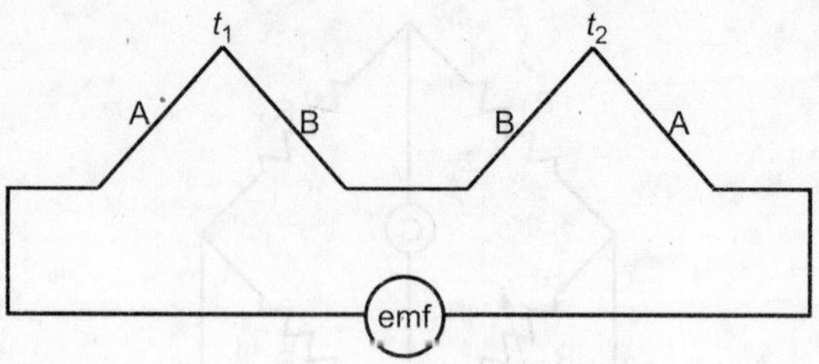

Fig. 3.5 A thermocouple circuit.

where a, b, c and d are calibration constants which are different for different thermocouple pairs. The most commonly used thermocouple pairs and the approximate maximum temperature limit up to which they can be used are given in Table 3.2.

Table 3.2 Common thermocouple pairs

S. No.	Thermocouple pair	Approx. max. temp. range in °C
1	Copper–Constantan	350
2	Iron–Constantan	450
3	Nickel–Nickel Chrome	1100
4	Chromel–Alumel	1200
5	Platinum–Platinum Rhodium	1800

Example 3.6

The emf, in millivolts (mV), in a thermocouple with the test junction at t°C on gas thermometer scale and reference junction at ice point is given by $e = 0.0367t + 1.33 \times 10^{-4}\ t^2$. The millivoltmeter is calibrated at ice and steam points. What will this thermometer read in a place where the gas thermometer reads 50°C.

▶ **Solution:**

At ice point, $t = 0$. Substituting for t in the given equation, $e = 0$

At steam point, $t = 100$

and hence,

$$e = 0.0367 \times 100 + 1.33 \times 10^{-4} \times 100^2 = 5.0 \text{ mV}$$

$$\text{At} \quad t = 50, e = 0.0367 \times 50 + 1.33 \times 10^{-4} \times 50^2 = 2.1675 \text{ mV}$$

As the millivoltmeter is calibrated only at ice and steam points, one degree temperature on this scale corresponds to a mV of $\dfrac{5}{100} = 0.05$. Therefore, 2.1675 mV corresponds to a temperature of $\dfrac{2.1675}{0.05} = 43.35$°C.

3.4 Heat

Heat is the interaction between a system and its surroundings that occurs solely by virtue of a temperature difference existing between the two. Alternatively, when two systems at different temperatures are brought in communication with each other, there occurs a spontaneous process of heat transfer, and its direction is from the high-temperature system to the low-temperature system. The driving potential for heat transfer is the temperature difference. Thus, heat is an *occurrence* and not something that is stored in a system.

When a hot piece of metal is dropped into oil, at ambient temperature, in a container, the process of heat transfer suddenly sets in. Heat is transferred from the metal to oil. During this process of heat transfer, the temperature of metal starts decreasing and the temperature of oil keeps increasing. Ultimately, a time will come when the two temperatures reach a common value. From this instant, heat transfer ceases to occur as there will be no driving force for the heat transfer to continue.

Heat transfer can occur by any of the following three modes, namely, (i) conduction, (ii) convection and (iii) radiation.

In *conduction mode*, heat is transferred from one particle to the next in contact without appreciable movement of the particles themselves. In *convection mode*, heat is transferred from one part of the system to another by the particles of a fluid moving from hotter regions to colder regions or the fluid to be heated is flown over a heated surface. In *radiation mode*, heat is transferred by electromagnetic waves between two bodies at different temperature separated by a distance either in vacuum or otherwise.

3.4.1 Sign Convention

Similar to work, heat is also a path function. While W is used to denote work, customarily, Q is used to denote heat transfer.

While heat transfer to a system is given a positive sign, heat transfer out of a system is given a negative sign. This is illustrated in Figure 3.6.

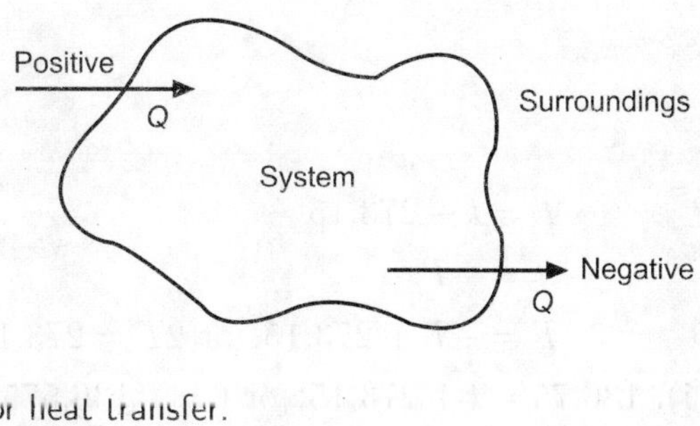

Fig. 3.6 Sign convention for heat transfer.

3.4.2 Unit of Heat Transfer

The unit of heat transfer is a calorie, which is defined as the magnitude of heat transfer required to rise the temperature of one gram of water through one degree from 14.5°C to 15.5°C. However, the unit of heat transfer in SI is joule, abbreviated as J. A joule is equal to 1/4.189 of a calorie. The various practical units of heat transfer in use are kJ (10^3 J) and MJ (10^6 J). The rate of heat transfer, measured in J/s, is a watt, abbreviated by W. The practical units of the rate of heat transfer are kW (10^3 J/s) and MW (10^6 J/s).

3.4.3 A Note of Caution on Heat Transfer

In general, heat transfer to a system causes a rise in temperature. This need not always be so. For example, when heat is transferred to melting ice, there will be no change in the temperature until the last particle of ice melts. Similarly, when heat is transferred to boiling water, there will be no change in temperature until the last drop of water evaporates. Therefore, it is to be noted that *heat transfer need not necessarily bring about a change in the temperature of a system.*

Similarly, whenever there is a rise in temperature, it need not necessarily be attributed to heat transfer. For example, when you rub your palms against each other vigorously, their temperatures increase. This is not because of heat transfer as, at any instant during the process of rubbing, both palms are at the same temperature and there is no driving potential for heat transfer. In fact, *the interaction that brings about the rise in temperature is work interaction* (movement against friction).

Illustrative Problems

3.1 The temperature of a system was measured using a celsius thermometer and a kelvin thermometer. Surprisingly, the numerical value on both the thermometers were the same except that the celsius thermometer gave a negative value. Find the temperature of the system in °C.

▶ **Solution:**

$$T = t + 273.15 \quad \text{(i)}$$

As per the data $t = -T$

and therefore $T = -T + 273.15, \text{ or } 2T = 273.15, \text{ or } T = 136.575 \text{ K}$

Substituting for T in (i), $136.75 = t + 273.15$, or $t = -136.575$°C.

3.2 In 1709, Sir Isaac Newton proposed a new temperature scale. On this scale, the temperature was a linear function of the temperature on celsius scale. The readings on this scale at the ice point (0°C) and the normal human body temperature (37°C) were 0°N and 12°N, respectively. Obtain the relation between the newton scale and the celsius scale.

► **Solution:**

It is given that t_N is a linear function of t_C. Accordingly, let $t_N = a + bt_C$

When $t_C = 0$ (ice point), $t_N = 0$
Substituting for t_C and t_N in (i) $a = 0$
Also, when $t_C = 37$ (human body temp), $t_N = 12$

Therefore, $12 = a + b \times 37$, or $b = 0.3243$.

Thus, the relation between t_N and t_C is $t_N = 0.3243\ t_C$

3.3 The temperature, t, on a certain scale is defined in terms of the thermometric property X by the relation $t = a\ln X + b$ where a and b are constants. On this scale the temperatures of ice and steam points are 0 and 100, respectively. Experiments reveal that $X_i = 1.86$ and $X_s = 6.81$. Find the temperature for an X value of 2.5 on this thermometer.

► **Solution:**

$$t = a \ln X + b$$

From the data, when $t = 0,\ X_i = 1.86$
and therefore ,

$$0 = a \ln 1.86 + b = 0.6206\ a + b \quad \text{(i)}$$

Similarly, $$100 = a \ln 6.81 + b = 1.9184\ a + b \quad \text{(ii)}$$

Solving for a and b, between (i) and (ii)

$$a = 77.053 \text{ and } b = -47.82$$

Therefore, $$t = 77.053 \ln X - 47.82 \quad \text{(iii)}$$

When $X = 2.5$, $t = 77.053 \ln 2.5 - 47.82 = 22.78°$.

3.4 State with reasons whether heat transfer Q, and work transfer W, are positive, negative or zero in each of the following cases. The systems to be considered are in italics:

(i) A rigid steel vessel containing *steam* at a temperature of 150°C is left standing in the atmosphere which is at a temperature of 25°C.

(ii) A *mixture of ice and water* is contained in an insulated vertical cylinder closed at the top by a non-conducting piston: the upper surface of the piston is exposed to atmosphere. The piston is held stationary while the mixture is stirred by a paddle-wheel, the shaft of which protrudes into the cylinder through the cylinder head. Some ice melts.

(iii) As given in (ii), except that the piston is allowed to move so as to keep the mixture pressure constant.

(iv) A well-insulated, sealed vessel contains some oil and oxygen gas. The oil is ignited by an infinitesimal spark from a spark plug. The temperature of the *contents of the vessel* increase.

(v) An electric current flows steadily through a *resistor* immersed in flowing water.

▶ **Solution:**

(i) As the vessel is rigid, there is no displacement work and as per the data there exists no other mode of work. Therefore, $W = 0$. The vessel is made of steel, a conducting material. Steam temperature is greater than the surrounding temperature. Therefore, heat is transferred from the steam to the surroundings and in conformity with the sign convention adopted, Q is negative .

(ii) For the mixture, $Q = 0$ as the cylinder and the piston are insulated. As per data the only work is that done by the stirrer on the mixture. The displacement work is zero, as the piston is held stationary. There are no other modes of work. Therefore, the net work for the mixture is negative, that is, W is negative.

(iii) As in (ii), $Q = 0$. Besides stirrer work, which is negative, there is also displacement work as the piston is free to move. As ice melts, the volume of the mixture decreases. Consequently, the piston moves down doing work on the mixture. There is no other work. Hence, the net work is negative, being equal to the sum of the stirrer work and the displacement work, that is, W is negative. (Note: The magnitude of $W_{(iii)} > W_{(ii)}$)

(iv) As the vessel is insulated, $Q = 0$. Electrical work due to the spark can be considered to be negligible. There is no displacement work, as the vessel is rigid. There are no other modes of work. Therefore, net $W = 0$.

(v) Work is done by the external source of electricity on the resistor. Therefore, W is negative for the resistor. The resistor gets heated up, which in turn will heat the water flowing over it. Therefore, Q is negative.

3.6 Comparison between Work and Heat

In this section, work and heat are compared to find the similarities and differences between them. There are a few similarities between work and heat, as listed below:

- Both are energies in transition across the system boundary.
- Both are transient in nature as they occur only when there exists required potential to cause the interaction.
- Both of them are not properties of the systems in question.
- Both are path-dependent.
- Both are inexact differentials.

The differences between the two are presented in Table 3.3.

Table 3.3 Differences between work and heat

S. No.	Work	Heat
1.	Positive work is defined in terms of the lifting of a weight in the surroundings as the sole effect.	Heat is defined as an interaction that occurs solely because of a temperature difference existing between the system and its surroundings.
2.	Can cause a rise or fall in temperature in an adiabatic system.	Cannot affect the temperature in an adiabatic system.
3.	Generally, work transferred out of a system or work done by a system is given a positive sign and work transferred into a system or work done on a system is given a negative sign.	Generally, heat transferred out of a system is given a negative sign and heat transferred into a system is given a positive sign.
4.	There are as many modes of work transfer as there are varieties of forces.	There are only three modes of heat transfer, namely, conduction, convection and radiation.
5.	As will be shown later, in the chapter on second law, complete conversion of work to other modes of work or heat is possible.	In a cyclic process complete conversion of heat to work is impossible.

Problems

3.1 The temperature of a fluid is measured with a celsius thermometer and a fahrenheit thermometer. Surprisingly, the numerical value on both the thermometers are the same. Find the temperature of the fluid.

3.2 A thermometer is calibrated using ice and steam points. The thermodynamic function chosen to establish the scale is $t = a \ln x + b$.

(i) Determine the constants a and b in terms of x_s and x_i.

(ii) Show that the new scale is given by $t = 100 \left[\frac{\ln\left(\frac{x}{xi}\right)}{\ln\left(\frac{xs}{x}\right)} \right]$.

3.3 The readings t_A and t_B of two celsius thermometers A and B agree well at ice and steam points and are related at other points in-between, by $t_A = p + qt_B + rt_B^2$ where p, q and r are constants. When both are immersed in a well-stirred oil bath, A registers 49°C, while B registers 50°C. (i) Determine the reading on A when B reads 25°C (ii) Discuss which thermometer is correct.

3.4 A constant volume gas thermometer containing helium gives readings of gas pressure p, of 1000 and 1366 mm Hg at ice and steam points, respectively. (i) Express the gas thermometer celsius temperature t in terms of gas pressure p. (ii) The thermometer when left exposed to atmosphere registers 1073 mm Hg. Determine the atmospheric temperature.

3.5 Two mercury in glass thermometers, A and B, are made of identical materials and are accurately calibrated at ice and steam points. Thermometer A has a capillary of constant diameter; B has a capillary of conical bore, 10% bigger in diameter at 100°C mark than that at the mark of 0°C. Both thermometers have the length between their 0 and 100 marks subdivided uniformly. When A and B are exposed to a system at an uniform temperature the latter reads 50°C. What will be the reading on A?

3.6 In testing an electrical apparatus, the average temperature of a coil of wire is often obtained by measuring the electrical resistance of the wire. The resistance of a certain magnet coil at room temperature (30°C) is 1240 ohms. After a certain testing operation, the coil resistance is 1435 ohms. The resistance R, in ohms, of the copper wire used in the coil is related to the temperature t, in degree C, by $R = R_0[1 + 0.00395(t - 20)]$, where R_0 is a constant. Find the average temperature of the coil at the end of the test.

3.7 A thermometer is calibrated using ice and steam points. The thermodynamic function chosen to establish the scale is $t = ax^2 + bx$. Calculate the relation between t and x, x_i and x_s.

3.8 A thermometer is constructed by arranging an aluminium rod such that its length can be measured accurately by a scale made of invar, a nickel alloy. The thermometer is calibrated at ice and steam points. The distance between these two points on the invar scale is uniformly divided into 100 parts, each mark being equal to one degree. Find the difference between the reading of this thermometer and actual temperature values of 0, 25, 50 and 100 degrees on the gas thermometer scale. The linear expansion of aluminium is represented by $L_t = L_0(1 + 0.2221 \times 10^{-4}t + 0.114 \times 10^{-7}t^2)$, where L_t and L_0 are the lengths of the aluminium rod at t°C and 0°C on gas thermometer scale. Neglect the linear expansion of invar.

3.9 A resistance thermometer is calibrated at ice (0°C) and sulphur points (444.6°C). The thermometric scale is then made assuming linear change in resistance between the calibration temperatures. Such a thermometer is brought in communication with boiling water at 1 atmosphere pressure. Will the thermometer read 100°C ? Discuss the problem.

3.10 A very-good-quality, closed, vacuum flask contains coffee at 85°C, in a place where the surroundings are at 30°C. Are the coffee and the surroundings in thermal equilibrium? Substantiate your answer.

3.11 State with reasons whether the heat transfer Q, and the work transfer W, are positive, negative or zero in each of the following cases. The systems to be considered are in italics.

(i) 0.1 kg of a *gas* contained in an insulated cylinder expands as the piston slowly moves outwards.

(ii) A *mixture of ice and water* is contained in a vertical closed at the top by a piston; the the upper surface of the is exposed to atmosphere. The piston is held stationary while a flame applied to the bottom of the cylinder causes some of the ice to melt.

(iii) As under (ii), but the piston is allowed to move so as to keep the mixture pressure constant.

(iv) A rigid vessel containing *gaseous ammonia* is connected through a valve to an evacuated rigid vessel. The vessels, the valve and the connecting pipe are all insulated. The valve is opened and after some time, the conditions throughout the two vessels become uniform.

(v) 1 kg *air* flows slowly but adiabatically from the atmosphere into a previously evacuated bottle through a valve.

(vi) A *sealed calorimeter* containing powdered coal and gaseous oxygen is immersed in a tank containing water. In the first half-minute, after ignition by an infinitesimal spark, the temperature of water increases lightly.

(vii) A rigid insulated vessel contains some fluid and an *electrical resistor*. When the resistor is connected to the electric mains for some time, the temperature of the resistor and the fluid increases.

(viii) *A lead-acid battery*, discharging adiabatically, supplies current to a resistor.

(ix) The *lead-acid battery* of (viii) undergoes the same change from the same initial state while standing on open-circuit.

(x) A combustible mixture of *hydrogen and oxygen* in a non-conducting and rigid envelope is ignited by a negligible spark as a consequence of which the two gases combine.

(xi) A *hydrogen and oxygen mixture* forming a bubble in a large reservoir of water is ignited by a minute spark.

(xii) A *liquid* in a non-conducting vessel comes to rest from an initial state of turbulent motion.

(xiii) When the plunger of a tyre pump is pushed down, the *air* it contains is forced into the tyre. Assume the tyre, the pump barrel and the connecting tubes to be non-conducting.

(xiv) A vessel with rigid non-conducting walls holds a complete electrical circuit consisting of a heating element and a charged battery. The temperature of the *contents of the vessel* increases.

(xv) A *copper block* slides down a rough, non-conducting inclined plane. All the sides of the block, except the one in contact with the plane, are insulated.

(xvi) An *electric battery*, discharging adiabatically, supplies current to an electric motor.

(xvii) An electric battery discharges slowly while in open circuit for a long time.

(xviii) A conducting vertical frictionless piston–cylinder device contains a *mixture of ice and water* in equilibrium. The mixture is heated by a Bunsen flame until half the inital amount of ice melts.

(xix) An infinitesimal spark ignites a *mixture of hydrogen and oxygen* in a rigid insulated cylinder. The gases combine to form water.

(xx) An infinitesimal spark ignites a large bubble of a *mixture of hydrogen and oxygen* in a large reservoir.

(xxi) A rigid vessel containing a certain *gas* has a heating element surrounding the vessel and it is wired to a battery. A paddle wheel protruding through the walls of the vessel churns the gas.

(xxii) A *gas* enclosed in a rigid vessel is stirred by a paddle wheel. As a consequence, the temperature of the gas increases.

(xxiii) A non-conducting cylindrical vessel is divided into two equal parts by a very thin diaphragm. While a *gas* at a certain pressure and temperature occupies the left half, the right half is evacuated. The gas is churned by a paddle wheel connected to an external motor until the diaphragm ruptures and the gas fills the entire cylinder.

THE FIRST LAW OF THERMODYNAMICS FOR SYSTEMS—PURE SUBSTANCES

4

4.1 Introduction

The first law of thermodynamics discusses the quantitative relationship that exists between heat and work. Consider a device that produces work using heat transfer. The simplest example of such a device is a steam power plant. In this power plant, fuel (coal, oil or gas) is burnt and the heat transfer from this burning fuel causes water to boil and form high-pressure steam in a boiler. Steam from the boiler goes to a turbine, where it expands to produce shaft power. If no heat transfer takes place in the boiler, there would be no generation of steam and consequently the turbine would not operate to produce work. It can hence be concluded that production of work is subject to heat transfer elsewhere in the system. In the present day energy scenario, this example has great economic and technical importance. Depleting fuel reserves, declining quality of fuel, spiralling cost of fuels, combustion-generated pollution and the like have forced man to find the most economical way of utilizing fuel (heat transfer) to produce work. In other words, heat transfer from a given mass of a burning fuel has to produce the maximum possible work. The knowledge of the 'quantitative relation' that exists between heat and work is essential for the 'performance analysis' of such processes of energy conversion. While the first law of thermodynamics confines itself to the former, namely, the quantitative relation between heat and work, the second law of thermodynamics, described in Chapter 6, deals with the latter, namely, the performance analysis of devices that convert heat to work or work to heat. Also, the first law of thermodynamics introduces an important property of a system, namely *energy*, which has many forms.

4.2 Joule's Experiments

There exists no logical relation between heat and work. The possibility of the existence of any such relation has to be found out though experiments. James Prescott Joule, between

AD 1840 and 1849, conducted several experiments to establish the quantitative relationship existing between heat and work. Explained below is one such simple experiment.

As shown in Figure 4.1, consider an insulated vessel which contains m kg water at a temperature of t_1°C. The water is now churned by a stirrer, whose shaft protrudes through an opening in the insulated cover of the vessel, until the temperature increases to t_2°C. The work done on the water can be calculated in terms of the torque in the stirrer shaft, speed of the shaft and the churning time. Let the magnitude of this work be W and is measured in Nm.

Consider now, another similar vessel, as shown in Figure 4.2, insulated all around but at the bottom and also containing m kg water at t_1°C as in the previous case. Let the water now be heated to the same final temperature of t_2°C, as in the previous experiment, by exposing the bottom of the vessel to a heater. Let the magnitude of heat transfer required to bring about the same change as in the previous experiment be Q calorie.

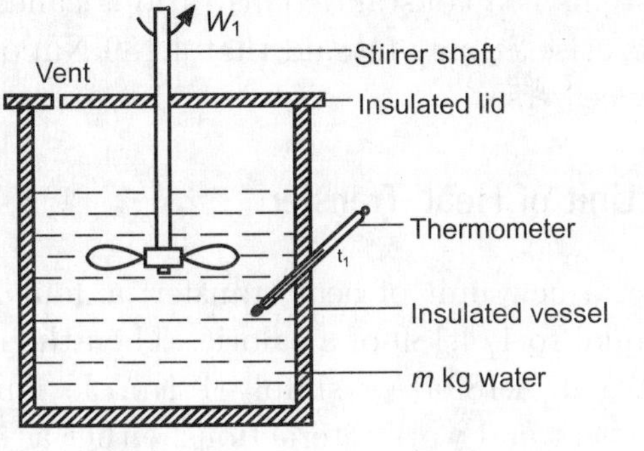

Fig. 4.1 Heating of water by work interaction.

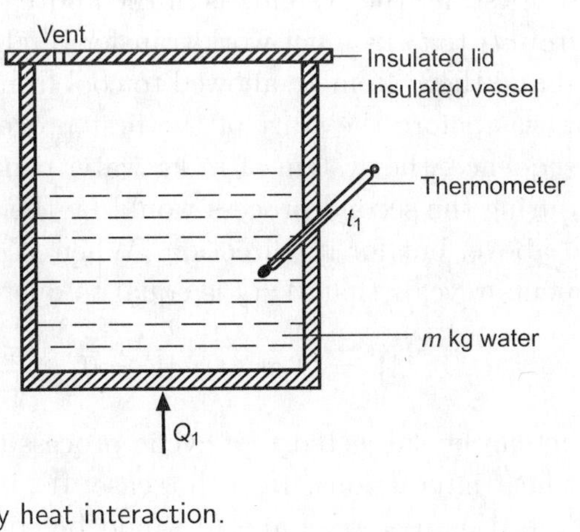

Fig. 4.2 Heating of water by heat interaction.

These variables W and Q are work and heat interactions which have brought about identical change in similar systems.

Suppose a number of such sets of experiments are conducted in which the state change of a system is brought about by work interaction and then the same change is brought about by heat transfer in a similar system. Let the magnitudes of work and heat interactions, in each such experiment, be $(W_1, Q_1), (W_2, Q_2), (W_3, Q_3)$ etc. The experimental results would confirm that the ratio of $W : Q$ in each experiment turns out to be the same and is equal to 4.189 Nm/cal. That is to say, $\dfrac{W_1}{Q_1} = \dfrac{W_2}{Q_2} = \dfrac{W_3}{Q_3} = 4.189$ Nm/cal.

4.2.1 Mechanical Equivalent of Heat

It is now evident that heat and work interactions can bring about identical changes in a system and that the ratio of the quantities of work and heat which bring about identical changes in similar systems is a constant. This ratio is called the *mechanical equivalent of heat*. To be more precise, it may be said that 4.189 Nm of work can do the same effect as 1 cal of heat, and vice versa.

4.2.2 Joule as the Unit of Heat Transfer

Suppose, at this stage, a new unit of heat transfer, a Joule, denoted by J, is introduced whose magnitude is equal to 1/4.189 of a calorie. Then the mechanical equivalent of heat turns out to be 4.189 Nm/4.189 J or simply 1 Nm/J. This concept facilitates using a common unit for both heat and work interactions, either a 'Nm' or a 'J', as one 'Nm' and one 'J' can bring about identical changes.

Let a system, say m kg water, be subjected to a cyclic process comprising two steps conducted in sequence. First, let the system, as in the above experiments, be subjected to a rise in temperature from t_1 to t_2 by a net work transfer W. Let the insulation around the vessel be then removed and the system be allowed to cool from t_2 to t_1, so that the system returns to the state it was before the start of the first process. After the completion of the said two steps in sequence, the system of m kg water completes a cyclic process. The net heat transfer, Q, during the second process would be the same as in the second set of experiments explained above, but for its direction. When W and Q are expressed in Nm (or J), Joule's experiments reveals that W/Q is equal to one. Therefore,

$$W = Q \text{ or } Q - W = 0$$

In general, a system can be subjected to a cyclic process through a series of processes involving work and/or heat interactions. In such a case, W, in the above expression, shall be the net work $= W_1 + W_2 + W_3 + \cdots$ and Q would be equal to the net heat transfer, equal to $Q_1 + Q_2 + Q_3 + \cdots$.

The above relation, as confirmed by experiments, may hence be written as follows:

$$(Q_1 + Q_2 + Q_3 + \cdots) - (W_1 + W_2 + W_3 + \cdots) = 0$$

or,

$$\Sigma Q - \Sigma W = 0$$

However, if the cyclic process were to be performed through a number of infinitesimal processes, each with a heat interaction δQ and/or a work interaction δW, it may then be generalized and stated that the integral of δQ around the cycle minus the integral around the cycle of δW is equal to zero.

Mathematically,

$$\oint \delta Q - \oint \delta W = 0$$

The above relations embody the statement of the first law of thermodynamics.

4.3 Statement of the First Law of Thermodynamics

When a system executes a cyclic process the net heat transfer during the process minus the net work transfer during the same process is always equal to zero, provided the magnitudes of heat and work interactions are expressed in the same units. Mathematically, the first law of thermodynamics may be stated as follows:

$$\Sigma Q - \Sigma W = 0 \text{ or } \Sigma (Q - W) = 0 \tag{4.1}$$

or, in integral form as,

$$\oint \delta Q - \oint \delta W = 0 \quad \text{or} \quad \oint (\delta Q - \delta W) = 0 \tag{4.2}$$

Example 4.1

A system executes a cyclic process during which there are four heat transfers of 900, 80, −800 and 150 J and four work interactions of 200, 150, 300 and W_4 J. Find the magnitude and direction of the fourth work interaction, W_4.

▶ Solution:

For a system undergoing a cyclic process, from the first law of thermodynamics,

$$\Sigma Q - \Sigma W = 0, \text{ or } \Sigma Q = \Sigma W$$

that is, $[Q_1 + Q_2 + Q_3 + Q_4] = [W_1 + W_2 + W_3 + W_4]$

Therefore, $900 + 80 + (-800) + 150 = 200 + 150 + 300 + W_4$.

Hence, $W_4 = -320$ J

4.4 Extension of the First Law of Thermodynamics to Non-Cyclic Processes—ENERGY

The first law of thermodynamics has been stated for a system undergoing a cyclic process. More often than not, one encounters non-cyclic processes rather than cyclic processes. It is, hence, essential to know the following form the first law takes for such processes.

When a system undergoes a non-cyclic process, the difference between the net heat during the process (Q) and the net work (W) during the same process is always equal to the change in the energy (ΔE) of the system. This can be stated mathematically as follows:

$$Q - W = \Delta E \tag{4.3}$$

where ΔE = (the energy at the final state – the energy at the initial state) = $(E_2 - E_1)$ or in differential form as,

$$\delta Q - \delta W = \mathrm{d}E \tag{4.4}$$

Note: While solving problems using the first law relations, it is *necessary* and *essential* to substitute for Q, W and ΔE, from data, their respective numerical values in *consistent units* along with their correct signs (– or +) as the case may be.

Example 4.2

(a) A system undergoes a change of state from 1 to 2 along path A, during which heat transferred from the system is 50 kJ and work done on it is 65 kJ. Find the change in energy of the system. (b) If the system were to return from state 2 to state 1 through an adiabatic process B, find the magnitude and direction of work during B.

▶ Data:

$Q_A = -50$ kJ, $W_A = -65$ kJ, $Q_B = 0$, $W_B = ?$

▶ Solution:

(a) Applying the first law to process A,

$$Q_A - W_A = E_2 - E_1 = \Delta E_A = -50 - (-65) = +15 \text{ kJ}$$

(b) Similarly, for process B, for which the final state is 1 and the initial state is 2,

$$Q_B - W_B = E_1 - E_2 = \Delta E_B = -(E_2 - E_1) = -\Delta E_A = -15 \text{ kJ}$$

Therefore, $0 - W_B = -15$ kJ, or, $W_B = 15$ kJ.

In other words, 15 kJ work is done by the system during the return adiabatic process.

Alternatively, processes A and B together make the system undergo a thermodynamic cycle.

Therefore, from the first law for a cyclic process, $\Sigma Q - \Sigma W = 0$

where, $\Sigma\, Q = Q_A + Q_B$, and $\Sigma W = W_A + W_B$

Therefore, $(Q_A + Q_B) - (W_A + W_B) = 0$

that is, $(-50 + 0) - (--65 + W_B) = 0$, or $W_B = 15$ kJ

4.5 More about Energy

4.5.1 Definition of Energy and Remarks

- Relation (4.3), $Q - W = E_2 - E_1 = \Delta E$, defines energy.
- It may be noticed that the above relation defines only a change in energy. This means E_2, the energy of the system at the end of a process, can be calculated with the knowledge of Q and W during the process if and only if E_1, the initial energy of the system, is known. This means that energy has to be measured with respect to an arbitrarily fixed datum.
- Energy, E, is an extensive property of a system, and its unit is either J or Nm.
- Energy per unit mass, E/m, denoted by e, the specific energy, has the unit of J/kg or Nm/kg.

4.5.2 Energy is a Property of the System

Any quantity related to a system whose change between two states is independent of the path and whose change in a cyclic process is zero is a property of the system. This concept is used to prove that energy is a property of the system.

Let us consider, as shown in Figure 4.3, two cyclic processes passing through two common state points 1 and 2 of a certain system.

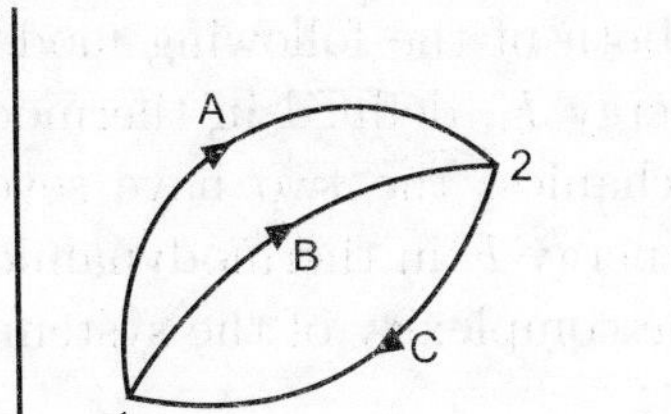

Fig. 4.3 Two cyclic processes through two common states.

The first cyclic process starts at state 1, proceeds along path A to state 2 and returns to state 1 via the process, C. For this cyclic process 1–A–2–C–1, from the first law,

$$\oint (\delta Q - \delta W) = 0$$

The cyclic integral can be split into two linear integrals corresponding to paths A and C.

Therefore, $$\int_{1-A}^{2} (\delta Q - \delta W) + \int_{2-C}^{1} (\delta Q - \delta W) = 0 \tag{i}$$

Similarly, for the second cyclic process, 1–B–2–C–1,

$$\int_{1-B}^{2} (\delta Q - \delta W) + \int_{2-C}^{1} (\delta Q - \delta W) = 0 \tag{ii}$$

(i) – (ii) gives, $$\int_{1-A}^{2} (\delta Q - \delta W) = \int_{1-B}^{2} (\delta Q - \delta W) \tag{iii}$$

Generalization of relation (iii) reveals that the integral of $(\delta Q - \delta W)$ between two states is the same for all paths between the two state points. In other words, as $\int \delta Q = Q$, the net heat during a process, and $\int \delta W = W$, the net heat transfer during the same process, $(Q - W)$ between any two states is independent of the path. But, $(Q - W) = \Delta E$, and hence ΔE between two states is independent of the path between the two states.

Also, from the first law for a cycle, the integral around the cycle of $(\delta Q - \delta W)$ is equal to zero. Thus, for a cycle, $(Q - W) = \Delta E = 0$.

Therefore, E is such that its change between two states is independent of the path and that its change in a cyclic process is zero. Thus, E has to be a property of a system.

4.5.3 The Nature of Energy

School-level physics and mechanics state that a system can posses energy, the capacity for doing work, in either or both of the following modes:(i) potential energy, (E_P), and (ii) kinetic energy (E_K). Energy E, defined in thermodynamics, though not always the same as that defined in mechanics, the two have several things in common. Besides comprising E_P and E_K, the energy E in thermodynamics may include many other modes of energy, depending upon the complexity of the system.

4.5.3.1 Potential Energy

Potential energy, E_P, is the energy possessed by a system by virtue of its position, and its magnitude is evaluated from

$$E_p = mgz$$

where m is the mass of the system in kg, g is the acceleration due to gravity in m/s^2, z is the vertical distance, in m from an arbitrary datum and E_p is in J.

4.5.3.2 Kinetic Energy

Kinetic energy, E_K, is the energy possessed by a system by virtue of its motion, and its magnitude is evaluated from

$$E_K = mV^2/2$$

where m is the mass of the system in kg, V is the velocity in m/s and E_k in J.

4.5.3.3 Internal Energy

Consider, for example, as shown in Figure 4.4, a system comprising an insulated hemispherical bowl with a spherical small marble at its middle. The broken line around the bowl indicates the boundary of the system considered. Assume the marble is slowly raised from its initial position to the rim of the bowl. This process requires work from outside. For the system considered, $Q = 0$ and W is negative. Therefore, from the first law, $\Delta E = Q - W =$ positive. This means $E_2 > E_1$. The same inference can be drawn from mechanics as the potential energy of the system has increased by mgz, where m is the mass of the marble and z is the height of the bowl rim to which the marble is raised from its initial position of rest. Hence, $E_2 > E_1$.
Suppose now the marble, with the system having an initial potential energy of mgz, is released to make to and fro oscillations inside the bowl surface.

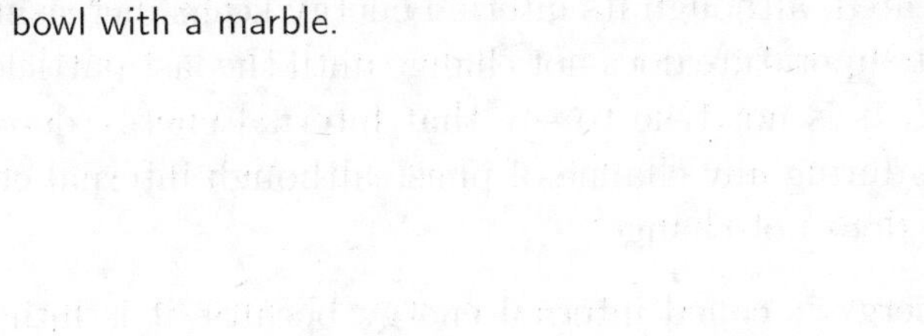

Fig. 4.4 Hemispherical bowl with a marble.

(a) If the friction between the marble and the bowl is neglected, the marble travels first to the bottom and then up to the right-side rim, returns to the bottom and raises to left-side rim to the same height z. This process of oscillation of the marble keeps happening forever in conformation with Newton's laws of motion. Applying the first law between any two locations of the marble, as $Q = 0$ and $W = 0$ (because nothing is crossing the system boundary), $\Delta E = 0$. The system possesses the same energy at every location of the marble. The same is true from concepts in mechanics also because, while the marble is at the rim, the system has a potential energy of mgz. When it is at the bottom centre of the bowl, its potential energy would be zero, but it possesses kinetic energy whose magnitude equals mgh. At any other location between the rim and the bottom of the bowl, the system possesses part of the energy in the form of potential energy and the rest in the form of kinetic energy. Thus, there is only a change of energy from one mode to another and no change in net energy.

(b) Suppose friction is considered, in each oscillation, the height to which the marble rises keeps decreasing, until, eventually, the marble settles at the bottom with no further movement possible. Even in such a case, as nothing crosses the system boundary, $Q = 0, W = 0$, and therefore $\Delta E = 0$. This observation is in contradiction to the mechanics concept, as both potential and kinetic energies have become zero and hence the energy of the system has apparently decreased instead of remaining constant. To overcome this problem it is said in mechanics that 'energy is dissipated' or 'energy is converted to heat'. Both statements are not acceptable from the thermodynamics standpoint, as energy is neither dissipated nor converted to heat. In reality, the marble would be hotter when it settles down at the middle. Thus, it is legitimate to infer that the system must now posses the same amount of energy but in a different form. This form of energy is the *internal energy*, U. More often than not internal energy is related to temperature. Though not always true, generally, higher the temperature of the system higher would be its internal energy. Temperature of a system can be altered by both heat and work interactions.

Thus, internal energy may be defined as that form of energy which could directly be altered by heat and/or work transfer.

A Note of Caution on Internal Energy: When a mixture of ice and water in equilibrium is heated, although its internal energy keeps increasing in conformity with the first law, its temperature does not change until the last particle of ice melts. Hence, strictly speaking, it is not true to say that internal energy depends on temperature only. In general, during any change of phase although internal energy keeps changing the temperature does not change.

This mode of energy is called internal energy because it is intimately related to the internal structure of the system in question, in terms of the internal microscopic energy modes. They include translational, rotational and vibrational energies of the molecules of the system. Thus, it is not possible to understand the concept of internal energy directly from macroscopic considerations.

4.5.3.4 Chemical Energy

Consider an insulated rigid vessel containing a mixture of some fuel and oxygen. Let a minute spark ignite the mixture. The mixture burns instantaneously, causing an increase in temperature and pressure of the contents of the vessel. During this process $Q = 0$ because the vessel is insulated and there is no work done either, neglecting the electrical work supplied through the spark as insignificant. From the first law, as $Q = 0$ and $W = 0$, $\Delta E = 0$. Though there is no change in energy, there has been a substantial increase in temperature after combustion. Thus, some form of energy which was in the fuel before burning has manifested itself in the form of internal energy of the combustion products. This is the *chemical energy*, the energy possessed by the fuel, denoted by E_{Che}.

4.5.3.5 Electrical Energy

If a system were to comprise electrically charged particles, because of the attractions and repulsions existing between these particles, a part of its energy will be in the form of *electrical energy*. This is denoted by E_{Ele}.

4.5.3.6 Magnetic Energy

Similar to electrical energy, a system comprising magnetic dipoles will have a part of its energy in the form of *magnetic energy*. This is denoted by E_{Mag}.

4.5.3.7 Surface Energy or Strain Energy

A compressed spring or a stretched membrane has part of its energy in the form of *strain energy*. A related form of energy is that associated with capillary or surface tension of the system. Systems exhibiting this phenomenon are supposed to have part of their energy in the form of *surface energy*. This is denoted by E_{Sur}.

4.5.3.8 Stored or Total Energy

The stored energy, or the total energy, E, of a system is the sum of all modes of energy the system may posses. Hence,

$$E = E_P + E_K + U + E_{Che} + E_{Elec} + E_{Mag} + E_{Sur} \tag{4.5}$$

For a system showing negligible or no chemical, electrical, magnetic and surface effects,

$$E = E_P + E_K + U \tag{4.6}$$

Example 4.3

(a) A system comprising 2.3 kg of a substance that shows negligible chemical, electrical, magnetic and surface effects has a specific internal energy of 21 kJ/kg. The system is moving with a velocity of 110 m/s at an elevation of 1500 m above the sea level. Evaluate the total energy of the system with respect to an observer at rest at sea-level.

▶ **Data:**

$m = 2.3$ kg, $E_{Che} = 0, E_{Elec} = 0, E_{Mag} = 0$ and $E_{Sur} = 0, u = 21 \times 10^3$ J/kg, $V = 110$ m/s, $z = 1500$ m. Assume potential energy is zero at sea level from where z is measured.

▶ **Solution:**

As the total energy of the system comprises of only potential, kinetic and internal energies

$$E = E_P + E_K + U \tag{4.6}$$

As energy is an extensive property, $E = me = m(e_p + e_k + u)$, where the lower-case letters denote the specific properties corresponding to various forms of energy.

$$\begin{aligned} E &= m\left(gz + \frac{V^2}{2} + u\right) \\ &= 2.3\left[9.81 \times 1500 + \frac{110^2}{2} + 21 \times 10^3\right] \\ &= 96{,}059.5 \text{ J} = 96.0595 \text{ kJ} \end{aligned}$$

4.6 Pure Substance

The first law of thermodynamics dictates that the net heat minus net work during a process is always equal to the change in stored energy of the system: however complicated the system may be with regard to the modes of energy it possesses. More often than not, the systems generally encountered in engineering practice happen to be simple substances which do not change chemically or exhibit significant electrical, magnetic and surface effects. Because of this fact such simple substances are referred to as 'pure substances'.

A pure substance can be defined as a system which is (i) homogeneous in chemical composition and (ii) homogeneous in chemical aggregation.

The term homogeneous in chemical composition means that the composition of each part of the system is the same as every other part; *homogeneous in chemical aggregation* means that the elements of the system are combined chemically in the same manner in all parts of the system.

Suppose we consider a system of an equilibrium mixture of steam and water in a container as shown in Figure 4.5*a*. The chemical composition of the system is homogeneous as it is the same everywhere in the substance. A molecule of water as well as a molecule of steam contain hydrogen atoms and oxygen atoms in the constant ratio of 2 : 1. Also, the hydrogen and oxygen atoms are held together chemically by identical bonds in both water and steam and hence the system is also homogeneous in chemical aggregation. Thus, an equilibrium mixture of water and steam can be considered as a pure substance.

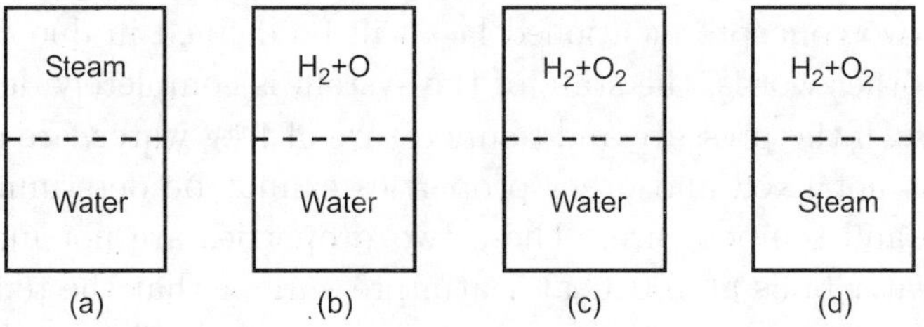

Fig. 4.5 Pure substance or not ?

Let us now consider, as shown in Figure 4.5*b*, a mixture of gaseous hydrogen and oxygen in the proportion of two atoms of hydrogen to one atom of oxygen and water in a vessel. Although the system is homogeneous in chemical composition, it is not homogeneous in chemical aggregation because while hydrogen and oxygen are not bonded chemically in the hydrogen–oxygen mixture, the two are bonded chemically in water. Thus, the system cannot be treated as a pure substance as it does not satisfy the second requirement.

Suppose we consider a random gaseous mixture of hydrogen and oxygen and water, as shown in Figure 4.5*c*, or a random gaseous mixture of hydrogen and oxygen and steam, as shown in Figure 4.5*d*. These systems are not homogeneous in chemical composition as the ratio of the number of hydrogen atoms and oxygen atoms is not 2 : 1 in the random gaseous mixture, whereas the ratio is 2 : 1 in both water and the steam. Also, as explained in Figure 4.5*b*, the system is also not homogeneous in chemical aggregation.

Thus, the two systems in Figures 4.5*c* and *d* cannot be treated as pure substances as they do not satisfy both requirements.

Some commonly encountered pure substances are gases, non-reactive gas mixtures, homogeneous mixtures of gaseous fuel and air, gaseous combustion products, refrigerants etc.

4.6.1 Two-Property Rule for Pure Substances

The two-property rule for a pure substance is stated as follows:

The state of a given mass of a pure substance can be fixed by specifying two independent properties of the system, provided the system is in equilibrium and that the effects due to gravity, motion, electricity, magnetism, chemical and surface actions are not significant.

The two properties selected to describe the state of a pure substance must be independent, which means the two properties must be such that it is possible to vary each property over a wide range holding the other constant and vice versa.

For example, suppose the pressure and volume of 1 kg of air is specified as p_1 and V_1. These two are independent properties. Air can be treated as a pure substance as it satisfies the requirements of a pure substance. Thus, according to the two property rule the state 1 of air is completely specified. The temperature of air at state 1, t_1, has a definite value as decided by the values of pressure and volume. Also, its energy, enthalpy and entropy (the two concepts mentioned last will be defined in due course) would have unique values. In other words, the state of the system is completely described.

On the contrary, if the pressure and temperature of 1 kg water, are given as 1 atm and 100°C, the state is not fixed and other properties cannot be determined from the given values of pressure and temperature. These two properties are not independent. It is a known fact that water boils at 100°C at 1 atm pressure or that the temperature must be 100°C if water were to boil at 1 atm pressure. As water is boiling at the given condition, the system may exist as either completely water, or steam, or a mixture of water and steam in varied proportions. Therefore, the volume of the system can be as small as the liquid volume or as high as the steam volume. As the volume can have different values for the same pressure and temperature, the state of the system cannot be fixed. This is because the given properties do not happen to be independent properties.

The two-property rule may be mathematically stated as

$$z = z(x, y)$$

where x and y are independent properties and z is any other third property.

4.6.2 Energy of a Pure Substance

When the effects of position and motion are negligible, the energy of a pure substance is in the form of internal energy only. Thus,

$$E_{puresub} = U \tag{4.7}$$

Hence, the first law of thermodynamics, applied to a pure substance takes the form,

$$Q - W = \Delta U = m\Delta u \tag{4.8}$$

4.6.3 Specific Heats of a Pure Substance

The *specific heat at constant volume of a pure substance*, c_v, is defined as the rate of change of specific internal energy with temperature at constant volume. that is

$$c_v = (\partial u/\partial t)_v \tag{4.9}$$

Before the specific heat at constant pressure, c_p, is defined, a new property will be introduced here. In thermodynamics, during several analyses, the term $(u + pv)$ is encountered very often. $(u + pv)$ is a property of the system under consideration as all entities in the bracket happen to be properties. For convenience this property is denoted by h and is called the specific enthalpy. Thus, $h = u + pv$. The total enthalpy, H, of a system is equal mh where m is the mass of the system.

The *specific heat at constant pressure of a pure substance*, c_p, is defined as the rate of change of specific enthalpy, h, with temperature at constant pressure. That is,

$$c_p = (\partial h/\partial t)_p \tag{4.10}$$

It can be seen that c_p and c_v are both properties of the system and that $c_p > c_v$, as in any state, $h > u$.

Example 4.4

The properties of a certain pure substance are related as follows:

$u = 196 + 0.718\ t$ and $pv = 287(t + 273)$, where u is the specific internal energy in kJ/kg, t is the temperature in °C, p is the pressure in N/m^2 and v is the specific volume in m^3/kg. Find c_v and c_p for this substance.

▶ Solution:

From the data, $u = 196 + 0.718\ t$

$$c_v = \left(\frac{\partial u}{\partial t}\right)_v = \left[\frac{\partial(196 + 0.718t)}{\partial t}\right]_v = 0.718 \text{ kJ/kg K.}$$

Again, from the data, $pv = 287(t + 273)$,

where the unit of pv = N/m^2 × m^3/kg = J/kg.

From the definition of enthalpy, $h = u + pv$

Substituting for u and pv in the same units of kJ/kg

$$h = (196 + 0.718t) + 287 \times 10^{-3} \times (t + 273)$$

$$c_p = \left(\frac{\partial h}{\partial t}\right)_p = \left\{\frac{\partial[(196 + 0.718t) + 287 \times 10^{-3} \times (t + 273)]}{\partial t}\right\}_p = 1.005 \text{ kJ/kg K}$$

Example 4.5

A pure substance is contained in a cylinder closed by a piston. A stirrer, rotated by means of a shaft protruding through the cylinder end plate, causes the substance to undergo adiabatically a fully resisted constant pressure expansion process as the piston moves outwards slowly. Show that the shaft work (stirrer work) done on the system is equal to the increase in enthalpy of the substance.

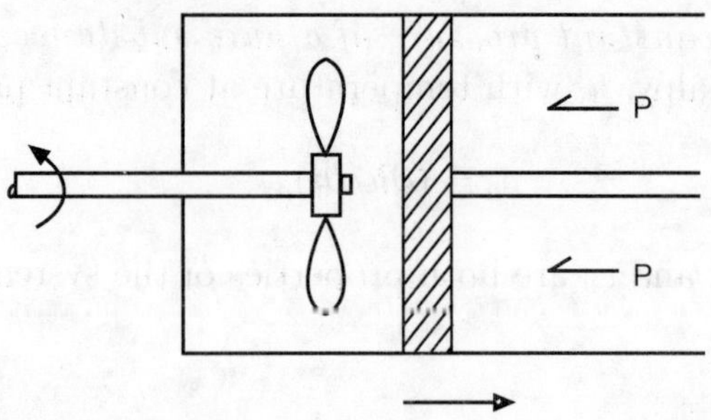

Fig. 4.6 Sketch for solving Example 4.5.

▶ Solution:

Applying the first law for the pure substance,

$$Q - W = U_2 - U_1 \tag{i}$$

From the data, $$Q = 0$$

hence, $$-W = U_2 - U_1 \tag{ii}$$

It can be noted from Figure 4.6, that the net work W is made up of two modes of work, (i) the displacement work, W_d, as the piston moves out at constant pressure and (ii) the stirrer work, W_s. Therefore,

$$W = W_d + W_s \tag{iii}$$

Here, $W_d = \int p\mathrm{d}V = p(V_2 - V_1)$, as the displacement occurs at constant pressure.

Substituting for W_d in (iii)

$$W = p(V_2 - V_1) + W_s \tag{iv}$$

Substituting for W in (ii) from (iv)

$$-[p(V_2 - V_1) + W_s] = U_2 - U_1 \tag{v}$$

Therefore, $-W_s = (U_2 - U_1) + p(V_2 - V_1)$

$p(V_2 - V_1)$ can be written as $(p_2V_2 - p_1V_1)$ as $p_1 = p_2 = p$. (v) now becomes,

$$-W_s = (U_2 - U_1) + (p_2V_2 - p_1V_1) = (U_2 + p_2V_2) - (U_1 + p_1V_1)$$

or, $$W_s = -[(U_2 + p_2V_2) - (U_1 + p_1V_1)] = -(H_2 - H_1)$$

Thus the stirrer work is equal to increase in enthalpy. The negative sign indicates that the stirrer work is negative for the substance.

Illustrative Problems

4.1 (a) A system consisting of a mixture of air and gasoline at an initial temperature of 15°C is contained in a rigid vessel. The following processes occur in sequence: (i) The mixture is heated to 200°C by a heat transfer of 3.0 kJ; (ii) The mixture is then ignited by a minute spark which causes the mixture to burn completely. This process is adiabatic and the final temperature is 1450°C. (iii) The products of combustion are cooled to 120°C by a heat transfer of 32 kJ. Evaluate the energy of the contents of the vessel after each of the above three process, assuming the initial energy of the system to be 10 kJ.

(b) An equal mass of the same mixture, as in (a), is contained in a cylinder closed by a piston. The following processes occur in sequence: (i) adiabatic compression to a temperature of 200°C; (ii) adiabatic complete combustion at constant volume; (iii) expansion to a temperature of 120°C, the work done by the system being 31 kJ and (iv) Cooling at constant volume until the temperature reaches 15°C. On the assumption that the energy of the system depends only upon its temperature and chemical aggregation, evaluate the work done during Process (i) and the heat transfer during Process (iii). Also, explain whether or not the system has executed a thermodynamic cycle.

► **Solution:**

(a) During Process (i), $Q = +3.0$ kJ. As the vessel is rigid, the mixture does not sweep any volume and hence $W = 0$. By the first law,

$$Q - W = E_{2(i)} - E_{1(i)} \qquad (1)$$

where $E_{1(i)}$ and $E_{2(i)}$ are, respectively, the energies of the system initially and finally during process (i). From data $E_{1(i)} = 10$ kJ. Therefore, substituting for Q, W and $E_{1(i)}$ in (1),

$$3 - 0 = E_{2(i)} - 10, \text{ or } E_{2(i)} = 13 \text{ kJ}.$$

During Process (ii), which is adiabatic, $Q = 0$. Because the electrical work supplied by the spark is negligible and that the displacement work is zero, the net $W = 0$. Applying the first law for this process

$$Q - W = E_{2(ii)} - E_{1(ii)} \qquad (2)$$

As the processes occur sequentially, $E_{2(i)} = E_{1(ii)} = 13$ kJ.
Substituting for Q, W and $E_{1(ii)}$ in (2),

$$0 - 0 = E_{2(ii)} - 13,$$

or

$$E_{2(ii)} = 13 \text{ kJ}.$$

During Process (iii), $Q = -32$ kJ and $W = 0$. Application of the first law gives,

$$Q - W = E_{2(iii)} - E_{1(iii)} \tag{3}$$

Also $E_{1(iii)} = E_{2(ii)} = 13$ kJ. Substituting for Q, W and $E_{1(iii)}$ in (3),

$$-32 - 0 = E_{2(iii)} - 13, \text{ or } E_{2(iii)} = -19 \text{ kJ}.$$

(b) The system is given to be the same as that in (a). Applying the first law for Process (i),

$$Q - W = E_{2(i)} - E_{1(i)} \tag{4}$$

Here, $Q = 0$, as the process is given to be adiabatic. Also, the energy of the system is said to depend only on temperature and chemical aggregation. The temperatures and chemical aggregations at the beginning and end of process (i) in both (a) and (b) are the same. Therefore, $[E_{2(i)} - E_{1(i)}]$ in both (a) and (b) have to be the same. Substituting for Q and the change in energy in (4),

$$0 - W = 13 - 10 = 3 \text{ kJ, or } W_{(i)} = -3 \text{ kJ}.$$

Applying the first law to the system during Process (ii)

$$Q - W = E_{2(ii)} - E_{1(ii)} \tag{5}$$

From the data, $Q = 0$ and $W = 0$. Also, based on the fact that the temperature (200°C) and chemical aggregation (unburnt mixture) are the same at the beginning of Processes (ii) in both (a) and (b), $E_{1(ii)}$ in the above relation is equal to $E_{2(i)}$ in (a) = 13 kJ. Substituting for Q, W and $E_{1(ii)}$ in (5)

$$0 - 0 = E_{2(ii)} - 13 \text{ kJ, or, } E_{2(ii)} = 13 \text{ kJ}.$$

Applying the first law for Process (iii),

$$Q - W = E_{2(iii)} - E_{1(iii)} \tag{6}$$

Here, $\quad W = +\,31$ kJ and $E_{1(iii)} = E_{2(ii)} = 13$ kJ.

At the end of Process (iii) the system comprises combustion products at a temperature of 120°C. These conditions are the same as those at the end of Process (iii) in (a). Thus, $E_{2(iii)}$ in (6) = $E_{2(iii)}$ in (a) = −19 kJ. Substituting for $W, E_{2(iii)}$ and $E_{1(iii)}$,

$$Q - 31 = -19 - 13 \text{ or } Q = -1 \text{ kJ}.$$

During Process (iv), $W = 0$, and Q is negative. Applying the first law for this process,

$$Q - W = E_{2(iv)} - E_{1(iv)} \tag{7}$$

Substituting for W, and $E_{1(iv)} = E_{2(iii)} = -19$ kJ,

$$Q - 0 = E_{2(iv)} - (-19) \text{ or } E_{2(iv)} = -19 \text{ kJ} + Q.$$

As Q is negative, $E_{2(iv)}$ is less than -19 kJ.

At the end of Process (iv), although the temperature reaches 15°C, which is the initial temperature of the system, its energy, which is less than -19 kJ, has not reached the initial value of 10 kJ. Thus, the state of the system at the beginning of Process, (i) and that at the end of Process (iv) are not the same. The system has not, hence, undergone a cyclic process.

4.2 A mass of 2.5 kg of a pure substance is initially at a pressure of 700 kPa, a temperature of 200°C and a specific volume of 0.2 m^3/kg. The system undergoes a process to a final state of 700 kPa and a specific volume of 0.2 m^3/kg. For the pure substance, it is given that pressure and specific volume are two independent properties. (a) Find the final temperature and the change in internal energy during the process. (b) If the work done by the system during the process is 12 kJ, determine the magnitude and direction of heat transfer.

▶ **Solution:**

(a) The properties p and V are given to be independent. From the two-property rule,

$$t = t(p, V)$$

The values of p and V at the initial and the final states are the same. Initially, $t = 200$°C, the final temperature also must be the same. For the same reason, the initial energy must be the same as the final energy. Hence, $\Delta U = 0$ during this process.

(b) Applying the first law for the process, $Q - W = \Delta U$, here $W = 12$ kJ and $\Delta U = 0$.

Therefore, $Q = 12 + 0 = 12$ kJ. Heat is transferred to the system

4.3 A certain mass of air entrapped in a piston–cylinder device is first compressed. The work done on the air during this process is 82 kJ and the heat transferred out is 45 kJ. The air is then brought back to its initial state by an expansion process. If the work output during this process is 10 kJ, find the magnitude and direction of heat transfer during this process.

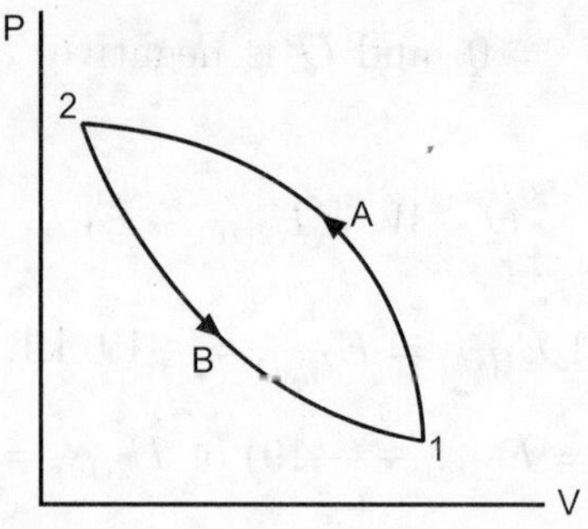

Fig. 4.7 Schematic for Illustrative Example 4.3.

► **Solution:**

The compression process 1–A–2 and the expansion process 2–B–1 are schematically shown, on the p– V plane, in Figure 4.7. The path 1–A–2–B–1 is a thermodynamic cycle.

Applying the first law for the compression process, 1–A–2,

$$Q_A - W_A = \Delta U = U_2 - U_1$$

From the data, $Q_A = -45$ kJ and $W_A = -82$ kJ. Therefore,

$$-45 - (-82) = U_2 - U_1 = 37 \text{ kJ}$$

Applying the first law for the expansion process 2–B–1,

$$Q_B - W_B = \Delta U = U_1 - U_2 = -(U_2 - U_1) = -37 \text{ kJ}$$

From the data, $W_B = 10$ kJ

and therefore,

$$Q_B - 10 = -37, \text{ or } Q_B = -27 \text{ kJ}$$

27 kJ of heat is transferred out of the system.

Alternatively, 1–A–2–B–1 is a cyclic process and therefore $\Sigma Q - \Sigma W = 0$

that is $$(Q_A + Q_B) - (W_A + W_B) = 0$$

$$(-45 + Q_B) - (-82 + 10) = 0, \text{ or } Q_B = -27 \text{ kJ}$$

4.4 A system comprising a certain mass of a gas is heated at constant volume. The heat transfer is 200 kJ. The gas is then cooled at constant pressure during which heat transferred out is 70 kJ and the work done on it is 50 kJ. If the system is then restored to its initial state through an adiabatic process, find the magnitude and direction of work during this adiabatic process.

▶ Solution:

Refer to Figure 4.8 which shows the three processes on a $p-V$ plane.

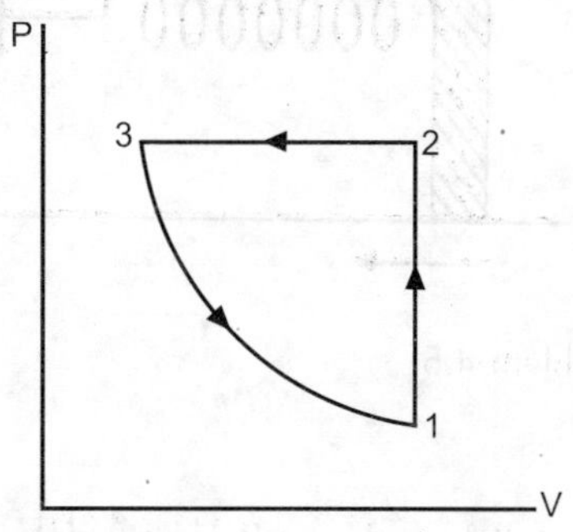

Fig. 4.8 $p-V$ diagram for solving Illustrative Problem 4.4.

From the data, during the constant volume process, 1–2, $Q_{1-2} = +200$ kJ, and $W_{1-2} = 0$.

During the next constant pressure process, 2–3, $Q_{2-3} = -70$ kJ and $W_{2-3} = -50$ kJ.

During the final adiabatic process, 3–1, $Q_{3-1} = 0$ and let the work $= W_{3-1}$.

The processes performed in the given sequence makes the system undergo a thermodynamic cycle. From the first law,

$$\Sigma Q - \Sigma W = 0$$

that is, $(Q_{1-2} + Q_{2-3} + Q_{3-1}) - (W_{1-2} + W_{2-3} + W_{3-1}) = 0$

Therefore, $(200 - 70 + 0) - (0 - 50 + W_{3-1}) = 0$ or $W_{3-1} = +180$ kJ.

4.5 The piston–cylinder device shown in Figure 4.9 contains initially 0.1 m^3 of air at 2 bar. During a certain process, the frictionless piston moves out slowly, against the spring pressure and the atmospheric pressure of 1 bar, until the volume and the pressure of air increase to 0.15 m^3 and 3 bar, respectively. During the process, the force exerted by the spring on the piston is proportional to the distance travelled by the piston from its position where the spring would be just touching the piston without exerting any pressure. For air, the pressure, volume and temperature are related by $pV = 40 \times (t + 273)$, where p is in N/m^2, V is in m^3 and t is in °C. Also, the internal energy of air is related to its temperature by $U = 0.1\,t$, where U is in kJ and t is in °C. Find the magnitude and direction of heat interaction for the air during the process.

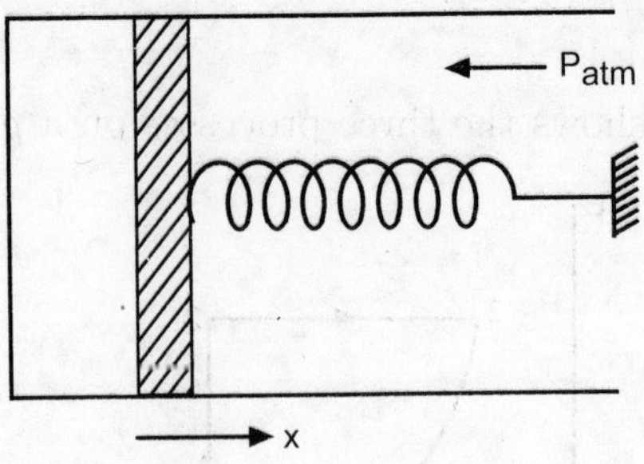

Fig. 4.9 Figure for Illustrative Problem 4.5.

▶ **Data:**

$P_1 = 2$ bar, $V_1 = 0.1$ m^3, $P_2 = 3$ bar, $V_2 = 0.15$ m^3, $PV = 40(t + 273)$, $U = 0.1t$ Spring force is proportional to distance moved by the piston from datum position.

▶ **Solution:**

The spring force is proportional to x, where x is measured from that position at which the spring is just touching the piston. Let V_0 be the volume of air in the cylinder when the spring is just touching the piston, that is, when $x = 0$. The spring pressure is proportional to force and also the distance is proportional to the volume. Hence, $P_{spr} = \mathrm{K}(V - V_0)$ where K is a constant.

The displacement work W is given by

$$W = \int p\mathrm{d}V = \int (p_{atm} + p_{spr})\mathrm{d}V$$

$$= \int p_{atm}\mathrm{d}V + \int p_{spr}\mathrm{d}V$$

As p_{atm} is a constant, and $p_{spr} = \mathrm{K}(V - V_0)$,

$$W = p_{atm}\int \mathrm{d}V + \int \mathrm{K}(V - V_0)\mathrm{d}V$$

$$= p_{atm}(V_2 - V_1) + \frac{\mathrm{K}[V_2^2 - V_1^2]}{2} - \mathrm{K}V_0(V_2 - V_1) \quad \text{(i)}$$

From the data, $p_1 = p_{atm} + p_{spr} = 1 + \mathrm{K}(V_1 - V_0)$ (ii)

And $p_2 = p_{atm} + p_{spr} = 1 + \mathrm{K}(V_2 - V_0)$ (iii)

(iii) – (ii) gives $\mathrm{K}(V_2 - V_1) = p_2 - p_1$ and therefore,

$$\mathrm{K}(0.15 - 0.1) = 3 - 2, \text{ or } \mathrm{K} = 20 \text{ bar/m}^3$$

Substituting for K and V_1 in (ii),

$$2 = 1 + 20 \times (0.1 - V_0)$$

Therefore, $V_0 = 0.05 \text{ m}^3$

Subtituting the numerical values for every quantity in (i)

$$W = 1 \times 10^5 \times (0.15 - 0.1) + \frac{20 \times 10^5 (0.15^2 - 0.1^2)}{2} - 20 \times 10^5 \times 0.05 \times (0.15 - 0.1)$$

$$= 0.05 \times 10^5 + 0.125 \times 10^5 - 0.05 \times 10^{5\prime}$$

$$= 12{,}500 \text{ J} = 12.5 \text{ kJ}.$$

As $pV = 40(t + 273)$, $t = \left(\dfrac{pV}{40}\right) - 273.$

Therefore, $t_1 = \left(\dfrac{p_1 V_1}{40}\right) - 273$ and $t_2 = \left(\dfrac{p_2 V_2}{40}\right) - 273$, or

$$(t_2 - t_1) = \left(\frac{1}{40}\right)(p_2 V_2 - p_1 V_1)$$

$$= \left(\frac{1}{40}\right) \times 10^5 \times (3 \times 0.15 - 2 \times 0.1) = 625°\text{C}$$

It is given that $U = 0.1t$ and hence $U_2 - U_1 = 0.1(t_2 - t_1) = 0.1 \times 625 = 62.5$ kJ. Applying the first law for the process,

$$Q - W = U_2 - U_1, \text{ or } Q = W + (U_2 - U_1)$$

Therefore, $Q = 12.5 + 62.5 = 75$ kJ (i.e., 75 kJ of heat is transferred to air)

Problems

4.1 Find out whether the change in the energy of the systems underlined in each of the cases in Problem 3.11 are positive, negative or zero.

4.2 Refer to Figure 4.10. When a system is taken from state ‘a’ to ‘b’ along ‘acb’, 100 kJ of heat is transferred to the system and the system does 35 kJ work. (a) Given that the work done by the system during the process ‘adb’ is 12 kJ, find the magnitude and direction of heat transfer during this process. (b) If during the process ‘ba’ work done on the system is 25 kJ, find the magnitude and direction of heat interaction during this process. (c) Assuming the energy of the system to be 0 and 50 kJ at states ‘a’ and ‘d’, respectively, find the direction and magnitude of heat transfers during processes ‘ad’ and ‘db’.

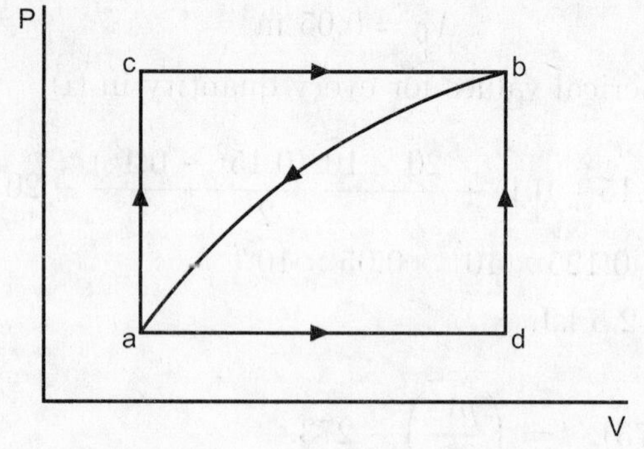

Fig. 4.10 Figure for Problem 4.2.

4.3 A certain system undergoes a cyclic process. The process starts at state 1, goes along path 'A' to state 2 and then returns to state 1 via the path 'B'. During process 'A', heat transferred to the system is 85 kJ and work done by the system is 40 kJ. Suppose the same system undergoes another cyclic process as follows: the process starts at state 2, goes along path 'C' to state 1 and then the system returns to state 2 via path 'D'. It is given that heat interactions during Processes B and C are same and each is equal to −55 kJ, and heat transferred to the system during process D is 75 kJ. Calculate the magnitude and directions of work interactions during C and D.

4.4 Consider the arrangement shown in Figure 4.11. A volume equal to 0.005 m^3 of a certain gas, initially at 1.2 bar (at the position of piston indicated in the figure) is heated until its pressure is 2.5 bar. Given that the internal energy of the gas during this process increases by 520 J, evaluate the heat and work transfers for the gas.

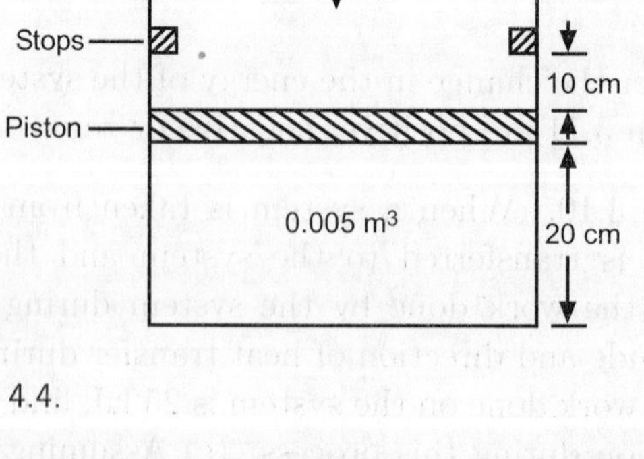

Fig. 4.11 Figure for Problem 4.4.

4.5 A system undergoes a change of state through a constant pressure process followed by a constant volume process. During the constant pressure process, 500 kJ of heat is transferred to the system and 200 kJ of work is done by the system. During the constant volume process, 500 kJ of heat is rejected from the system. If an adiabatic process restores the system to its initial state, what would be the direction and magnitude of work during this process. Assume all are resisted processes and show them on a $p-V$ diagram.

4.6 A sealed vessel containing certain chemicals is completely immersed in a tank of water which is open to the atmosphere. When the chemicals react, heat is transferred from the vessel to the water, causing the temperature of water to increase. A stirring device is used to churn the water and the power input to the shaft driving the stirrer is 30 W. In a 15-min period, the heat transferred from the vessel is 60 kJ and the heat transferred from the tank to the surrounding air 4 kJ. Assuming that there is no evaporation of water, determine the change in internal energy of the water.

4.7 A fluid is confined in a cylinder by a spring-loaded frictionless piston such that the pressure of the fluid is a linear function of volume given by $p = a + bV$. The internal energy of the fluid is given by the relation $U = 34 + 3.15\,pV$. In these relations, a and b are constants and p is in kPa, V is in m^3 and U is in kJ. If the fluid changes from an initial state of 170 kPa, 0.03 m^3 to a final state of 400 kPa, 0.06 m^3, with no other work interaction than that done on the piston, find the magnitude and direction of work and heat transfer for the fluid.

4.8 A system undergoes a state change from 1 to 2 during which heat transferred to the system is 200 kJ. Internal energy of the system when it is at state 2 is 150 kJ higher than that at state 1. Suppose the system is brought back to its initial state 1 by a process during which work done on the system is 80 kJ. Calculate the work transfer during the process 1–2 and heat transfer during the process 2–1.

4.9 A dashpot is a device for controlling the motion of mechanisms. It consists of a cylinder, closed at both ends, fitted with a piston. A rod which protrudes through one of the covers of the cylinder connects the piston to the mechanism. Both ends of the cylinder contain oil and are in communication via a small hole drilled through the piston. As the piston is pushed inwards, the oil at one end of the cylinder is forced through the hole to the other end. In a particular dashpot, the mechanism connected to the piston has a kinetic energy of 4000 Nm, initially; finally, the mechanism comes to rest. Assuming the heat transfers are negligible, evaluate the increase in energy of the following systems: (a) the mechanism–piston combination, (b) the oil and (c) the mechanism–dashpot combination.

4.10 (a) An insulated rigid vessel contains some powdered coal and air at 10 bar, 25°C. The coal is ignited by an infinitesimal spark and at the end of the combustion process the temperature of the contents of the vessel increases to 540°C. Consider the vessel with its contents as the system and evaluate the increase in energy of the system.

(b) The insulation is now removed, and after a time, the vessel with its contents attains thermal equilibrium with the surroundings at 25°C. Find the decrease in energy of the system during this process if the heat transferred out of the system is 50 kJ.

(c) If the initial energy of the system is 30 kJ find the values of the energies at the end of processes (a) and (b).

4.11 A vertical cyclinder whose diameter is 10 cm and which is closed by a piston contains a combustible mixture at 25°C. The piston is free to move and its weight is 0.2 kg. The upper surface of the piston is exposed to atmosphere at 1 bar. The mixture is ignited. Because of combustion the piston moves up slowly. By the time the piston moves up a distance of 7.5 cm it is found that not only the combustion is complete but also the contents in the cylinder have been cooled to 25°C. The heat transfer during this process is 4 kJ. Evaluate the work interaction and decrease in energy of the contents.

4.12 On a warm summer day, Sheela decides to beat the heat by closing the windows and doors of the kitchen and opening the refrigerator door. At first, she feels cool and refreshed but after a while the effect begins to wear off. Analyse the situation in the light of the first law of thermodynamics.

4.13 A household refrigerator is loaded with fresh food and closed. Consider the whole refrigerator and contents as the system. The refrigerator uses 1 kWh electrical energy in cooling the food and the internal energy of the system decreases by 5000 kJ as the temperature drops. Find the direction and magnitude of heat transfer for the system.

4.14 A cylinder with a vertical axis has a frictionless piston at its upper end. The piston weight is such as to exert a pressure of 3 bar on 0.5 kg gas enclosed. When this gas is cooled, its volume decreases from 0.03 to 0.01 m^3 and its specific internal energy decreases from 200 to 80 kJ/kg. For the gas, calculate (a) the heat interaction (b) work interaction and (c) the change in internal energy.

4.15 A mass of 0.3 kg of air at 4 bar, 60°C is contained in a cylinder closed by a piston. As a result of expansion and heat transfer the state of air changes to 3 bar, 55°C. Assume that, for air, the enthalpy and internal energy are functions of temperature only and that c_p and c_v of air are 1005 and 718 J/kg K, respectively. Evaluate (i) the increase in internal energy of air and (ii) the increase in enthalpy. Can the work and heat transfers be evaluated for the process. If the answer is 'yes' calculate their directions and magnitudes, if the answer is 'no', explain why it is so?

FIRST LAW OF THERMODYNAMICS FOR CONTROL VOLUMES

5

5.1 Introduction

Most energy conversion devices fall under the category of control volumes, across whose control surfaces flow of material is mandatory. During most of their operating periods these devices work under steady conditions. Chapter 4 elaborated how the first law of thermodynamics is stated for a system and how it is used to analyse several processes. This chapter explains how the first law can be extended to control volumes operating under both steady and unsteady conditions. Several processes of practical importance are also analysed.

5.2 Extension of the First Law for a Control Volume

Generally speaking, when a control volume (e.g., turbine, rotary compressor, nozzle, diffuser, heat-exchanger, throttling device etc.) operates, there is material flow into and out of the control volume through the inlet and exit manifolds, respectively, and heat and work interactions across the control surface.

Consider, as shown in Figure 5.1, a control volume, CV, bound by the control surface C. In a small time interval $d\tau$, let the mass of material entering the control volume through the inlet manifold at cross-section 1–1 be dm_1 and the mass of material leaving the control volume at cross-section 2–2 in the exit manifold be dm_2. At the same time, let δQ and δW_x be the heat and work interactions across the control surface.

As the size, shape and orientation of a control surface with respect to a stationary observer cannot change, W_x includes all modes of work (mostly shaft work and/or electrical work) other than displacement work. Since the first law is known to be applicable to a system only, an attempt is now made to imagine the happening in a control volume as an equivalent happening in a suitably imagined system.

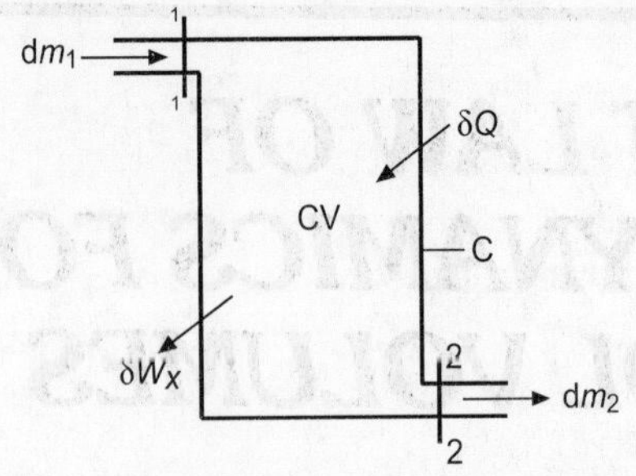

Fig. 5.1 Schematic for the extension of the first law of thermodynamics to a control volume.

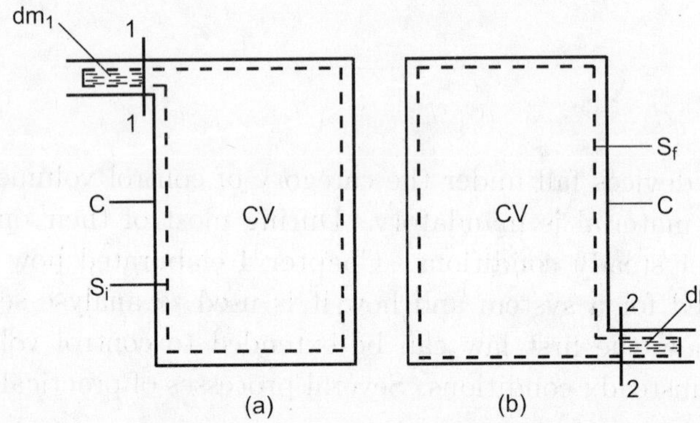

Fig. 5.2*a* Initial system boundary; *b* Final system boundary.

The initial system boundary, S_i, is shown in Figure 5.2*a* by the broken line. This comprises the control surface C and the boundary in the inlet manifold that contains the material dm_1. Similarly, the final system boundary, S_f, is shown in Figure 5.2*b* by the broken line again. This boundary, at the end of the considered small interval of time $d\tau$, comprises the control surface C and the material dm_2 in the exit manifold. It is necessary to point out here that the material content in S_i and S_f are the same. The process undergone by the imagined system is the displacement of the system boundary at the inlet, causing material dm_1 to enter the CV and the displacement of the system boundary at the exit manifold, causing material dm_2 to leave the CV. During the said process there could be heat and work interactions across the system boundary.

Let the properties of the fluid at the entry be p_1, v_1, e_1, h_1 and V_1 and the corresponding properties at the exit be p_2, v_2, e_2, h_2 and V_2, where p is the pressure, v the specific volume, e the net specific energy, h the specific enthalpy and V the velocity.

Considering dm_1 and dm_2 to be very small, the areas of the system boundary in the inlet and exit manifolds become negligible, thereby tending the area of the system towards the control surface area. Thus, the net heat transfer across the system boundary would be the same as δQ, the heat transfer across the control surface.

However, the net work interaction, δW, across the system boundary is equal to δW_x, the work interaction across the control surface plus the displacement works at the inlet and the exit manifolds. Displacement work at the inlet manifold $= p\mathrm{d}V$, where p is the pressure at entry and $\mathrm{d}V$ is the volume swept by the part of the system boundary in the entry manifold. Thus $p = p_1$ and $\mathrm{d}V = (\mathrm{d}m_1 v_1)$. Also $\mathrm{d}V$ is negative as the system volume in the inlet manifold keeps decreasing as the material flows into the CV. Therefore, the displacement work at the inlet manifold $= -\mathrm{d}m_1\, p_1\, v_1$. The negative sign indicates that the work is done on the system at the inlet.

Similarly, the displacement work at the exit manifold $= +\mathrm{d}m_2\, p_2\, v_2$.

Therefore,
$$\delta W = \delta W_x + \mathrm{d}m_2\, p_2\, v_2 - \mathrm{d}m_1\, p_1\, v_1. \tag{i}$$

Suppose, E_{c1} and E_{c2} are, respectively, the initial and final energies of the material in the control volume and that E_1 and E_2 are the initial and final energies of the system. Then,
$$E_1 = E_{c1} + \mathrm{d}m_1\, e_1 \quad \text{and} \quad E_2 = E_{c2} + \mathrm{d}m_2\, e_2,$$

or,
$$E_2 - E_1 = (E_{c2} + \mathrm{d}m_2\, e_2) - (E_{c1} + \mathrm{d}m_1\, e_1) = \mathrm{d}E_c + (\mathrm{d}m_2\, e_2 - \mathrm{d}m_1\, e_1) \tag{ii}$$

where, $\mathrm{d}E_c = (E_{c2} - E_{c1})$ = change in energy of the contents of the control volume.

Application of the first law to the system gives,
$$\delta Q - \delta W = E_2 - E_1$$

Substituting for δW and $(E_2 - E_1)$ from (i) and (ii),
$$\delta Q - [\delta W_x + \mathrm{d}m_2\, p_2\, v_2 - \mathrm{d}m_1\, p_1\, v_1] = \mathrm{d}E_c + (\mathrm{d}m_2\, e_2 - \mathrm{d}m_1\, e_1)$$

that is,
$$\delta Q - \delta W_x = \mathrm{d}E_c + [\mathrm{d}m_2\,(e_2 + p_2 v_2) - \mathrm{d}m_1\,(e_1 + p_1 v_1)] \tag{5.1}$$

The differential form of the first law applicable to control volumes is as in (5.1).

In the integrated form, applicable over a certain time, τ, it is
$$\int \delta Q - \int \delta W_x = \int \mathrm{d}E_c + \int [\mathrm{d}m_2\,(e_2 + p_2 v_2) - \mathrm{d}m_1\,(e_1 + p_1 v_1)]$$

that is,
$$Q - W_x = \Delta E_c + m_2\,(e_2 + p_2 v_2) - m_1\,(e_1 + p_1 v_1) \tag{5.2}$$

where, ΔE_c = the difference in the energies of the control volume after and before the process in the time interval $\tau = (E_{c,f} - E_{c,i})$, and m_1 and m_2 are, respectively, the masses entering and leaving the control volume in the same time interval, τ. These interactions are shown schematically in Figure 5.3.

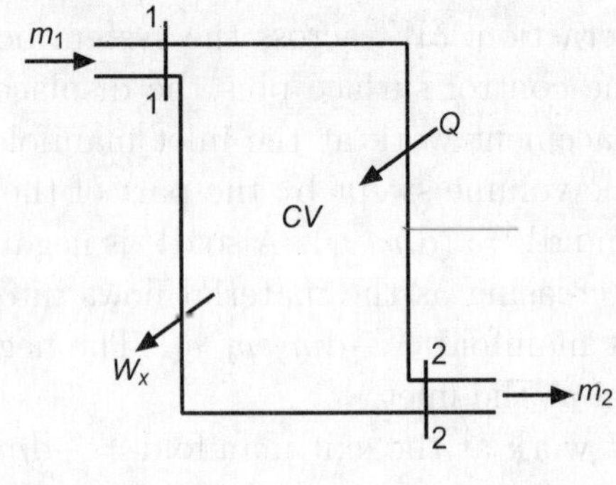

Fig. 5.3 Interactions across a control volume.

5.3 Steady-State, Steady-Flow Energy Equation

As stated before, more often than not, control volumes operate under steady conditions. Conditions are said to be steady if the properties of the fluid at entry and exit are steady, and that in the given time interval there is no change in the properties of the contents of the control volume. This means $\Delta E_c = 0$. To satisfy this condition, the flow into and out of the CV must be steady. Flow is said to be steady if m_2 equals m_1. If not, there could be either accumulation or depletion of material in the control volume, leading to a change in the control volume energy. Let $m_2 = m_1 = m$.

The steady-state, steady-flow energy equation (SS, SFEE), is obtained by incorporating the steady-state, steady-flow conditions in (5.1) and (5.2). Hence,

$$\delta Q - \delta W_x = \mathrm{d}m\,[(e_2 + p_2 v_2) - (e_1 + p_1 v_1)] \tag{5.3}$$

or

$$Q - W_x = m\,[(e_2 + p_2 v_2) - (e_1 + p_1 v_1)] \tag{5.4}$$

(***Note***: Although it is not strictly correct, in most books, SS, SFEE is also abbreviated as SFEE.)

5.3.1 SS, SFEE based on Unit Mass

In general, e, the specific energy, may comprise u, the internal energy, $V^2/2$, the kinetic energy, and gz, the potential energy. Hence, (5.4) can be rewritten as

$$Q - W_x = m\left[\left(u_2 + \frac{V_2^2}{2} + gz_2 + p_2 v_2\right) - \left(u_1 + \frac{V_1^2}{2} + gz_1 + p_1 v_1\right)\right]$$

that is, $$Q - W_x = m\left\{\left[(u_2 + p_2 v_2) + \frac{V_2^2}{2} + gz_2\right] - \left[(u_1 + p_1 v_1) + \frac{V_1^2}{2} + gz_1\right]\right\}$$

By definition, $h = (u + pv)$

and therefore, $(u_2 + p_2 v_2) = h_2$ and $(u_1 + p_1 v_1) = h_1$

$$\text{Therefore,}\quad Q - W_x = m\left[\left(h_2 + \frac{V_2^2}{2} + gz_2\right) - \left(h_1 + \frac{V_1^2}{2} + gz_1\right)\right]$$

$$\text{or,}\quad Q - W_x = m\left[(h_2 - h_1) + \left(\frac{V_2^2}{2} - \frac{V_1^2}{2}\right) + g(z_2 - z_1)\right] \tag{5.5}$$

$$\text{that is,}\quad Q - W_x = m\Delta\left[h + \frac{V^2}{2} + gz\right] \tag{5.6}$$

If $m = 1$ kg, (5.6) reduces to

$$Q - W_x = \Delta\left[h + \frac{V^2}{2} + gz\right] \tag{5.7}$$

The SS, SFEE based on unit mass is represented in (5.7). Here, Δ stands for 'the change in'. For example, Δh = change in specific enthalpy = $(h_2 - h_1)$. The relation (5.7) is such that while the left-hand side has only the interaction across the control surface, the right-hand side contains just the properties of the fluid at entry and exit.

The units of the quantities in (5.7) have to be as follows: If Q and W_x are in J/kg material, then each property within parenthesis must also be in J/kg. Therefore, V must be in m/s, g in m/s^2 and z in m. If Q and W_x are in kJ/kg, then, while h must also be in kJ/kg, the terms $\frac{V^2}{2}$ and gz must be divided by 1000 to convert each of them to kJ/kg.

Example 5.1

A fluid flows steadily through a rotary device. For a kilogram of fluid, the heat transfer out of the device is 24 kJ. The fluid properties at the entry are 5 bar, 227°C, 50 m/s, and 0.78 m^3/kg. The corresponding properties at exit are 1 bar, 57°C, 100 m/s and 0.97 m^3/kg. The inlet manifold is 5 m above the exit manifold. For the fluid, the internal energy is a function of temperature only and that its c_v = 0.7 kJ/kg K. Find the work output in kJ/kg.

▶ **Data:**

The schematic of the device will be as in Figure 5.3.

$$Q = -24 \text{ kJ/kg},\ p_1 = 5\times10^5 \text{ N/m}^2,\ t_1 = 227°\text{C},\ V_1 = 50 \text{ m/s},\ v_1 = 0.78 \text{ m}^3\text{/kg},$$
$$p_2 = 1\times10^5 \text{ N/m}^2,\quad t_2 = 57°\text{C},\ V_2 = 100 \text{ m/s},\ v_2 = 0.97 \text{ m}^3\text{/kg},$$
$$z_2 - z_1 = -5 \text{ m},\quad u = u(t) \quad\text{and}\quad c_v = 0.7 \text{ kJ/kg K}.$$

▶ **Solution:**

SS, SFEE on unit-mass basis is,

$$Q - W_x = \Delta\left[h + \frac{V^2}{2} + gz\right] \tag{5.7}$$

$$= (h_2 - h_1) + \left(\frac{V_2^2 - V_1^2}{2}\right) + g\,(z_2 - z_1) \qquad \text{(i)}$$

But, $$(h_2 - h_1) = (u_2 + p_2 v_2) - (u_1 + p_1 v_1)$$
$$= (u_2 - u_1) + (p_2 v_2 - p_1 v_1) = c_v\,(t_2 - t_1) + (p_2 v_2 - p_1 v_1)$$
$$= 0.7 \times (57 - 227) + 10^5 \times (1 \times 0.97 - 5 \times 0.78) \times 10^{-3}$$
$$= -412 \text{ kJ/kg}$$

Now, $$\left(\frac{V_2^2}{2} - \frac{V_1^2}{2}\right) = \frac{(100^2 - 50^2)}{2000} = 3.75 \text{ kJ/kg}$$

And $$g\,(z_2 - z_1) = 9.81 \times \left(\frac{-5}{1000}\right) = -0.049 \text{ kJ/kg}$$

Substituting the values of various quantities in (i),

$$-24 - W_x = -412 + 3.75 - 0.049, \quad \text{or} \quad W_x = 384.3 \text{ kJ/kg}$$

5.3.2 SS, SFEE Based on Unit Time

When (5.6) is divided by time, τ, we get the SS, SFEE equation based on unit time as,

$$(Q/\tau - W_x/\tau) = (m/\tau)\Delta\left[h + \frac{V^2}{2} + gz\right], \quad \text{or}$$

$$\dot{Q} - \dot{W}_x = \dot{m}\Delta\left[h + \frac{V^2}{2} + gz\right] \qquad (5.8)$$

While the units of all the fluid properties within parentheses on the right-hand side would have to be necessarily the same as those used in (5.7), the units of $\dot{Q}$ and $\dot{W}_x$ will have to be in J/s or kJ/s and $\dot{m}$ in kg/s.

Example 5.2

Steam flows into a turbine, at a flow rate of 5000 kg/h. The turbine develops a power of 550 kW. The heat loss from the casing of the turbine and the bearings is negligible. (a) Find the change in enthalpy across the turbine, if the inlet velocity is negligible and the exit velocity is 360 m/s and the change in potential energy is negligible. (b) Find the change in enthalpy across the turbine, if the velocity at entry is 66 m/s (exit velocity being the same as given above) and the inlet pipe is 3 m above the exit pipe.

▶ **Data:**

Referring to Figure 5.4,

$$\dot{m} = \frac{5000}{3600} = 1.389 \text{ kg/s}, \quad \dot{W}_x = \dot{W}_T = 550 \text{ kJ/s}, \quad \dot{Q} = 0,$$

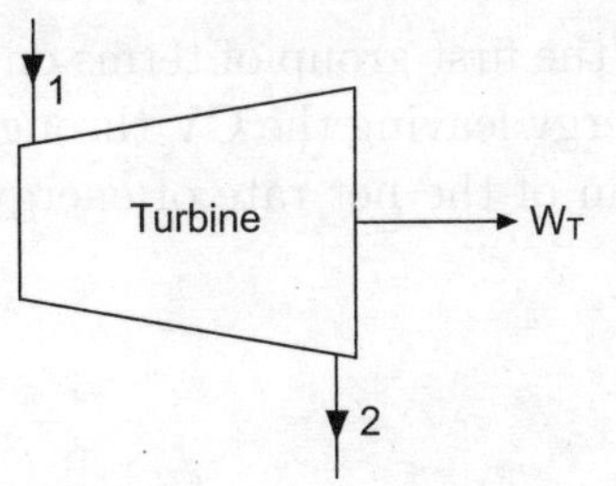

Fig. 5.4 Schematic for analysing Example 5.2.

(a) $V_1 = 0$, $V_2 = 360$ m/s, $\Delta gz = 0$ and (b) $V_1 = 60$ m/s, $V_2 = 360$ m/s, $(z_2 - z_1) = -3$ m.

► **Solution:**

SS, SFEE on unit time basis is

$$\dot{Q} - \dot{W}_x = \dot{m}\Delta\left[h + \frac{V^2}{2} + gz\right]. \tag{5.8}$$

$$= \dot{m}\left[\Delta h + \frac{(V_2^2 - V_1^2)}{2} + g(z_2 - z_1)\right]$$

(a) Substituting the various values from data,

$$0 - 550 = 1.389\left[\Delta h + \frac{360^2}{2000}\right].$$

Therefore, $\Delta h = -460.77$ kJ/kg.

(b) Substituting the relevant values in (5.8),

$$0 - 550 = 1.389\left[\Delta h + \left(\frac{360^2 - 60^2}{2000}\right) - 9.81 \times \frac{3}{1000}\right]$$

Therefore, $\Delta h = -458.94$ kJ/kg.

5.3.3 SS, SFEE for Control Volumes with Multi-Streams

In many situations, there could be more than one stream at entry or exit, or both at entry and exit. As the fluid properties have been exclusively brought on to the right side of the SS, SFEE relation in (5.8), it would not be difficult to visualise that, in such cases, the equation, based on unit time, could be modified to read as follows:

$$\dot{Q} - \dot{W}_x = \Sigma_{\text{exit}}\dot{m}\left(h + \frac{V^2}{2} + gz\right) - \Sigma_{\text{entry}}\dot{m}\left(h + \frac{V^2}{2} + gz\right) \tag{5.9}$$

In the above relation, while the first group of terms on the right-hand side corresponds to the sum of the net rate of energy leaving the CV through each stream, the second group of terms corresponds to the sum of the net rate of energy entering the CV through each stream.

Example 5.3

Two air streams A ($\dot{m}_A = 0.8$ kg/s, $p_A = 15$ bar and $t_A = 250$°C) and B ($\dot{m}_B = 0.5$ kg/s, $p_B = 15$ bar and $t_B = 200$°C) are mixed in a large chamber and the chamber supplies the mixed air, at 10 bar, to a turbine. The turbine exhausts the air at 30°C to the atmosphere. Assume steady adiabatic flow all through. Neglect kinetic and potential energies. Assume that enthalpy and internal energy of air are functions of temperature only and that $c_p = 1$ kJ/kg K and $c_v = 0.718$ kJ/kg K. Determine (a) the temperature of air at inlet to the turbine and (b) the power developed by the turbine.

▶ Data:

Refer to Figure 5.5.

$$\dot{m}_A = 0.8 \text{ kg/s}, \quad p_A = 15 \times 10^5 \text{ N/m}^2, \quad t_A = 250°\text{C},$$
$$\dot{m}_B = 0.5 \text{ kg/s}, \quad p_B = 15 \times 10^5 \text{ N/m}^2, \quad t_B = 200°\text{C}, \quad Q = 0 \text{ althrough},$$
$$p_1 = 10 \text{ bar}, \quad t_1 = ? \; t_2 = 30°\text{C}.$$
$$h = h(t) \text{ and } u = u(t), \quad c_v = 0.718 \text{ kJ/kg deg}, \quad c_p = 1 \text{ kJ/kg deg}.$$

▶ Solution:

SS, SFEE for a CV with multi-streams is

$$\dot{Q} - \dot{W}_x = \Sigma_{\text{exit}}\dot{m}\left(h + V^2/2 + gz\right) - \Sigma_{\text{entry}}\dot{m}\left(h + V^2/2 + gz\right) \tag{5.9}$$

For the mixing chamber, as per data, $\dot{Q} = 0$, $\dot{W}_x = 0$ and kinetic and potential energies are negligible. Therefore, 5.9 reduces to,

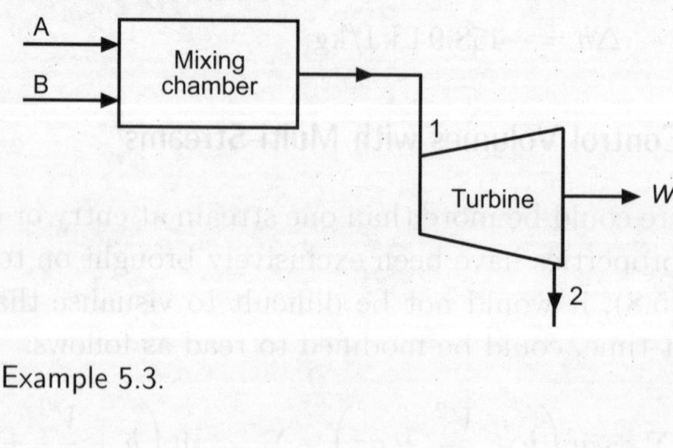

Fig. 5.5 Schematic for Example 5.3.

$$0 - 0 = \Sigma_{\text{exit}}\dot{m}(h + 0 + 0) - \Sigma_{\text{entry}}\dot{m}(h + 0 + 0)$$

that is, $$\Sigma_{\text{exit}}\dot{m}h - \Sigma_{\text{entry}}\dot{m}h = 0$$

Therefore, $(\dot{m}_A + \dot{m}_B)\, h_1 - \dot{m}_A h_A - \dot{m}_B h_B = 0$, where suffix 1 indicates the condition at the exit of the mixing chamber and also at the inlet to the turbine.

Rearranging the above relation,

$$\dot{m}_A\,(h_1 - h_A) + \dot{m}_B\,(h_1 - h_B) = 0,$$

As enthalpy is a function of temperature only and c_p is a constant,

$$\dot{m}_A c_p\,(t_1 - t_A) + \dot{m}_B c_p\,(t_1 - t_B) = 0$$

Substituting the numerical values,

$$0.8 \times 1 \times (t_1 - 250) + 0.5 \times 1 \times (t_1 - 200) = 0, \quad \text{and hence} \quad t_1 = 230.77°\text{C}.$$

Applying SS, SFEE to turbine, for which, from data, $Q = 0$, and kinetic and potential energies are negligible and $W_x = W_T$, the turbine work output.

Therefore $$-\dot{W}_T = (\dot{m}_A + \dot{m}_B)\,(h_2 - h_1)$$
$$= (\dot{m}_A + \dot{m}_B)\,c_p\,(t_2 - t_1) = (0.8 + 0.5) \times 1 \times (30 - 230.77)$$

Therefore, $\dot{W}_T = 261$ kW.

5.3.4 Continuity Equation

An essential condition for a control volume to operate under steady conditions is that the mass flow rate at entry equals that at exit. If at the entry to the control volume, the area of cross-section is A_1, the velocity is V_1 and the specific volume is v_1, and the corresponding characteristics at the exit are, respectively A_2, V_2 and v_2, then the continuity equation can be written as

$$\dot{m} = A_1V_1/v_1 = A_2V_2/v_2 \tag{5.10}$$

Example 5.4

Steam enters a rotary turbine steadily through an inlet pipe of 15 cm diametre at a rate of 4000 kg/h. The specific volume at entry is 0.285 m^3/kg. The exit pipe diametre is 25 cm and that the specific volume of steam at exit is 15.0 m^3/kg. Find the inlet and exit velocities.

► **Data:**

$d_1 = 0.15$ m, $\dot{m} = 4000/3600 = 1.11$ kg/s, $v_1 = 0.285$ m^3/kg, $d_2 = 0.25$ m, $v_2 = 15$ m^3/kg, $V_1 =$? and $V_2 =$?

► **Solution:**

From the continuity equation,

$$\dot{m} = A_1V_1/v_1 = A_2V_2/v_2. \tag{5.10}$$

Here, $A_1 = \pi d_1^2/4 = \pi \times 0.15^2/4 = 0.01767$ m^2

And $A_2 = \pi d_2^2/4 = \pi \times 0.25^2/4 = 0.0491$ m^2

Substituting the numerical values in (5.10),

$$1.11 = 0.01767 \times V_1/0.285 = 0.0491 \times V_2/15,$$

and therefore, $V_1 = 17.9$ m/s and $V_2 = 339$ m/s

5.4 Application of SS, SFEE to some Specific Cases

5.4.1 Adiabatic Throttling Process

A throttling process is one in which a fluid flowing through a duct experiences a heavy resistance. Flow of a fluid across an almost closed valve, or through a pipe plugged with porous material such as glass wool or through a long capillary are examples of throttling processes.

Shown in Figure 5.6 is the schematic of the flow of a fluid through a porous plug in a pipe. There is a very-high-pressure drop during the process, which serves to overcome the friction for the flow. Such a process is called a *throttling process*. Indeed, a throttling process is an extremely irreversible process. Often, the velocities at entry and exit are negligible and the process can be assumed to be adiabatic. Being a simple flow process, there are no work interactions. Thus, in the SS, SFEE equation,

$$Q - W_x = \Delta\left[h + \frac{V^2}{2} + gz\right] \tag{5.7}$$

$Q = 0$, $W_x = 0$, $\Delta V^2 = 0$ and Δz is generally so small that it is neglected.

Therefore, the SS, SFEE equation for a throttling process reduces to

$$\Delta h = 0, \quad \text{or} \quad h_2 = h_1 \tag{5.11}$$

The SS, SFEE applicable to a throttling process is as in (5.11). A throttling process is thus an *isenthalpic process*.

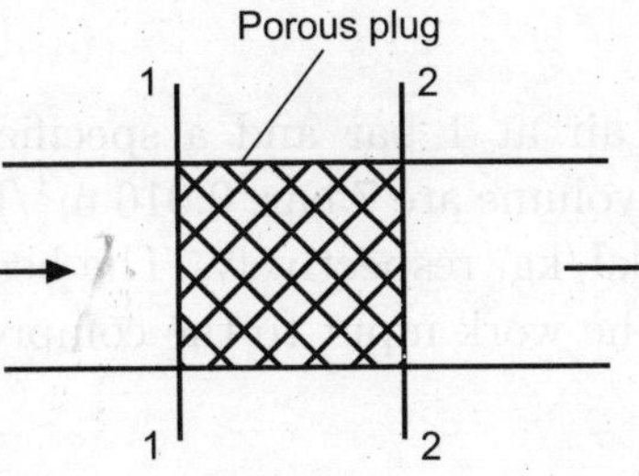

Fig. 5.6 Flow through a porous plug.

Example 5.5

Air is throttled steadily from 10 bar, 300°C to 0.1 bar. The enthalpy of air is a function of temperature only and that its $c_p = 1$ kJ/kg K. Find the temperature of air after throttling.

► **Data:**

Refer to Figure 5.6,

$p_1 = 10$ bar, $t_1 = 300$°C, $p_2 = 0.1$ bar, $h = h(t)$, $c_p = 1$ kJ/kg K.

► **Solution:**

SS, SFEE as applicable to an adiabatic throttling process is,

$$\Delta h = 0 \tag{5.11}$$

Therefore, $c_p\,(t_2 - t_1) = 0$, or $t_2 = t_1 = 300$°C.

5.4.2 Turbines and Rotary Compressors

Unless otherwise mentioned, the rotary machines are considered to be adiabatic with negligible change in potential energy. Thus, with $Q = 0$ and $\Delta z = 0$, the SS, SFEE for rotary machines reduces to

$$-W_x = \Delta\left(h + \frac{V^2}{2}\right) \quad \text{or}$$

$$W_x = (h_1 - h_2) + \left(\frac{V_1^2}{2} - \frac{V_2^2}{2}\right) \tag{5.12}$$

And, if the changes in kinetic energies are negligible,

$$W_x = (h_1 - h_2) \tag{5.13}$$

In a turbine, $h_1 > h_2$ and hence W_x would be positive.
In a compressor, $h_2 > h_1$ and hence W_x would be negative.
Thus, (5.12) or (5.13) is the SS, SFEE as applicable to rotary machines.

Example 5.6

A rotary compressor takes in air at 1 bar and a specific volume of 0.08 m^3/kg. The discharge pressure and specific volume are 7 bar, 0.016 m^3/kg. The entry and exit internal energies of air are 48 and 200 kJ/kg, respectively. The heat loss from the turbine casing is 120 kJ/kg of air. Calculate the work input to the compressor neglecting the changes in kinetic and potential energies.

▶ **Data:**

$p_1 = 10^5$ N/m^2, $v_1 = 0.08$ m^3/kg, $p_2 = 7 \times 10^5$ N/m^2, $v_2 = 0.016$ m^3/kg, $u_1 = 48$ kJ/kg, $u_2 = 200$ kJ/kg, $Q = -120$ kJ/kg, $\Delta V^2/2 = 0$, $\Delta gz = 0$.

▶ **Solution:**

The SS, SFEE on unit mass basis is,

$$Q - W_x = \Delta\left[h + \frac{V^2}{2} + gz\right] \tag{5.7}$$

From the data, $\quad Q = -120$ kJ/kg, $\quad \Delta\dfrac{V^2}{2} = 0, \quad \Delta gz = 0$ and $W_x = W_c$.

$$\text{Therefore,} \quad -120 - W_c = \Delta h = \Delta(u + pv) = (u_2 - u_1) + (p_2v_2 - p_1v_1)$$
$$= (200 - 48) + (7 \times 10^5 \times 0.016 - 1 \times 10^5 \times 0.08) \times 10^{-3}$$

$$\text{Hence,} \quad W_c = -275.2 \text{ kJ/kg}.$$

Work is supplied to the compressor.

5.4.3 Heat Transfer Equipment

In devices like the boiler in a steam power plant or an evaporator in a refrigerator, heat is continuously transferred to a steadily flowing working substance to change it from liquid to vapour. Whereas in a condenser of either a steam power plant or a refrigerator, heat is transferred out of the working substance to condense it from vapour to liquid and in domestic geysers, heat is transferred to a steadily flowing water to rise its temperature. In all such devices, there is no work involved and the changes in kinetic and potential energies can be neglected.

$$Q - W_x = \Delta\left[h + \frac{V^2}{2} + gz\right] \tag{5.7}$$

with $W_x = 0$, $\Delta\dfrac{V^2}{2} = 0$ and $\Delta gz = 0$, equation (5.7) reduces to

$$Q = \Delta h \tag{5.14}$$

The SS, SFEE for a heat transfer equipment is thus as in (5.14).

5.4.4 Heat Exchangers

Heat exchangers are devices in which, generally, heat is transferred from one flowing fluid to another without physical mixing of the two. During the process of heat exchange, a hot fluid entering the heat exchanger transfers heat to another cold fluid. As a sequel to the energy interaction between the two fluids, while the hot fluid leaves the exchanger with a reduced temperature, the cold fluid leaves the heat exchanger with an increased temperature. Generally, there would be no appreciable change in the velocities of fluid flow. In a well-designed heat exchanger there is no heat loss to the surroundings from its exterior surfaces. Thus, an ideal heat exchanger is adiabatic with energy interaction occurring only internally. A simplest heat exchanger is a double-pipe heat exchanger which consists of two concentric pipes. While the hot fluid flows through the central pipe, the cold fluid flows in the annulus between the two pipes. The exterior surface of the heat exchanger is insulated to reduce heat losses. The two fluids may flow in the same direction or in opposite directions. While the former is called a parallel-flow heat exchanger, the latter is called a counterflow heat exchanger. Figure 5.7 shows the schematic of a parallel-flow heat exchanger. The SS, SFEE applicable for a heat exchanger is of the form,

$$\dot{Q} - \dot{W}_x = \Sigma_{\text{exit}} \dot{m} \left(h + \frac{V^2}{2} + gz \right) - \Sigma_{\text{entry}} \dot{m} \left(h + \frac{V^2}{2} + gz \right) \tag{5.9}$$

In an ideal case, $Q = 0$. There is no work interaction and the changes in kinetic and potential energies are negligible. Thus, equation (5.9) reduces to,

$$\Sigma_{\text{exit}} \dot{m} h - \Sigma_{\text{entry}} \dot{m} h = 0$$

Using suffixes h and c to denote, respectively, the hot and cold fluids, and suffixes 1 and 2 to denote properties of each fluid at its respective entry and exit,

$$\begin{aligned} (\dot{m}_h h_{h2} + \dot{m}_c h_{c2}) - (\dot{m}_h h_{h1} + \dot{m}_c h_{c1}) &= 0, \text{ or} \\ \dot{m}_h (h_{h2} - h_{h1}) + \dot{m}_c (h_{c2} - h_{c1}) &= 0 \end{aligned} \tag{5.15}$$

The SS, SFEE applicable to an ideal heat exchanger is as in (5.15).

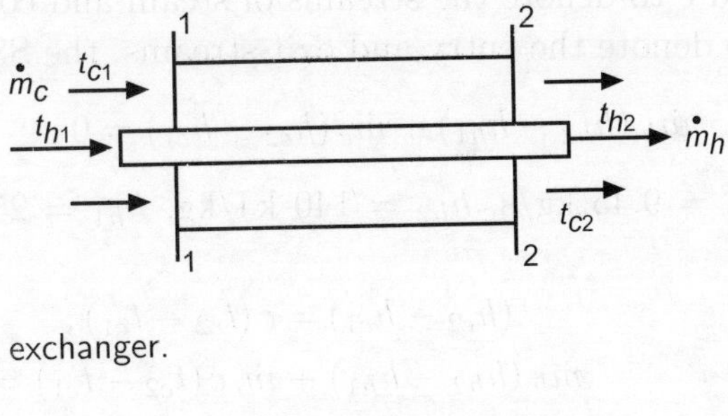

Fig. 5.7 Parallel-flow heat exchanger.

Alternatively SS, SFEE can be applied separately to each fluid flowing through the heat exchanger and yet obtain the same result as in 5.15. Applying SS, SFEE to the hot and cold fluids, respectively,

$$\dot{Q}_h = \dot{m}_h (h_{h2} - h_{h1}),$$

and
$$\dot{Q}_c = \dot{m}_c (h_{c2} - h_{c1}),$$

where $\dot{Q}_h$ is the rate of heat transfer from the hot fluid and $\dot{Q}_c$ is the rate of heat transfer to the cold fluid. In conformation with the sign convention used for heat transfer, while $\dot{Q}_h$ is negative, $\dot{Q}_c$ is positive. In an ideal heat exchanger, (100% efficient heat exchanger), as already explained, $\left|\dot{Q}_h\right| = \left|\dot{Q}_c\right|$

Hence,
$$-\dot{m}_h (h_{h2} - h_{h1}) = \dot{m}_c (h_{c2} - h_{c1}),$$

or,
$$\dot{m}_h (h_{h2} - h_{h1}) + \dot{m}_c (h_{c2} - h_{c1}) = 0 \tag{5.15}$$

If the heat exchanger is not adiabatic, and that the heat loss to the surroundings is, $\dot{Q}_l$,

then,
$$\left|\dot{Q}_h\right| - \left|\dot{Q}_l\right| = \left|\dot{Q}_c\right|.$$

If the efficiency of the heat exchanger is given to be η, then

$$\eta \times \left|\dot{Q}_h\right| = \left|\dot{Q}_c\right|$$

Example 5.7

A steam condenser receives 9.45 kg/s of steam with an enthalpy of 2570 kJ/kg. The condensate leaves with an enthalpy of 140 kJ/kg. Cooling water enters the condenser at a temperature of 25°C and leaves at 36°C. Assuming the specific heat of water to be 4.189 kJ/kg deg, find the mass flow rate of cooling water.

▶ Solution:

A steam condenser is a heat exchanger in which the hot fluid is the steam and the cold fluid is the cooling water. Heat is transferred from the steam to the cooling water. Consequently, steam gets condensed.

Using suffixes h and c to denote the streams of steam and cooling water, respectively, and suffixes 1 and 2 to denote the entry and exit streams, the SS, SFEE is,

$$\dot{m}_h (h_{h2} - h_{h1}) + \dot{m}_c (h_{c2} - h_{c1}) = 0 \tag{5.15}$$

From the data, $\dot{m}_h$ = 9.45 kg/s, h_{h2} = 140 kJ/kg, h_{h1} = 2570 kJ/kg, $\dot{m}_c$ =?, t_{c2} = 36°C, t_{c1} = 25°C

Also,
$$(h_{c2} - h_{c1}) = c\,(t_{c2} - t_{c1}),$$

Therefore,
$$\dot{m}_h (h_{h2} - h_{h1}) + \dot{m}_c c\,(t_{c2} - t_{c1}) = 0$$

Substituting the various numerical values,

$$9.45 \times (140 - 2570) + \dot{m}_c \times 4.189(36 - 25) = 0, \qquad \text{(i)}$$

and hence

$$\dot{m}_c = 498.35 \text{ kg/s}.$$

Example 5.8

Redo Example 5.7, if the heat loss to the surroundings is 10% of heat transferred from steam to cooling water.

▶ **Solution:**

Here, $\dot{Q}_h - \dot{Q}_l = \dot{Q}_c$, where $\dot{Q}_l$ = rate of heat loss

$\therefore \quad \dot{Q}_h - 0.1\dot{Q}_h = 0.9\dot{Q}_h = \dot{Q}_c$

Heat transfer to cooling water in this case = 0.9 of that in Example 5.7.

$\therefore \quad 0.9 \times 9.45(140 - 2570) + \dot{m}_c \times 4.189(36 - 25) = 0$, or $\dot{m}_c = 448.5$ kg/s.

5.4.5 Nozzles and Diffusers

A nozzle is a variable-area duct or a passage. It is used to increase the velocity of a fluid flowing through it. In the direction of flow, the pressure of the fluid decreases continuously and the velocity increases. Figure 5.8 shows a converging nozzle. The fluid enters the nozzle at section 1–1 and leaves at 2–2 .

Applying SS, SFEE to the nozzle,

$$Q - W_x = \Delta\left[h + \frac{V^2}{2} + gz\right] \qquad (5.7)$$

Unless otherwise mentioned, the flow through a nozzle is treated to be adiabatic. There is no work interaction and the change in potential energy is neglected. Thus the SS, SFEE reduces to

$$\Delta\left[h + \frac{V^2}{2}\right] = 0, \text{ or } h_1 - h_2 = \frac{(V_2^2 - V_1^2)}{2} \qquad (5.16)$$

$$\therefore \qquad V_2 = \left[2(h_1 - h_2) + V_1^2\right]^{1/2} \qquad (5.17)$$

Fig. 5.8 A converging nozzle.

In (5.17), while V_1 and V_2 are in m/s, h_1 and h_2 are in J/kg.

From the continuity equation, as $\dot{m} = \dfrac{A_1V_1}{v_1} = \dfrac{A_2V_2}{v_2}$, $V_2 = V_1\left(\dfrac{A_1}{A_2}\right)\left(\dfrac{v_2}{v_1}\right)$.

Further, if the fluid is *incompressible* its specific volume does not change substantially with change in either pressure or temperature. Hence, $v_2 = v_1$. Therefore, the exit velocity of the fluid from a nozzle is given by $V_2 = V_1\left(\dfrac{A_1}{A_2}\right)$, which means that, for a given V_1 and A_1, the exit velocity is inversely proportional to the exit area. Thus the area of the nozzle decreases in the direction of flow as shown in Figure 5.8.

For a *compressible fluid*, the specific volume changes considerably with both pressure and temperature. It is for this reason that the shape of the nozzle for a compressible fluid flow can either be converging or converging–diverging as shown in Figure 5.9.

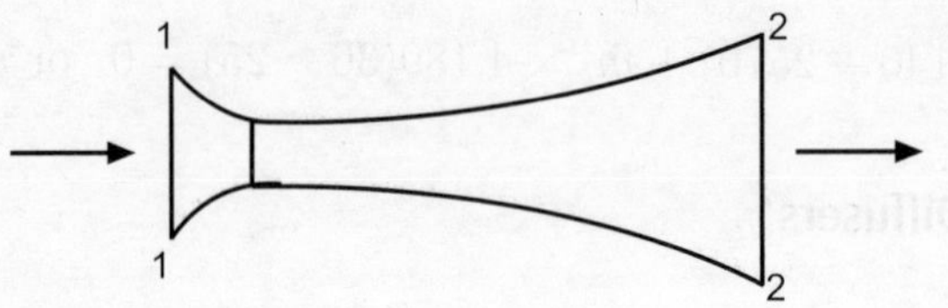

Fig. 5.9 Converging–diverging nozzle.

The maximum attainable compressible fluid velocity in the converging nozzle is limited to sonic velocity. If velocities greater than the sonic velocity are desired, the nozzle has to be converging–diverging. This is because higher velocity requires a greater expansion in the nozzle. Thus, the exit pressure would be very low. At such low pressures, the specific volume of the compressible fluid is very large. In conformation with the continuity equation to accommodate the required flow rate, the area has to be large towards the exit.

A diffuser is also a duct or passage of variable area; however, a diffuser does exactly opposite to that of a nozzle. In the direction flow, in a diffuser, while the pressure gradually but continuously increases, the flow velocity keeps decreasing.

Example 5.9

Air enters an adiabatic diffuser at t_1°C and 300 m/s and leaves at 100°C and 15 m/s. Find the inlet air temperature. Assume for air $h = a + t$, where h is in kJ/kg, t is in °C and a is a constant.

► **Solution:**

The SS, SFEE for an adiabatic diffuser is,

$$\Delta\left[h + \frac{V^2}{2}\right] = 0, \text{ or } h_1 - h_2 = \frac{(V_2^2 - V_1^2)}{2} \tag{5.16}$$

Here, $h_1 - h_2 = (a + t_1) - (a + t_2) = (t_1 - t_2)$ kJ/kg $= (t_1 - t_2) \times 1000$ J/kg.

Therefore, $$(t_1 - t_2) \times 1000 = \frac{(V_2^2 - V_1^2)}{2}$$

Substituting the various numerical values from the data,

$$(t_1 - 100) \times 1000 = \frac{(15^2 - 300^2)}{2}, \text{ hence } t_1 = 55.1°\text{C}.$$

5.5 Unsteady Flow

The flow through a control volume is said to be unsteady if the properties of the control volume were to be changing with time. This can happen for two reasons, mainly: (i) the flow rates into and out of the control volume are not the same or (ii) the flow rates may be same, but the control volume has not reached steady operating conditions. In such situations, the analysis needs to consider the change in energy of the CV. The popular examples of such flow processes are: (i) evacuating a gas cylinder, (ii) filling a gas cylinder, (iii) an accelerating or decelerating turbine or a compressor, (iv) an unsteady heat exchanger etc. Thus the energy equation to be considered is (5.1). To make the energy equation more general, the suffixes 2 and 1 in (5.1) are replaced by out and in, respectively.

$$\delta Q - \delta W_x = \mathrm{d}E_c + [\mathrm{d}m_{out}(e + pv)_{out} - \mathrm{d}m_{in}(e + pv)_{in}] \tag{5.18}$$

or, $$\delta Q - \delta W_x = \mathrm{d}E_c + \left[\mathrm{d}m_{out}\left(h + \frac{V^2}{2} + gz\right)_{out} - \mathrm{d}m_{in}\left(h + \frac{V^2}{2} + gz\right)_{in}\right] \tag{5.19}$$

Example 5.10

An insulated rigid vessel containing 0.8 kg air at 150 kPa, 300 K is connected via a valve to a pipe in which air is flowing steadily at 600 kPa and 330 K. The valve is opened. Find the mass of air entering the vessel to bring the pressure of air in the vessel to that in the pipe. Also, find the final temperature of air in the vessel. Assume, for air, $h = 1.4\ u = 1.005\ T = 0.0035$ pv, where h and u are in kJ/kg, T is in K, p is in N/m^2 and v is in m^3/kg.

▶ **Data:**

With suffix p denoting the properties of air in the pipe, 1 and 2 denoting the initial and final properties of air in the vessel: $m_1 = 0.8$ kg, $p_1 = 150$ kPa, $T_1 = 300$ K, $p_p = 600$ kPa, $T_p = 330$ K, and the properties of air are related by:

$$h = 1.4u = 1.005T = 0.0035\ pv.$$

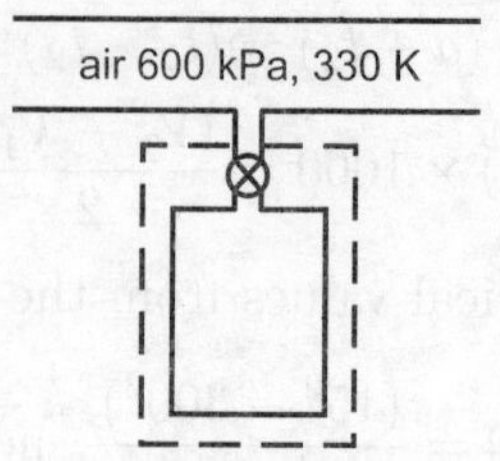

Fig. 5.10 Schematic for Example 5.10.

► **Solution:**

The process of filling the vessel with air from the pipe is not a steady-flow process. Considering the CV (vessel), shown in Figure 5.10, enclosed by the control surface marked by the broken line, the energy equation for the filling process is given by,

$$\delta Q - \delta W_x = \mathrm{d}E_c + \left[\mathrm{d}m_{out}\left(h + \frac{V^2}{2} + gz\right)_{out} - \mathrm{d}m_{in}\left(h + \frac{V^2}{2} + gz\right)_{in} \right] \quad (5.19)$$

Here, $\delta Q = 0$, $\delta W_x = 0$ and $\mathrm{d}m_{\text{out}} = 0$.

Therefore, (5.19) reduces to, $\mathrm{d}E_c - \mathrm{d}m_{in}\left(h + \frac{V^2}{2} + gz\right)_{in} = 0.$

Neglecting kinetic energy and potential energy terms and noting that E_c comprises only internal energy, (5.19) further reduces to,

$$\mathrm{d}U_c - \mathrm{d}m_{in}h_{in} = 0$$

that is $\quad \mathrm{d}(mu) - \mathrm{d}m_{in}h_{in} = 0.$

Here, $\quad h_{in} = h_p = 1.005T_p = 1.005 \times 330 = 331.65$ kJ/kg

Therefore, $\quad \mathrm{d}(mu) - \mathrm{d}m_{\text{in}} \times 331.65 = 0.$

Integrating this equation between the initial and final states of air in the vessel,

$$(m_2u_2 - m_1u_1) - 331.65\,(m_2 - m_1) = 0$$

Dividing throughout by m_1,

$$\left[\left(\frac{m_2}{m_1}\right)u_2 - u_1\right] - 331.65\left(\frac{m_2}{m_1} - 1\right) = 0 \quad \text{(i)}$$

From the given property relation, $u = 1.005T/1.4 = 0.718T$

Substituting for u_2 and u_1 in (i) in terms of temperatures,

$$\left[\left(\frac{m_2}{m_1}\right)0.718T_2 - 0.718T_1\right] - 331.65\left(\frac{m_2}{m_1} - 1\right) = 0$$

that is,

$$0.718\left[\left(\frac{m_2}{m_1}\right)T_2 - T_1\right] - 331.65\left(\frac{m_2}{m_1} - 1\right) = 0 \quad \text{(ii)}$$

From property relation, $1.005\ T = 0.0035\ pv$. Multiplying both sides by the mass m, $1.005\ mT = 0.0035\ pmv = 0.0035\ pV$, where V is the total volume in m^3.

Therefore, at the final and initial states,

$$1.005\ m_2T_2 = 0.0035\ p_2V_2 \text{ and} \qquad \text{(iii)}$$

$$1.005\ m_1T_1 = 0.0035\ p_1V_1 \qquad \text{(iv)}$$

In (iii) and (iv) $V_2 = V_1$ = vessel volume. Diving (iii) by (iv),

$$\frac{m_2T_2}{m_1T_1} = \frac{p_2}{p_1}, \text{ or}$$

$$\frac{m_2T_2}{m_1} = \left(\frac{p_2}{p_1}\right) T_1 = \left(\frac{600}{150}\right) 300 = 1200 \qquad \text{(v)}$$

Substituting for $\left(\dfrac{m_2T_2}{m_1}\right)$ from (v), and the value of T_1 from data, in (ii)

$$0.718\ [1200 - 300] - 331.65 \left(\frac{m_2}{m_1} - 1\right) = 0 \text{ and therefore } \frac{m_2}{m_1} = 2.948.$$

From data, $m_1 = 0.8$ kg, and therefore, $m_2 = 0.8 \times 2.948 = 2.3584$ kg.

The mass entering the vessel = $(m_2 - m_1) = 2.3584 - 0.8 = 1.5584$ kg.

Substituting for m_2 and m_1 in (v), $2.3584\ \dfrac{T_2}{0.8} = 1200$ and hence, $T_2 = 407$ K.

Example 5.11

A well-insulated tank initially contains m_1 kg air at a pressure p_1. A small valve mounted on the tank is opened to let some air escape into the atmosphere. Assume that the conditions are uniform throughout the tank at any given instant. Show that $\left(\dfrac{m_2}{m_1}\right) = \left(\dfrac{p_2}{p_1}\right)^{0.714}$, where m_2 and p_2 are, respectively, the mass and pressure of air finally existing in the tank. Assume for air, $h = 1.4\ u = 0.0035\ pv$, where, u and h are in kJ/kg, p is in N/m^2 and v is in m^3/kg.

▶ **Solution:**

Emptying the tank is an unsteady process. The energy equation for the process is

$$\delta Q - \delta W_x = \mathrm{d}E_c + \left[\mathrm{d}m_{out}\left(h + \frac{V^2}{2} + gz\right)_{out} - \mathrm{d}m_{in}\left(h + \frac{V^2}{2} + gz\right)_{in}\right] \qquad (5.19)$$

Here, $\delta Q = 0$, $\delta W = 0$ and $\mathrm{d}m_{in} = 0$. And, neglecting the kinetic energy and potential energy terms, the above energy equation reduces to

$$\mathrm{d}E_c + \mathrm{d}m_{out}h_{out} = 0$$

The only mode of energy the control volume (air in the tank) can have is the internal energy.

Thus, $$\mathrm{d}E_c = \mathrm{d}U = m(\mathrm{d}u).$$

Therefore, $$\mathrm{d}(mu) + \mathrm{d}m_{out}h_{out} = 0$$

that is, $$(m\mathrm{d}u + u\mathrm{d}m) + \mathrm{d}m_{out}h_{out} = 0,$$

where, for an infinitesimal process at any instant, dm is the change in mass of air in the tank and u is the specific internal energy of the air remaining in the tank. Also, dm_{out} = −dm (because, mass leaking out = reduction in mass of air in the tank) and $h_{\text{out}} = h$ of air in the tank at the same instant. Therefore, the above equation reduces to

$$(m\mathrm{d}u + u\mathrm{d}m) - \mathrm{d}mh = 0, \text{ or } \mathrm{d}m(h - u) = m\mathrm{d}u$$

From data, $(h - u) = 0.4u$ and hence

$$0.4\ \mathrm{d}mu = m\mathrm{d}u, \text{ or } 0.4\frac{\mathrm{d}m}{m} = \frac{\mathrm{d}u}{u}$$

Therefore, $$0.4\int_1^2 \frac{\mathrm{d}m}{m} = \int_1^2 \frac{\mathrm{d}u}{u}$$

that is, $$\left(\frac{m_2}{m_1}\right)^{0.4} = \frac{u_2}{u_1}$$

From data, 1.4 u = 0.0035 pv or $u = \left(\dfrac{0.0035}{1.4}\right) pv$

Therefore, $\dfrac{u_2}{u_1} = \dfrac{p_2v_2}{p_1v_1}$.

That is, $\left(\dfrac{m_2}{m_1}\right)^{0.4} = \dfrac{p_2v_2}{p_1v_1}$. Here, $v = \dfrac{V}{m}$, where V, the volume of the tank, is a constant.

Therefore, $$\left(\frac{m_2}{m_1}\right)^{0.4} = \left(\frac{p_2}{p_1}\right)\frac{\left(\frac{V}{m_2}\right)}{\left(\frac{V}{m_1}\right)} = \left(\frac{p_2}{p_1}\right)\left(\frac{m_1}{m_2}\right)$$

that is, $$\left(\frac{m_2}{m_1}\right)^{0.4}\left(\frac{m_2}{m_1}\right) = \left(\frac{m_2}{m_1}\right)^{1.4} = \left(\frac{p_2}{p_1}\right)$$

or $$\left(\frac{m_2}{m_1}\right) = \left(\frac{p_2}{p_1}\right)^{1/1.4} = \left(\frac{p_2}{p_1}\right)^{0.714}$$

Illustrative Problems

5.1 Steam at a rate of 0.42 kg/s and an enthalpy of 2785 kJ/kg and a velocity of 33.3 m/s is supplied to a steadily operating turbine. The steam leaves the turbine at 100 m/s and an enthalpy of 2512 kJ/kg. The inlet pipe is 3 m above the exit pipe. Rate of heat loss from the turbine casing is 0.29 kJ/s. What is the power output of the turbine?

▶ **Data:**

$\dot{m} = 0.42$ kg/s, $h_1 = 2785$ kJ/kg, $V_1 = 33.3$ m/s, $V_2 = 100$ m/s, $h_2 = 2512$ kJ/kg, $(z_2 - z_1) = -3$ m, $Q = -0.29$ kJ/s.

▶ **Solution:**

The SS, SFEE on unit time basis is,

$$\dot{Q} - \dot{W}_x = \dot{m}\Delta\left[h + \frac{V^2}{2} + gz\right]. \tag{5.8}$$

that is,
$$\dot{Q} - \dot{W}_x = \dot{m}\left[(h_2 - h_1) + \left(\frac{V_2^2}{2} - \frac{V_1^2}{2}\right) + g(z_2 - z_1)\right]$$

Substituting the given numerical values,

$$-0.29 - \dot{W}_T = 0.42\left[(2512 - 2785) + \frac{(100^2 - 33.3^2)}{2000} + \frac{(9.81 \times -3)}{1000}\right]$$

or
$$\dot{W}_T = 112.51 \text{ kJ/s} = 112.5 \text{ kW}$$

5.2 A nozzle is a device to increase the velocity of a steadily flowing fluid. At the inlet to a certain horizontal adiabatic nozzle, the enthalpy of the fluid is 3000 kJ/kg and the velocity is 60 m/s. The enthalpy at discharge is 2762 kJ/kg. The area of the nozzle at inlet is 0.1 m^2 and the specific volume at entry is 0.187 m^3/kg. The specific volume at exit is 0.498 m^3/kg. Find (a) the velocity of the fluid at exit, (b) mass flow rate through the nozzle and (c) the exit area of the nozzle.

▶ **Data:**

$h_1 = 3000$ kJ/kg, $V_1 = 60$ m/s, $h_2 = 2762$ kJ/kg, $A_1 = 0.1$ m^2, $v_1 = 0.187$ m^3/kg and $v_2 = 0.498$ m^3/kg.

▶ **Solution:**

(a) The SS, SFEE for an adiabatic nozzle is

$$\Delta\left[h + \frac{V^2}{2}\right] = 0, \text{ or } h_1 - h_2 = \frac{(V_2^2 - V_1^2)}{2} \tag{5.16}$$

or $$V_2 = \left[2\,(h_1 - h_2) + V_1^2\right]^{1/2} \tag{5.17}$$

Substituting the various numerical values,

(b)
$$V_2 = \left[2(3000 - 2762) \times 10^3 + (60^2)\right]^{1/2} = 692.5 \text{ m/s}.$$

$$\dot{m} = \frac{A_1 V_1}{v_1} = 0.1 \times \frac{60}{0.187} = 32.08 \text{ kg/s}.$$

(c)
$$\dot{m} = \frac{A_2 V_2}{v_2}$$

and therefore, $$A_2 = \frac{\dot{m} v_2}{V_2} = 32.08 \times \frac{0.498}{692.5} = 0.023 \text{ m}^2.$$

5.3 In an oil cooler, oil flows steadily through a bundle of metal tubes submerged in a steady stream of cooling water. Oil enters the cooler at 90°C and leaves at 30°C. Water enters at 25°C and leaves at 70°C. The enthalpy of oil in kJ/kg, at any temperature t in °C, is given by $h = 1.68t + 10.5 \times 10^{-4}\, t^2$. What is the required cooling water flow rate for an oil flow rate of 2.78 kg/s.

► **Data:**

Using suffixes 1 and 2 to denote the properties of oil and water at inlet and exit, respectively, $t_{\text{oil},1} = 90$°C, $t_{\text{oil},2} = 30$°C, $t_{\text{wat},1} = 25$°C, $t_{\text{wat},2} = 70$°C, $m_{\text{oil}} = 2.78$ kg/s and $h_{\text{oil}} = 1.68\, t + 10.5 \times 10^{-4}\, t^2$.

► **Solution:**

The SS, SFEE for an adiabatic heat exchanger is given by

$$\dot{m}_h\,(h_{h2} - h_{h1}) + \dot{m}_c\,(h_{c2} - h_{c1}) = 0 \tag{5.15}$$

where h denotes the hot fluid (in this case the oil) and c denotes the cold fluid (here it is the water).

$$\dot{m}_{\text{oil}}\,(h_2 - h_1)_{\text{oil}} + \dot{m}_{\text{wat}}\,(h_2 - h_1)_{\text{wat}} = 0 \tag{i}$$

From data, $$(h_2 - h_1)_{\text{oil}} = 1.68\,(t_2 - t_1) + 10.5 \times 10^{-4}\,(t_2^2 - t_1^2)$$
$$= 1.68(30 - 90) + 10.5 \times 10^{-4}\,(30^2 - 90^2) = -108.36 \text{ kJ/kg}$$

Assuming the specific heat water as 4.189 kJ/kg deg,

$$(h_2 - h_1)_{\text{wat}} = 4.189(70 - 25) = 188.05 \text{ kJ/kg}$$

Therefore, substituting the values of enthalpy changes and the mass flow rate of oil in (i),

$$2.78(-108.36) + \dot{m}_{\text{wat}}(188.05) = 0, \text{ or } \dot{m}_{\text{wat}} = 1.598 \text{ kg/s}.$$

5.4 A mass of air in an insulated rigid tank is stirred by means of a paddle wheel whose shaft protrudes into the tank and is externally connected to an electric motor. Air is allowed to escape out of the tank through a carefully controlled valve such that the temperature of air remaining in the tank at any instant has the same value. At a certain instant the temperature of air in the tank is 160°C and the current drawn by the motor connected to the mains at 220 V is 0.505 amps. Given that the electro-mechanical efficiency of the motor is 90% find the rate of flow air out of the tank in kg/h. Assume for air (i) $u = u_0 + 0.718t$, where u is in kJ/kgK and t is in °C and $pv = 0.287(t + 273)$, where p is in kPa and v is in m^3/kg.

▶ **Data:**

$t_1 = t_2 = 160°\text{C}$, $V = 220$, $I = 0.505A$, $u = u_0 + 0.718t$, $pv = 0.287(t + 273)$

▶ **Solution:**

Figure 5.11 shows the schematic of the arrangement. Broken line shows the surface of the control volume considered.

The general energy equation for a control volume is

$$\delta Q - \delta W_x = \text{d}E_c + \left[\text{d}m_{out}\left(h + \frac{V^2}{2} + gz\right)_{out} - \text{d}m_{in}\left(h + \frac{V^2}{2} + gz\right)_{in}\right] \quad (5.19)$$

Here, $\delta Q = 0$, $\text{d}m_{in} = 0$. Also, neglecting the kinetic and potential energy terms as negligible, the above equation reduces to:

$$-\delta W_x = \text{d}E_c + \text{d}m_{out}h_{out}$$

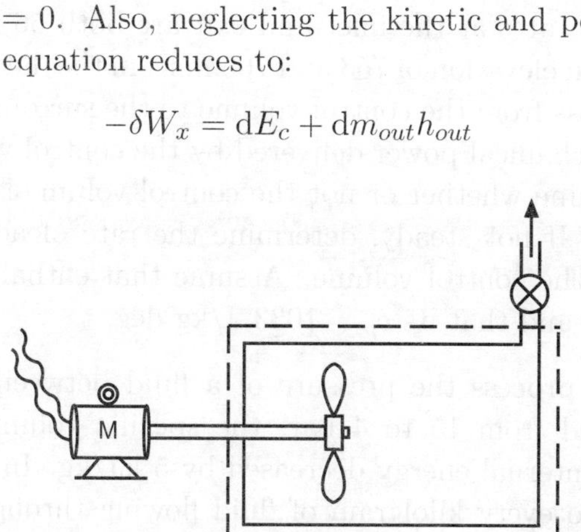

Fig. 5.11 Schematic for solving Illustrative Problem 5.4.

In the control volume only the internal energy of air is changing and h_{out} at any instant is equal to h of air at the same instant. Therefore,

$$-\delta W_x = \mathrm{d}(mu) + \mathrm{d}m_{out}h \text{ or } -\delta W_x = m\mathrm{d}u + u\mathrm{d}m + \mathrm{d}m_{out}h,$$

Also, $\mathrm{d}m_O = -\mathrm{d}m$ and therefore

$$-\delta W_x = m\mathrm{d}u + u\mathrm{d}m - \mathrm{d}mh$$

that is, $-\delta W_x = m\mathrm{d}u - \mathrm{d}m(h-u) = m\mathrm{d}u - \mathrm{d}m(pv)$

As the temperature of air in the tank does not change, $\mathrm{d}u = 0$, as from data it can be seen that u is a function of temperature only. Therefore,

$$\delta W_x = \mathrm{d}m(pv) = \mathrm{d}m[0.287(t+273)]$$

At the given instant, δW_x = power supplied to air from the paddle wheel

= (electrical power supplied to the motor) $\times \eta$

$= IV\eta = 0.505 \times 220 \times 0.9$

= 100 J/s = 0.1 kW

Therefore, $(-0.1) = \mathrm{d}m[0.287(160+273)]$

And hence, $\mathrm{d}m = \dfrac{0.1}{[0.287(160+273)]} = 8.047 \times 10^{-4}$ kg/s = 2.897 kg/h.

Problems

5.1 At a particular instant, air enters a control volume at, 15 MPa, 490°C and 350 m/s. The exit conditions at the same instant are, 5 MPa, 210°C, and 50 m/s. The flow rates at the inlet and exit are both 56 kg/s. The inlet and exit manifolds are at an elevation of 753 and 705 m from the datum. At the same instant, the rate of heat loss from the control volume to the surroundings is estimated to be 4 kW and the mechanical power delivered by the control volume is estimated to be 6000 kW. Determine whether or not the control volume is operating under steady state conditions. If not steady, determine the rate of accumulation or depletion of energy within the control volume. Assume that enthalpy of air is a function of temperature only and that its $c_p = 1033$ J/kg deg.

5.2 In a steady flow process the pressure of a fluid between the inlet and exit of a machine decreased from 15 to 4 bar, the specific volume increased from 0.1 to 0.33 m^3/kg, the internal energy decreased by 5 kJ/kg. In the machine 1.3 kJ heat was transferred to every kilolgram of fluid flowing through it. Entrance and exit pipes were at the same elevation and entrance and exit velocities were 3 and 5 m/s, respectively. Determine the magnitude and direction of shaft work per kilogram of fluid.

5.3 A steady flow of water at a temperature of 60°C and a pressure of 3 bar ($h = 248$ kJ/kg) enters a section of a heating plant of a building in which there are no pumps. The water leaves the section at a temperature of 48°C and a pressure of 2.6 bar ($h = 200$ kJ/kg). The exit-pipe is 25.5 m above the entry-pipe. Evaluate the heat transfer from the water assuming the change in kinetic energy to be negligible.

5.4 In a steady flow device, 135 kJ of work is done by each kilogram of fluid. For the fluid the specific volume, pressure and velocity at the inlet are 0.37 m^3/kg, 600 kPa and 16 m/s, respectively. The corresponding properties of the fluid at the exit are 0.62 m^3/kg, 100 kPa and 270 m/s. The inlet is 32 m above the floor level and the discharge pipe is at the floor level. The heat transfer from the device is 9 kJ/kg of fluid. What is the change in specific internal energy of the fluid as it passes through the device.

5.5 (a) Water flows through a shower head steadily at a rate of 10 lit/min. An electric resistance placed around the water pipe heats the water from 16°C to 43°C. Taking the density of water to be 1 kg/l, determine the electric power input to the heater, in kW. (b) In an effort to conserve energy, it is proposed to pass the drained warm water at a temperature of 39°C through a heat exchanger to preheat the incoming cold water. If the heat exchanger has an effectiveness of 0.50 (i.e., it recovers only half of the energy which can possibly be transferred from the drained water to incoming cold water), determine the electric power input required in this case. If the price of the electric energy is Rs.4.5/unit, determine how much money is saved during a 10-min shower as a result of installing this heat exchanger.

5.6 In a test of water cooled air compressor, it is found that the shaft work required to drive the compressor is 175 kJ/kg of air delivered and that the enthalpy of air leaving is 72 kJ/kg greater than that entering. Heat transfer to cooling water is 96 kJ/kg air. Neglect the changes in kinetic energy. How much heat is transferred from the compressor to the atmosphere.

5.7 An air compressor takes in air at a pressure of 1 bar and a specific volume of 0.8 m^3/kg. The discharge pressure and specific volume are 7 bar and 0.2 m^3/kg. The initial and final internal energies of air are 48 and 200 kJ/kg, respectively. The heat transferred from air is 132 kJ/kg. Neglect changes in kinetic and potential energies and calculate the shaft work.

5.8 An adiabatic blower handles 1 kg/s of air at 20°C and consumes a power of 15 kW. The inlet and outlet velocities of air are 100 and 150 m/s, respectively. Find the exit air temperature. Assume for air enthalpy is a function of temperature only and that c_p for air 1005 J/kg deg.

5.9 A jet engine, propelled by a hot jet leaving the engine, has a gas turbine power plant whose major four components are a compressor, a heater, a turbine and a nozzle. Air at 1.013 bar, 15°C (density = 1.22 kg/m^3) is taken into the engine at a velocity of 130 m/s through an opening of 0.7 m^2 area. The air is compressed, heated and expanded through the turbine and is discharged through the nozzle at 1.5 bar, 60°C (density = 1.57 kg/m^3) through an opening of 0.3 m^2 area. Work produced by the turbine is completely used by the compressor. Calculate the heat transfer to the air assuming the enthalpy of air is a function of temperature only and that for air c_p is 1005 J/kg deg.

5.10 The gas leaving a turbine of an aircraft jet engine flows steadily into the engine jet-pipe at 900°C, 192 kPa and a velocity of 300 m/s relative to the pipe. The gas leaves the jet-pipe at 820°C, 105 kPa. Heat transfer from the gas is negligible. Evaluate the velocity of the gas leaving the jet-pipe. The enthalpy of the gas is given to depend only on temperature. The following information is obtained from the gas tables: enthalpy of the gas is 862.6 kJ/kg at 820°C and 951.7 kJ/kg at 900°C.

5.11 Air at a temperature of 15°C passes through a heat exchanger at a velocity of 30 m/s where its temperature is raised to 800°C. It then enters a turbine with the same velocity of 30 m/s and expands until the temperature falls to 650°C. Heat transfer from the turbine casing is 30 kW. Air leaving the turbine at 60 m/s is then passed through an adiabatic nozzle until the temperature falls to 500°C. The flow rate of air through all the components is steady at 2 kg/s. Assume that enthalpy h, in kJ/kg, of air is given by $h = 1.005\, t$, where t the temperature is in °C. Calculate the following: (a) rate of heat transfer in the heat exchanger, (b) power output of the turbine and (c) velocity of air at the exit of nozzle.

5.12 An adiabatic steam turbine receives a steam flow of 25,000 kg/h and its power output is 500 kW. Find the change in specific enthalpy of steam as it flows through the turbine for the following conditions: (a) neglect changes in kinetic and potential energies, (b) inlet and exit velocities are, respectively, 80 and 400 m/s and change in potential energy is small, (c) neglect change in kinetic energy and assume that inlet pipe is 5 m above the exhaust pipe and (d) inlet and exit velocities are same as in (b) and that the inlet pipe is 5 m above the exhaust pipe.

5.13 Steam enters a turbine at 4800 kPa and leaves at 20 kPa. The specific internal energy and specific enthalpy of steam at entry are 2960 and 3260 kJ/kg, respectively. The corresponding values at the exit are 2100 and 2230 kJ/kg. The velocity of steam at entry is negligible. The exit area of the turbine is 0.464 m^2. Heat loss from turbine casing is 24 kJ/kg steam. Determine the following: (i) specific volume of steam at exit, (ii) velocity of steam at exit and (iii) the power developed by the turbine.

5.14 A fluid enters an adiabatic nozzle with an enthalpy of 3000 kJ/kg and a velocity of 60 m/s and is discharged with an enthalpy of 2500 kJ/kg. (a) Calculate the velocity with which the fluid leaves the nozzle. (b) If the inlet area is 1000 cm^2 and the specific volume at entry is 0.187 m^3/kg find the mass flow rate of the fluid through the nozzle. (c) If the specific volume of the fluid at exit is 0.5 m^3/kg find the exit area of the nozzle.

5.15 A long insulated horizontal pipeline consists of two pipes connected in series, the diametres of which are 7.5 and 2.5 cm, respectively. A steady flow of steam enters the 7.5 cm pipe with an enthalpy of 2990 kJ/kg and a specific volume of 0.0829 m^3/kg. At a point downstream in the 2.5 cm pipe, the enthalpy is 2931 kJ/kg and the specific volume is 0.838 m^3/kg. Determine the velocities of steam at these two locations and the mass flow rate of steam through the pipeline.

5.16 Oil flows into a cooler at a temperature of 88°C with a flow rate of 4.5 kg/min and flows out at 38°C. Cooling water enters the cooler at 15°C and leaves at 26.5°C. The cooler is adiabatic. Changes in kinetic and potential energies of the two streams are negligible. Calculate the mass flow rate of water required to accomplish the given cooling of oil. For both oil and water the enthalpy h is related to temperature t by $h = c\,(t - t_0)$ where t is in °C and c and t_0 are constants. Assume that $c = 4.2$ kJ/kg deg for water and 1.9 kJ/kg deg for oil.

5.17 Air and fuel enter a furnace used for house heating. At entry air has an enthalpy of 300 kJ/kg and the fuel has an enthalpy of 43,000 kJ/kg. The gases leaving the furnace have an enthalpy of 620 kJ/kg. 17 kg air is required for every kg of fuel supplied to the furnace. Water circulating around the furnace is heated and sent to the house and the house requires 18 kW heat. What is the fuel consumption per day. Assume heat loss from the furnace to the atmosphere is negligible and that any changes in potential and kinetic energies are negligible.

5.18 The steam supply to an engine comprises two streams which mix before entering the engine. One stream is supplied at a rate of 0.5 kg/min with an enthalpy of 2950 kJ/kg and a velocity of 20 m/s. The other stream is supplied at the rate of 5 kg/min with an enthalpy of 2575 kJ/kg and a velocity of 135 m/s. At the exit from the engine the fluid leaves as two streams, one as water at a rate of 0.05 kg/min with an enthalpy of 420 kJ/kg and the other as steam. The fluid velocities at the exit are negligible. The engine develops 16 kW power. The heat loss from the engine is small. Evaluate the enthalpy of the exit stream of steam.

5.19 A water heater operates under steady flow conditions, receiving water at 75°C (enthalpy = 314 kJ/kg) with a flow rate of 4.2 kg/s. The water is heated by mixing it with steam which is supplied to the heater at a temperature of 120°C (enthalpy = 2676 kJ/kg). The mixture leaves the water heater as liquid water at a temperature of 100°C (enthalpy = 420 kJ/kg). How much steam is to be supplied to the heater every hour.

5.20 A mixture of air and water vapour with an enthalpy of 108 kJ/kg enters the dehumidifying section of an air-conditioning system at a rate of 360 kg/h. Liquid water drains out of the dehumidifier with an enthalpy of 36 kJ/kg at a rate of 8 kg/h. An air vapour mixture leaves the dehumidifier with an enthalpy of 40 kJ/kg. Determine the rate of heat transfer from the dehumidifier.

5.21 A stream of air and gasoline vapour, in the ratio of 14:1 by mass, enters a water cooled engine at 30°C. The combustion products leave the engine at 750°C. The engine develops a shaft power of 26 kW at a specific fuel consumption of 0.3 kg/kWh. The rate of heat transfer to the cooling water is 35 kW. Find the change in specific enthalpy between the entering air-gasoline mixture and the exiting combustion products per kilogram of the latter.

***Note*:** In problems 5.22–5.25, if and when required, assume for air, $h = 1.4\ u = 1.005\ T = 0.0035\ pv$, where h and u are in kJ/kg, $T = (273 + t)$ with t in °C, p is in N/m^2 and v is in m^3/kg.

5.22 A small leak develops in a 0.6 m^3 insulated tank which initially contains air at 3 bar. What fraction of the air escapes by the time the pressure drops to 1.5 bar?

5.23 A well-insulated rigid tank initially contains some air at 40°C. A paddle wheel in the tank is turned by an external motor, and air is bled from the tank through a small valve to keep the temperature constant at 40°C. Assume that the pressure and temperature of air is uniform throughout the tank at any one instant. Determine the mass rate of flow of air out of the tank at an instant when the pressure is 400 kPa and the power input is 0.04 kW.

5.24 An insulated tank contains air initially at 150 kPa, 17°C. An electrical heating coil in the tank steadily dissipates 100 W to the air in the tank while a pressure regulating valve bleeds air from the tank to the atmosphere at such a rate as to maintain the pressure constant in the tank. The tank volume is 0.12 m^3. Determine the temperature of air in the tank 2 min after the initial conditions.

5.25 Air at 750 kPa and 100°C flows through a large pipeline. Connected to the line through a valve is an insulated tank with a volume of 0.2 m^3 containing 0.23 kg air at 100 kPa, 30°C, while the valve is closed. The valve is then opened and air flows into the tank until the pressure in the tank is 750 kPa. Compute the final mass and the final temperature of air in the tank. Assume that no heat transfer occurs between the air in the tank and any part of the surroundings and that conditions are uniform throughout the tank at any instant. Neglect changes in kinetic and potential energy.

HEAT ENGINES—THE SECOND LAW OF THERMODYNAMICS

6

6.1 Introduction

Although the first law of thermodynamics explains clearly the relationship that exists between heat and work, it does not deal with the aspects of actual conversion of work to heat and vice versa which are of paramount importance in engineering practice. In this chapter, the principle of working of a few practical devices that convert heat to work and work to heat are explained. The performance characteristics of such devices are also elaborated. That the performance limits are bound by the constraints imposed by the second law of thermodynamic is explained further.

6.2 Limitations of the First Law of Thermodynamics

The first law of thermodynamics has certain limitations. Although it gives the relationship that exists between heat and work it does not say anything about the actual conversion of heat to work or work to heat. For example, a thermodynamic cycle in which the heat transferred to the system is completely converted to an equal amount of work does not violate the first law of thermodynamics. But, experience shows that such a cycle is an impossibility.

A second point is, suppose several hypothetical processes have to be analysed for their feasibility or otherwise, and also the direction of such feasible processes has to be determined, this task cannot be comprehensively and conclusively accomplished with the help of the first law of thermodynamics alone. As an example, let two systems at different temperatures be brought in communication with each other. As a result of this communication, there occurs a non-cyclic process of heat interaction between the two systems. This is a natural process that occurs to bring about thermal equilibrium between the two communicating systems. It is spontaneous in nature and the direction of the process is fixed, heat being transferred from the high-temperature system to the low-temperature system. However, in a situation where two systems at different temperatures are brought in communication with each other if it is said that heat is transferred from the low-temperature system to the high-temperature system it does not violate the first law. The first law is only concerned with the fact that the magnitude of heat transferred from the low-temperature system is equal to the magnitude of heat transferred to the high-temperature system so that energy is conserved.

To cite another example, consider air flowing steadily through a horizontal, insulated, variable area duct. Suppose measurements made at two different locations A and B along the length of the duct are as given in Table 6.1. What would be the direction of flow of air? Is it from A to B or from B to A?

Table 6.1 Properties of fluid at two locations in a duct

Location	Pressure (kPa)	Enthalpy (kJ/kg)	Velocity (m/s)
A	120	50	150
B	100	30	250

Either direction of flow is possible according to the first law of thermodynamics. This is elaborated below.

Let it be assumed that the flow is from A to B. From the first law,

$$Q - W_x = \Delta\left(h + \frac{V^2}{2} + gZ\right)$$

Here, $Q = 0, W_x = 0 \text{ and } \Delta Z = 0$

and therefore, $\left(h + \dfrac{V^2}{2}\right) = 0,$

or
$$(h_B - h_A) = \frac{(V_A^2 - V_B^2)}{2} \tag{i}$$

Substituting the given values for each term in the above relation,

the left-hand side $= h_B - h_A = 30 - 50 = -20$ kJ/kg, and

the right-hand side $= \dfrac{(V_A^2 - V_B^2)}{2} = \dfrac{(150^2 - 250^2)}{1000} = -20$ kJ/kg.

The first law is thus satisfied as the energy terms on both sides of the relation happen to be the same. Does this mean that in actual practice the flow is from A to B ? Not necessarily, because even if the flow is assumed to be from B to A we arrive at the same result that the first law is satisfied as the value of each side of (i) becomes equal to + 20 kJ/kg. Thus, according to the first law of thermodynamics, both directions of flow are possible. In reality, the flow is possible under the given conditions in only one particular direction. The possible direction is only from B to A and not from A to B. This can be confirmed only by the second law analysis using the *principle of increase of entropy*, discussed in Chapter 8, which is on entropy. According to this principle, the flow of energy is possible only in the direction of increase of entropy.

In many chemical reactions, it is very essential to know the direction of occurrence. Suppose the simple chemical reaction represented by $2CO + O_2 \rightarrow 2CO_2$ is considered as an example. Experience shows that this reaction cannot take place in the forward direction beyond a certain pressure and temperature. Yet, the first law of thermodynamics does not impose any such restriction on the process.

Many such hypothetical processes can be cited as examples which, although they do not violate the first law of thermodynamics in principle, cannot either occur in practice or proceed in the presumed direction.

It is the second law of thermodynamics, and its corollaries, which not only deal with all aspects of conversion of heat to work and vice versa but also throw light on the criteria for the feasibility of processes and their possible directions of occurrence. For a process, cyclic or non-cyclic, to be feasible, in addition to the first law of thermodynamics, the second law of thermodynamics must also be satisfied.

Before the second law of thermodynamics is stated, a few practical devices that convert heat to work and vice versa are considered here.

6.3 A Steam Power Plant

A schematic of a steam power plant is given in Figure 6.1. A steady flow of water, at high pressure, enters the boiler. Heat transfer from the combustion products (generated by burning fuel and air in a furnace and led to the boiler) causes the water in the boiler to boil and generate steam. Let the magnitude of this heat transfer be Q_1. The steam generated at high pressure leaves the boiler steadily to enter the turbine. This steam expands in the turbine to deliver shaft work, W_T.

The low-pressure steam leaving the turbine enters a water-cooled condenser where it is condensed by rejecting heat to the cooling water. Let the magnitude of this heat transfer be Q_2. A steady flow of condensate (water) enters the pump. To complete the cycle of events, the pump delivers water at high pressure to the boiler. The work W_P supplied to the pump is usually very small in comparison to other interactions and, hence, unless otherwise mentioned, it is customary to neglect the magnitude of this work during the analysis of the working of a steam power plant.

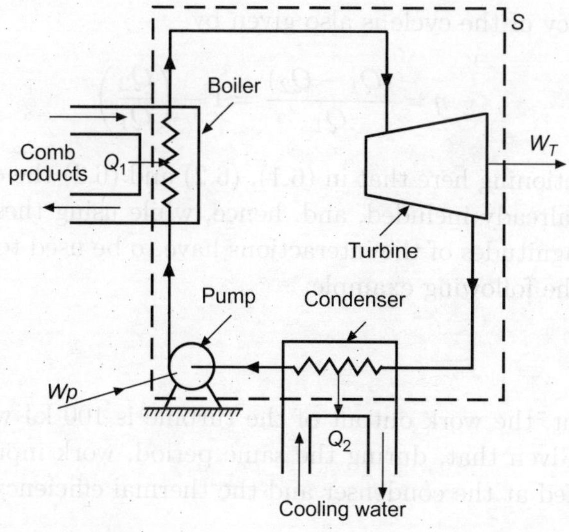

Fig. 6.1 Schematic of a steam power plant.

Although each of the four devices in the plant (boiler, turbine, condenser and pump) is a steady-flow device, all four together enable the water to undergo a thermodynamic cycle. The system (water) that undergoes a thermodynamic cycle is shown enclosed by the boundary S in Figure 6.1. A careful study of the system reveals that during the cycle there are four interactions across the system boundary, namely (i) Q_1, the heat transfer to the system (ii) W_T, the work output from the system (iii) Q_2, the heat transfer from the system and (iv) W_P, the work input to the system. By the sign convention employed in this book, while Q_1 and W_T are positive, Q_2 and W_P are negative.

Applying the first law of thermodynamics, $\Sigma Q = \Sigma W$, to the system executing the cycle,

$$Q_1 + (-Q_2) = W_T + (-W_p)$$

or

$$Q_1 - Q_2 = W_T - W_p \tag{6.1}$$

The performance of any device is usually evaluated by its efficiency, which is defined as the ratio of the output to input (or the ratio of what one wants to what one pays for). The net work output from the plant is $(W_T - W_p)$ and the input is the heat transfer Q_1 in the boiler. Hence, the thermal efficiency or the cycle efficiency of the plant is given by,

$$\eta = \frac{(W_T - W_p)}{Q_1} \tag{6.2}$$

Also, in a thermodynamic cycle, from the first law of thermodynamics, the net work $(W_T - W_p)$ is equal to net heat transfer $(Q_1 - Q_2)$. Replacing $(W_T - W_p)$ in (6.2) by

$(Q_1 - Q_2)$, the efficiency of the cycle is also given by

$$\eta = \frac{(Q_1 - Q_2)}{Q_1} = 1 - \left(\frac{Q_2}{Q_1}\right) \tag{6.3}$$

***Note*:** It is worth mentioning here that in (6.1), (6.2) and (6.3) the directions of heat and work interactions are already included, and, hence, while using these equations for computations, only the magnitudes of the interactions have to be used to avoid any confusion. This is illustrated in the following example:

Example 6.1

In a steam power plant, the work output of the turbine is 100 kJ while heat supplied at the boiler is 300 kJ. Given that, during the same period, work input to the pump is 0.5 kJ find the heat rejected at the condenser and the thermal efficiency of the plant.

► **Data:**

$Q_1 = 300$ kJ, $W_T = 100$ kJ and $W_P = 0.5$ kJ.

► **Solution:**

Referring to Figure 6.1,

$$Q_1 - Q_2 = W_T - W_p \tag{6.1}$$

Substituting the magnitudes of Q_1, W_T and W_p, in (6.1)

$$300 - Q_2 = 100 - 0.5$$

Therefore, Q_2 = Heat rejected at the condenser = 200.5 kJ

$$\eta = \frac{(W_T - W_p)}{Q_1} \tag{6.2}$$

$$= \frac{(100 - 0.5)}{300} = 0.33 \text{ or } 33\%$$

Alternatively, $$\eta = 1 - \left(\frac{Q_2}{Q_1}\right) \tag{6.3}$$

$$= 1 - \frac{200.5}{300} = 0.33 \text{ or } 33\%$$

In a steam power plant, while the work output is around $\frac{1}{3}Q_1$ the heat rejected in the condenser is around $\frac{2}{3}Q_1$. The heat rejected goes as waste to the cooling water which cannot be recovered. Hence, it can be said that all the heat transferred (Q_1) to this thermodynamic cycle is not completely converted to work. In other words, it may be stated that no steam power plant can be 100% efficient. This fact is highly unpalatable to an engineer who would want to have all the heat transfer converted to work.

On second thought, one realizes that in a steam power plant the heat transferred to the system could have been completely converted to work provided no heat was rejected in the condenser. This would be possible if the steam discharged from the turbine, without being condensed in the condenser, is directly pumped by some means and made to enter the boiler as high-pressure liquid. Such a process is practically impossible. Theoretically, it can be shown that, in such a hypothetical process, work required would be equal to or more than the work output of the turbine and hence the plant, as a whole, cannot produce any net positive work at all. And, experience has shown that the plant will never operate without the condenser. Hence rejection of an amount of heat, Q_2, is unavoidable. Before generalizing the observations made on the steam power plant, the working of another practical thermodynamic cycle that delivers net work is considered. It is the closed-cycle gas turbine power plant.

6.4 Closed-Cycle Gas Turbine Power Plant

The schematic of a closed-cycle gas turbine power plant is shown in Figure 6.2. Compressed air from the centrifugal compressor is heated steadily in the heater by heat transfer Q_1 from the combustion products. High-temperature, high-pressure air leaving the heater is then expanded steadily as it flows through a centrifugal turbine. As a result of this expansion process, the turbine delivers a shaft work W_T and air leaves the turbine at low pressure and a relatively low temperature. This low-pressure air from the turbine is then cooled in the cooler by transferring heat to the cooling water from the surroundings. Let the magnitude of this heat transfer be Q_2. Air from the cooler, in turn, is sucked by the centrifugal compressor to be compressed and delivered to the heater again. The air undergoes a thermodynamic cycle as it sequentially passes through the four components. The turbine and the compressor are normally mounted on the same shaft so the work required for compression W_c is provided directly by the turbine. The net work output of the plant is thus $(W_T - W_c)$.

Here also the air (system enclosed by boundary S in Figure 6.2) in the plant undergoes a cyclic process with two heat interactions and two work interactions similar to the happenings in the steam power plant. Hence, relations (6.1)–(6.3) with W_p being replaced by W_c hold good in this case also. However, in this power plant, unlike in the steam power plant, W_c is of considerable magnitude. The magnitude of W_c can never be neglected in computations.

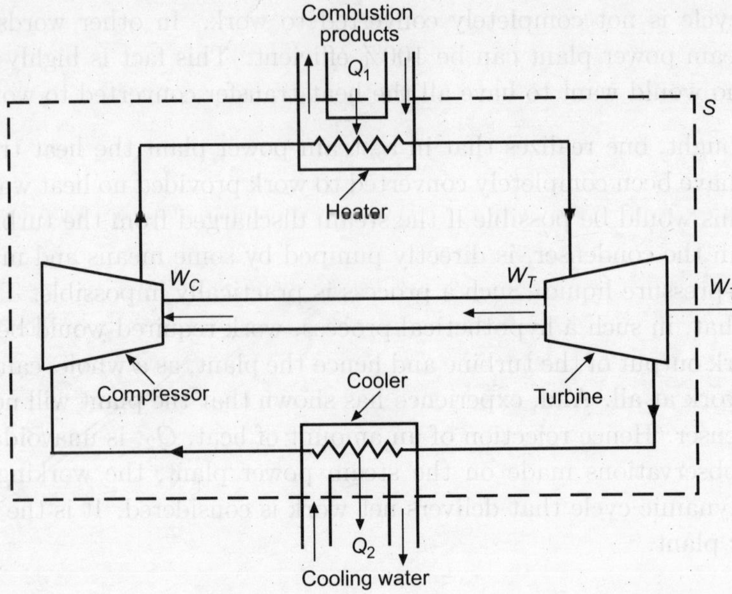

Fig. 6.2 Schematic of a closed gas turbine power plant.

It is evident that the closed-cycle gas turbine power plant, working in a thermodynamic cycle, is also unable to convert all the heat transferred to the system, Q_1, to the network $(W_T - W_C)$. Heat, Q_2, is always rejected as waste to the surroundings. To avoid Q_2, if the cooler is physically eliminated from the plant, experience shows that the plant would not operate. This is because W_c, in such a case, would either be equal to or greater than W_T, making the net work output zero or negative. When Q_2 is zero, the efficiency of the plant, from (6.3), would be 1.0 or 100%. Thus, it can be said that it is impossible for a system operating in a thermodynamic cycle to have 100% efficiency.

All thermodynamic cycles, delivering net work, give the same information and hence it can be generalized that it is impossible to have a thermodynamic cycle in which all the heat transferred to the system is completely converted to work. In other words, it is impossible for any device operating in a thermodynamic cycle to operate with 100% efficiency. This fact forms the basis for the Kelvin-Planck statement of the second law of thermodynamics.

6.5 Refrigerator

The schematic of a domestic refrigeration system that operates in a thermodynamic cycle is shown in Figure 6.3. The working substance that undergoes the cycle is called a refrigerant. Heat Q_2 is transferred to the refrigerant in the evaporator to maintain the refrigerator at a low temperature (in reality, this is the heat transferred from the articles stored in the refrigerator to the refrigerant flowing through the coils around the freezer situated inside the cabinet at the top).

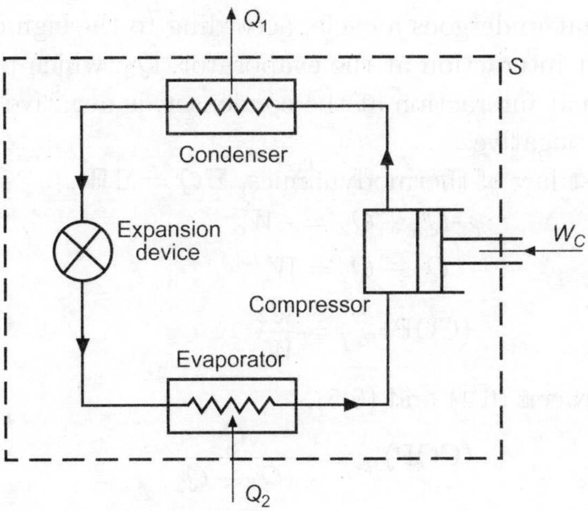

Fig. 6.3 Schematic of a refrigerator/heat pump.

As a consequence of this heat transfer the liquid refrigerant coming from the expansion device gets vapourized in the evaporator and the low-pressure low-temperature vapour leaves the evaporator to enter the compressor. Work W_c supplied to the compressor from electrical mains, to which the refrigerator is connected to, increases the pressure and temperature of the refrigerant vapour as it passes through the compressor. This compressed vapour passes to the air-cooled condenser (situated behind but outside the refrigerator cabinet) where it is steadily condensed to a liquid at high pressure. In the condenser, heat Q_1 is transferred from the refrigerant to the surrounding air which is at a temperature higher than the temperature inside the refrigerator cabinet. The high-pressure liquid refrigerant coming from the condenser is then throttled as it passes through an expansion device. Low-pressure, low-temperature liquid (actually, after throttling, the refrigerant will be a mixture of liquid and vapour with a high percentage of the former) from the expansion device enters the evaporator to complete the cycle. Note that heat is transferred to the refrigerant at a low temperature while heat is rejected from the refrigerant at a higher temperature and work is supplied to the system. Thus, all interactions are opposite in direction to those in either a steam power plant or a closed cycle gas turbine power plant.

In the refrigerator, as the refrigerant undergoes a cycle, it has been possible to transfer heat from a low- to a high-temperature region only because of the work input to the compressor. In the absence of this work input, it would be impossible to carry out the above heat transfer. This fact embodies the Clausius statement of the second law of thermodynamics.

It has been explained earlier that the performance of a work-producing cycle is evaluated by its efficiency. However, it is customary to evaluate the performance of a work-absorbing cycle, such as the refrigerator, by its coefficient of performance, COP. Similar to efficiency, COP is also the ratio of what one wants to what one pays for.

For the system S that undergoes a cycle, according to the sign convention of heat and work transfers, the heat interaction at the evaporator, Q_2, which provides refrigeration is positive, and Q_1, the heat interaction at the condenser, is negative and the work input to the compressor, W_c, is negative.

Hence, from the first law of thermodynamics, $\Sigma Q = \Sigma W$,

$$-Q_1 + Q_2 = -W_c$$

or,

$$Q_1 - Q_2 = W_c \tag{6.4}$$

By definition,

$$(\text{COP})_{Ref} = \frac{Q_2}{W_c} \tag{6.5}$$

Eliminating W_c between (6.4) and (6.5),

$$(\text{COP})_{Ref} = \frac{Q_2}{(Q_1 - Q_2)} \tag{6.6}$$

***Note*:** In (6.4), (6.5) and (6.6) the directions of heat and work interactions are already incorporated. Thus, while using these equations for computations, only the magnitudes of the interactions will have to be substituted.

6.6 Heat Pump

A refrigerator creates a low-temperature region by continuously transferring heat from the region being cooled to the surroundings which are at a relatively high temperature. A refrigerator or a window air conditioner which works according to the above cycle is a necessity in hot climates. In cold climates, it may be necessary to provide a relatively high-temperature region such as a heated room. This can be achieved by a heat pump, whose operation is exactly similar to that of a refrigerator. Figure 6.3 can also be considered as the schematic of a heat pump. In a heat pump, one is interested in Q_1, the heat transfer to the high-temperature region. To maintain its high temperature, heat is continuously transferred to this region from the low-temperature surroundings with the help of work supplied to the compressor. It may be imagined that a domestic refrigerator can be used as a heat pump by installing it in such a manner that, with its door open, the cabinet (the evaporator) is exposed to the cold weather outside the room and the condenser coil is exposed to the room to be heated. It is needless to mention again that a heat pump is also a work-absorbing cycle.

While (6.4) also holds good for a heat pump, its *coefficient of performance* (COP) is given by,

$$(\text{COP})_{\text{hp}} = \frac{Q_1}{W_c} \tag{6.7}$$

as, what is expected of the heat pump is the transfer of heat to the room to be heated.

As
$$W_c = (Q_1 - Q_2)$$
$$(\text{COP})_{\text{hp}} = \frac{Q_1}{(Q_1 - Q_2)} \tag{6.8}$$

Subtracting 1 from both sides of (6.8), we get,

$$(\text{COP})_{\text{hp}} - 1 = \frac{Q_1}{(Q_1 - Q_2)} - 1$$
$$= \frac{Q_2}{(Q_1 - Q_2)}$$

But,
$$(\text{COP})_{\text{Ref}} = \frac{Q_2}{(Q_1 - Q_2)} \tag{6.6}$$

Therefore,
$$(\text{COP})_{\text{hp}} - 1 = (\text{COP})_{\text{Ref}}$$

or,
$$(\text{COP})_{\text{hp}} = 1 + (\text{COP})_{\text{Ref}} \tag{6.9}$$

Example 6.2 is meant to explain how computations have to be made while analysing the performances of refrigerators and heat pumps.

Example 6.2

(a) A refrigerator with a COP of 4.0 transfers heat at a rate of 0.5 kJ/s at the condenser. Find the rate of heat transfer in the evaporator and the power input to the compressor. (b) Also, calculate the COP if the refrigerator were to operate as a heat pump with the same magnitudes of heat and work interactions.

▶ Data:

$\text{COP}_{\text{ref}} = 4, Q_1 = 0.5$ kJ/s.

▶ Solution:

(a) Referring to Figure 6.3

from the definition,
$$\text{COP}_{\text{ref}} = \frac{Q_2}{W_c} \tag{6.5}$$

from data,
$$\text{COP}_{\text{ref}} = 4.0$$

therefore,
$$4 = \frac{Q_2}{W_c}, \text{ or } Q_2 = 4W_c \tag{i}$$

From the first law,
$$Q_1 - Q_2 = W_c \tag{ii}$$

Eliminating Q_2 Between (i) and (ii),

$$Q_1 - 4W_c = W_c$$

or,
$$W_c = \frac{Q_1}{5}$$

from the data
$$Q_1 = 0.5 \text{ kJ/s}$$

therefore,
$$W_c = \frac{0.5}{5} = 0.1 \text{ kJ/s}$$

Substituting for W_c in (i), $\quad Q_2 = 4 \times 0.1 = 0.4$ kJ/s.

Rate of heat transfer in the evaporator $= 0.4$ kJ/s

and the power input to the compressor $= 0.1$ kJ/s

(b)
$$(\text{COP})_{\text{hp}} = \frac{Q_1}{W_c} \tag{6.7}$$

Substituting the values for Q_1 and W_c

$$(\text{COP})_{\text{hp}} = \frac{0.5}{0.1} = 5$$

Alternatively, we know that $\quad (\text{COP})_{\text{hp}} = 1 + (\text{COP})_{\text{ref}} \quad$ (6.9)

Therefore, $\quad (\text{COP})_{\text{hp}} = 1 + 4 = 5$

Example 6.3

(a) In winter, a person decides to beat the cold in a room by switching on an electric resistance heater connected to the mains at 220 V. The current drawn by the heater is 4.5 amp. If one unit of electricity costs Rs 5.00 and the heater is on for 8 h per day find the additional monthly electricity bill the person has to pay for using the heater. (b) Instead of using the electric heater, had the person installed a heat pump of COP = 4 to heat the room, what would be the saving in the monthly electricity bill.

▶ Solution:

(a) Rate of electricity consumption $= IV = 4.5 \times 220 = 990$ W

One unit of electricity = 1 kWh

Therefore, units of electricity per day $= \left(\frac{990}{1000}\right) \times 8 = 7.92$ kWh.

Electricity bill per month $= (7.92 \times 5.00) \times 30$ days $=$ Rs 1188.00

(b) When the heat pump is used, and the heating is identical to that in (a), the rate of heating demanded from the heat pump is equal to the rate of electrical energy consumption by the electric heater.

Therefore, the rate of heat transfer from the heat pump,

$$Q_1 = 990 \text{ W} = 0.99 \text{ kW}.$$

As

$$\text{COP}_{\text{hp}} = 4 = \frac{Q_1}{W},$$

$$W = \frac{Q_1}{4} = \frac{0.99}{4} = 0.2475 \text{ kW}.$$

This will be the rate at which electric energy is consumed by the heat pump.

Therefore, the electric energy consumed per day = 0.2475 × 8 = 1.98 kWh.

The monthly electricity bill =(1.98 × 5.00) × 30 days = Rs 297.00

Saving in monthly electricity bill = Rs 1188.00 − Rs 297.00 = Rs 891.00

Note: This example brings out clearly the advantage of using a heat pump for heating rooms in place of simple electric heaters. There is a substantial saving in energy and thus saving in money as well.

6.7 Heat Engine

Many practitioners of thermodynamics prefer to define a *heat engine* as a continuously operating thermodynamic system across the boundary of which there are only heat and work interactions. In other words, a heat engine is nothing but a system undergoing a thermodynamic cycle. According to this understanding, a steam power plant and a closed-cycle gas turbine power plant are examples of *heat engines*. However, this concept of a heat engine does not put any restriction on the directions of heat and work transfers. In the refrigerator and heat pump, which are also systems undergoing thermodynamic cycles, the heat and work interactions are in opposite directions to those in a steam power plant and a closed-cycle gas turbine power plant. Yet, by definition, they are also heat engines.

The terms direct heat engine and reversed heat engine are generally used to distinguish between the above two categories of heat engines.

6.7.1 Direct Heat Engine

A system operating in a thermodynamic cycle which delivers work by receiving heat from a high-temperature region and rejecting heat to a low-temperature region is called a *direct heat engine*. Accordingly, the steam power plant, the closed-cycle gas turbine power plant and similar other cycles are direct heat engines.

6.7.2 Reversed Heat Engine

A system operating in a thermodynamic cycle which transfers heat from a low-temperature region to a high-temperature region with the assistance of work from the surroundings, is a *reversed heat engine*. Accordingly, the refrigerator, the heat pump and similar other cycles are reversed heat engines.

6.7.3 Symbolic Representation of Heat Engines

By now, it is clear that a direct heat engine is one in which the working substance undergoes a thermodynamic cycle to produce work while heat is transferred to the working substance from a high-temperature system and the working substance, in turn, transfers out heat to a low-temperature system. Thus, it can be said that a heat engine operates between two systems at different temperatures. The high-temperature system at a fixed temperature with which the heat engine engages in heat transfer is usually referred to as *high-temperature reservoir* (or a *source* by a few authors). Similarly, the low-temperature system at a fixed temperature with which the heat engine engages in heat transfer is usually called a *low temperature reservoir* (or a *sink* by a few authors).

Figure 6.4 shows how a direct heat engine can be represented symbolically. In this figure, the circle with the letter E at the centre represents the heat engine. The top block represents the high-temperature reservoir at temperature t_1. The bottom block represents the low-temperature reservoir at t_2. The arrows indicate the directions of heat and work interactions. While Q_1 is the magnitude of heat interaction with the high-temperature reservoir, Q_2 is the magnitude of heat interaction with the low-temperature reservoir and W is the magnitude of net work interaction. Similarly, Figure 6.5 shows the symbolic representation of a reversed heat engine.

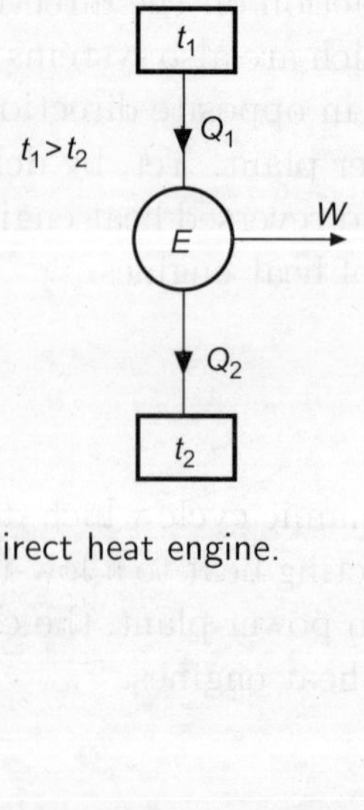

Fig. 6.4 Symbolic representation of a direct heat engine.

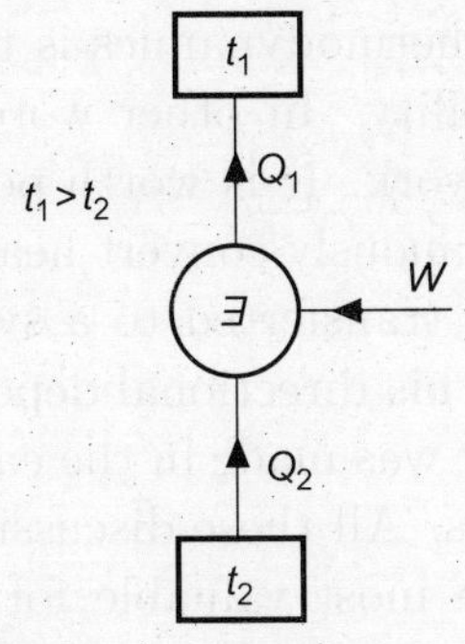

Fig. 6.5 Symbolic representation of a reversed heat engine.

6.8 The Second Law of Thermodynamics

The two classical statements of the second law of thermodynamics, namely, the Kelvin–Planck statement and the Clausius statement, emerge when hitherto-made observations on direct and reversed heat engines are generalized.

6.8.1 The Kelvin–Planck Statement of the Second Law

It is impossible for a device operating in a thermodynamic cycle to produce work while having heat interaction with a single system at a fixed temperature.

Mathematically, it may be stated as follows,

$$\oint_{\substack{\text{single}\\ \text{temp.system}}} \delta W \leq 0$$

This statement of the second law is made based on the observations on the direct heat engines. In brief, it means that a direct heat engine cannot operate in the absence of heat rejection. Such an impossible engine, which is called the *perpetual motion machine of the second kind* (PMM II), is shown symbolically in Figure 6.6. In PMM II, there is no heat rejection and, hence, according to the first law, $Q_1 = W$. Therefore, the efficiency of the engine, which is given by W/Q_1, is 100%.

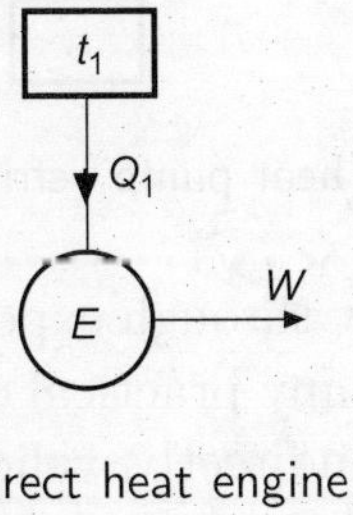

Fig. 6.6 The impossible (100% efficient) direct heat engine, PMM II.

The gist of the second law of thermodynamics is that a PMM II, or a 100% efficient direct heat engine, is an impossibility. In other words, no heat engine can completely and continuously convert heat to work. It is worth pointing out here that although it is impossible to completely and continuously convert heat to work, experiments have shown that it is possible to convert work transferred to a system to an equal quantity of heat transfer out of it. It is because of this directional dependence of the extent of conversion of one to another that a great effort was made in the earlier chapters to clearly distinguish between heat and work interactions. All these discussions establish for certain that work is a refined form of energy or the most valuable form of energy or high-grade energy compared to heat.

(Incidentally, a perpetual motion machine of the first kind, PMM I, is a device that contradicts the first law of thermodynamics. It is a system operating in a cycle producing positive work with no heat interaction across the system boundary.)

6.8.2 The Clausius Statement of the Second Law of Thermodynamics

It is impossible for any device operating in a thermodynamic cycle to transfer heat from a low-temperature region to a high-temperature region without the help of work interaction from the surroundings.

This statement is made with respect to the observations made earlier, that a reversed heat engine cannot operate unless work is supplied to it from outside. Thus, a heat pump shown symbolically in Figure 6.7 is an impossibility. If such a heat pump were to be possible, its COP ($= Q_1/W$) would be infinite, as $W = 0$. Thus, the gist of the Clausius statement is that no reversed heat engine (a refrigerator or a heat pump) can operate without work being supplied to it. In other words, it is impossible to have a heat pump of infinite COP.

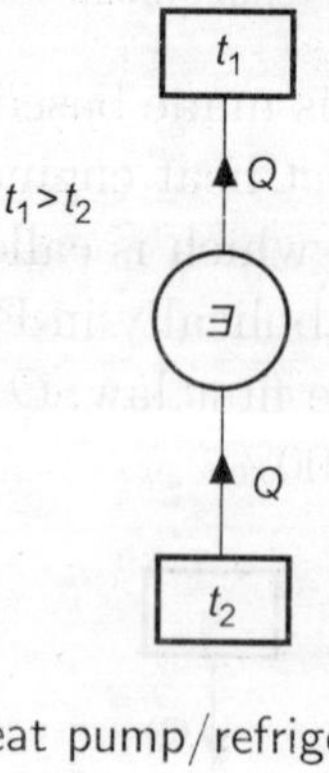

Fig. 6.7 Impossible reversed heat engine (heat pump/refrigerator with infinite COP).

The second law derives its validity through experimental evidence only. Until this day, it has not been possible to construct any practical device that contradicts the second law. Thus, every experiment, directly or indirectly, validates the second law.

The two statements of the second law, although appearing to convey different things, are, in reality, equivalent. As a matter of fact, they have to be so. This is elaborated below.

6.8.3 Equivalence of the Kelvin–Planck and Clausius Statements

A simple method by which the equivalence can be shown is by making use of the fact that any two statements are equivalent if the violation of one statement implies the violation of the other and vice versa.

Consider a hypothetical heat pump that violates the Clausius statement. Let this heat pump, HP, as shown in Figure 6.8, transfer heat from a low-temperature reservoir at t_2 to a high-temperature reservoir at t_1 without any work being supplied to it. Let the heat interaction with the reservoir at t_2 be Q_2 and that with the reservoir at t_1 be Q_1. HP is working in a cycle. Hence, by the first law, $\Sigma Q = \Sigma W$. As per the assumption made, $W = 0$. Therefore, $\Sigma Q = 0$, hence, $Q_2 - Q_1 = 0$, or in other words, $Q_1 = Q_2$. Assume now that a direct heat engine E, which works in accordance with the Kelvin–Planck statement, operates between the same two temperature limits and that heat supplied to it, from the reservoir at t_1, is Q_1.

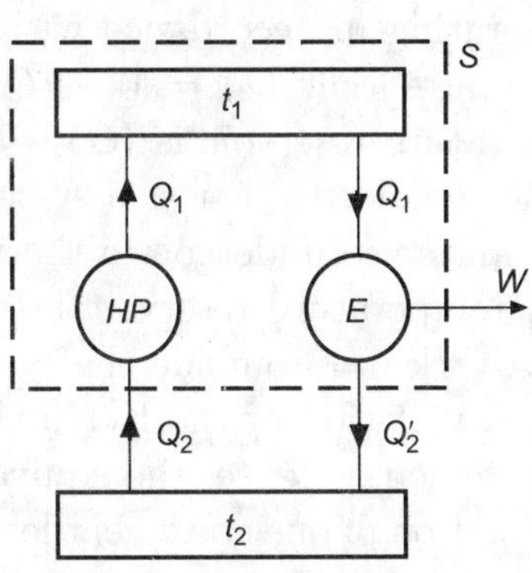

Fig. 6.8 Schematic for showing the equivalence of the Kelvin–Planck and Clausius statements.

Let the work produced by this engine be W and the heat rejected by it to the low-temperature reservoir at t_2 be Q_2'. Now, consider the system S enclosed by the boundary shown in Figure 6.8 by the broken line. This system must also be undergoing a cyclic process as it consists of two sub-systems, E and HP, each of which is undergoing a cycle. It may be stated here that inclusion of the high-temperature reservoir inside the boundary does not pose any problem as this reservoir experiences no change whatsoever with time because the heat transfer from it to E is equal to heat transfer to it by HP, both being equal to Q_1. The net work output of this system is W. Applying the first law to E, $W = (Q_1 - Q_2')$. The system, identified by the broken-line boundary, while operating in a cycle, has heat interaction with only the low-temperature reservoir. It can also be seen from Figure 6.8 that the net heat transfer to the system from the reservoir at t_2 is equal to $(Q_2 - Q_2')$. It is already shown that $Q_2 = Q_1$. Hence,

net heat transfer, $\quad Q = Q_1 - Q_2', \quad$ and

net work produced, $\quad W = Q_1 - Q_2'.$

Therefore, $$\eta = \frac{W}{Q} = 100\%$$

Hence, this system working in a thermodynamic cycle, converts the net heat transferred to it from a single reservoir to an equal quantity of work. This amounts to the violation of the Kelvin–Planck statement, as it turns out to be a 100% efficient engine (or a PMM II). Thus, it can be concluded that a violation of the Clausius statement leads to the violation of the Kelvin–Planck statement.

To complete the exercise, let it now be assumed that there exists a heat engine E that operates violating the Kelvin–Planck statement. As shown in Figure 6.9, let the heat transfer from the high-temperature reservoir to this heat engine be Q_1. As no heat is rejected by this engine to the low-temperature region, according to the first law, the work output of the engine $W = Q_1$. Let this work drive a heat pump HP which receives Q_2 from the low-temperature and transfers Q_3 to the high-temperature reservoir. It can be seen that the heat pump is operating in accordance with the Clausius statement. The first law of thermodynamics dictates that $Q_3 = W + Q_2 = Q_1 + Q_2$. Thus, the net heat transfer to the high-temperature reservoir is $(Q_3 - Q_1) = (Q_1 + Q_2) - Q_1 = Q_2$. Consider the heat pump and the engine together as a system. This system, S, enclosed in the boundary shown by broken lines also undergoes a thermodynamic cycle during which Q_2 is transferred from the low-temperature reservoir to the high-temperature reservoir without the requirement of any work from outside the system. This is a violation of the Clausius statement. Therefore, a violation of the Kelvin–Planck statement also leads to the violation of the Clausius statement. Hence, the equivalence of the two statements is proved as it is shown that the violation of one statement leads to the violation of the other.

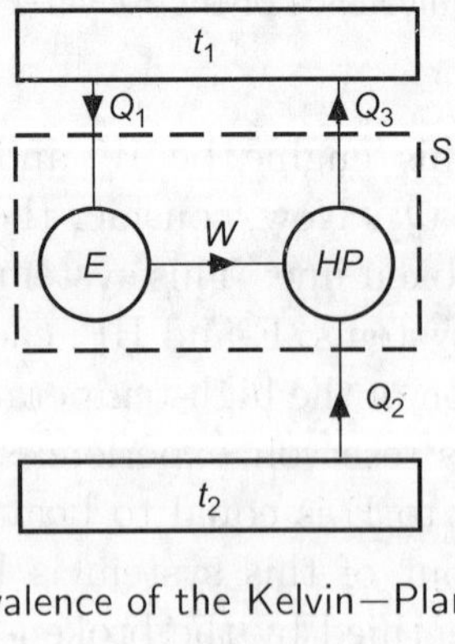

Fig. 6.9 Sketch for showing the equivalence of the Kelvin–Planck and Clausius statements.

6.9 Key Points Discussed so Far

- A heat engine is a system undergoing a thermodynamic cycle with heat and work interactions.
- In a direct heat engine, heat (Q_1) is transferred to the system from a high-temperature reservoir at t_1, work W is transferred out of the system and heat, Q_2, is rejected by the system to a low-temperature reservoir at t_2.
- The efficiency of a direct heat engine is given by

$$\eta = \frac{(Q_1 - Q_2)}{Q_1} = \frac{W}{Q_1}$$

- In a reversed heat engine the heat and work interactions are in opposite directions compared to those in a direct heat engine.
- Heat pump and refrigerator are reversed heat engines.
- If Q_1 and Q_2 are the heat transfers at the high temperature reservoir and the low temperature reservoir, respectively, then the coefficient of performance of a reversed engine is given by

$$(\mathrm{COP})_{\mathrm{hp}} = \frac{Q_1}{(Q_1 - Q_2)} = \frac{Q_1}{W}, \text{ and}$$

$$(\mathrm{COP})_{\mathrm{ref}} = \frac{Q_2}{(Q_1 - Q_2)} = \frac{Q_2}{W}.$$

- Also, $(\mathrm{COP})_{\mathrm{hp}} = 1 + (\mathrm{COP})_{\mathrm{ref}}$.
- The gist of the Kelvin–Planck statement of the second law of thermodynamics is that it is impossible for a direct heat engine to work with 100% efficiency.
- According to the Clausius statement of the second law it is impossible for a reversed heat engine to work with infinite COP.

6.10 Illustrative Problems

6.1 A heat engine performs many cycles during which it develops 21 kJ work and receives 85 kJ heat from the source. Evaluate the thermal efficiency of and the heat rejected by the engine.

▶ **Data:**

$Q_1 = 85$ kJ, $W = 21$ kJ.

▶ **Solution:**

Refer to Figure.6.4.

$$\eta = \frac{W}{Q_1} = \frac{21}{85} = 0.247$$

also, $$\eta = 0.247 = 1 - \frac{Q_2}{Q_1}$$

therefore, $$\frac{Q_2}{Q_1} = 1 - 0.247 = 0.753,$$

Or, $$Q_2 = 0.753Q_1 = 0.753 \times 85 = 64 \text{ kJ}.$$

Alternatively, from the first law, $Q_1 - Q_2 = W$, and hence, $Q_2 = Q_1 - W = 85 - 21 = 64$ kJ.

6.2 In a reversed heat engine, the work done on the engine is 75 kJ and the heat transferred to the engine is 220 kJ. Evaluate the heat transfer from the engine and the COP of the engine (a) as a refrigerator and (b) as a heat pump.

▶ **Data:**

Refer to Figure 6.5.
$W = 75$ kJ, $Q_2 = 220$ kJ.

▶ **Solution:**

Applying the first law, $$(Q_1 - Q_2) = W$$

Therefore, $$(Q_1 - 220) = 75, \text{ or } Q_1 = 295 \text{ kJ}.$$

$$(\text{COP})_{\text{Ref}} = \frac{Q_2}{W} = \frac{220}{75} = 2.93$$

$$(\text{COP})_{\text{hp}} = \frac{Q_1}{W} = \frac{295}{75} = 3.93$$

6.3 A heat engine, X, operates between a reservoir at 1000°C and a body at 600°C. Heat transfer to the heat engine is 1000 kJ and work output is 220 kJ. Another engine, Y, operates between the body at 600°C and the atmosphere at 27°C. Heat rejected to the atmosphere is 400 kJ. The body at 600°C maintains its temperature steady and has interactions only with the two engines, X and Y. Calculate the work output and thermal efficiency of engine, Y.

▶ **Data:**

$Q_{1\text{X}} = 1000$ kJ, $W_{\text{X}} = 220$ kJ, $Q_{2\text{Y}} = 400$ kJ .

► **Solution:**

Refer to Figure 6.10

Applying the first law to engine X , $Q_{1X} - Q_{2X} = W_X$

Therefore, $Q_{2X} = Q_{1X} - W_X = 1000 - 220 = 780$ kJ.

As the body at 600°C maintains its temperature steady, heat transfer to it must be equal to heat transfer from it. Therefore,

$$Q_{2X} = Q_{1Y} = 780 \text{ kJ}$$

Applying the first law to engine Y,

$$Q_{1Y} - Q_{2Y} = W_Y = 780 - 400 = 380 \text{ kJ.}$$

$$\eta_Y = \frac{W_Y}{Q_{1Y}} = \frac{380}{780} = 0.487.$$

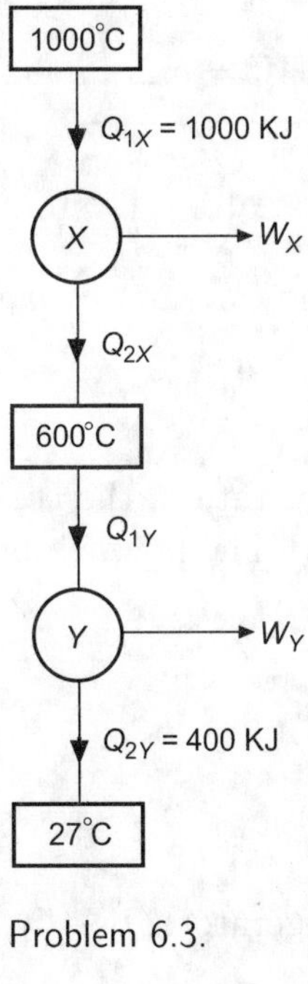

Fig. 6.10 Schematic to solve Illustrative Problem 6.3.

6.4 The COP of a heat pump is 5 when the power supplied to drive it is 35 kW. (a) Evaluate the heat transfers from and to the working fluid. (b) The heat transfer from the heat pump is used to heat water flowing through the radiators of a building. Evaluate the flow rate of radiator water in kg/s, given that its temperature increases by 20°C . Neglect the change in velocity of water as it flows through the condenser.

▶ Data:

$(\text{COP})_{\text{hp}} = 5, W = 35 \text{ kW}, (\Delta t)_{\text{radiator water}} = 20°\text{C}$.

▶ Solution:

Refer to Figure 6.5.

By definition, $(\text{COP})_{\text{hp}} = \dfrac{\dot{Q}_1}{\dot{W}} = \dfrac{\dot{Q}_1}{35} = 5.$

Therefore, the heat transfer from the heat pump $= \dot{Q}_1 = 175$ kW.

From the first law, $(\dot{Q}_1 - \dot{Q}_2) = \dot{W}$

Hence, $175 - \dot{Q}_2 = 35$, or

$\dot{Q}_2 =$ heat transfer to the heat pump

$= 175 - 35 = 140$ kW.

Applying the first law to water flowing through the radiator

$$\dot{Q} - \dot{W} = \dot{m}\Delta\left(h + \frac{V^2}{2} + gz\right)$$

Here $\dot{Q} = \dot{Q}_1 = 175$ kW, $W = 0, \Delta\left(\dfrac{V^2}{2}\right) = 0$ and assume $\Delta(gz) = 0$.

Therefore, $175 = \dot{m}\Delta h_{water} = \dot{m}c_{water}(\Delta t) = m \times 4.2 \times 20$, or

$$m = \frac{175}{(4.2 \times 20)} = 2.083 \text{ kg/s}.$$

6.5 Find the COP and heat transfer rate in the condenser of a refrigerator which has a refrigeration capacity of 7.5 kW. The power input to the refrigerator is 1 kW.

▶ Data:

Refer to Figure 6.5

$\dot{Q}_2 = 7.5$ kW, $\dot{W} = 1$ kW.

▶ Solution:

Applying the first law to the refrigerator, $\dot{Q}_1 - \dot{Q}_2 = \dot{W}$, where $\dot{Q}_1$ is the heat transfer rate in the condenser.

Therefore, $\dot{Q}_1 = \dot{W} + \dot{Q}_2 = 1 + 7.5 = 8.5$ kW.

$$(\text{COP})_{\text{Ref}} = \frac{\dot{Q}_2}{\dot{W}} = \frac{7.5}{1.0} = 7.5.$$

6.6 The properties of the working substance (air) at different locations in a closed-cycle gas turbine power plant are as shown in the table below:

The heat transfer to air in the heater is 640 kJ/kg air. Assume no heat loss from any component to atmosphere. Calculate (a) t_1, (b) heat transfer in the cooler, (c) net work output and (d) thermal efficiency of the plant. Assume that enthalpy of air is a function of its temperature only and that $c_p = 1.005$ kJ/kg deg.

Location	Temperature (°C)	Velocity (m/s)
Between cooler and compressor	35	95
Between compressor and heater	240	35
Between heater and turbine	t_1	220
Between turbine and cooler	525	110

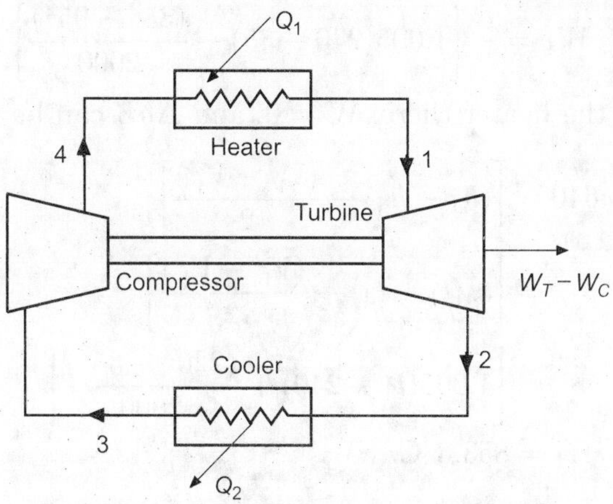

Fig. 6.11 Schematic to solve Illustrative Problem 6.6.

▶ **Data:**

The schematic of the plant is shown in Figure 6.11;

$$t_3 = 35°\text{C}, \quad V_3 = 95 \text{ m/s},$$
$$t_4 = 240°\text{C}, \quad V_4 = 35 \text{ m/s},$$
$$V_1 = 220 \text{ m/s}, \quad t_2 = 525°\text{C}$$
$$V_2 = 110 \text{ m/s}, \quad \text{and} \quad Q_1 = 640 \text{ kJ/kg air.}$$

Heat loss from all components to the atmosphere = 0. As h is a function of temperature only and that c_p is a constant, for air flowing through each component, $h_{out} - h_{in} = c_p(t_{out} - t_{in})$.

► **Solution:**

Applying SS, SFEE to the compressor,

$$Q - W_c = \left[(h_4 - h_3) + \frac{(V_4^2 - V_3^2)}{2} + g(Z_4 - Z_3)\right]$$

Here, $Q = 0$ (no heat losses) and ΔgZ can be neglected. Hence,

$$-W_c = \left[(h_4 - h_3) + \frac{(V_4^2 - V_3^2)}{2}\right]$$

$$= \left[c_p(t_4 - t_3) + \frac{(V_4^2 - V_3^2)}{2}\right]$$

Therefore, $$Wc = -\left[1.005(240 - 35) + \frac{(35^2 - 95^2)}{2000}\right] = -202 \text{ kJ.}$$

Applying SS, SFEE to the heater (here, W = 0, and ΔgZ can be neglected),

$$Q_1 = 640 = \left[(h_1 - h_4) + \frac{(V_1^2 - V_4^2)}{2}\right]$$

$$= \left[c_p(t_1 - t_4) + \frac{(V_1^2 - V_4^2)}{2}\right]$$

$$= \left[1.005(t_1 - 240) + \frac{(220^2 - 35^2)}{2000}\right]$$

Therefore, $$t_1 = 853.3°\text{C.}$$

Applying SS, SFEE to the turbine (where $Q = 0$ and ΔgZ can be neglected),

$$-W_T = \left[(h_2 - h_1) + \frac{(V_2^2 - V_1^2)}{2}\right]$$

$$= \left[c_p(t_2 - t_1) + \frac{(V_2^2 - V_1^2)}{2}\right]$$

Hence, $$W_T = -\left[1.005(525 - 853.3) + \frac{(110^2 - 220^2)}{2000}\right] = 348 \text{ kJ.}$$

Applying SS, SFEE to the cooler (here, $W = 0$ and ΔgZ can be neglected),

$$Q_2 = \left[(h_3 - h_2) + \frac{(V_3^2 - V_2^2)}{2}\right] = \left[c_p(t_3 - t_2) + \frac{(V_3^2 - V_2^2)}{2}\right]$$

$$= \left[1.005(35 - 525) + \frac{(95^2 - 110^2)}{2000}\right] = -494 \text{ kJ.}$$

(a) t_1 = temperature of air after the heater = 853.3°C.

(b) Q_2 = heat rejected at the condenser = 494 kJ.

(c) net work output = $W_T - W_c$ = 348 – 202 = 146 kJ.

(d) Thermal efficiency $= \dfrac{(W_T - W_c)}{Q_1} = \dfrac{146}{640} = 0.228.$

Alternatively, efficiency $= \dfrac{(Q_1 - Q_2)}{Q_1} = \dfrac{(640 - 494)}{640} = 0.228$

6.7 A heat engine is used to drive a heat pump. The heat transfers from the heat engine and the heat pump are used to heat the water circulating through the radiators of a building. It is given that the efficiency of the heat engine is 30% and the COP of the heat pump is 4. How much heat is transferred to the radiator water, for every kJ heat transferred to the heat engine.

▶ **Solution:**

Refer to Figure 6.12.

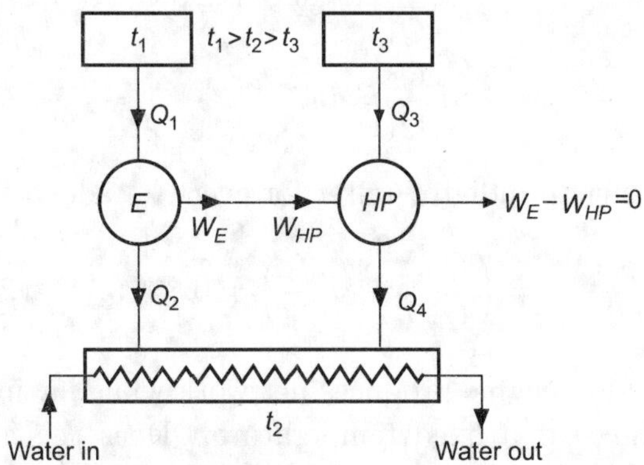

Fig. 6.12 Heat pump driven by the heat engine.

$$\text{Efficiency of the heat engine} = \frac{(Q_1 - Q_2)}{Q_1} = 1 - \frac{Q_2}{Q_1} = 0.3, \text{ or, } \frac{Q_2}{Q_1} = 0.7$$

Here, $W_E = W_{\text{hp}}$ because the heat engine is driving the heat pump

$$(\text{COP})_{\text{hp}} = 4 = \frac{Q_4}{W_{\text{hp}}}, \text{ where } W_{\text{hp}} = W_E = \eta \times Q_1 = 0.3\ Q_1$$

$$\text{Therefore,} \quad \frac{Q_4}{W_{hp}} = \frac{Q_4}{0.3\ Q_1} = 4, \text{ hence, } \frac{Q_4}{Q_1} = 0.3 \times 4 = 1.2.$$

Total heat transfer to the radiator water $= Q_4 + Q_2$.
Heat transfer to the heat engine $= Q_1$.
Therefore, heat transfer to radiator water for every kJ added to the heat engine

$$= \frac{(Q_2 + Q_4)}{Q_1} = \left(\frac{Q_2}{Q_1}\right) + \left(\frac{Q_4}{Q_1}\right) = 0.7 + 1.2 = 1.9 \text{ kJ}.$$

6.8 Redo Problem 6.7 given that the engine–pump combination delivers a work output of 10% of heat transferred to the heat engine.

► **Solution:**

Refer to Figure 6.12 again. In this problem also $W_E = \eta \times Q_1 = 0.3\ Q_1$. However, W_{hp} is not the same as W_E, because the work developed by the engine goes to run the heat pump and also to deliver shaft work. Therefore,

$$W_{hp} = W_E - W_{out} = 0.3Q_1 - 0.1Q_1 = 0.2Q_1$$

$$(\text{COP})_{\text{hp}} = 4 = \frac{Q_4}{W_{\text{hp}}} = \frac{Q_4}{0.2Q_1}$$

Therefore,
$$\frac{Q_4}{Q_1} = 0.2 \times 4 = 0.8.$$

As in Problem 6.7,
$$\frac{Q_2}{Q_1} = 0.7.$$

Therefore, heat transfer to radiator water for every kJ added to heat engine is:

$$\frac{(Q_2 + Q_4)}{Q_1} = \left(\frac{Q_2}{Q_1}\right) + \left(\frac{Q_4}{Q_1}\right) = 0.8 + 0.7 = 1.5 \text{ kJ}$$

6.9 An internal combustion engine produces net work while having heat interaction only with the surroundings at a fixed temperature. Does this mean that the internal combustion engine violates the second law of thermodynamics?

► **Solution:**

In an internal combustion engine, a steady flow of fuel and air enters the engine. The fuel–air mixture is burnt in the piston–cylinder device of the engine and subsequently the combustion products expand to do work on the piston. This causes the engine to deliver shaft work. Because of the combustion process, there will be an increase in temperature of the engine and hence the engine exchanges heat with the surroundings. There is no heat input in the internal combustion engine. A steady flow of combustion products leaves the engine through the exhaust pipe. From the operation of the internal combustion engine, it becomes evident that it does not work in a thermodynamic cycle as the working substance is not returned to its original state after each cycle of mechanical events. In fact, there is a steady flow of material into and out of the engine. Hence, an internal combustion engine is not a system undergoing a thermodynamic cycle. In other words, an internal combustion engine is not a heat engine. The Kelvin–Planck statement of the second law of thermodynamics is applicable only for heat engines. It cannot, therefore, be applied here to verify the violation or otherwise of the second law. An internal combustion is a practically feasible engine. It could not have been so if it were to violate the second law of thermodynamics. It shall be explained later that processes such as those occurring in an internal combustion engine are also bound by the corollaries of the second law of thermodynamics.

Problems

6.1 A heat engine developing a power of 100 kW has an efficiency of 20%. Evaluate the rates of heat transfer to and from the working fluid.

6.2 It is given that in a steam power plant the condition of water at various locations are as follows:

Location	Enthalpy (kJ/kg)	Velocity (m/s)
Between pump and boiler	115.15	6.5
Between boiler and turbine	3020.10	50
Between turbine and condenser	2330.20	310
Between condenser and pump	113.40	6.5

Find the thermal efficiency of the plant, neglecting heat loss in various pipes between the components. Neglect also the changes in kinetic and potential energies of water.

6.3 Shown in Figure 6.13 are heat engines, X and Y, operating having heat interactions with three reservoirs at t_1, t_2 and t_3. The reservoir at t_2 has heat interactions with only engines X and Y. Given that the work output of Y is half that of X and that the efficiencies of the two engines are the same, show that the ratio of heat transfer to engine X to the heat transfer from engine Y is 4.

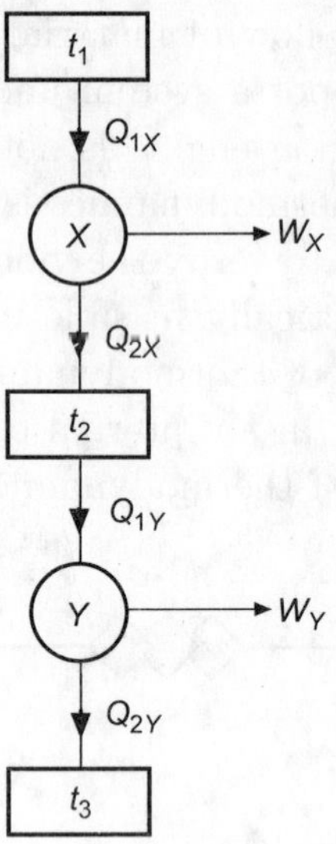

Fig. 6.13 Sketch for Problem 6.3.

6.4 A dry-cell battery can produce work in the form of electricity while exchanging heat only with the constant temperature atmosphere. Is this a violation of the second law of thermodynamics?

6.5 An open-cycle gas turbine power plant produces work while having heat interaction only with the atmosphere. Does this violate the second law of thermodynamics?

6.6 A household refrigerator is in thermal communication solely with the atmosphere in a room at constant temperature. The refrigerator operates in a thermodynamic cycle receiving work from the external electrical mains, and exchanges heat only with the atmosphere. Is this a violation of the second law of thermodynamics?

REVERSIBILITY AND THE THERMODYNAMIC TEMPERATURE SCALE

7

7.1 Introduction

The second law of thermodynamics is on overt statement that the most cherished wish of an engineer to convert heat completely and continuously to work is unattainable. Laws of thermodynamics can never be violated because they are the laws of nature. Hence, with no alternative but to accept the second law of thermodynamics, an inquisitive engineer would probe deeper in search of answers to several logical questions that arise as a sequel to the restrictions imposed by the second law. Some of these interesting and important questions are:

- According to the Kelvin–Planck statement of the second law of thermodynamics, as it is not possible for a system operating in a thermodynamic cycle to convert heat transfer to it to an equal amount of work, what is the maximum conversion possible? In other words, what is the maximum possible efficiency of a heat engine? Can this limit of conversion be predicted? If the answer is 'yes', can the predicted limit of conversion be realized in practice? If not, why? What are the precautions to be taken in the design of an actual engine so that its efficiency tends towards that of an ideal engine—an engine that converts maximum percentage of heat to work?

- According to the Clausius statement of the second law of thermodynamics, as it is impossible to transfer heat from low- to high-temperature regions without external work being employed, what is the minimum work that can accomplish the task of transferring heat in the said direction under given conditions, theoretically? In other words, what is the maximum possible COP of a reversed heat engine? Can this maximum COP be predicted?

- If the answer is 'yes', can this predicted minimum work do the job in actual practice? If not why? What are the precautions to be taken in the design of a heat pump or a refrigerator to ensure that its performance tends towards the predicted maximum performance of an ideal reversed heat engine?

It is needless to emphasize here that the answers to the above questions have a very high economic and practical significance. This chapter is aimed at providing answers to some of these questions. Chapter 8 on 'entropy' provides answers to the remaining ones.

7.2 Reversible Heat Engine

The concept of a *reversible heat engine* is highly theoretical and imaginary in nature. A reversible heat engine is one which works most efficiently under given conditions. This means, while a reversible direct heat engine converts heat input to it to maximum work, a reversible reversed heat engine requires minimum work to transfer a given amount of heat from a low-to a high-temperature region. It is to be noted here that a reversible heat engine is only a mathematical model. No practical engine can actually be reversible, but as the engine is made to perform better with improved design and operating conditions, its performance tends towards the ideal performance.

What are the specialties of a reversible engine? To discuss this point, a reader must have a thorough understanding of the concept of reversibility. The following example, which may appear to be crude, is aimed at helping the reader to understand the concept.

Consider a centrifugal gas turbine. High-pressure, high-temperature gas entering through the inlet pipe expands in the turbine to provide shaft work and then it leaves the turbine as low-pressure low-temperature gas through the exit pipe. Let the efficiency of the turbine be η_T. Assume that this turbine is made to run in the opposite direction, that is, as a centrifugal compressor. This is practically possible with minimal modifications to the machine. In this case, the low-pressure, low-temperature gas enters the compressor (the entry and exit locations get interchanged—the exit of turbine becoming the inlet and the inlet of the turbine becoming the exit of the compressor) and gets compressed because of shaft work input (the direction of work is opposite to that in the turbine) and leaves it as a high-pressure high-temperature gas. When such an experiment is conducted, one finds that the efficiency of the machine while operating as a compressor, η_C, is always less than its efficiency, η_T, when it operates as a turbine. In other words, the work required to drive 1 kg air through the compressor under the given conditions is much more than the work delivered by the same kilogram of air as it flows through the device when it operates as a turbine. It may be generalized here that devices designed to operate in a particular direction when made to operate in the opposite direction can only work with an efficiency less than its value in the direction for which they have been designed. Suppose an unique device is imagined, which works with identical efficiencies in both the designed direction and its opposite. Such a device is called a *reversible device*.

With the above understanding, one can now imagine a heat engine, direct or reversed (i.e., in general, a system operating in a thermodynamic cycle), which works equally well in both the forward and the reverse directions. Such a heat engine is a reversible heat engine. The performance of a direct heat engine is measured by its η which is the ratio of work output to heat input. The performance of a heat pump is measured by its COP which is the ratio of heat output to work input. Before an attempt is made to compare the η with COP, the two must be reduced to the same ratio of either heat to work or work to heat. Thus, η is to be compared with the reciprocal of COP and not directly with COP.

The precise definition of a reversible heat engine is stated below:

A heat engine having heat interactions with two different temperature reservoirs is said to be reversible if its efficiency, η_R, while operating as a direct heat engine is equal to the reciprocal of its coefficient of performance $COP_{hp,R}$, when it works as a heat pump between the same two temperatures.

That is,
$$\eta_R = \frac{1}{(\text{COP})_{hp,R}} \tag{7.1}$$

Consider a reversible engine, E_R, shown in Figure 7.1*a*, operating between two constant temperature reservoirs at t_1 and t_2. While Q_{1R} and Q_{2R} are, respectively, the heat transfers to and from the heat engine, W_R is the work output of the engine. Suffix R indicates that all interactions are with the reversible heat engine.

Efficiency of this engine is given by
$$\eta_\text{R} = \frac{W_R}{Q_{1R}} \tag{i}$$

Let E_R be now made to work as a heat pump, HP_R, between the same two reservoirs, as shown in Figure 7.1*b*. Let the magnitude of heat transfer by this heat pump to the high-temperature reservoir be Q_{1R}. To make the comparison between the engine and the heat pump meaningful, the magnitudes of heat interactions with the high-temperature reservoir in both the cases are considered to be the same (Note, however, that Q_{1R} is negative for the heat pump, whereas it was positive for the direct heat engine). To accomplish this task, let the work input to the heat pump be $W_{R'}$ (while $W_{R'}$ is negative, W_R in the direct heat engine is positive) and heat transfer to it from the low-temperature reservoir be $Q_{2R'}$ (***Note:*** While $Q_{2R'}$ is positive, Q_{2R} in the direct heat engine is negative).

By definition,
$$(\text{COP})_{hp,R} = \frac{Q_{1R}}{W_{R'}} \tag{ii}$$

Substituting for η_R and $(\text{COP})_{hp,R}$ from (i) and (ii) in (7.1),
$$\frac{W_R}{Q_{1R}} = \frac{1}{Q_{1R}/W_{R'}} = \frac{W_{R'}}{Q_{1R}}$$

Therefore,
$$W_R = W_{R'} \tag{iii}$$

Applying the first law to HP_R,
$$-Q_{1R} + Q_{2R'} = -W_{R'}, \text{ or}$$
$$Q_{1R} - Q_{2R'} = W_{R'} \tag{iv}$$

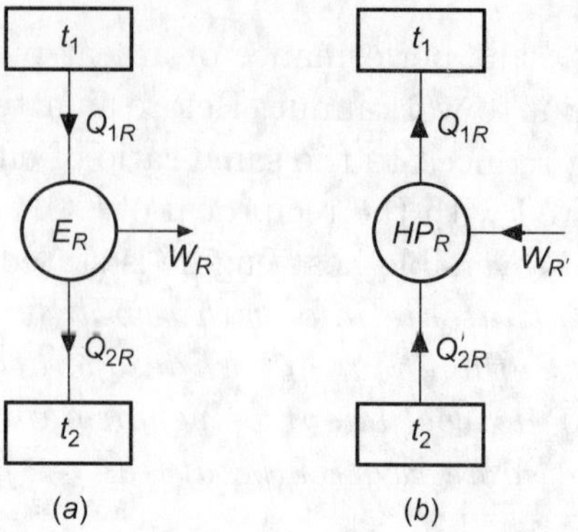

Fig. 7.1*a* Reversible heat engine; *b* heat engine working as a heat pump.

But, from (iii), $W_{R'} = W_R$, and hence (iv) can be written as,

$$Q_{1R} - Q_{2R'} = W_R \tag{v}$$

Applying the first law to E_R,

$$Q_{1R} - Q_{2R} = W_R \tag{vi}$$

Equation (v) minus (vi) gives

$$Q_{2R} = Q_{2R'} \tag{vii}$$

Therefore, the magnitudes of Q_{1R}, Q_{2R} and W_R remain the same for a reversible engine in both the directions of its operation—either as an engine or a heat pump.

This analysis clearly demonstrates the uniqueness of the reversible heat engine which may be generalized as follows: if a heat engine operating between two reservoirs at different temperatures is reversible, then the magnitudes of heat interactions it will have with each reservoir, and also the magnitude of work interaction it will have with the surroundings, are the same irrespective of whether the heat engine works as a direct heat engine or as a reversed heat engine (heat pump or refrigerator). But the directions of all the three interactions in a reversed heat engine are opposite to those in a direct heat engine.

Example 7.1

A reversible heat engine operates between two reservoirs at t_1 and t_2. A heat transfer of 500 kJ occurs from the reservoir at t_1 to the engine which rejects 187.5 kJ to the one at t_2. (a) Find the work output of the engine. (b) If this engine were to work as a heat pump between the same temperature limits and that it transfers out 500 kJ to the reservoir at t_1 find the magnitude of heat interaction with the reservoir at t_2 and the work input to the heat pump.

▶ **Data:**

(a) $Q_1 = 500$ kJ, $Q_2 = 187.5$ kJ.

▶ **Solution:**

(a)

$$\eta_R = \frac{(Q_l - Q_2)}{Q_1} = \frac{(500 - 187.5)}{500} = 0.625$$

$$\eta_R = 0.625 = \frac{W}{Q_1} = \frac{W}{500}$$

Therefore, $W = 500 \times 0.625 = 312.5$ kJ

▶ **Data:**

(b) $Q_1 = 500$ kJ.

When the reversible direct engine operates as a heat pump between the same two temperatures,

$$\eta_\mathrm{R} = \frac{1}{(\mathrm{COP})_\mathrm{HP}} \tag{7.1}$$

Therefore, $(\mathrm{COP})_\mathrm{HP} = \dfrac{1}{\eta_\mathrm{R}} = \dfrac{1}{0.625} = 1.6$

Also, $(\mathrm{COP})_\mathrm{HP} = \dfrac{Q_1}{W} = 1.6 = \dfrac{500}{W}$

Therefore, $W = 312.5$ kJ

From the first law, $Q_1 - Q_2 = W = 500 - Q_2 = 312.5$

Therefore, $Q_2 =$ Heat transfer from $t_2 = 500 - 312.5 = 187.5$ kJ

Alternatively The heat pump is operating between the same two temperature as the direct heat engine does. From the concept of reversible engine, the magnitudes of heat and work interactions in both cases have to be the same. Therefore, for the heat pump, $W = W$ in (a) $= 312.5$ kJ and $Q_2 = Q_2$ in (a) $= 187.5$ kJ.

7.2.1 Importance and Superiority of a Reversible Heat Engine

The following exercise demonstrates clearly the superiority of a reversible heat engine.

Consider a reversible heat engine R and any other heat engine X operating between the same two reservoirs as shown in Figure 7.2. Let the heat transfer to each engine from the high-temperature reservoir at t_1 be Q_1. The proposition is to analyse which engine is more efficient under the given conditions.

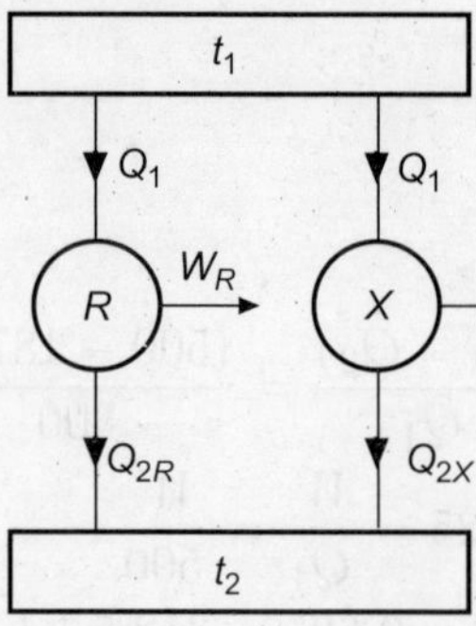

Fig. 7.2 Heat engines R and X operating between t_1 and t_2.

As it is not possible to guess which engine operates better, let it be assumed that engine X is more efficient than engine R.

By definition,
$$\eta_X = \frac{W_X}{Q_1} \text{ and } \eta_R = \frac{W_R}{Q_1}$$

As per the assumption made,
$$\eta_X > \eta_R.$$

Therefore,
$$\frac{W_X}{Q_1} > \frac{W_R}{Q_1}, \text{ or } W_X > W_R.$$

As shown in Figure 7.3, let engine R be made to function as a heat pump between the same two temperatures. R is a reversible engine, and, hence, as a heat pump, HP_R, it will have the same magnitudes of heat and work interactions with the reservoirs as it had while working as direct heat engine (Fig.7.2).

Engine X can drive the heat pump HP_R as the work delivered by X is more than the work required by the heat pump. A careful study of Figure 7.3 reveals that the combined system of the two engines and the reservoir at t_1 (shown by the dotted boundary) undergoes a cyclic process with $(Q_{2R} - Q_{2X})$ as the only heat interaction with the low-temperature reservoir at t_2, and producing a positive work of $(W_X - W_R)$. Inclusion of the high-temperature reservoir in the system boundary does not pose any problem as this reservoir experiences no change because the heat transfer to it by HP_R is equal to the heat transfer by the reservoir to X. Thus, the system, shown enclosed by the boundary S, executes a cycle whose efficiency, η, is given by,

$$\eta = \frac{\text{net work output}}{\text{heat input}}$$
$$= \frac{(W_X - W_R)}{(Q_{2R} - Q_{2x})} \qquad \text{(i)}$$

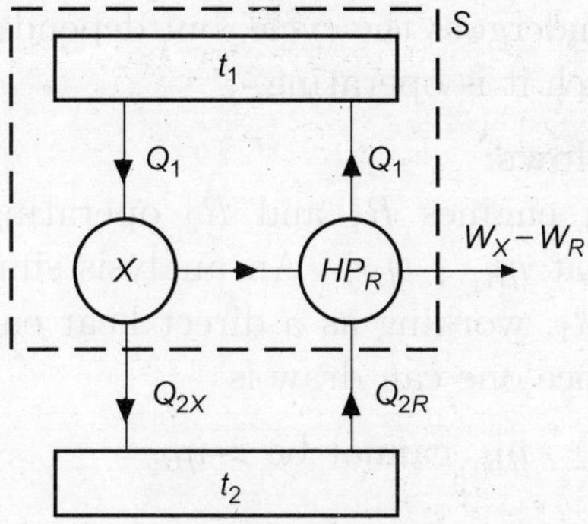

Fig. 7.3 Engine X driving engine R.

and, η as given by (i) = 100% because,

By applying the first law to X and HP_R,

$$W_X = Q_1 - Q_{2X}$$

and $$W_R = Q_1 - Q_{2R}, \text{ therefore}$$

$$W_X - W_R = (Q_{2R} - Q_{2X}).$$

That is, $$\frac{(W_X - W_R)}{(Q_{2R} - Q_{2X})} = 1 \qquad \text{(ii)}$$

Therefore, $$\eta = \frac{(W_X - W_R)}{(Q_{2R} - Q_{2X})} = 100\%$$

This is impossible according to the second law. There has been no flaw in the above analysis. Yet, it has led to this impossible situation. Hence, the one and only assumption made in arriving at this impossible observation, namely η_X ¿ η_R, has to be essentially wrong. Therefore, η_X cannot be greater than η_R. This fact, generally known as Carnot's theorem, is stated as follows:

7.2.1.1 Carnot's Theorem 1

No heat engine operating between two given temperature limits can be more efficient than a reversible engine operating between the same two temperatures.

This theorem provides answer to one of the questions posed earlier, namely, what is the maximum possible conversion of heat to work by a direct heat engine? It is only a reversible engine that can convert heat to maximum work. In other words, among all engines operating between two given temperatures, the reversible engine gives the maximum efficiency. Thus the superiority of the reversible engine is established.

A point will now be made that *all reversible heat engines operating between the given two temperatures have the same efficiency.* This means that the efficiency of a reversible heat engine does not depend on the mode of operation of the engine or on the nature of the working substance that undergoes the cycle, but depends only upon the temperatures of the reservoirs between which it is operating.

This can be proved as follows:
Consider two reversible heat engines R_1 and R_2 operating between two reservoirs at t_1 and t_2. Let us assume that η_{R_1} ¿ η_{R_2}. An analysis similar to the previous one can now be effected by making R_1, working as a direct heat engine, drive R_2, working as a heat pump. Then, the inference one can draw is

$$\eta_{R_1} \text{ cannot be } > \eta_{R_2} \tag{i}$$

As η_{R_1} cannot be ¿ η_{R_2} let it now be assumed that η_{R_2} ¿ η_{R_1}. An identical analysis as the previous one can now be effected, making R_2 drive R_1 working as a heat pump. The inference, now, would be the following:

$$\eta_{R_2} \text{ cannot be } > \eta_{R_1} \tag{ii}$$

The only possibility for (i) and (ii) to be simultaneously satisfied is that $\eta_{R_1} = \eta_{R_2}$. This fact is generally known as either *Carnot's Theorem 2* or *Corollary to Carnot's Theorem 1.*

7.2.1.2 Carnot's Theorem 2

This theorem, which has been proved above, is stated as follows:

All reversible heat engines operating between two given temperatures have the same efficiency.

Example 7.2

A reversible heat engine operating between two reservoirs at temperatures of t_1 and t_2 has an efficiency of 70%. A manufacturer of heat engines claims that he has designed an engine, X, which when operating between the above two temperatures develops 80 kJ work when supplied with 100 kJ of heat from the reservoir at t_1. Evaluate the validity of his claim.

▶ Solution:

From the data, $\eta_R = \eta_{max} = 0.7.$

And, $\eta_x = \dfrac{W}{Q_1} = \dfrac{80}{100} = 0.8.$

Therefore, $\eta_X > \eta_R.$

Engine R and engine X are operating between the same temperature limits of t_1 and t_2. According to Carnot's theorem, η_X cannot be greater than η_R. Therefore, engine X is not a viable proposition.

7.3 Reversible Process

Let us now examine the features of a reversible heat engine. A heat engine is a system operating in a cycle. If a heat engine is reversible, it means the thermodynamic cycle on which it operates has to be reversible. A thermodynamic cycle consists of more than one non-cyclic process. Hence, it is a logical conclusion that for a cycle to be reversible, each of its processes must be reversible. This makes it necessary to define and understand a reversible process.

7.3.1 Definition of a Reversible Process

A process is said to be reversible if, after the process is completed, means can be found to restore the system and all parts of the surroundings to the states they were in at the beginning of the process.

In other words, a process, 1–2, is said to be reversible when the changes in the system and all parts of the surroundings during this process are completely undone when process 2–1 is carried out after 1–2. In such a case, there will be no evidence left either in the system or in the surroundings to indicate later that the system had actually been subjected to two processes in opposite directions. In such a case, it is obvious that the heat and/or work interactions between the system and its surroundings have to be the same in magnitudes during both the forward and reverse directions, but for their directions which have to be opposite. Only when it is so do the changes in the system and the surroundings during one direction of the process get totally undone in the opposite direction.

Hence, if a process is reversible, it is possible to change the direction of the process by making infinitesimal changes in the conditions that control the process. It means to emphasize that it is necessary that the reversible process occurs very slowly with no dissipative effects and that the gradients which control the process are infinitesimally small so that the system is always in equilibrium with its surroundings. These explanations indicate that a reversible process is a quasi-static process or a fully resisted process.

The conditions under which a process can be declared a reversible process are theoretical in nature. It can hence be said that reversible processes are only ideal processes and hence can be treated as only mathematical models. All natural processes and almost all man-made processes do not fall under this category.

7.4 Irreversible Process

A process which is not reversible, is called an irreversible process. It is necessary to understand the factors that make a process irreversible. Once these factors are identified, means can be found in practice to either minimize or eliminate them, if possible, so that in the limit, the process tends towards a reversible process. Existence of dissipative effects or lack of equilibrium between the system and the interacting surroundings are responsible for making a process irreversible. Therefore, in a general context, the factors that make a process irreversible are:

- Friction (solid or fluid)
- Unresisted expansion
- Heat transfer with a finite temperature difference
- Mixing
- Combustion

The above list is not exhaustive. These are factors that are very commonly encountered, especially in mechanical engineering. There are several other factors that can make a process irreversible. For example, magnetic hysteresis, flow of current through a resistor, plastic deformation of materials, hydraulic jump in flow of water in channels etc are a few other factors.

Some authors advocate the use of a general method, as stated below, to show that each of the factors listed above actually makes a process irreversible: (i) Describe a process P, which has one of the factors in the above list. Note the initial and final states of the system and the magnitudes and directions of heat and/or work interactions during process. (ii) Assume P is reversible. Therefore, during ꟼ the heat and work interactions between the system and the surroundings are the same as in P but are opposite in direction, and at the end of the process ꟼ, the system and all parts of the surroundings are brought to the states they were in at the beginning of the process P. (iii) Devise a mechanism to work in a thermodynamic cycle using ꟼ as one of the processes. (iv) Show that such a device turns out to be a 100% efficient engine and so it is impossible. (v) Infer, hence, that the assumption that P is reversible has to be wrong.

The readers may look into such qualitative proofs in some books on thermodynamics to satisfy their curiosity. It will, however, be explained later, in Chapter 8, as to how the property *entropy* can be used to quantitatively assess the irreversibility, or otherwise, of a process.

A very brief description, based purely on common sense, is given below to impress upon the reader that the above-listed factors really make a process irreversible.

7.4.1 Friction

Consider a rough inclined plane and a block at the top of its inclined surface as a system. Assume that both the block and the plane are good conductors of heat. Let the process considered be the sliding of the block down the inclined plane against friction. At the end of the process, the block will be at a lower level and the temperature of both the block and the plane would have increased because of the friction between the sliding block and the inclined plane. During this process, the interaction is only between the block and the plane. As the plane and the block together constitute the system, there can be no heat and work interactions across the boundary of the considered system. If this process of sliding of the block down the plane were to be reversible, it should be possible, in the return process, to push the block up the inclined plane to its original position, using no work from outside, and that, at the end of the process, the block and the plane should cool down to the original temperature without any heat transfer to the surroundings. Common sense dictates that such a process is impossible. Hence, this process is irreversible as it is not possible to find means to bring the system and surroundings to the states they were in at the start of the process. It can thus be inferred that friction makes a process irreversible. Dissipative effects associated with friction is the cause for making the process irreversible. If friction were not to be present during the above-cited process it would be possible to bring the system and all parts of the surroundings to their initial states by simply guiding the block at the bottom of the plane to rise up to the same top of the plane.

7.4.2 Unresisted Expansion

Consider an insulated rigid vessel divided into two parts by a thin membrane. Presume that the left-hand compartment is filled with a gas at a certain pressure and temperature and that the right-hand side compartment is evacuated. Let the membrane be now ruptured instantaneously. Considering the gas as the system, it undergoes an unresisted expansion as the gas expands suddenly, with no resistance, to fill the entire vessel. The system does not experience any heat (because of insulation) or work interaction (unresisted expansion) but its volume increases and its pressure decreases. It is impossible that this process is reversible because without work interaction it is not possible in the reverse direction to restore the system to its initial volume and pressure. The system can be restored to its initial state only with the help of work interaction from the surroundings. In such a case, the surroundings will undergo a state change which is not in accordance with the concept of a reversible process. Hence, unresisted expansion makes a process irreversible. This has to be so as in an unresisted expansion there exists no mechanical equilibrium between the system and the surroundings.

7.4.3 Heat Transfer with Finite Temperature Difference

Suppose a body, A, at a temperature t_A, is brought in communication with another body, B, at a temperature t_B. Let $t_A > t_B$. A spontaneous natural process of heat transfer occurs from A to B. This process of heat transfer cannot be reversible as, in the reverse direction, it is impossible to transfer heat from B to A (as $t_B < t_A$), spontaneously. The irreversibility, in this case, is because of the nonexistence of thermal equilibrium between the interacting bodies.

7.4.4 Combustion

Consider a mixture of fuel and air in a rigid insulated container as the system. Ignite the mixture with the help of an infinitesimal spark. Combustion of the fuel occurs instantaneously. There is generally a rise in pressure and temperature of the contents of the container. There is no heat or work interaction with the surroundings. If this process were to be reversible, it should be possible in the reverse direction to get back the fuel-air mixture at its original pressure and temperature from the hot combustion products, with no heat and work interactions from the surroundings. This is an impossibility, and, hence, combustion makes a process irreversible. The irreversibility, here, is due to the nonexistence of chemical equilibrium during the process.

7.4.5 Mixing

Gases in different conditions may be mixed with no heat or work interactions but, in the reverse direction, they cannot be separated out to their initial states without help from outside. As the system and the surroundings cannot be returned to their initial states, a process involving mixing is always irreversible.

The above examples and explanations lead one to think that almost all processes are irreversible in nature. However, if one is interested in making a model of a reversible heat engine for achieving maximum efficiency, the person must work on a cycle consisting of only reversible processes. Hence, from this point of view, it is necessary to identify some processes which are reversible.

A few examples of a reversible processes are:

(i) Frictionless motion of solids.

(ii) Slow, frictionless, adiabatic expansion, or compression, of gases and vapours.

(iii) Slow, frictionless, isothermal compression, or expansion, of gases and vapours.

7.5 Carnot Heat Engine

By now, it is very clear that all processes that constitute a reversible cycle have to be individually reversible. In other words, no process can have in it any factor that makes it irreversible. A reversible cycle was conceived as early as 1824 by Sadi Carnot and is appropriately named, after him, the *Carnot cycle* or *Carnot heat engine*. The Carnot heat engine is only a mathematical model, as it comprises ideal reversible processes. Hence, no practical engine can reach the performance of this engine in reality. How close can an engine be made to reach Carnot's performance? This is dictated by economic and practical constraints. The details of the Carnot heat engine are given below.

As shown in Figure 7.4, consider a horizontal cylinder comprising an insulated barrel and a frictionless insulated piston. The left-face of the cylinder is a plate made of a good conducting material. Let the piston–cylinder mechanism contain, when the piston is at its left extreme position, a certain mass of gas at a temperature t_a, and a pressure p_a indicated by state 'a' on the corresponding p–V diagram shown below the piston–cylinder mechanism. Assume that a high-temperature reservoir at t_1 (where t_1 is infinitesimally higher than t_a so that for all practical purposes it may be assumed that $t_1 = t_a$) is readily available and that it is communicated with the left face of the cylinder. Heat is transferred at a slow rate from the reservoir to the gas in the cylinder. This is because of the infinitesimal temperature difference between the two. The gas expands because of heat and the piston moves out slowly such that the work done by the gas equals the heat transfer to the gas. This maintains the gas at the same temperature. Under the circumstances, it may be said that the gas undergoes a reversible isothermal expansion process, the temperature remaining constant at t_1. Let this expansion be continued until the gas reaches a predetermined state 'b'. Let the heat transfer to the gas during this process be Q_1. When the gas reaches state 'b' let the high-temperature reservoir be removed and in its place let an insulating material be brought in instantaneously. The expansion of the gas continues with no heat transfer and consequently the pressure and temperature of the gas decrease. This expansion process is reversible and adiabatic and let it continue until the piston reaches the right extreme position. At this instant, the gas is at state 'c' with temperature t_c and pressure p_c. Assume that a low-temperature reservoir at temperature t_2 (where t_2 is infinitesimally less than t_c so that for all practical purposes it may be assumed that $t_2 = t_c$) is available and when the gas reaches state 'c' the insulator is removed and in its place the low-temperature reservoir is brought in, instantaneously.

As the piston moves slowly to the left, the gas is compressed and to maintain the gas at the same temperature t_2, heat is slowly transferred out to the reservoir. This process is reversible and isothermal. Let it continue until the gas reaches the predetermined state 'd'. Let the heat rejected during this process be Q_2. When the gas reaches the state 'd' the low-temperature reservoir is suddenly removed and the insulator is brought in communication with the cylinder head. The compression now continues with no heat interaction. During this reversible adiabatic compression process, the temperature and pressure of the gas increase. The cycle is completed when the gas reaches its initial state 'a' and the piston is at its extreme left position. The gas is now ready to perform the next cycle in the same sequence.

Thus, the Carnot cycle consists of two reversible isotherms (a–b, the expansion process at t_1, and c–d, the compression process at t_2) and two reversible adiabatics (b–c, the expansion process, and d–a, the compression process). The engine operating on this cycle between two reservoirs at different temperatures t_1 and t_2 must have the highest efficiency as it operates on a reversible cycle.

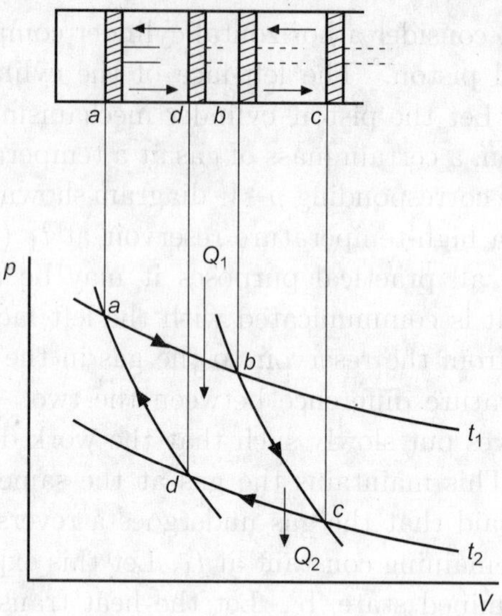

Fig. 7.4 Carnot heat engine (carnot cycle).

7.5.1 Remarks on Carnots Engine

Carnot's engine is only a mathematical model to which all engines must tend for best performance. A practical realization of the Carnot engine is difficult because of several constraints. They are:

(i) The restriction that the piston should move slowly makes the power output from such an engine very small and of no practical use.

(ii) During isothermal processes, not only must the temperature of the working substance be uniform throughout but also the difference between it and the two reservoirs must be infinitesimal. The constraint of infinitesimal temperature difference demands large engines even for small heat transfer rates. For a given heat transfer rate across such small temperature differences, the required contact areas are large. Economic constraints may not permit such large engines for small power output.

(iii) It is assumed that, in a Carnot engine, there is no friction between the piston and cylinder. This condition is highly impossible to comply with in practice. Friction can be decreased by using precise machining processes and a proper lubrication system but can never be eliminated totally.

(iv) Perfect conductors and insulators of heat are not available in practice. This poses a problem of choice of materials in making the cylinder and piston.

(v) A careful study of the Carnot cycle reveals that the changes from the isothermal expansion process to adiabatic and from the isothermal compression process to adiabatic have to be carried out instantaneously. Any delay may not take the working substance through the required cycle. Practical limitations do not permit such instantaneous changes.

7.6 Internal and External Reversibilities

During a given process, the factors that make the process irreversible may exist either only in the system, or only in the surroundings or in both the system and surroundings.

(i) If the factors exist only in the system and not in the surroundings, the process is said to be internally irreversible but externally reversible.

(ii) If the factors exist only in the surroundings and not in the system, the process is internally reversible but externally irreversible.

(iii) If the factors exist in both the system and the surroundings the process is said to be totally irreversible.

For example, consider a system undergoing a Carnot cycle explained above. During the isothermal expansion process, the system is not only at a steady temperature but also has heat interaction with a reservoir which is at the same temperature as its own. Similarly, during the isothermal compression process the system is at a steady temperature and is transferring heat to the reservoir at a temperature equal to its own temperature. In other words, the heat transfers in a Carnot cycle occur practically with no temperature difference between the system and the reservoirs. Hence, the Carnot heat engine or the Carnot cycle is both internally and externally reversible.

Let us now discuss another example. Consider a system undergoing a cycle consisting of two reversible isotherms and two reversible adiabatics. The system may be said to be undergoing a Carnot cycle. During the isothermal expansion process, although the system temperature is steady and uniform at t_1 let the high-temperature reservoir transferring heat to the system be at t_H where $t_H > t_1$. Similarly, during the isothermal compression process let the system be at a steady and uniform temperature of t_2 but let it transfer heat to the low-temperature reservoir at t_L where $t_L < t_2$. In such a cycle, although there are no factors internally that make the cycle irreversible, there are such factors outside the system (heat transfer with finite temperature difference). Hence, although the cycle undergone by the system is internally reversible, it is externally irreversible.

Example 7.3

One kilogram of water at atmospheric pressure and 100°C is evaporated completely, without a change in either its pressure or temperature, by heating it over an electric heater maintained at a temperature of 700°C. Simultaneously, the vapour is collected and condensed in a vessel cooled by the surroundings at 27°C such that the water condenses at 100°C to bring it to its initial state. As the water has been restored to its initial state after two processes in opposite directions, the process it has undergone is reversible. Comment on (a) this statement and (b) on internal and external reversibility.

▶ Solution:

In the forward direction, water is evaporated at a constant temperature of 100°C. In the reverse direction, water is condensed back to liquid state at the same temperature of 100°C. After the two process, the system has returned to its initial state. Therefore, it can be said that the system has undergone an *internally reversible cycle*.

But, during the forward process, heat is transferred to water from the heater at 700°C. Thus, the heat transfer is across a finite temperature difference. Similarly, in the return process, heat is transferred from the water vapour at 100°C to the surroundings at 27°C.

Thus, the two heat transfers occur with a finite temperature difference. Hence, although the process undergone by water is internally reversible, it cannot be externally reversible.

7.7 Thermodynamic Temperature Scale

7.7.1 The Concept

It is only a reversible engine that can convert heat transferred to it to maximum work output. In other words, efficiency of a reversible engine is the maximum. However, it has not been made known quantitatively as to what this maximum conversion is. The concept of thermodynamic temperature scale provides the answer for this important question. In addition to this, it also eliminates the lacunae in the Celsius temperature scale. The thermometers with Celsius temperature scale are not calibrated at each 'degree' marking on them. Generally, they are calibrated at certain reproducible temperatures in the required range, such as the ice point, the steam point etc. What the thermometer reads in between two calibrated markings depends to a large extent on the properties of the thermometric substance and the mode of operation of the thermometer. Hence it is unlikely that two thermometers of different types, or for that matter even of the same type, inserted into a system at a particular temperature would give the same reading. There is thus a need for developing a temperature scale obtained through a device whose performance neither depends on the properties of the working substance nor on its mode of operation. A reversible heat engine, such as the Carnot engine, provides such a device as its performance neither depends on the properties of the working substance nor on the mode of operation but only on the temperature limits between which it operates. For example, under identical conditions, a reversible engine operating with steam as the working substance will have the same efficiency as the one operating on air or any other working substance.

Let an externally reversible engine operate between a high-temperature reservoir at a temperature t_H and a low-temperature reservoir at a temperature t_L. Let the heat transfer to the engine be Q_H and heat transferred out of the engine be Q_L. According to Carnot's theorems, the efficiency of such an engine is the maximum and its value depends only on the temperatures between which the engine is operating and not on the properties of the substance that undergoes the cycle. Hence, the efficiency has to be a function, say ψ, of the two temperatures only..

That is, $$\eta_R = 1 - \left(\frac{Q_L}{Q_H}\right) = \psi(t_H, t_L) \text{ or}$$

$$\left(\frac{Q_H}{Q_L}\right) = [1 - \psi(t_H, t_L)]^{-1} = \phi(t_H, t_L) \tag{7.2}$$

where ϕ is another function related to ψ.

This means that the ratio of heat transfer to and from a reversible heat engine is a function of the two reservoir temperatures only.

Consider, three reversible heat engines A, B and C working between a high-temperature reservoir at t_1 and a low-temperature reservoir at t_2, as shown in Figure 7.5.

Engine A operates between the reservoir at t_1 and an intermediate system at a fixed temperature of t_X, where $t_1 > t_X > t_2$. Let the heat transfer from the reservoir at t_1 to engine A be Q_1 and the heat transferred by it to the system at t_X be Q_X. Let the work produced by A be W_A. Engine B operates between the system at t_X and the reservoir at t_2. With the assumption that all the heat transferred by engine A goes to engine B through the system at constant temperature, t_X, heat transferred to B is Q_X. Let heat transferred by engine B to the reservoir at t_2 be Q_2, and the work produced by it be W_B. Engine C operates between the two reservoirs at t_1 and t_2. Let heat transferred to engine C be Q_1, which is equal to the heat transferred to engine A. The heat transferred out of this engine has to be Q_2 as, according to Carnot's second statement, the efficiency of engine C must be same as the combined efficiencies of engine A and B. (Note that engines A and B together with the constant temperature system at t_X is equivalent to another reversible heat engine operating between t_1 and t_2 and hence if this combination transfers out Q_2 to the reservoir at t_2, then engine C must also transfer out Q_2 to the same reservoir.)

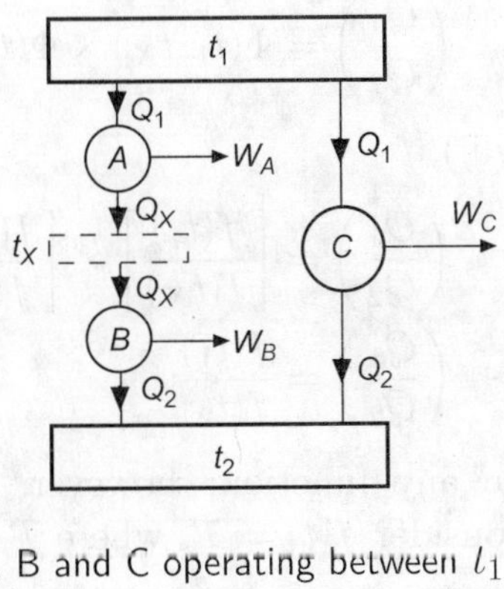

Fig. 7.5 Reversible heat engines A, B and C operating between t_1 and t_2.

According to (7.2), the following relations hold good for the three reversible engines A, B and C, respectively:

$$\left(\frac{Q_1}{Q_X}\right) = \Phi(t_1, t_X) \tag{i}$$

$$\left(\frac{Q_X}{Q_2}\right) = \Phi(t_X, t_2) \tag{ii}$$

$$\left(\frac{Q_1}{Q_2}\right) = \Phi(t_1, t_2) \tag{iii}$$

But,
$$\left(\frac{Q_1}{Q_X}\right) \times \left(\frac{Q_X}{Q_2}\right) = \left(\frac{Q_1}{Q_2}\right) \tag{iv}$$

Replacing the ratios of heat transfers in (iv) by the corresponding functions of temperatures from (i), (ii) and (iii),

$$\Phi(t_1, t_X) \times \Phi(t_X, t_2) = \Phi(t_1, t_2) \tag{v}$$

While the left-hand side of (v) is a function of three temperatures t_1, t_X and t_2 the right-hand side is a function of two temperatures, t_1 and t_2, only. Hence, the value of the left-hand side product has to be independent of the value of temperature t_X. This gives an insight into the nature of the function Φ which can satisfy such a condition. Therefore,

$$\Phi(t_1, t_X) = \frac{f(t_1)}{f(t_X)}, \text{ and} \tag{vi}$$

$$\Phi(t_X, t_2) = \frac{f(t_X)}{f(t_2)} \tag{vii}$$

where f is some other function.

The right-hand side of (v), from (iii), is equal to $\left(\frac{Q_1}{Q_2}\right)$.

Therefore,
$$\left(\frac{Q_1}{Q_2}\right) = \Phi(t_1, t_X) \times \Phi(t_x, t_2) \tag{viii}$$

Combining (vi), (vii) and (viii),

$$\left(\frac{Q_1}{Q_2}\right) = \left[\frac{f(t_1)}{f(t_X)}\right] \times \left[\frac{f(t_X)}{f(t_2)}\right]$$

that is,
$$\left(\frac{Q_1}{Q_2}\right) = \frac{f(t_1)}{f(t_2)} \tag{ix}$$

Here, f, in general, can be any function; however, it is universally accepted, as suggested by Lord Kelvin, to consider $f(t) = T$, where T is in thermodynamic or kelvin temperature scale.

Therefore,
$$\left(\frac{Q_1}{Q_2}\right) = \frac{T_1}{T_2} \tag{7.3}$$

7.7.2 Definition of the Thermodynamic Temperature Scale

The thermodynamic temperature scale is defined by (7.3). The ratio of two temperatures on thermodynamic temperature scale is equal to the ratio of heat transfers to and from an externally reversible heat engine operating between the two given temperatures.

7.7.3 Unit of Thermodynamic Temperature

Although (7.3) defines the thermodynamic temperature scale, it does not indicate how big or small one degree on this scale is. It is internationally accepted that the kelvin, the unit of thermodynamic temperature, is $\frac{1}{273.16}$ of the thermodynamic temperature of the triple-point of water, T_{tp}.

For a reversible heat engine operating between any temperature, T, and the triple-point temperature T_{tp}, according to (7.3),

$$\left(\frac{Q}{Q_{tp}}\right) = \frac{T}{T_{tp}} \text{ or } T = T_{tp}\left(\frac{Q}{Q_{tp}}\right)$$

Or,
$$T = 273.16\left(\frac{Q}{Q_{tp}}\right) \tag{7.4}$$

Thus, (7.4) gives the numerical value in kelvin of any temperature, T, as a function of the heat transfer, Q, to a reversible heat engine at this temperature and the heat transfer from the engine, Q_{tp}, at the triple-point temperature, T_{tp}, of water.

To get a value of T on thermodynamic temperature scale, as indicated by (7.4) a reversible heat engine must be made to operate between T and T_{tp} and the magnitudes of heat transfer to and from the engine are measured. This appears to pose problems, as a reversible heat engine is only a mathematical model and hence is not possible to measure the heat transfers in actual practice. Chapter 10 shows that Q and Q_{tp} can actually be estimated using the properties of steam available in the saturated steam tables.

It is clear from (7.4) that the thermodynamic temperature scale agrees well with the ideal gas thermometer scale, or the absolute temperature scale. Hereafter, while lower-case 't' is used to indicate the temperature on celsius scale, upper-case 'T' is used to denote the temperature on thermodynamic temperature scale.

The triple-point temperature of water is arbitrarily assigned a value of 273.16 on thermodynamic temperature scale (also called kelvin scale) or 0.01° on celsius scale (for details about the triple-point of water, refer to Chapter 10). Hence, the relation between the temperatures on celsius scale and kelvin scale is the following:

$$T = t + 273.15 \tag{7.5}$$

For example, the boiling point of water at standard atmospheric pressure on celsius scale is 100°C. Hence, on kelvin scale, it is equal to 100 + 273.15 = 373.15 K.

(***Note*:** While, in celsius scale, the temperature is indicated by °C, on kelvin scale, it is just represented by K and not by °K. For example, 500 kelvin is represented as 500 K and not as 500°K)

In engineering applications,

$$T = t + 273 \tag{7.6}$$

is considered to be sufficiently accurate.

The thermal efficiency of a heat engine in general is given by

$$\eta = 1 - \frac{Q_2}{Q_1}$$

If the engine is externally reversible, according to (7.3),

$$\frac{Q_2}{Q_1} = \frac{T_2}{T_1}$$

and hence the efficiency of a reversible heat engine is given by

$$\eta_R = 1 - \frac{T_2}{T_1}$$

It may also be noted that

$$\eta_R \neq 1 - \frac{t_2}{t_1}, \text{ but } \eta_R = 1 - \frac{(t_2 + 273)}{(t_1 + 273)}$$

Carnot's statement dictates that no engine can be more efficient than a reversible heat engine, which operates between the same two temperatures. In general, if Q_1 and Q_2 are the heat transfers to and from any heat engine operating between T_1 and T_2 then,

$$1 - \frac{Q_2}{Q_1} \leq 1 - \frac{T_2}{T_1} \tag{7.7}$$

When the equality holds, the engine must be a reversible engine, otherwise the engine is irreversible. However, the magnitude of the left-hand side quantity can never be greater than that the right-hand side. If so, such an engine cannot operate. Equation (7.7) can thus be used to identify whether a given engine is reversible, irreversible or impossible.

The kelvin temperature scale provides answer to yet another important question posed at the beginning, namely, the maximum possible conversion of heat to work by a heat engine quantitatively. Supposing the engine operates between two reservoirs at T_1 and T_2, ($T_1 > T_2$), and the heat transfer to the engine at the temperature T_1 is Q_1, then the maximum possible work output from the engine is limited to $Q_1\left(1 - \frac{T_2}{T_1}\right)$. No engine can give more work than this under given conditions.

Let us now consider a reversible reversed heat engine (either a refrigerator or a heat pump) operating between T_1 and T_2, ($T_1 > T_2$), with heat interactions of Q_1 and Q_2, respectively, at these two temperatures.

By definition,
$$(\text{COP})_{ref} = \frac{Q_2}{(Q_1 - Q_2)} \tag{6.6}$$

Dividing the numerator and denominators on the right–hand side of (6.6) by Q_2,

$$(\text{COP})_{ref} = \frac{1}{\left[\left(\frac{Q_1}{Q_2}\right) - 1\right]}$$

As the engine is reversible, the ratio of heat transfer is equal to the ratio of temperatures.

Therefore,
$$\text{COP}_{rev,ref} = \text{COP}_{max,ref} = \frac{1}{\left[\left(\frac{T_1}{T_2}\right) - 1\right]} = \frac{T_2}{(T_1 - T_2)} \tag{7.8}$$

It can similarly be shown that,

$$\text{COP}_{rev,hp} = \text{COP}_{max,hp} = \frac{1}{\left[1 - \left(\frac{T_2}{T_1}\right)\right]} = \frac{T_1}{(T_1 - T_2)} \tag{7.9}$$

Example 7.4

A heat engine operates between 527°C and 27°C. Find its efficiency, if (a) the engine is reversible and (b) the engine is irreversible.

▶ **Data:**

$T_1 = (527 + 273) = 800$ K, $T_2 = (27 + 273) = 300$ K

▶ **Solution:**

(a) The engine is given to be reversible. Therefore, its efficiency is given by

$$\eta_R = 1 - \frac{Q_2}{Q_1} = 1 - \frac{T_2}{T_1} = 1 - \frac{300}{800} = 0.625$$

(b) The engine is irreversible. Therefore its efficiency is given by

$$\eta = 1 - \frac{Q_2}{Q_1} = \frac{W}{Q_1}, \text{ and } \eta \neq \left(1 - \frac{T_2}{T_1}\right)$$

As the values of at least any two from among Q_1, Q_2 and W are not given the efficiency of the engine cannot be evaluated. However, according to Carnot's statement, the efficiency will be less than 0.625.

Example 7.5

A reversible reversed heat engine operates between −13°C and 37°C. Find its COP as (a) a refrigerator and (b) as a heat pump.

▶ Data:

$T_1 = 273 + 37 = 310$ K and $T_2 = 273 + (-13) = 260$ K and the reversed engine is given to be reversible.

▶ Solution:

$$\text{COP}_{rev,ref} = \text{COP}_{max,ref} = \frac{T_2}{(T_1 - T_2)} = \frac{260}{(310 - 260)} = 5.2 \qquad (7.8)$$

$$\text{Similarly,} \quad \text{COP}_{rev,hp} = \text{COP}_{max,hp} = \frac{T_1}{(T_1 - T_2)} = \frac{310}{(310 - 260)} = 6.2 \qquad (7.9)$$

Illustrative Problems

7.1 Check for the presence, or otherwise, of the factors of irreversibility in the following cases and classify them as either reversible or irreversible. Also, wherever possible, comment on the internal and external reversibility. The system to be considered in each case is underlined. Assume, in all cases, the properties are uniform throughout the system at any given instant.

(a) *A mass of gas* is compressed slowly in a frictionless leak-proof non-conducting piston–cylinder mechanism.

(b) *A block* slides down a rough inclined plane slowly.

(c) One kilogram of *water* is heated very slowly from 30°C to 40°C by an inflow of heat.

(d) One kilogram of *water* is heated slowly but adiabatically from 30°C to 40°C by churning with a paddle wheel mounted on a frictionless pulley.

(e) One kilogram of *water* is heated rapidly from 30°C to 40°C.

(f) One kilogram of a *mixture of petrol and oxygen* contained in a sealed vessel immersed in a large pool of water at 27°C reacts spontaneously.

(g) A stream of *steam* at 100°C mixes with liquid water at 25°C in a non-conducting vessel.

(h) A heavy piston confining a gas in a vertical cylinder is held in position by stops. When the stops are removed the piston falls rapidly because it has much more weight than the initial pressure of the gas can support.

► **Solution:**

(a) The process is reversible as there are no factors of irreversibility (no friction, slow and hence not an unresisted process, no heat transfer because of insulation, no mixing, no combustion). It is worth recalling here that this process is actually one of the four processes of the Carnot cycle.

(b) The process is irreversible. The factor that makes the process irreversible is the solid friction between the sliding block and the inclined plane.

(c) This process can be either reversible or irreversible. It depends upon how slowly the water is heated.

(i) Let it be assumed that the water is heated slowly, step by step, by bringing it in communication, successively, with several reservoirs, each with a temperature infinitesimally higher than the previous one starting from the one at (30 + dT)°C to the last one at (40 + dT)°C. When heated in this manner the factor heat transfer with finite temperature difference, which makes the process otherwise irreversible, is almost eliminated. In such a case, the process is reversible.

(ii) Let it now be assumed that the water is in contact with a single reservoir at 40°C or slightly higher. The heating would still be slow under this condition. In such a case, at the beginning of the process the irreversibility is more as the temperature difference is more and the irreversibility keeps decreasing as the temperature difference reduces with the heating of water. The process is thus irreversible.

(d) The process is irreversible. The temperature of water gets increased only because of the fluid friction between the paddle wheel and the water. The process is internally irreversible but is externally reversible as there are no factors in the surroundings that makes the process irreversible.

(e) Rapid heating requires a large temperature difference between the water being heated and the external system supplying heat transfer, such as an electric heater or a burner. This heat transfer with finite temperature difference makes the process irreversible.

(f) This is an irreversible process for two factors.

(i) Combustion makes the process irreversible, and (ii) Because of the spontaneous combustion there will be a rise in temperature of the vessel containing the combustion products. The vessel transfers heat to the pool of water due to finite temperature difference between the vessel and the water pool. The process is internally irreversible due to combustion and heat transfer with finite temperature difference. However, the process is externally reversible as during the heating process the temperature of the large pool of water hardly changes.

(g) This process is irreversible for two reasons namely mixing and heat transfer with finite temperature difference between the steam and water.

(h) The process is possibly irreversible as the piston moves down rapidly. The system is not in mechanical equilibrium with the surroundings during the process as outside pressure is high compared to the system pressure.

7.2 Consider three hypothetical heat engines, A, B and C, each operating between 1000 and 300 K. When each engine involves itself with a heat interaction of 1000 kJ with the high-temperature reservoir, it is claimed that while A develops a work of 600 kJ, B and C develop 700 and 800 kJ, respectively. Use the Carnot statement and identify the engines A, B and C as reversible, irreversible or impossible.

▶ **Solution:**

As shown in Figure 7.6 the three engines A, B and C operate between the same temperature limits of $T_1 = 1000$ K and $T_2 = 300$ K. No engine operating between these two temperature can be more efficient than a reversible engine whose efficiency is given by

$$\eta_{\text{max}} = 1 - \left(\frac{T_2}{T_1}\right) = 1 - \left(\frac{300}{1000}\right) = 0.7$$

Considering engine A, $\quad Q_1 = 1000 \text{ kJ}, W_A = 600 \text{ kJ}$

Therefore, $\quad \eta_A = \frac{W_A}{Q_1} = \frac{600}{1000} = 0.6$

Alternatively

Applying the first law to engine A,

$$Q_{2A} = Q_1 - W_A = 1000 - 600 = 400 \text{ kJ}$$

Therefore, $\quad \eta = 1 - \left(\frac{Q_{2A}}{Q_1}\right) = 1 - \left(\frac{400}{1000}\right) = 0.6$

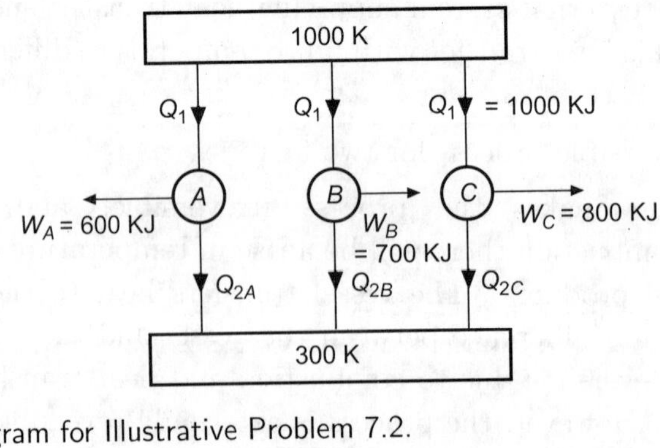

Fig. 7.6 Schematic diagram for Illustrative Problem 7.2.

As $\eta_A < \eta_{max}$, the heat engine A is irreversible.

Similarly, for engine B,

$$Q_{2B} = Q_1 - W_B = 1000 - 700 = 300 \text{ kJ}$$

and hence,
$$\eta_B = 1 - \left(\frac{300}{1000}\right) = 0.7$$

As $\eta_B = \eta_{max}$, engine B is reversible.

Considering engine C,

$$Q_{2C} = Q_1 - W_C = 1000 - 800 = 200 \text{ kJ}$$

Therefore,
$$\eta_C = 1 - \left(\frac{200}{1000}\right) = 0.8$$

As $\eta_C > \eta_{max}$ the heat engine C is an impossibility.

7.3 A direct heat engine A and a reversed heat engine B operate between 177°C and 27°C. The COP of B as a heat pump is 2.5. A drives B. The magnitudes of heat interaction of A and B with the reservoir at 27°C are 200 and 50 kJ, respectively. The combined work output of A and B is 20 kJ. Identify whether the heat engine A is reversible or irreversible.

▶ **Data:**

Refer to Figure 7.7, where $T_1 = 273 + 177 = 450$ K, $T_2 = 273 + 27 = 300$ K, $Q_{2A} = 200$ kJ, $Q_{2B} = 50$ kJ, $\text{COP}_{\text{HP,B}} = 2.5$

▶ **Solution:**

$$\text{COP}_{\text{HP,B}} = 2.5 = \frac{Q_{1B}}{W_B}$$

Therefore,
$$Q_{1B} = W_B \times 2.5 \tag{i}$$

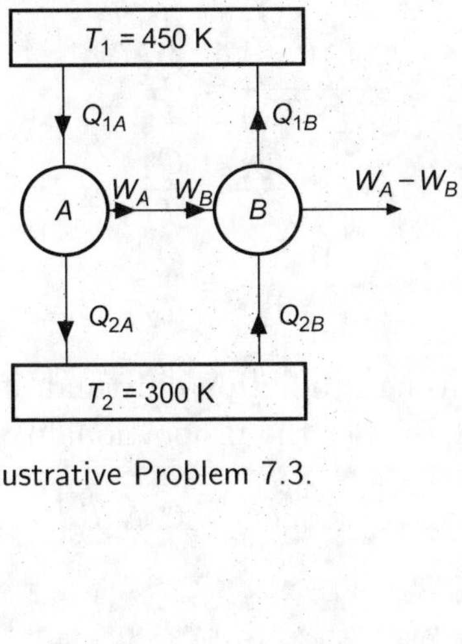

Fig. 7.7 Schematic diagram for Illustrative Problem 7.3.

where W_B is the work required by the reversed heat engine B.
Applying the first law to engine B,

$$Q_{1B} = W_B + Q_{2B} \tag{ii}$$

Eliminating Q_{1B} between (i) and (ii), and substituting the value for Q_{2B}

$$W_B \times 2.5 = W_B + 50,$$

that is,
$$W_B = 33.3 \text{ kJ}$$

From data, $W_A = W_B + W_{output} = 33.3 + 20 = 53.3$ kJ.
Applying the first law to engine A,

$$Q_{1A} = Q_{2A} + W_A = 200 + 53.3 = 253.3 \text{ kJ}$$

$$\eta_A = \frac{W_A}{Q_{1A}} = \frac{53.3}{253.3} = 0.21$$

But,
$$\eta_{max} = 1 - \frac{T_2}{T_1} = 1 - \frac{300}{450} = 0.33$$

As $\eta_A < \eta_{max}$, engine A is irreversible.

7.4 Consider a reversible engine in the outer space which operates between T_1 and T_2 (temperatures are in kelvin and $T_1 > T_2$). The only way in which heat can be transferred out from the engine is by radiation. The rate at which heat is radiated is directly proportional to fourth power of kelvin temperature and the area of the radiating surface. Show that for a given power output and a given T_1, the area of the radiator is minimum when $\frac{T_2}{T_1} = 0.75$.

Let Q_1 be the heat transfer to the reversible engine from the reservoir at T_1 K.

Then,
$$\eta = \frac{W}{Q_1} = \frac{(T_1 - T_2)}{T_1}$$

Or,
$$\frac{W}{(T_1 - T_2)} = \frac{Q_1}{T_1} \tag{i}$$

For a reversible engine
$$\frac{Q_1}{T_1} = \frac{Q_2}{T_2} \tag{ii}$$

Between (i) and (ii),
$$\frac{W}{(T_1 - T_2)} = \frac{Q_2}{T_2} \tag{iii}$$

The heat rejected, Q_2, is through radiation only and, it is given to be proportional to area A and fourth power of T_2. Let the proportionality constant be C. Then,

$$Q_2 = \text{C A } {T_2}^4 \tag{iv}$$

Eliminating Q_2 between (iii) and (iv)

$$\frac{W}{(T_1 - T_2)} = \text{C } A\ \frac{{T_2}^4}{T_2} = C\ A\ {T_2}^3 \tag{v}$$

Therefore,
$$A = \frac{W}{C}\left(\frac{1}{{T_2}^3 T_1 - {T_2}^4}\right) \tag{vi}$$

For a given W and T_1, A is minimum when $\dfrac{dA}{dT_2} = 0$. Hence, from (vi),

$$\frac{dA}{dT_2} = \left(\frac{W}{C}\right)\left[\left(\frac{3{T_2}^2 T_1 - 4{T_2}^3)}{(T_1 {T_2}^3 - {T_2}^4}\right)^2\right] = 0$$

Here, $(T_1 {T_2}^3 - T_2^4)$ cannot be zero as $T_1 > T_2$

Hence, $$3{T_2}^2 T_1 - 4{T_2}^3 = 0$$

Thus, $$3T_1 = 4T_2 \text{ or } \frac{T_2}{T_1} = \frac{3}{4} = 0.75.$$

7.5 Two Carnot engines are operating in series between two reservoirs at T_1 and T_2 as follows: heat rejected from engine A operating between the reservoir at T_1 and an intermediate body at a constant temperature of T_3 is picked up by engine B operating between T_3 and T_2, where $T_2 < T_3 < T_1$. Show that T_3 is the arithmetic mean of T_1 and T_2 if the work output of each engine were to be the same

► **Solution:**

Figure 7.8 shows the schematic of the two engines A and B operating in series. As A and B are reversible engines,

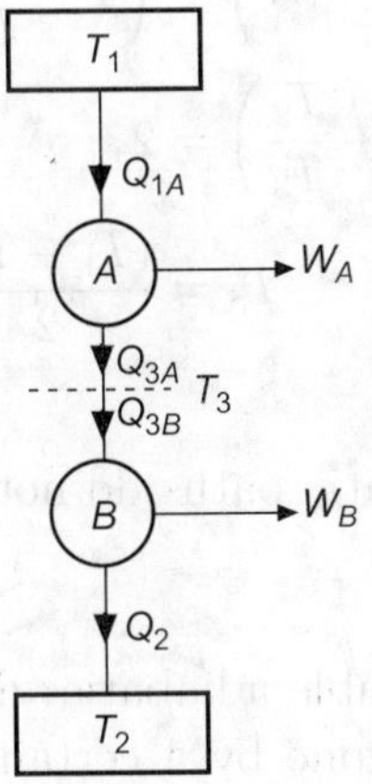

Fig. 7.8 Carnot engines A and B in series.

$$\eta_A = \left(1 - \frac{T_3}{T_1}\right) = \frac{W_A}{Q_{1A}}$$

and therefore $$W_A = Q_{1A}\left(1 - \frac{T_3}{T_1}\right)$$

Similarly, $$W_B = Q_{3B}\left(1 - \frac{T_2}{T_3}\right)$$

From data, $$W_A = W_B.$$

Therefore, $$Q_{1A}\left(1 - \frac{T_3}{T_1}\right) = Q_{3B}\left(1 - \frac{T_2}{T_3}\right)$$

that is, $$\left(\frac{Q_{1A}}{Q_{3B}}\right)\left(1 - \frac{T_3}{T_1}\right) = \left(1 - \frac{T_2}{T_3}\right)$$

Heat rejected by A is picked up by engine B to maintain T_3 constant. Therefore, $Q_{3A} = Q_{3B}$. Substituting for Q_{3B} in the above relation,

$$\left(\frac{Q_{1A}}{Q_{3A}}\right)\left(1 - \frac{T_3}{T_1}\right) = \left(1 - \frac{T_2}{T_3}\right)$$

From the definition of thermodynamic temperature scale,

$$\left(\frac{Q_{1A}}{Q_{3A}}\right) = \frac{T_1}{T_3}$$

Therefore, $$\left(\frac{T_1}{T_3}\right)\left(1 - \frac{T_3}{T_1}\right) = \left(1 - \frac{T_2}{T_3}\right)$$

i.e., $$\left(\frac{T_1}{T_3} - 1\right) = \left(1 - \frac{T_2}{T_3}\right)$$

or, $$\left(\frac{T_1}{T_3} + \frac{T_2}{T_3}\right) = 2$$

Hence, $$T_3 = \frac{(T_1 + T_2)}{2}$$

7.6 Show that two reversible adiabatic paths do not intersect on a state diagram.

► **Solution:**

Let it be assumed that two reversible adiabatics do intersect on a state diagram. Consider now a cyclic process undergone by a certain system with these two intersecting adiabatics and a reversible isothermal process. This cycle is shown on a $p-V$ diagram in Figure 7.9.

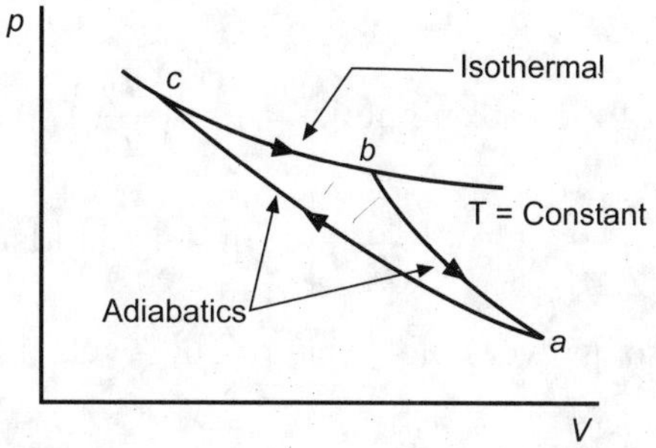

Fig. 7.9 Assumed cycle with two reversible adiabatics and an isothermal.

In Figure 7.9, 'ac' and 'ba' are two reversible adiabatics intersecting at 'a' and 'cb' is an isothermal. The area of the diagram, *cbac*, gives the net work output of the cycle. Heat is transferred to the cycle only during the isothermal process at a constant temperature of T. There is no heat rejection during the cycle as the other two processes happen to be adiabatics. Hence, this turns out be a PMM II as all the heat transferred to the system at a fixed temperature is converted to work during this cycle.

This is impossible according to the Kelvin–Planck statement of the second law. Therefore, the assumption made that two reversible adiabatics do intersect on a state diagram must be wrong.

SUMMARY

- A reversible heat engine is one whose efficiency while operating directly between two given reservoirs at fixed but different temperatures is equal to the reciprocal of the COP when it operates as a heat pump between the same two reservoirs.

- According to Carnot's theorems: (a) no engine can be more efficient than a reversible engine, when both are operating between given two temperatures and (b) all reversible engines operating between given two temperature reservoirs have the same efficiency.

- Friction, un-resisted expansion, heat transfer with finite temperature difference, combustion and mixing make a process irreversible.

  (In the following explanations suffixes 1 and 2 stand for high-temperature and low-temperature reservoir, respectively.)

- The thermodynamic temperature scale is defined with respect to a reversible heat engine by $\dfrac{T_1}{T_2} = \left(\dfrac{Q_{1R}}{Q_{2R}}\right)$.

- Maximum efficiency of a heat engine is $= 1 - \left(\frac{T_2}{T_1}\right)$.

  In general $\left[1 - \left(\frac{Q_2}{Q_1}\right)\right] \leq \left[1 - \left(\frac{T_2}{T_1}\right)\right]$, equality holds good for reversible cycle and the inequality to irreversible cycle. The cycle in which $\left[1 - \left(\frac{Q_2}{Q_1}\right)\right] > \left[1 - \left(\frac{T_2}{T_1}\right)\right]$ is ar impossible cycle.

- $\text{Max(COP)}_{ref} = \frac{T_2}{(T_1 - T_2)}$ and $\max(\text{COP})_{Hp} = \frac{T_1}{(T_1 - T_2)}$.

Problems

7.1 Check for the presence, or otherwise, of the factors of irreversibility in the following cases and classify them as reversible, irreversible or might-be-reversible. Also, wherever possible, comment on the internal and external reversibilities.

(a) One kilogram of water is evaporated at constant temperature by an inflow of heat.

(b) One kilogram of water is evaporated at constant temperature by an inflow of work.

(c) A mass of air in a constant volume container is heated slowly from 50 to 100°C by an inflow of heat.

(d) Hot stagnant gases in a non-conducting frictionless piston–cylinder mechanism expand as the piston moves out slowly.

(e) Hot turbulent gases in the cylinder of a water-cooled automobile engine expand as the piston moves out rapidly.

(f) A mass of air contained in a rigid vessel is sprayed with some quantity of petrol. After sometime the vessel contains a homogeneous mixture of air and petrol vapour. Consider the contents as the system.

(g) A compressed mixture of air and petrol vapour confined in a cylinder by a piston is ignited by a negligible electric spark. After sometime, the cylinder contains a homogeneous mixture of CO_2, CO, O_2 and N_2.

7.2 A heat engine develops a work of 21.5 kJ when supplied with a heat transfer of 90 kJ from a reservoir at 227°C. The sink temperature is 27°C. What is the efficiency of the engine. Is the engine reversible or irreversible?

7.3 A refrigerator operates between 38°C and −7°C. It requires twice as much work for a given refrigerating effect as required by a reversible refrigerator operating between the same two temperature limits. How much power does the refrigerator need to produce a cooling at a rate of 420 kW.

7.4 A series combination of two reversible engines operate between 180°C and 20°C. What is the intermediate temperature if (a) the engines develop same work and (b) the engines perform with the same efficiency.

7.5 Shown in Figure 7.10 is a heat engine 'A' operating between T (in kelvin) and 300 K and a heat pump 'B' operating between 250 and 300 K. Find the value of T given that $\eta_{\text{A}} = 0.7\ \eta_{\text{max}}$, $Q_1 = 1000$ kJ, $(Q_2 + Q_4) = 1590$ kJ, $(\text{COP})_B = 0.6 \times (\text{COP})_{max}$ and $\text{W}_{net} = 190$ kJ.

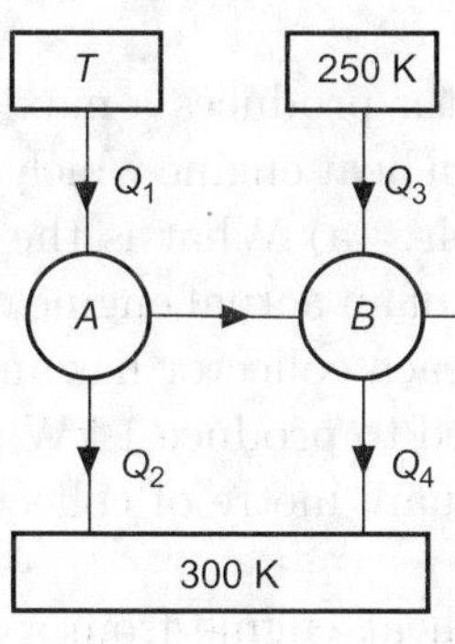

Fig. 7.10 Diagram for Problem 7.5.

7.6 A reversed heat engine is used for winter heating and summer cooling to maintain the house-interior temperature steady at 25°C through the year. The average heat loss from the roof and the walls of the house is estimated to be 5 kW per degree temperature difference between the inside and the outside of the house. While during winter the outside temperature is 5°C, it is 35°C in summer. Find the minimum power required to run the device in summer and in winter.

7.7 The COP of a heat pump is 5. (a) Evaluate the rates of heat transfer to and from the pump if the power supplied to it is 50 kW. (b) The heat transfer from the pump is used to heat water flowing through the radiators of a building. Calculate the mass flow rate of water if its temperature increases from 50°C to 70°C as it flows through the radiator. Assume that there is no change in velocity of water and that the specific heat of water is constant at 4.189 kJ/kg K.

7.8 Which of the following is the more effective way of increasing the efficiency of a Carnot engine? (i) Increasing T_1 and keeping T_2 constant or (ii) decreasing T_2 and keeping T_1 constant, where T_1 and T_2 are, respectively, the temperatures of high-and low-temperature reservoirs.

7.9 A Carnot engine operates between two reservoirs at temperatures T_1 K and T_2 K, where $T_1 > T_2$. The work output of the engine is 0.6 times the heat rejected. Given that the difference between the temperatures of the two reservoirs is 200 degrees calculate (a) the temperatures of the two reservoirs and (b) efficiency of the engine.

7.10 (a) A reversible heat engine E operates between two reservoirs at 600°C and 40°C. The engine E drives a reversible refrigerator R which operates between 40°C and −20°C. The heat transfer to E is 2000 kJ and the network output of the combined E and R plant is 350 kNm. Evaluate the heat transfer to the refrigerant and the net heat transfer to the reservoir at 40°C. (b) Redo the above problem assuming E and R are irreversible and that the efficiency of E and the COP of R are each 40% of their maximum values.

7.11 A certain flat plate solar collector produces a maximum temperature of 90°C. The energy collected is to be used in a heat engine which employs the atmosphere at 27°C as the low-temperature reservoir. (a) What is the maximum possible efficiency of the engine? (b) If the efficiency of an actual engine operating under these conditions is 40% of that in (a), and that each collector has an exposed area of 2 m^2, calculate the number of collectors required to produce 1 kW power. Assume the average rate of solar energy collected per square metre of collector area is 200 W.

7.12 For a given heat transfer to a heat engine from a high-temperature reservoir at a given temperature, the work output from the engine increases with a decrease in the low-temperature reservoir. To take advantage of this fact, a person proposes to run his engine between reservoirs at 500 K and 250 K at a place where the surroundings are at 300 K. He also proposes to maintain the low-temperature reservoir at 250 K by running a refrigerator. Assess his proposal if (a) the heat engine and the refrigerator are both reversible and (b) the performance of both the engine and the refrigerator are 80% of their theoretical maximum values.

7.13 A heat engine drives a heat pump. The heat transfer from the heat engine and the pump is used to heat water flowing through the radiators of a building. The efficiency of the heat engine is 27% and the COP of the heat pump is 4. Calculate the heat transfer to the radiator water for every kJ of heat transfer to the engine.

7.14 Two Carnot engines are operating in series between two reservoirs at T_1 and T_2 as follows: heat rejected from engine A operating between the reservoir at T_1 and an intermediate body at a constant temperature of T_3 is picked up by engine B operating between T_3 and T_2, where $T_2 < T_3 < T_1$. Show that T_3 is the geometric mean of T_1 and T_2 if the efficiency of each engine were to be the same.

7.15 An inventor claims that his heat engine operating between 750°C and 25°C develops a power of 75 kW. To maintain the temperature of the source steady at 750°C, a fuel of calorific value (heating value)42 000 kJ/kg is burnt at a rate of 0.12 kg/min. Assess the validity of the claim.

7.16 A reversible heat engine operates between two reservoirs at 700°C and 50°C. The engine drives a reversible refrigerator operating between 50°C and −25°C. The heat transfer to the engine is 2500 kJ. The net work output of the engine–refrigerator combination is 400 kJ. Find the heat transfer to the refrigerant and the net heat transfer to the reservoir at 50°C.

7.17 Redo Problem 7.16 assuming that the engine and the refrigerator are irreversible and that the efficiency of the engine and the COP of the refrigerator are each 45% of their maximum values.

ENTROPY - AVAILABLE AND UNAVAILABLE ENERGY

8

8.1 Introduction

The second law of thermodynamics and its corollaries have been introduced in Chapters 6 and 7 for systems that execute thermodynamic cycles. These chapters also elaborate on the ways in which the law can be used for several purposes such as (i) assessing the feasibility or otherwise of thermodynamic cycles, (ii) classifying the thermodynamic cycles as reversible, irreversible or impossible and (iii) for evaluating the maximum performance of a cycle operating between given two reservoirs. The subject does not end here because in their stated forms the second law and its corollaries cannot be used to analyse non-cycli processes. It is therefore necessary to know the form the law takes when extended to non-cyclic processes and how it can be used to assess quantitatively the feasibility or otherwise of these processes and to classify them as reversible, irreversible or impossible. The necessary extension of the second law to non-cyclic processes introduces an important property, 'entropy', and hence the second law for non-cyclic processes is stated using this property.

8.2 Some Important Statements on $\dfrac{\delta Q}{T}$

A few important statements are first made before defining and explaining entropy and how it is used to analyse processes from the stand-point of the second law of thermodynamics. These statements are then proved with the knowledge gained thus far in thermodynamics. They are then used to define and explain entropy. The first statement in the series of three is Clausius's inequality statement.

8.2.1 Statement 1: Clausius Inequality

When a system executes a cyclic process, the integral around the cycle of $\frac{\delta Q}{T}$ is less than or equal to zero. This statement, popularly known as Clausius's inequality statement, is symbolically written as

$$\oint \frac{\delta Q}{T} \leq 0$$

Let the system enclosed by the system boundary S, in Figure 8.1, undergo a cyclic process. At a given instant, let δQ and δW be the heat and work interactions the system has with its surroundings.

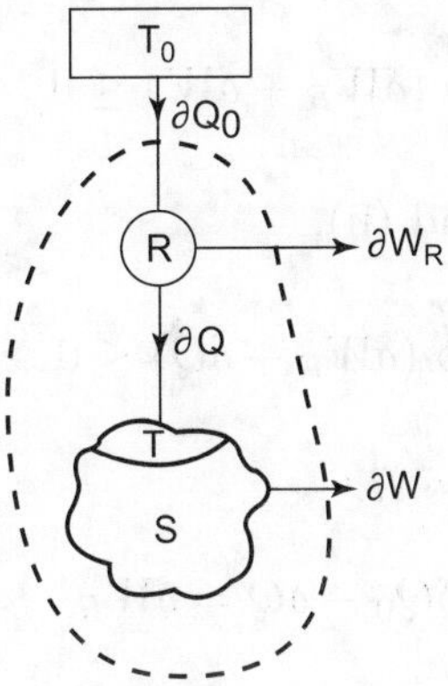

Fig. 8.1 Diagram to prove the Clausius inequality.

Let T be the temperature of the system at this instant. It is necessary to note here that the directions of heat and work interactions (δQ and δW) are taken to be positive, only to get a generalized result and it need not be misconstrued that the system considered is already violating the second law. Although the system can exchange δQ directly with a reservoir at T_0, let it be assumed that this heat interaction occurs through a reversible heat engine R operating between the reservoir at T_0 and that part of the system boundary at temperature T such that, while δQ_0 is the heat interaction between R and the reservoir, δQ is the heat interaction between R and the system. Under these conditions, let the work output of R be δW_R. The reversible engine R is considered to be so small that it performs a finite number of cycles in the time taken by the system S to undergo one cycle. Thus, δQ is only an infinitesimal heat interaction between R and the system at temperature T in the period during which R undergoes one cycle. Conceptually, R cannot be a single heat engine but a series of several small reversible engines. Each engine comes into effect sequentially. Imagine that the first engine has a heat interaction of δQ_0 with the reservoir at T_0 and δQ with the system at temperature T. The second one, while having the same heat interaction of δQ_0 with T_0, has an interaction of δQ with the system whose temperature could have been changed to $T + \mathrm{d}T$. Similarly, the third engine operates between T_0 and $T + 2\mathrm{d}T$ with δQ_0 and δQ, respectively, and so on until S undergoes a full cycle.

Applying the first law to the system, S,

$$\oint \delta Q = \oint \delta W \tag{i}$$

Consider, now, the system inside the boundary shown by broken lines. This system also undergoes a cyclic change (engine R undergoing a finite number of cycles for each cycle undergone by S) having heat interaction with only the reservoir at T_0. In accordance with the Kelvin–Planck statement of the second law, the net work output of this system cannot be positive, and, hence,

$$\oint (\delta W_R + \delta W) \le 0 \tag{ii}$$

Eliminating δW between (i) and (ii),

$$\oint (\delta W_R + \delta Q) \le 0 \tag{iii}$$

Applying the first law to engine R,

$$\delta Q_0 - \delta Q = \delta W_R \tag{iv}$$

As R is a reversible engine, by the definition of thermodynamic temperature scale,

$$\frac{T_0}{T} = \frac{\delta Q_0}{\delta Q} \quad \text{or } \delta Q_0 = \delta Q\left(\frac{T_0}{T}\right) \tag{v}$$

Eliminating δQ_0 between (iv) and (v),

$$\delta Q\left(\frac{T_0}{T}\right) - \delta Q = \delta W_R \tag{vi}$$

Substituting for δW_R from (vi) in (iii)

$$\oint \left\{\left[\delta Q\left(\frac{T_0}{T}\right) - \delta Q\right] + \delta Q\right\} \le 0 \text{ or}$$

$$\oint \delta Q\left(\frac{T_0}{T}\right) \le 0 \tag{vii}$$

In (vii), T_0 is always a positive number and, therefore,

$$\oint \frac{\delta Q}{T} \le 0 \tag{8.1}$$

8.2.2 Statement 2

When a system undergoes an internally reversible cycle the integral around the cycle of $\left(\frac{\delta Q_R}{T}\right)$ is always equal to zero. This may be symbolically written as

$$\oint\left(\frac{\delta Q_R}{T}\right) = 0$$

As shown in Figure 8.2, let a system execute an internally reversible cycle 1–A–2–B–1. From Clausius, inequality,

$$\oint\left(\frac{\delta Q_R}{T}\right) \leq 0 \tag{i}$$

Suppose the system now undergoes the cycle 1–B–2–A–1, in the opposite direction. Applying Clausius inequality to this cycle,

$$\oint\left(\frac{\delta Q'_R}{T}\right) \leq 0 \tag{ii}$$

In (i), δQ_R represents the heat interaction while the system executes the cycle in the given direction, and in (ii) $\delta Q'_R$ represents the heat interaction while the system undergoes the cycle in the opposite direction. Both cycles are reversible and one is the reversed cycle of the other. By the concept of reversible cycles, the magnitudes of $\delta Q'_R$ and δQ_R have to be necessarily the same but their directions are opposite and hence,

$$\delta Q'_R = -\delta Q_R \tag{iii}$$

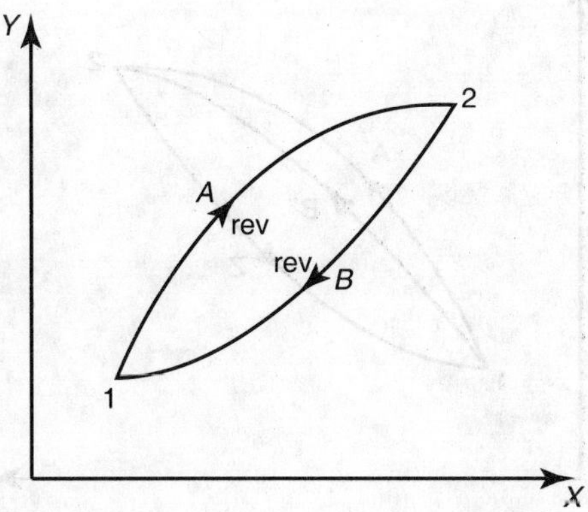

Fig. 8.2 Diagram to show $\oint\left(\frac{\delta Q_R}{T}\right) = 0$.

Substituting for $\delta Q'_R$ from (iii) in (ii),

$$\oint\left(\frac{-\delta Q_R}{T}\right) \leq 0 \text{ or } \oint\left(\frac{\delta Q_R}{T}\right) \geq 0 \quad \text{(iv)}$$

For (i) and (iv) to be simultaneously true the only possibility is that

$$\oint\left(\frac{\delta Q_R}{T}\right) = 0 \quad (8.2)$$

8.2.3 Statement 3

When a system undergoes a change of state, the value of the integral $\left(\frac{\delta Q_R}{T}\right)$ is independent of the path between the given two states. In other words, suppose a system changes from state 1 to 2 through several reversible processes A, B, C etc. The value of $\int\left(\frac{\delta Q_R}{T}\right)$ between states 1 and 2 is the same for each reversible path.

$$\text{That is,} \quad \int_A\left(\frac{\delta Q_R}{T}\right) = \int_B\left(\frac{\delta Q_R}{T}\right) = \int_C\left(\frac{\delta Q_R}{T}\right) \cdots \text{ etc.}$$

As shown in Figure 8.3, consider a system that changes state from 1 to 2 as the system is subjected to reversible process A. Assume that there exists another reversible process Z which can restore the system back to its initial state.

1–A–2–Z–1 is a reversible cycle. Therefore,

$$\oint\left(\frac{\delta Q_R}{T}\right) = 0$$

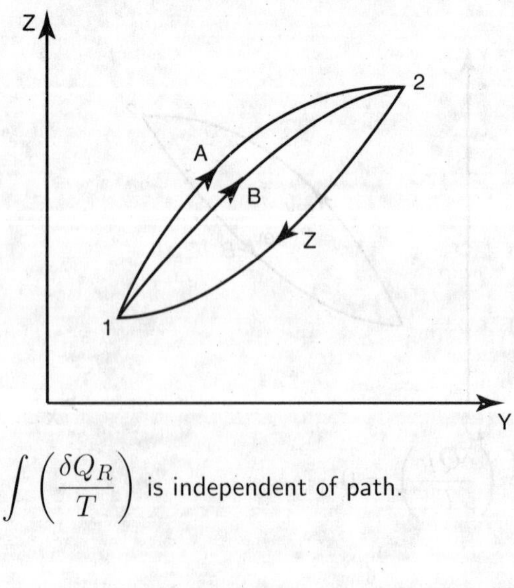

Fig. 8.3 Diagram to show $\int\left(\frac{\delta Q_R}{T}\right)$ is independent of path.

This cyclic integral can now be replaced by two linear integrals, one for processes A and the other for Z. Hence,

$$\int_{1}^{2-A}\left(\frac{\delta Q_R}{T}\right)+\int_{2}^{1-Z}\left(\frac{\delta Q_R}{T}\right)=0 \qquad \text{(i)}$$

Now, consider another reversible process B which causes the same state change as in process A. When process Z is carried out, after process B, the system will have executed a reversible cycle 1–B–2–Z–1.

Therefore, for this cycle,

$$\int_{1}^{2-B}\left(\frac{\delta Q_R}{T}\right)+\int_{2}^{1-Z}\left(\frac{\delta Q_R}{T}\right)=0 \qquad \text{(ii)}$$

Equation (i) minus (ii) gives

$$\int_{1}^{2-A}\left(\frac{\delta Q_R}{T}\right)-\int_{1}^{2-B}\left(\frac{\delta Q_R}{T}\right)=0 \text{ or } \int_{1}^{2-A}\left(\frac{\delta Q_R}{T}\right)=\int_{1}^{2-B}\left(\frac{\delta Q_R}{T}\right) \qquad (8.3)$$

Thus, $\int\left(\frac{\delta Q_R}{T}\right)$ is the same for the two reversible processes A and B between states 1 and 2. It can similarly be proved that for any other reversible path between 1 and 2 the value of the integral remains the same.

8.3 Entropy

It is evident from the above statements that $\left(\frac{\delta Q_R}{T}\right)$ is unique in the sense that not only its change in a cyclic process is zero (8.2) but also its change between two state points is independent of the path between them (8.3). By definition, a magnitude related to a system can be a property of a system if its change in a cyclic process is zero and that its change between any two states is independent of the path between them. Therefore, $\left(\frac{\delta Q_R}{T}\right)$ has to be a property of a system and this property is named entropy. Thus, entropy can be defined as follows: entropy, S, of a system is a property whose change between two states 1 and 2 is given by,

$$\Delta S = S_2 - S_1 = \int_{1}^{2}\left(\frac{\delta Q_R}{T}\right) \qquad (8.4)$$

or

$$\mathrm{d}S = \left(\frac{\delta Q_R}{T}\right) \qquad (8.5)$$

8.3.1 Remarks on Entropy

(i) Equation (8.4), or (8.5) in differential form, defines entropy. To be precise, it defines the change in entropy.

(ii) Similar to energy, a datum state is required to measure or evaluate the value of entropy at any other state of a system. If entropy is assigned a value (generally it is assigned as zero) at the selected datum, the entropy at any other state, relative to the datum state, can be calculated using the definition. The entropy at a given state can either be higher or lower than the assigned value of entropy at the datum state.

(iii) In any process, the magnitude of heat transfer required to bring about a certain change depends on the mass of the system. Hence, it is easy to infer from the definition of entropy that entropy is an extensive property. Therefore, specific entropy 's' is equal to S/m, where m is the mass of the system.

(iv) In accordance with its definition, the unit of entropy is J/K or kJ/K. Hence, the unit of specific entropy has to be J/kg K or kJ/kg K.

8.3.2 Similarities between Energy and Entropy

1. It is indeed useful to recall here that the first law of thermodynamics was stated first with reference to the systems undergoing thermodynamic cycles only. It was then extended to systems undergoing non-cyclic processes with the help of energy, the newly introduced property. Similarly, the second law has also been stated first with reference to systems undergoing cyclic processes only. It will be shown in this chapter that the extension of the second law to non-cyclic processes is made using another important property, entropy, introduced for the purpose. It may be said that, while the concept of energy was needed to extend the first law to non-cyclic processes the concept of entropy is required to extend the second law to non-cyclic processes. The property entropy is essential and so important that, in almost all subsequent chapters, this property is used for several meaningful and essential analyses.

2. The word energy is used freely in everyday life, although may not always be with the same meaning as in science or engineering. This familiarity perhaps makes one to feel that energy is something easy to comprehend, though in reality it need not necessarily be so. On the contrary, entropy is an unfamiliar word which is not used as frequently as energy in everyday life. This may, hence, lead to an uneasy feeling that it is something difficult to understand.

3. Similar to energy, entropy is also not a physical entity. It is also abstract in nature and is only a calculable property.

4. Energy is a property whose change in any process is equal to the difference between the magnitudes of the two interactions—net heat and net work. The mathematics required to understand this is very simple. A higher level concept in mathematics, namely integration, is essential to define entropy as in (8.4). This might further complicate matters to make the reader fear that entropy is something very difficult to comprehend. In reality, it is not so. It is as difficult or as easy as understanding energy.

   The process of study can be simplified, if, instead of struggling to know what entropy is, attention is focused on understanding what it can do. These two properties, energy and entropy, are important and essential to understand and analyse several happenings in science and technology to the extent that it will not be an exaggeration to state, 'Thermodynamics is the science of Energy and Entropy'.

Example 8.1

According to the Clausius inequality statement $\oint \frac{\delta Q}{T} < 0$ for an irreversible thermodynamic cycle. Therefore, there is a decrease in entropy of the system undergoing an irreversible cycle. Comment on this statement.

▶ Solution:

By definition, $ds = \frac{\delta Q_R}{T}$ and not $\frac{\delta Q}{T}$. Hence, when $\oint \frac{\delta Q}{T}$ is less than zero, it does not necessarily mean that $\oint \frac{\delta Q_R}{T} < 0$. Entropy, being a property, always shows zero change in a cycle, reversible or irreversible. The Clausius statement does not deal with entropy.

Example 8.2

10 kJ of heat is transferred from a reservoir at 300°C to a gas confined in a frictionless piston–cylinder mechanism, and as a result, the gas expands isothermally at 100°C. Is ΔS for the gas equal to $\int \frac{\delta Q}{T}$? Can ΔS for the reservoir be evaluated? If so, what is its value?

▶ Solution:

The process is externally irreversible because of heat transfer with finite temperature difference. However, it is internally reversible. Hence, ΔS for the system is equal to $\int \frac{\delta Q}{T}$ where $T = (100 + 273) = 373$ K. Therefore,

$$\Delta S_{sys} = \int \frac{\delta Q}{T} = \frac{Q}{T} = \frac{10}{373} = 2.68 \times 10^{-2} \text{ kJ/K}$$

Heat is transferred out of the reservoir at a fixed temperature and, hence, the process is also internally reversible as far as the reservoir is considered. Therefore, $\int \frac{dQ}{T}$ can also be used to calculate the entropy change of the reservoir. In such a case, T is to be taken as 300°C. Here Q is negative as it is from the reservoir. Therefore,

$$\Delta S_{res} = \int \frac{-\delta Q}{T} = \frac{-Q}{T} = \frac{-10}{(273+300)} = -1.7452 \times 10^{-2}\ \text{kJ/K}$$

Example 8.3

Reanalyse Example 8.2 if the reservoir temperature is also equal to 100°C.

▶ **Solution:**

In this case, the process is externally and internally reversible as the heat transfer occurs without finite temperature difference. Also there are no factors which otherwise would make the process irreversible. As in Example 8.2,

$$\Delta S_{sys} = \int \frac{\delta Q}{T} = \frac{Q}{T} = \frac{10}{373} = 2.68 \times 10^{-2}\ \text{kJ/K and}$$

$$\Delta S_{res} = \int \frac{-\delta Q}{T} = \frac{-Q}{T} = \frac{-10}{373} = -2.68 \times 10^{-2}\ \text{kJ/K.}$$

Example 8.4

Reanalyse Example 8.3 given that during the process, the gas is stirred using a paddle wheel.

▶ **Solution:**

In this case, while the process is externally reversible it is not internally reversible because of the stirring process which involves fluid friction between the blades of the paddle wheel and the gas. Therefore,

$$\Delta S_{res} = \int \frac{-\delta Q}{T} = \frac{-Q}{T} = \frac{-1}{373} = -2.68 \times 10^{-3}\ \text{kJ/K}$$

and

$$\Delta S_{sys} \neq \int \frac{\delta Q}{T}$$

as the gas undergoes an irreversible process.

8.3.3 Entropy as a Coordinate

Entropy, being a property, can be used as one of the coordinates in representing state diagrams. It shall be seen in Chapters 9 and 10 that $T-S$ and $h-s$ diagrams are widely used in analysing several processes of interest in thermodynamics.

In the state diagram, in Figure 8.4, while temperature is marked on the Y axis, entropy is marked on the X axis. This is known as the $T-S$ diagram. Let 1–2 be a reversible process undergone by a system between an initial state, 1, and a final state, 2. $1-a$ and $2-b$ are the ordinates at the initial and final states, respectively. Let us consider the hatched area $1-2-b-a-1$ bound by the process 1–2, the extreme ordinates and the X axis.

This area, $1-2-b-a-1$, according to mathematics, is equal to $\int y\mathrm{d}x$. As the temperature is represented on the y axis and entropy on the x axis, $\int y\mathrm{d}x = \int T\mathrm{d}S$ and, hence,

$$\text{Area} \quad 1-2-b-a-1 = \int y\mathrm{d}x = \int T\mathrm{d}S$$

By the definition of entropy given in (8.5), $T\mathrm{d}S = \delta Q_R$ and hence,

$$\int T\mathrm{d}S = \int \delta Q_R = Q_R = \text{net heat interaction} = \text{area} \quad 1-2-b-a-1$$

Hence, on a $T-S$ diagram, the area bound by a reversible or a quasi-static process, the two extreme ordinates and the X axis, is equal to the net heat interaction the system has during the process.

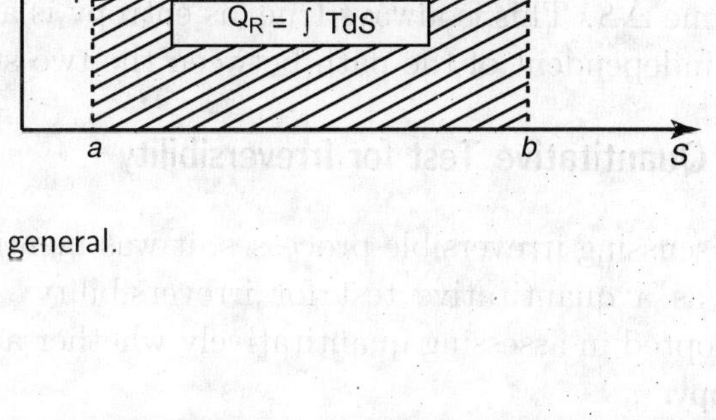

Fig. 8.4 $T-S$ diagram in general.

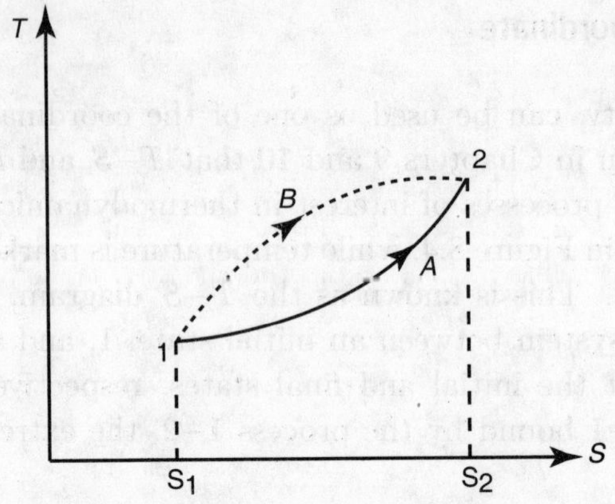

Fig. 8.5 Reversible and irreversible processes.

It is worth emphasizing here that $\mathrm{d}S = \dfrac{\delta Q_R}{T}$ and that $\mathrm{d}S$ is not equal to $\dfrac{\delta Q}{T}$ in general. This fact is further clarified below.

Consider a reversible process 1–A–2 on the T–S diagram in Figure 8.5. The area below 1–A–2 is equal to the net heat transfer during the reversible process A and the entropy change, ΔS, during this process is given by

$$S_2 - S_1 = \int_1^2 \frac{\delta Q_R}{T}$$

Now, consider an irreversible process, B, between the same two states 1 and 2. In Figure 8.5 it can be seen that the change in entropy in this process is also equal to $(S_2 - S_1)$, although it cannot be evaluated from $\dfrac{\delta Q}{T}$ along the irreversible process B, as $\mathrm{d}S \neq \dfrac{\delta Q}{T}$. This fact reveals a method by which entropy change in any irreversible process can also be evaluated.

Assume any convenient reversible process connecting the same end states as the irreversible process in question does. Evaluate $\displaystyle\int \frac{\delta Q_R}{T}$ along this assumed reversible process. Let this be equal to ΔS. The entropy change in the irreversible process should also be equal to the same ΔS. This is always true, as entropy is a property and its change between two states is independent of the path between the two states.

8.3.4 Entropy as a Quantitative Test for Irreversibility

In Chapter 7, while discussing irreversible processes, it was mentioned in Section 7.4 that entropy can be used as a quantitative test for irreversibility. Explained below is the methodology to be adopted in assessing quantitatively whether a process is irreversible or otherwise, using entropy.

Consider the thermodynamic cycle 1–A–2–B–1 shown on the T–S diagram of Figure 8.6. This is an irreversible cycle because of process A.

According to the Clausius inequality, for this irreversible cycle,

$$\oint \frac{dQ}{T} < 0.$$

Splitting the cyclic integral to linear integrals corresponding to the two processes,

$$\int_{1}^{2-A} \frac{\delta Q}{T} + \int_{2-B}^{1} \frac{\delta Q}{T} < 0 \qquad \text{(i)}$$

Process B is reversible and, hence, the second term on the right-hand side of (i), by the definition of entropy, has to be equal to $(S_1 - S_2)$ as, for this, process 1 is the end state and 2 is the initial state. Therefore,

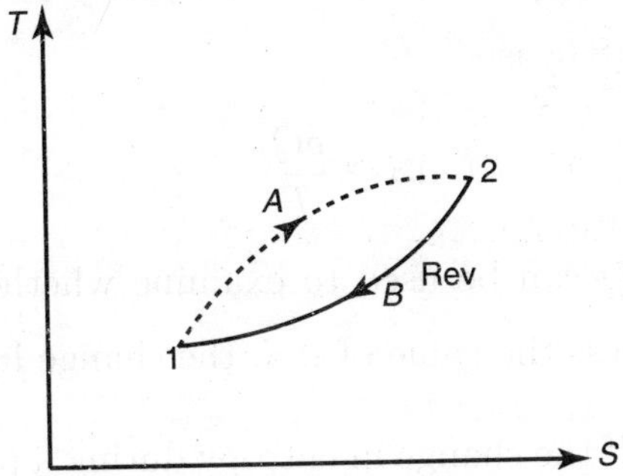

Fig. 8.6 T–S diagram to show entropy as quantitative test for irreversibility.

$$\int_{1}^{2-A} \frac{\delta Q}{T} + (S_1 - S_2) < 0 \qquad \text{(ii)}$$

or,

$$S_2 - S_1 > \int_{1}^{2-A} \frac{\delta Q}{T} \qquad \text{(iii)}$$

Process A is one among various irreversible processes that can occur between states 1 and 2. Thus (i) is valid for all irreversible processes between states 1 and 2. It can now be generalized and stated that for all irreversible processes between two given states, 1 and 2,

$$S_2 - S_1 > \int_{1}^{2} \frac{\delta Q}{T} \qquad \text{(iv)}$$

However, by the definition of entropy,

$$S_2 - S_1 = \int_1^2 \frac{\delta Q_R}{T} \tag{v}$$

Equations (iv) and (v) can be combined and written as

$$S_2 - S_1 \geq \int_1^2 \frac{\delta Q}{T} \tag{8.6}$$

where the equality holds good for a reversible process and the inequality for any irreversible process. Equation (8.6) is the mathematical representation of the second law for systems undergoing non-cyclic processes. This can be elaborated as follows:

When a system executes a non-cyclic process between two given state points the change in entropy during the process is either equal to or greater than $\int \frac{\delta Q}{T}$.

The differential form of (8.6) is

$$\mathrm{d}S \geq \frac{\delta Q}{T} \tag{8.7}$$

Equations (8.6) and (8.7) can be used to examine whether a process is reversible or irreversible. If during a process the value of ΔS, the change in entropy, is equal to $\int \frac{\delta Q}{T}$, such a process is reversible. If the change in entropy during a process is greater than $\int \frac{\delta Q}{T}$ such a process is irreversible.

A hypothetical process can be reversible or irreversible or impossible. Therefore, a process is impossible if

$$S_2 - S_1 < \int_1^2 \frac{\delta Q}{T}$$

8.3.5 Reversible and Irreversible Adiabatic Processes

An adiabatic process is one during which there are no heat interactions, that is, $Q = 0$. Substituting 0 for δQ in (8.6),

$$S_2 - S_1 \geq 0$$

As explained earlier, the equality holds good for a reversible process and the inequality for an irreversible process. Therefore, a reversible adiabatic process is one during which not only is the heat interaction zero but also the change in entropy during the process is equal to zero. Hence, a reversible adiabatic process is also an *isentropic process.* On the contrary, in an irreversible adiabatic process, although the heat interaction is zero the change in entropy is not equal to zero, but is greater than zero. In other words, entropy of a system can never decrease in an adiabatic process.

Thus, in general,

$$Q_{adiabatic} = 0, \text{ and } \Delta s_{adiabatic} \geq 0 \tag{8.8}$$

the equality holding good for a reversible adiabatic process and the inequality for the irreversible process. This fact is embodied in the principle of increase of entropy.

Example 8.5

Shown on the T–S diagram of Figure 8.7 are three different adiabatic expansion processes, A–B_1, A–B_2 and A–B_3 said to have undergone by a system. Each process starts at the same initial state A and ends at the same pressure p_2. Identify, giving reasons, possible and impossible processes.

▶ Solution:

All the processes are given to be adiabatic. In an adiabatic process, d$s \geq 0$ and hence entropy cannot decrease. Therefore, process A–B_1 is irreversible and possible as d$S = S_{B1} - S_A > 0$. Process A–B_2 is reversible because d$s = 0$ and hence may be possible. Process A–B_3 is impossible because d$S < 0$.

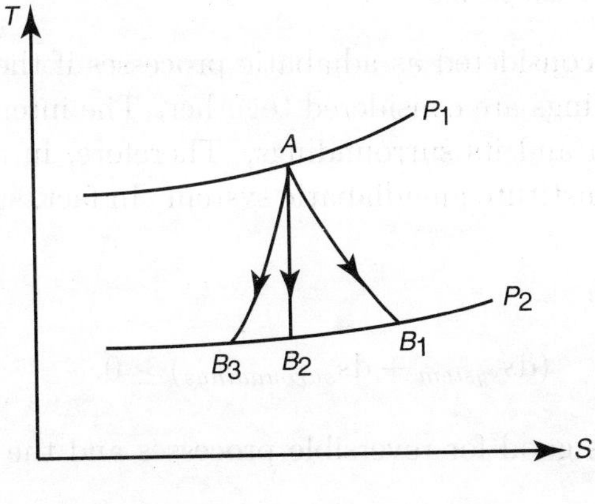

Fig. 8.7 T–S diagram for Example 8.5.

Example 8.6

Shown on a T–S diagram, Figure 8.8, are three adiabatic compression processes A–B_1, A–B_2 and A–B_3. Identify them as irreversible, reversible or impossible.

► Solution:

A quick perusal of the figure reveals that process A–B_1 is irreversible as in this process $dS > 0$. Process A–B_2 is reversible as $dS = 0$ and process A–B_3 is an impossible one as $dS < 0$.

It has been made clear that the change in entropy in an adiabatic process is always greater than zero, or, in the limiting case of a reversible adiabatic process, it tends to zero.

In other words, entropy can never decrease in an adiabatic process.

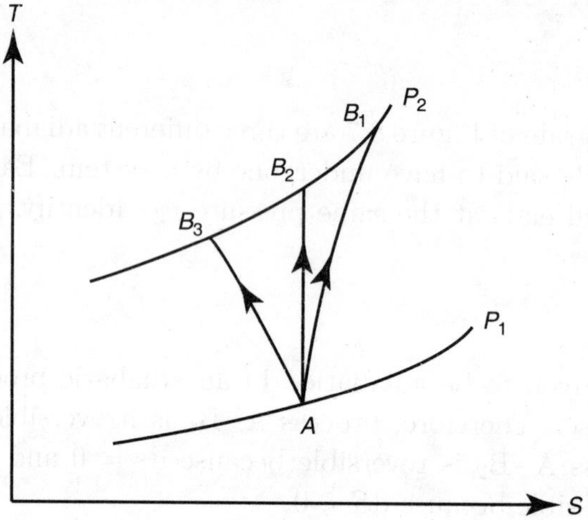

Fig. 8.8 $T–S$ diagram for Example 8.6.

All processes can be considered as adiabatic processes if the changes occurring in the system and the surroundings are considered together. The interactions in any process are only between the system and its surroundings. Therefore, in any situation, system and surroundings together constitute an adiabatic system. In fact, such a system is an isolated system.

Therefore, from (8.8),

$$(ds_{system} + ds_{suroundings}) \geq 0, \tag{8.9}$$

where, the equality holds good for reversible processes and the inequality for irreversible processes.

It is evident from (8.9) that, in a reversible process, the magnitude of the change in entropy of the system and that of the surroundings are equal, but their directions are opposite. In other words, in a reversible process, if the entropy of the system increases, the entropy of the surroundings has to decrease by the same extent, and vice versa.

In an irreversible adiabatic process the entropy change of a system and its surroundings cannot be the same. Therefore, the sum of these changes will always be positive, in accordance with (8.9).

All processes occur inside the universe. Thus, (8.8), when applied to the universe, takes the form:

$$\Delta S_{universe} > 0 \tag{8.10}$$

and not $\Delta S_{universe} \geq 0$ because almost all processes occurring in the universe, man-made or natural, are generally irreversible in nature and hence $\Delta S_{universe} \neq 0$.

Equations (8.9) and (8.10) lead to the concept of the principle of increase of entropy.

8.3.6 Principle of Increase of Entropy

The principle of increase of entropy states that for a system undergoing any process, the net change in entropy of the system and its surroundings is always greater than or equal to zero as in (8.9).

Alternatively, it may also be stated that the entropy of the universe is always on the increase as in (8.10).

8.3.7 Entropy as a Quantitative Test of Irreversibility

The fact that d$S \geq 0$ during an adiabatic process can be used to advantage in assessing whether a given process is reversible or irreversible. The quantitative assessment is not restricted to adiabatic processes alone. It can also be applied to all processes, reversible or irreversible, adiabatic or otherwise.

One of the important questions raised in Chapter 6 was how to quantitatively decide the feasibility or otherwise of a process. Although (8.6) can provide the required answer, it is not in a convenient form, as it may not always be possible to evaluate the right-hand side of this equation.

Whenever (8.6) cannot be used, for want of sufficient data, (8.7)–(8.9) can conveniently be used to provide the answer to this question. Thus,

(a) A process is irreversible and possible if

$$S_2 - S_1 > \int \frac{\delta Q}{T} \quad \text{or}$$
$$\Delta S_{system} + \Delta S_{surroundings} > 0$$

(b) A process is reversible and theoretically possible if

$$S_2 - S_1 = \int \frac{\delta Q}{T} \quad \text{or}$$
$$\Delta S_{system} + \Delta S_{surroundings} = 0$$

(c) A hypothetical process is impossible if

$$S_2 - S_1 < \int \frac{\delta Q}{T} \text{ or}$$

$$\Delta S_{system} + \Delta S_{surroundings} < 0$$

Example 8.7

A rigid non-insulated vessel containing a mixture of fuel and air is surrounded by a large body of water at a particular temperature. The fuel is burnt. State whether the entropy change of the universe is positive, negative or zero.

▶ **Solution:**

The process involving combustion is an irreversible process. Hence, the entropy change of the universe is positive.

8.4 Evaluation of Entropy Change

Equation (8.4), which defines entropy, cannot be used to easily calculate entropy change in each and every process. This equation can only be used for either a reversible isothermal process or any other reversible process during which the magnitude of heat transfer is known as a function of temperature. Hence, the utility of this relation is limited. Entropy is a property and, from the two-property rule for a pure substance, it must be possible to calculate the entropy at any state in terms of any two convenient independent properties. The following text is an explanation of such useful relations among properties.

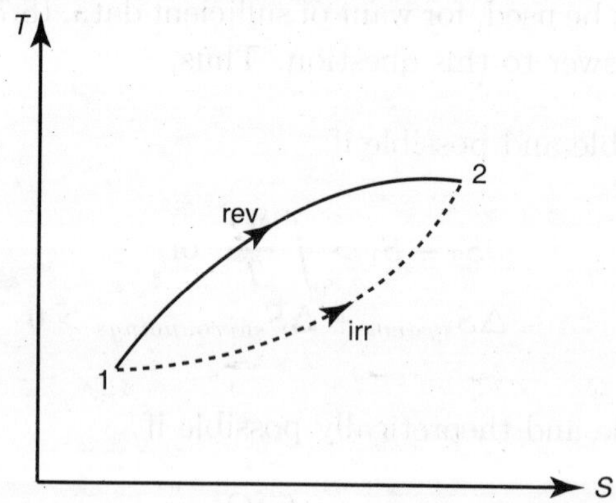

Fig. 8.9 Reversible and irreversible processes on a T–S diagram.

Shown on the T–S plane in Figure 8.9 is a reversible process undergone by a pure substance between states 1 and 2. Applying the first law in differential form to this process,

$$\delta Q - \delta W = \mathrm{d}U$$

As the process is reversible, $\delta Q = \delta Q_R$, and $\delta W = p\mathrm{d}V$ (displacement work is the only possible reversible work) and therefore,

$$\delta Q_R - p\mathrm{d}V = \mathrm{d}U$$

By the definition of entropy, $\delta Q_R = T\mathrm{d}S$ and hence,

$$T\mathrm{d}S = \mathrm{d}U + p\mathrm{d}V$$

This relation, when written in terms of specific properties, takes the form,

$$T\mathrm{d}s = \mathrm{d}u + p\mathrm{d}v \tag{8.11}$$

By the definition of enthalpy, $h = u + pv$ and hence,

$$\mathrm{d}h = \mathrm{d}u + p\mathrm{d}v + v\mathrm{d}p \text{ or } \mathrm{d}u = \mathrm{d}h - p\mathrm{d}v - v\mathrm{d}p$$

Substituting for $\mathrm{d}u$ in (8.11),

$$T\mathrm{d}s = \mathrm{d}h - p\mathrm{d}v - v\mathrm{d}p + p\mathrm{d}v, \text{ or } T\mathrm{d}s = \mathrm{d}h - v\mathrm{d}p \tag{8.12}$$

Equations (8.11) and (8.12) are called the $T\mathrm{d}s$ relations and are very important and useful relations. They are very frequently used in calculating the entropy changes in majority of the cases. For the reversible process 1–2 in Figure 8.9, using (8.11),

$$\Delta s = \int \mathrm{d}s = s_2 - s_1 = \int \frac{\mathrm{d}u}{T} + \int \left(\frac{p}{T}\right)\mathrm{d}v \tag{8.13}$$

Similarly, using (8.12),

$$\Delta s = \int \mathrm{d}s = s_2 - s_1 = \int \frac{\mathrm{d}h}{T} - \int \left(\frac{v}{T}\right)\mathrm{d}p \tag{8.14}$$

Now, consider the irreversible process shown in Figure 8.9. This process also occurs between the same two end states as that of the reversible process. Hence, $s_2 - s_1$ has to be the same for both processes. Although (8.11)–(8.14) are derived for a reversible process, these relations can, hence, also be used for calculating entropy changes in irreversible processes. As a further proof of this inference it is worth noting that there are no factors in (8.11)–(8.14) which are path-dependent. T, p, v, s, u and h are all properties of the substance. Hence, the $T\mathrm{d}s$ relations can be used to find the entropy change between two states in reversible as well as irreversible processes. However, to do so, the following information have to be available: (i) p–v–T relation for the system in question, (ii) u and/or h as a function of temperature.

Example 8.8

A system comprising m kg of a certain gas undergoes a change of state from 1 to 2. Given that (i) the $p-v-T$ relation for the gas is $pv = RT$, where R is a constant, (ii) its internal energy and enthalpy are functions of temperature only and (iii) the specific heats c_p and c_v are constants, determine the change in entropy of the gas in terms of (a) T and v, (b) T and p and (c) p and v.

► **Solution:**

(a) In general, $$\Delta s = \int \mathrm{d}s = s_2 - s_1 = \int \frac{\mathrm{d}u}{T} + \int \left(\frac{p}{T}\right) \mathrm{d}v \quad (8.13)$$

Here, $$pv = \mathrm{R}T, \quad \text{and therefore} \quad \frac{p}{T} = \frac{\mathrm{R}}{v} \quad \text{(i)}$$

By the definition for a pure substance, $c_v = \left(\dfrac{\partial u}{\partial T}\right)_v$

As it is given that u is a function of temperature only, $c_v = \dfrac{\mathrm{d}u}{\mathrm{d}T}$ and hence,

$$\mathrm{d}u = c_v \mathrm{d}T \quad \text{(ii)}$$

Substituting for $\dfrac{p}{T}$ and $\mathrm{d}u$, from (i) and (ii), in (8.13)

$$\Delta s = \int \frac{c_v \mathrm{d}T}{T} + \int \left(\frac{\mathrm{R}}{v}\right) \mathrm{d}v = c_v \int \frac{\mathrm{d}T}{T} + \mathrm{R} \int \frac{\mathrm{d}v}{v}$$

Therefore, $$\Delta s = c_v \ln \left(\frac{T_2}{T_1}\right) + \mathrm{R} \ln \left(\frac{v_2}{v_1}\right)$$

And $$\Delta S = m\Delta s = m \left[c_v \ln \left(\frac{T_2}{T_1}\right) + \mathrm{R} \ln \left(\frac{v_2}{v_1}\right)\right]$$

(b) In general, $$\Delta s = \int \mathrm{d}s = s_2 - s_1 = \int \frac{\mathrm{d}h}{T} - \int \left(\frac{v}{T}\right) \mathrm{d}p \quad (8.14)$$

Here, $$pv = \mathrm{R}T, \quad \text{and therefore} \quad \frac{v}{T} = \frac{\mathrm{R}}{p} \quad \text{(i)}$$

By the definition for a pure substance, $c_p = \left(\dfrac{\partial h}{\partial T}\right)_p$

As it is given that h is a function of temperature only, $c_p = \dfrac{dh}{dT}$ and hence,

$$dh = c_p dT \tag{ii}$$

Substituting for $\dfrac{v}{T}$ and dh in (8.14) from (i) and (ii)

$$\Delta s = \int c_p \frac{dT}{T} - \int \left(\frac{R}{p}\right) dp = c_p \int \frac{dT}{T} - R \int \frac{dp}{p}$$

Therefore, $\Delta s = c_p \ln \left(\dfrac{T_2}{T_1}\right) - R \ln \left(\dfrac{p_2}{p_1}\right)$

And $\Delta S = m\Delta s = m \left[c_p \ln \left(\dfrac{T_2}{T_1}\right) - R \ln \left(\dfrac{p_2}{p_1}\right)\right]$

(c) From (a), $\Delta S = m\Delta s = m \left[c_v \ln \left(\dfrac{T_2}{T_1}\right) + R \ln \left(\dfrac{v_2}{v_1}\right)\right]$

From data, $p_1 v_1 = RT_1$ and $p_2 v_2 = RT_2$

Therefore, $\dfrac{T_2}{T_1} = \left(\dfrac{p_2}{p_1}\right)\left(\dfrac{v_2}{v_1}\right)$

Substituting for $\dfrac{T_2}{T_1}$ in the expression for ΔS,

$$\Delta S = m \left\{ c_v \ln \left[\left(\frac{p_2}{p_1}\right)\left(\frac{v_2}{v_1}\right)\right] + R \ln \left(\frac{v_2}{v_1}\right)\right\}$$
$$= m \left[c_v \ln \left(\frac{p_2}{p_1}\right) + (c_v + R) \ln \left(\frac{v_2}{v_1}\right)\right]$$

Example 8.9

A solid body of mass m kg is heated from T_1 to T_2. Find the change in entropy.

▶ Solution:

From the Tds relation

$$Tds = dh - vdp$$

The heating of a solid is at constant pressure and hence $dp = 0$

Therefore $$Tds = dh$$

And by definition of the specific heat, $c = \dfrac{dh}{dT}$ or $dh = cdT$.

Therefore, $$T\mathrm{d}s = c\mathrm{d}T \text{ or } \mathrm{d}s = c\frac{\mathrm{d}T}{T}$$

that is $$\Delta s = \int \mathrm{d}s = \int \frac{c\mathrm{d}T}{T} = c \ln \frac{T_2}{T_1}$$

Or $$\Delta S = m\Delta s = mc \ln \frac{T_2}{T_1}$$

(The same relation as above can be used to determine the change in entropy when a certain mass of liquid undergoes a state change. Liquids and solids are treated as incompressible mediums. The change in volume and pressure, with temperature, is so negligible for these substances that $\mathrm{d}h = \mathrm{d}u + \mathrm{d}(pv) = \mathrm{d}u$. Thus, solids and liquids have only one specific heat, c, such that $\mathrm{d}h = \mathrm{d}u = c\mathrm{d}T$)

Example 8.10

At 0°C, m kg of ice is melted completely into water. Find the change in entropy during this melting process.

► **Solution:**

During the process, there occurs a change of phase from solid to liquid but the temperature does not change. The process of melting is thus an internally reversible isothermal process. Therefore,

$$\mathrm{d}s = \frac{\mathrm{d}Q}{T} \text{ or } \Delta s = \int \frac{\mathrm{d}Q}{T} = \frac{Q}{T}$$

where Q is the heat transfer required to completely melt 1 kg ice from solid to liquid at the same temperature. This heat is called *latent heat of melting* or *enthalpy of melting.* Let L be the enthalpy of melting. Then

$$\Delta s = \frac{L}{T} \text{ and } \Delta S = m\Delta s = m\frac{L}{T}.$$

Example 8.11

Can a system undergo an isentropic process which is not reversible and adiabatic?

► **Solution:**

The answer to this question is 'yes'. An isentropic process is one in which the final and initial entropies are the same. Such a process can be a combination of two or more processes occurring in such a manner that any increase in entropy anywhere is compensated by the corresponding decrease in entropy somewhere else. A reversible and adiabatic process is one in which the entropy of the system remains steady at its initial value all through the process.

Example 8.12

Air flows steadily through a horizontal, insulated, variable-area duct. The measurements made at two locations, A and B, along the duct gave the following information. Find the direction of flow of air. Assume for air, which follows the equation of state $pv = RT$, enthalpy is a function of temperature only and that $c_p = 1.005$ kJ/kg and R = 0.287 kJ/kg K.

Location	Pressure (kPa)	Temp. (°C)	Velocity (m/s)
A	120	50	150
B	100	30	250

***Note*:** While discussing the limitations of the first law of thermodynamics in Section 6.2, this example of flow through a variable-area duct was cited to evaluate the direction of flow of air through it. It was then mentioned that the flow direction could only be decided by the application of the second law. The example is now taken up for analysis here.

According to the principle of increase of entropy, the flow is possible only in that direction in which $\Delta S_{system} + \Delta S_{surroundings} > 0$.

The flow is such that the air has no heat and work interaction with the surroundings. Hence $\Delta S_{surroundings} = 0$. Therefore, in the direction in which follow is possible $\Delta S_{system} > 0$.

Assume the flow is from A to B:

$$\Delta s_{sys} = s_\text{B} - s_\text{A} = \int \frac{\text{d}h}{T} - \int \left(\frac{v}{T}\right) \text{d}p$$

From data, $h = h(T)$ and c_p is given to be a constant, hence, $\text{d}h = c_p \text{d}T$.

Also as $pv = \text{R}T$, $\dfrac{v}{T} = \dfrac{\text{R}}{p}$, therefore,

$$\begin{aligned}\Delta s_{sys} = s_\text{B} - s_\text{A} &= \int c_P \frac{\text{d}T}{T} - \int \text{R}\frac{\text{d}p}{p} \\ &= c_P \ln \frac{T_\text{B}}{T_\text{A}} - \text{R} \ln \frac{p_\text{B}}{p_\text{A}} \\ &= 1.005 \ln \frac{(273+30)}{(273+50)} - 0.287 \ln \left(\frac{100}{120}\right) \\ &= -0.0119 \text{ kJ/kg K}\end{aligned}$$

As $\Delta s_{adiabatic} = \Delta s_{sys} < 0$ the flow cannot be from A to B.

Assume now the flow is from B to A. Similar calculation results in $\Delta s_{adiabatic} = \Delta s_{sys} = +0.0119$ kJ/kg K which is > 0. Therefore, according to the principle of increase of entropy, the flow is possible only from B to A.

Example 8.13

5 kg of ice at −10°C is mixed with 20 kg water at 80°C in an insulated vessel. (a) What is the temperature of the mixture when equilibrium is established between ice and water. Find the change in entropy during this process for (b) ice, (c) water and (d) the universe. Assume specific heat of ice = 2.1 kJ/kg K and enthalpy of melting of ice (latent heat) is 330 kJ/kg.

▶ Data:

$$m_i = 5 \text{ kg}, \; T_i = 273 - 10 = 263 \text{ K}, \; c_i = 2.1 \text{ kJ/kg K}, L = 330 \text{ kJ/kg}$$

$$m_w = 20 \text{ kg}, \; T_w = 273 + 80 = 353 \text{ K}, \; c_w = 4.2 \text{ kJ/kg K (assumed)}$$

▶ Solution:

When ice at −10°C is mixed with water at 80°C, the following changes occur in sequence until thermal equilibrium is reached between the two.

(i) Temperature of ice increases from −10°C to 0°C, during which time, the temperature of water decreases because of the heat transfer between the two.

(ii) When ice reaches 0°C, it starts melting, and until the last particle of ice melts its temperature remains constant. But the temperature of water decreases as it supplies the required latent heat to ice.

(iii) At the end of the process of melting, the temperature of water from the ice keeps increasing and the temperature of original water keeps further decreasing until the two reach a common temperature at thermal equilibrium. Let this final equilibrium temperature be T.

(a) Considering ice plus water as the system, there are no heat and work interactions with the surroundings. Therefore, from the first law, as $Q = 0$ and $W = 0$,

$$\Delta U_{sys} = \Delta U_{ice} + \Delta U_{wat} = 0 \tag{1}$$

Here, $\Delta U_{ice} = \Sigma\, \Delta U_{ice}$ for the above phenomenon for ice.

that is

$$\begin{aligned}\Delta U_{ice} &= \Delta U_{ice(i)} + \Delta U_{ice(ii)} + \Delta U_{ice(iii)} \\ &= m_i\,[c_i(273 - 263) + L + c_w(T - 273)] \\ &= 5[2.1(273 - 263) + 330 + 4.2(T - 273)]\end{aligned} \tag{2}$$

and

$$\begin{aligned}\Delta U_{wat} &= m_w[c_w(T - 353)] \\ &= 20[4.2(T - 353)]\end{aligned} \tag{3}$$

Substituting for ΔU_{ice} and ΔU_{wat} in (1) from (2) and (3)

$$5[2.1(273 - 263) + 330 + 4.2(T - 273)] + 20[4.2(T - 353)] = 0$$

Solving for T,

$$T = 320\ \text{K}$$

(b) Now, $\Delta S_{ice} = \Delta S_{ice(i)} + \Delta S_{ice(ii)} + \Delta S_{ice(iii)}$

$$= m_i \left[c_i \ln \frac{273}{263} + \frac{330}{273} + c_w \ln \frac{320}{273} \right] \quad \text{(as in Examples 8.9 and 8.10)}$$

$$= 5 \left[2.1 \ln \frac{273}{263} + \frac{330}{273} + 4.2 \ln \frac{320}{273} \right]$$

$$= 9.77\ \text{kJ/K}$$

(c) $\Delta S_{wat} = m_w c_w \ln \dfrac{T}{T_w} = 20 \times 4.2 \ln \dfrac{320}{353} = -8.24\ \text{kJ/K}$

(d) $\Delta S_{univ} = \Delta S_{ice} + \Delta S_{wat} = 9.77 - 8.24 = 1.53\ \text{kJ/K}$

Note: As ΔS_{univ} is positive the process is irreversible.

8.5 Carnot Cycle on the slant 200 T--S Diagram

In Figure 8.10, the Carnot cycle is shown on $p-V$ and $T-S$ diagrams. The $p-V$ diagram has already been explained in Chapter 7. An attempt is now made to explain the Carnot cycle on the $T-S$ diagram. Corresponding state points on both the diagrams are indicated by the same alphabets. In the following analysis, only the $T-S$ diagram is considered.

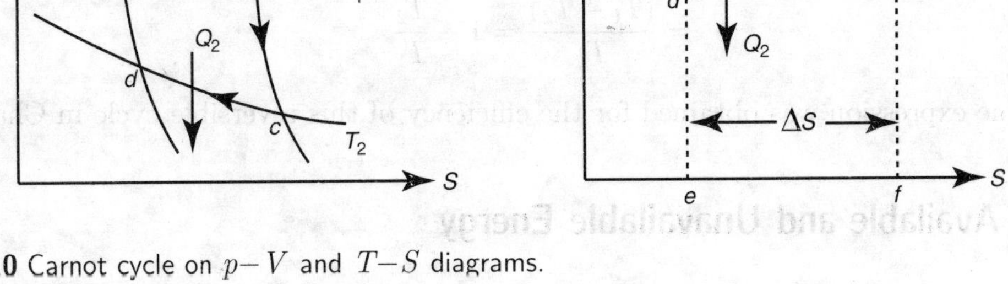

Fig. 8.10 Carnot cycle on $p-V$ and $T-S$ diagrams.

During a–b, the system undergoes a reversible isothermal expansion process. Hence, a–b is along a horizontal line with $T_a = T_1 = T_b$. During this process net heat transfer to the system at a constant temperature of T_1 is Q_1. Therefore,

$$S_b - S_a = \int_a^b \frac{\delta Q_R}{T} = \frac{Q_1}{T_1}$$

or,
$$Q_1 = T_1(S_b - S_a) = T_1 \Delta S = \text{ area abfea} \qquad \text{(i)}$$

As process b–c is reversible and adiabatic, $S_b = S_c$ and hence b and c must lie on a vertical line. c–d is the reversible isothermal compression or heat rejection process during which the system transfers out Q_2 at a steady temperature of T_2, such that $T_d = T_2 = T_c$. Hence,

pt

$$S_d - S_c = \int \frac{\delta Q_R}{T} = \frac{Q_2}{T_2}$$

or
$$Q_2 = T_2(S_d - S_c) = \text{area cdefc} \qquad \text{(ii)}$$

In (ii) the entropy change is negative as heat is transferred out of the system. Thus, the magnitude of Q_2 is given by

$$Q_2 = T_2(S_c - S_d) = T_2 \Delta S. \qquad \text{(iii)}$$

Process d–a is reversible and adiabatic and hence it is along a vertical line and $S_d = S_a$. Thus, the Carnot cycle on a T–S diagram is bound by two horizontal and two vertical lines, and it is always a rectangle.

From (i) and (iii) net work W for the cycle is given by

$$W = Q_1 - Q_2 = (T_1 \Delta S - T_2 \Delta S) = (\text{area abfea} - \text{area cdefc}).$$
$$= \text{area of the diagram abcda}$$
$$\eta = \frac{(Q_1 - Q_2)}{Q_1} = \frac{(T_1 \Delta S - T_2 \Delta S)}{T_1 \Delta S}$$
$$= \frac{(T_1 - T_2)}{T_1} = 1 - \frac{T_2}{T_1}$$

The same expression was obtained for the efficiency of this reversible cycle in Chapter 7.

8.6 Available and Unavailable Energy

Consider a heat transfer of magnitude Q_1 from a high-temperature reservoir at T_1 K to a heat engine. The magnitude of work produced by the heat engine depends upon: (i) T_2, the temperature of the low-temperature reservoir to which the engine rejects heat, and (ii) the nature of the engine—whether it is reversible or irreversible. The work produced increases with a decrease in T_2. The naturally available lowest temperature is the temperature of the surroundings at any place, say T_0. Thus, an engine rejecting heat to the surroundings at T_0 gives maximum work. Also, among all engines operating between T_1 and T_0, the reversible engine gives the maximum work. Thus, only a reversible engine operating between T_1, the temperature of the high-temperature reservoir, and T_0, the temperature of the surroundings, can convert heat transfer to it to maximum work, W_{max}.

Under these conditions W_{max} is called the available energy of Q_1. Q_0, the heat rejected to the surroundings at T_0 is called the unavailable energy of Q_1. These nomenclatures are quite logical and meaningful as W_{max}, the maximum work realizable is the available or the useful part of Q_1, while Q_0, the heat rejected to the surroundings, is lost forever and hence is the unavailable part, or the useless part, of Q_1.

From the first law, the unavailable energy is also equal to $(Q_1 - W_{max})$. It can now be generalized and stated as follows: Energy from any reservoir at a given temperature, or, for that matter, energy from even a finite system whose temperature keeps decreasing with the supply of energy, is made up of two parts, namely, the available energy (useful part) and the unavailable energy (useless part).

Available energy is that part of the energy supplied to an engine which can be converted to work by a reversible engine operating between the temperature of the energy supply system and a low-temperature reservoir at the surrounding temperature.

Unavailable energy is the energy rejected to the surroundings by this reversible heat engine.

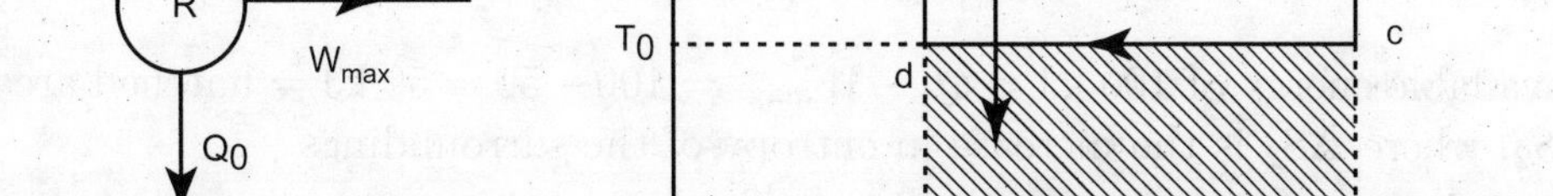

Fig. 8.11 Available and unavailable energy.

Figure 8.11 shows, as an example, the schematic and the T–S diagrams of a reversible engine operating between T_1 and T_0. Referring to this figure,

$$\text{Available energy of } Q_1 = W_{max} = \eta_{max}\ Q_1 = \left[1 - \frac{T_0}{T_1}\right] Q_1 \tag{8.15}$$

$$= \text{Area of the cycle abcda on } T\text{–}S \text{ diagram.}$$

$$\text{Unavailable Energy of } Q_1 = Q_0 = Q_1 - W_{max} \tag{8.16}$$

$$= \text{Area cdefc} = T_0 \times \Delta S_1,$$

where ΔS_1 = increase in entropy of the surroundings, which, in magnitude, is also equal to the decrease in entropy of the high-temperature reservoir.

Example 8.14

Find the available and unavailable parts of 100 kJ of heat from (a) a reservoir at 1000 K and (b) a reservoir at 600 K. Assume the surroundings are at 300 K.

► **Solution:**

(a) From (8.15), available energy of 100 kJ =

$$Q_1 \left[1 - \frac{T_0}{T_1}\right] = 100 \left[1 - \left(\frac{300}{1000}\right)\right] = 70 \text{ kJ}$$

And, from (8.16), unavailable energy of 100 kJ = $Q_1 - W_{max}$ = 100 – 70 = 30 kJ = hatched area cfedc in Figure 8.11 = $T_0 \Delta S_1$.

► **Solution:**

(b) Referring to Figure 8.12, available energy of 100 kJ =

$$Q_1 \left[1 - \frac{T_0}{T_1}\right] = 100 \left[1 - \left(\frac{300}{600}\right)\right] = 50 \text{ kJ}$$

Unavailable energy of 100 kJ = $Q_1 - W_{\text{max}}$ = 100 – 50 = 50 kJ = hatched area nqpon = $T_0 \Delta S_2$, where ΔS_2 is the increase in entropy of the surroundings.

***Note*:** It is worth observing from the two examples, 8.14(a) and (b), that while the available energy of 100 kJ from the reservoir at 1000 K is 70 kJ, the available energy of the same 100 kJ from another reservoir at 600 K, a lower temperature, is only 50 kJ. Therefore, available energy of a given magnitude of heat transfer decreases with a decrease in temperature. It is for this reason that energy at a higher temperature is said to be more valuable.

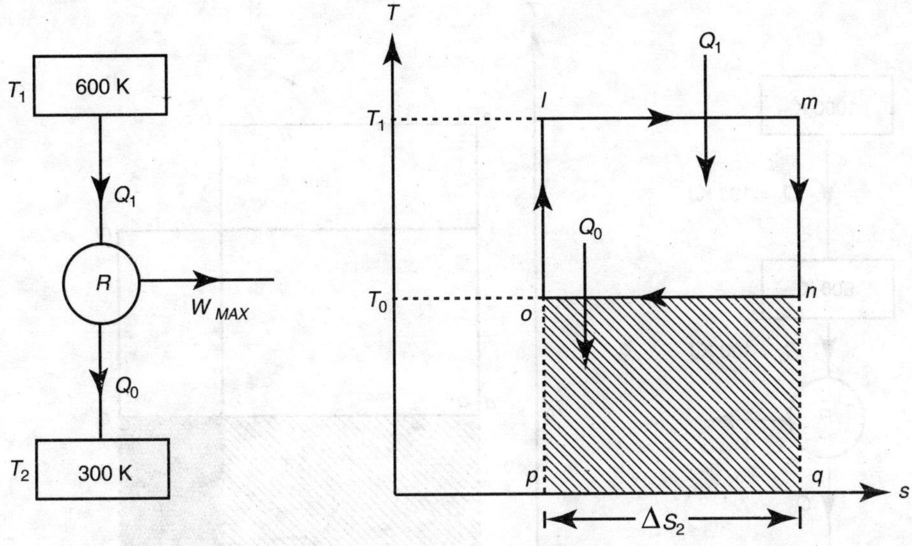

Fig. 8.12 Diagram for solving Example 8.14 (b).

Example 8.15

At 1000 K, 100 kJ of heat is transferred from a reservoir to a reversible engine through an intermediate reservoir at 600 K. The surroundings are at 300 K. Evaluate the increase in unavailable energy due to irreversible heat transfer between the two reservoirs.

▶ Solution:

This example is a combination of Examples 8.14 (a) and (b). Consider the schematic and the $T-S$ diagrams in Figure 8.13. This diagram comprises the $T-S$ diagram of Example 8.14 (b) superimposed on that of Example 8.14 (a).

Although the input to the engine is from the reservoir at 1000 K, the engine operates between the reservoirs at 600 K and 300 K. Hence, the work output of the engine is $= 100\left[1-\left(\dfrac{300}{600}\right)\right] = 50$ kJ which is equal to that in Example 8.14(b) and not 70 kJ as in Example 8.14 (a). This reduction is because the 100 kJ from 1000 K is first transferred to the reservoir at 600 K before it is transferred to the engine. The heat transfer between the two reservoirs is irreversible because of the finite temperature difference between the two. This irreversibility reflects in the decrease of the available energy by 20 kJ (= 70−50). This reduction in available energy makes the unavailable energy increase by the same extent, 20 kJ = (50 − 30) = difference between the unavailable energies in Example 8.14(a) and Example 8.14(b). Therefore, irreversibility always increases the unavailable energy.

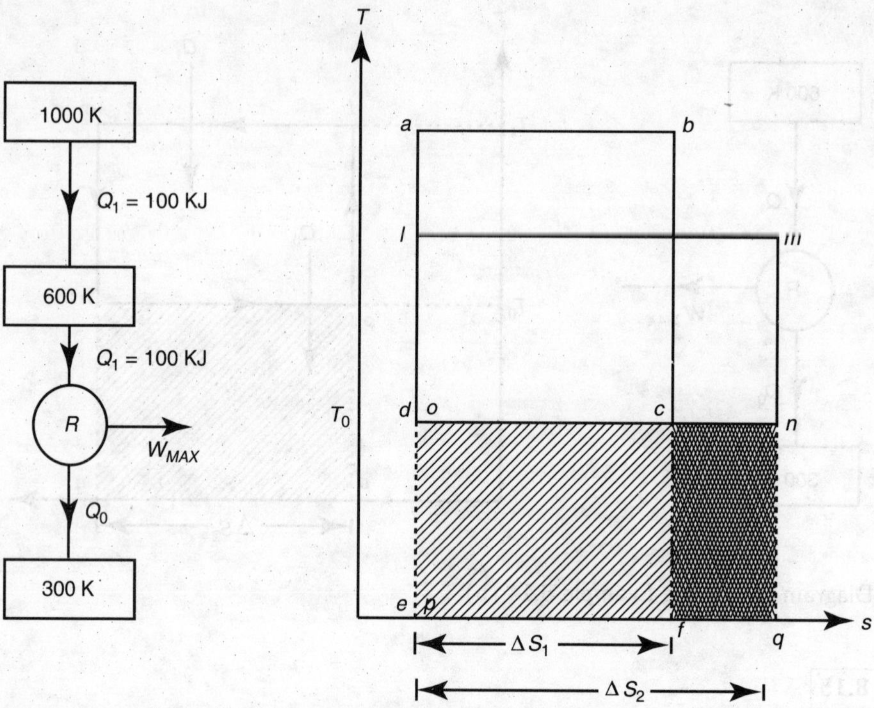

Fig. 8.13 Schematic and T–S diagrams for solving Example 8.12.

8.6.1 Relation between Increase in Unavailable Energy and Increase in Entropy

An important relation between the increase in unavailable energy and entropy can now be deduced based on observations on the results of the three examples discussed above. In the T–S diagram of Figure 8.13, the unavailable energy, in the absence of the intermediate reservoir, as in Example 8.14(a), is equal to 30 kJ which is represented by the single hatched area cfedc (same area shown in Figure 8.11). Because of the introduction of the intermediate reservoir, the unavailable energy, according to Example 8.14(b), has increased to 50 kJ. This is represented by the area nqpon (same area shown in Figure 8.12). The increase in unavailable energy is thus equal to the double-hatched area nqfcn which is equal to area nqpon ($= T_0\Delta S_2$) minus the area cfedc ($= T_0\Delta S_1$).

that is $$\text{Area nqfcn} = T_0[\Delta S_2 - \Delta S_1] = T_0[\Delta S_{net}] \quad (8.17)$$

where, ΔS_{net} = entropy increase of the universe due to irreversibility in the process. With reference to the schematic and T–S diagrams of Figure 8.13, entropy increase of the universe, ΔS_{net}, is given by

$$\begin{aligned}\Delta S_{net} &= \Delta S_{res\ at\ 1000\ k} + \Delta S_{res\ at\ 600\ k} + \Delta S_{engine} + \Delta S_{res\ at\ 300\ k}. \\ &= \frac{-100}{1000} + \left(\frac{100}{600} - \frac{100}{600}\right) + 0 + \frac{50}{300} \\ &= -0.1 + 0.1667 = 0.0667\end{aligned}$$

Therefore, increase in unavailable energy $= T_0[\Delta S_{net}] = 300 \times 0.0667 = 20$ kJ.

The above analysis has great practical importance. High-temperature reservoirs are made available in practice by burning fuel, fossil or nuclear. The heat transferred from the burning fuel is used to run a heat engine. During the heat addition process the working substance that undergoes a thermodynamic cycle is usually at a temperature lower than the reservoir temperature. This is a necessity because (i) in the absence of such temperature difference the heat transfer rates are small and thus the power from the engine is bound to be very small being of no practical use; (ii) if the substance were to be at the same temperature as that of the high-temperature reservoir, the material of the container of the working substance may not withstand this high-temperature and may thus lead to engine damage. Hence, although the reservoir that supplies energy is at a high temperature (say 1000 K, as in the above examples) the heat addition to the heat engine occurs at a low temperature (say 600 K, as in the above examples). As a ready example, the reader may recall here that the water in a steam power plant boils at a low temperature in the boiler, although the furnace, where the fuel is burnt, is at a very high temperature.

The definitions of available and unavailable energies are not restricted to only heat transfers from reservoirs (constant-temperature systems). They are general in nature and can also be used for energies from finite systems whose temperature may decrease with the supply of heat. In such a case, it is not possible to imagine a single externally reversible engine that can take heat from the finite system reversibly as its temperature keeps decreasing. Instead, it can be assumed that heat is supplied to a series of tiny Carnot engines, each engine taking heat from the system at one unique temperature between its initial and final temperatures.

As an example, suppose a finite system changes from state 1 to 2 as it supplies heat Q reversibly. This process of heat supply is represented by the process 1–2 in the T–S diagram of Figure 8.14 and Q is equal to the area below 1–2.

To find the available and unavailable parts of Q, it is necessary to imagine an engine that takes in Q reversibly and also rejects heat to the surroundings at T_0. If the heat transfer to the engine is reversible then the temperature of the system supplying heat must be equal to the temperature of the working substance of the engine at every instant. Thus, in Figure 8.14, while 1–2 represents the process of heat transfer from the finite system, 2–1 will represent the process of heat addition to the cycle. Let this be assumed to be possible through successive heat transfers to a series of Carnot engines. As shown in Figure 8.14 , let δQ be the heat transfer at temperature T to one such tiny Carnot cycle. Then $\delta Q = T\mathrm{d}s$. Unavailable energy of $\delta Q = T_0\mathrm{d}S$ = area single-hatched, and available energy of $\delta Q = \delta Q - T_0\mathrm{d}S$ = double-hatched area. As net $\delta Q = \int \delta Q$= area below 2–1, unavailable energy of $Q = \int T_0\mathrm{d}S = T_0\Delta S$ = area abcda and available energy of $Q = Q - T_0\Delta S$= area 1ad21.

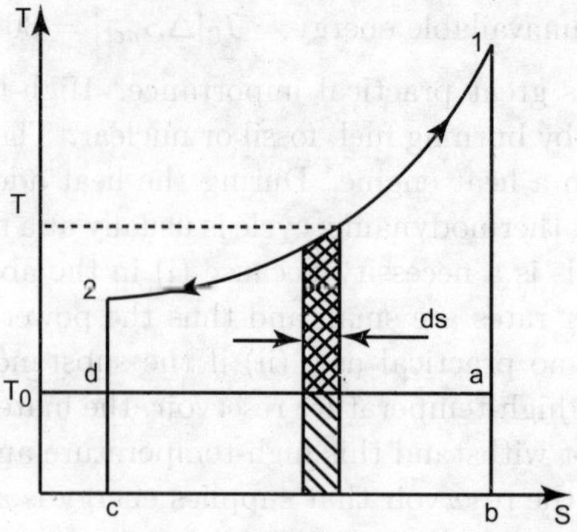

Fig. 8.14 Available and unavailable energies—finite systems.

Example 8.16

A heat engine operates between 500 K and 350 K at a place where the surroundings are at 300 K. The engine delivers a work of 250 kJ when supplied with $Q_1 = 1000$ kJ from the high-temperature reservoir. It is said that the available energy of Q_1 is 250 kJ and its unavailable part is 750 kJ. Do you agree with this ? If not, what are the available and unavailable energies of Q_1?

▶ **Solution:**

Figure 8.15 shows the schematic of the engine.

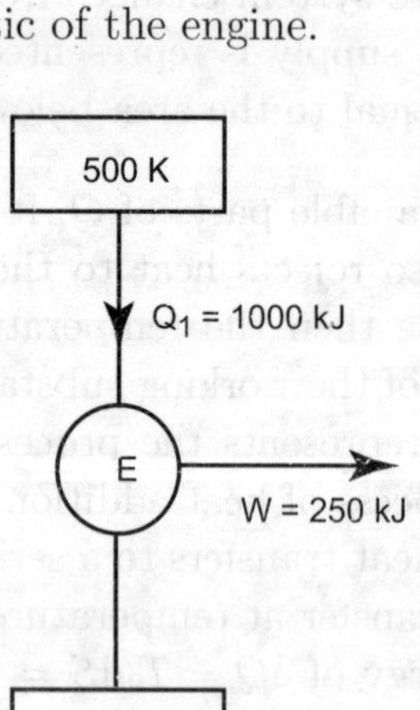

Fig. 8.15 Heat engine for Example 8.16.

Available energy is defined as that part of the energy which is converted to work by a reversible engine operating between the given high-temperature reservoir and the low-temperature reservoir at the surroundings temperature. Efficiency of such an engine operating between 500 K and 300 K = $\left(1 - \frac{300}{500}\right) = 0.4$. Thus, available energy of Q_1 is equal to $\eta Q_1 = 0.4 \times 1000 = 400$ kJ. The unavailable energy $= Q_1 - W_{max} = 1000 - 400 = 600$ kJ.

The engine shown in Figure 8.15 converts 1000 kJ supplied to it to 250 kJ work. Also the engine operates between 500 and 350 K and not between 500 and 300 K. In addition, the engine is not reversible. If it were to be reversible, the work output should have been equal to $1000\left(1 - \frac{300}{500}\right) = 300$ kJ and not 250 kJ. Hence, 250 kJ is not the available energy but it is only the actual work produced by the engine under the given conditions.

SUMMARY

(1) The Clausius inequality dictates that

(a) for a reversible cycle, $\oint \frac{\partial Q}{T} = 0$

(b) for an irreversible cycle, $\oint \frac{\partial Q}{T} < 0$

(c) for an impossible cycle, $\oint \frac{\partial Q}{T} > 0$.

(2) Entropy is a property of a system, which is defined by

(a) $\Delta S = \int \frac{\partial Q_R}{T}$

(3) $\int \frac{\delta Q}{T}$ and $\int \frac{\partial Q_R}{T}$ for a given change of state are, in general, not the same. While the former is a path function, the latter is independent of the path between the two given states.

(4) The relation $ds = \frac{\partial Q}{T}$ is valid for an internally reversible process and it is not in contravention with the second law of thermodynamics.

(5) For a reversible non-cyclic process $\Delta S = \frac{\partial Q_R}{T}$ (by definition), while for an irreversible process $\Delta S > \frac{\partial Q}{T}$.

(6) It follows from (5) that while $\Delta S = 0$ for a reversible adiabatic process, $\Delta S > 0$ for an irreversible adiabatic process.

(7) The $T\mathrm{d}s$ relations that are extensively used in evaluating the changes in entropy of pure substances are:

(a) $T\mathrm{d}s = \mathrm{d}u + p\mathrm{d}v$

(b) $T\mathrm{d}s = \mathrm{d}h - v\mathrm{d}p$

(8) In any process, the net entropy change of the system and surroundings is > or = 0, and this is the gist of the principle of increase of entropy

(9) When heat transfer occurs across a finite temperature difference, there is a decrease in available energy and this decrease manifests itself as the increase in unavailable energy.

Illustrative Problems

8.1 Analyse Illustrative Problem 7.2 using the Clausius inequality statement.

► **Solution:**

Refer to Figure 7.6. For engine A, $Q_1 = 1000$ kJ, and from the first law

$$Q_2 = Q_1 - W = 1000 - 600 = 400 \text{ kJ}.$$

With respect to the system undergoing thermodynamic cycle, while Q_1 is positive, Q_2 is negative. Also, heat addition and rejection occur reversibly with the reservoirs at $T_1 = 1000$ K and $T_2 = 300$ K, respectively. Hence, for each process $\int \frac{\delta Q}{T} = \frac{Q_{net}}{T}$.

Therefore,
$$\begin{aligned}\oint \frac{\delta Q}{T} &= \int \frac{\delta Q}{T} \text{ at } T_1 + \int \frac{\delta Q}{T} \text{ at } T_2.\\ &= \left(\frac{Q_1}{T_1}\right) + \left(\frac{Q_2}{T_2}\right)\\ &= \frac{1000}{1000} + \left(\frac{-400}{300}\right) = -0.333 \text{ kJ/K}.\end{aligned}$$

that is $\oint \frac{\delta Q}{T} < 0$ and, hence, according to the Clausius inequality, the heat engine A is irreversible.

Similarly, for engine B,
$$\oint \frac{\delta Q}{T} = \left(\frac{Q_1}{T_1}\right) + \left(\frac{Q_2}{T_2}\right)$$
$$= \frac{1000}{1000} + \left(\frac{-300}{300}\right) = 0$$

and, hence, according to the Clausius inequality, engine B is reversible.

For engine C,
$$\oint \frac{\delta Q}{T} = \left(\frac{Q_1}{T_1}\right) + \left(\frac{Q_2}{T_2}\right)$$
$$= \frac{1000}{1000} + \left(\frac{-200}{300}\right) = +0.33 > 0$$

and, hence, according to the Clausius inequality, engine C is impossible.

8.2 A reversible heat engine receives heat from three high-temperature reservoirs at 1000 K, 800 K and 600 K and rejects 200 kJ heat to the lone sink at 300 K. The work output of the engine is 300 kJ. The heat transfer from the reservoir at 600 K is twice that from the reservoir at 1000 K. Calculate the heat transfers to the engine from each of the high-temperature reservoirs.

► Solution:

Figure 8.16 shows the arrangement for the problem.

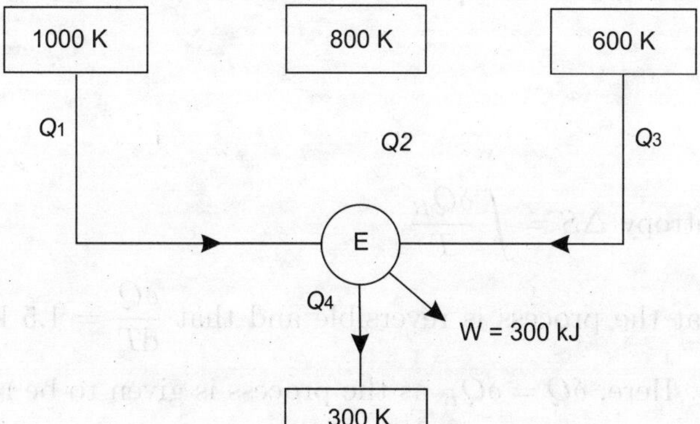

Fig. 8.16

From the first law of thermodynamics for the engine,
$$Q_1 + Q_2 + Q_3 - Q_4 = W$$

Here, from data, $Q_3 = 2Q_1, Q_4 = 200$ kJ and $W = 300$ kJ

and therefore, $$Q_1 + Q_2 + 2Q_1 - 200 = 300 \text{ or}$$
$$3Q_1 + Q_2 = 500 \tag{i}$$

Also, as the cycle is reversible, from the Clausius inequality the cyclic integral of $\dfrac{\delta Q}{T} = 0$. Taking the directions of heat transfer with respect to the engine which is undergoing a cycle,

$$\oint \frac{\delta Q}{T} = \frac{Q_1}{1000} + \frac{Q_2}{800} + \frac{Q_3}{600} - \frac{Q_4}{300} = 0$$

Therefore, $$\frac{Q_1}{1000} + \frac{Q_2}{800} + \frac{2Q_1}{600} = \frac{200}{300} \tag{ii}$$

Solving for Q_1 and Q_2, simultaneously, between relations (i) and (ii)

$$Q_1 = 72 \text{ kJ}, \; Q_2 = 284 \text{ kJ and } Q_3 = 144 \text{ kJ}$$

8.3 (a) In a certain reversible process, the rate of heat transfer to the system per unit temperature rise is constant and is given by $\delta Q/\mathrm{d}T = 1.5$ kJ/deg. Find the increase in entropy of the system as its temperature increases from 300 K to 500 K. (b) In a second process between the same end states, the temperature rise is accomplished by stirring plus addition of heat equal to half that in (a). What is the entropy increase in this case

► **Solution:**

(a) Increase in entropy $\Delta S = \displaystyle\int \frac{\delta Q_R}{T}$.

It is given that the process is reversible and that $\dfrac{\delta Q}{\mathrm{d}T} = 1.5$ kJ/deg and therefore $\dfrac{\delta Q}{T} = 1.5\,\dfrac{\mathrm{d}T}{T}$. Here, $\delta Q = \delta Q_R$ as the process is given to be reversible.

Therefore, $$\Delta S = \int 1.5\,\frac{\mathrm{d}T}{T} = 1.5 \ln\left(\frac{T_2}{T_1}\right).$$

From data, $T_1 = 300$ K and $T_2 = 500$ K, and therefore

$$\Delta S = 1.5 \ln\left(\frac{500}{300}\right) = 0.766 \text{ kJ/K}$$

(b) In this case, the process occurs between the same states as in (a). The system is also the same. But, the manner in which the state change occurs is different. Actually the process here is irreversible because of stirring work. Entropy, being a property, its change is independent of the path between two given states. Thus the entropy change in this case is also the same as that in (a), being equal to 0.766 kJ/K.

8.4 A certain pure substance of constant c_p (= 0.6 kJ/kg K) undergoes a change of state at constant pressure from 1 to 2 because of heating accompanied by stirring. The mass of the system is 2.5 kg and the temperatures at the two states are 30°C and 80°C, respectively. At the end of this process, the substance is subjected to a reversible constant pressure cooling process which restores it to the initial state. Find the increase in entropy in the first process.

► **Solution:**

The first and the second process occur between the same two state points. Hence, the entropy change in both processes must be the same but for their direction. The process from 1 to 2 is irreversible because of stirring work. The process from 2 to 1 is given to be reversible and hence entropy change can be evaluated along this path.

$$T\mathrm{d}s = \mathrm{d}h - v\mathrm{d}p$$

Therefore, $$\mathrm{d}s = \frac{\mathrm{d}h}{T} - \left(\frac{v}{T}\right)\mathrm{d}p$$

For the process 2–1, p is constant and therefore

$$\mathrm{d}s = \frac{\mathrm{d}h}{T}$$

For the substance, c_p is a constant and therefore $\mathrm{d}h = c_p\mathrm{d}T$

Therefore, $\mathrm{d}s = c_p\dfrac{\mathrm{d}T}{T}$ or

$$\Delta s = s_1 - s_2 = \int c_p \frac{\mathrm{d}T}{T} = c_p \ln\left(\frac{T_1}{T_2}\right)$$

Here, $T_1 = 30 + 273 = 303$ K and $T_2 = 80 + 273 = 353$ K.

Therefore, $s_1 - s_2 = 0.6 \ln\left(\dfrac{303}{353}\right) = -0.0916$ kJ/kg K

Now, entropy increase in the first process $= s_2 - s_1$

$= -(s_1 - s_2) = 0.0916$ kJ/kg K.

Therefore, $(S_2 - S_1) = m \times (s_2 - s_1) = 2.5 \times 0.0916 = 0.229$ kJ/K.

8.5 (a) one kilogram of water in an open vessel at 30°C is brought in contact with a reservoir at 100°C. What are the entropy changes for water, the reservoir and the universe, when the water temperature just reaches the temperature of the reservoir? Assume, for water, $dh = c dT$ where $c = 4.2$ kJ/kgK. (b) What would be the entropy increase of the universe had the water been heated from 30°C to 100°C by first bringing it in contact with a reservoir at 65°C and then with the reservoir at 100°C ?

▶ **Solution:**

(a) For a pure substance $ds = \dfrac{dh}{T} - \left(\dfrac{v}{T}\right) dp$

Heating of water in an open vessel is a constant-pressure process as the pressure is steady at the atmospheric value and hence in the equation $dp = 0$

Therefore,
$$ds = \frac{dh}{T} = c\frac{dT}{T} \text{ or}$$

$$\Delta s_{water} = s_2 - s_1 = c \ln\left(\frac{T_2}{T_1}\right)$$
$$= 4.2 \ln\left[\frac{(273+100)}{(273+30)}\right]$$
$$= 0.873 \text{ kJ/kg K.}$$

Heat transfer, Q, to water, by the first law is given by:

$Q = \Delta u + W$, where the work $W = 0$ as the process is also at constant volume (water is an incompressible substance), and $\Delta u = c(t_2 - t_1)$.

Therefore, $Q = 4.2(100 - 30) = 294$ kJ/kg. This heat is transferred from the reservoir to water. Heat transfer from the reservoir is internally reversible as its temperature remains the same.

Hence,
$$\Delta s_{reservoir} = \int \frac{\delta Q_{rev}}{T} = \frac{Q}{T} = \frac{-294}{(273+100)}$$
$$= -0.788 \text{ kJ/kg water K.}$$

Now,
$$\Delta s_{universe} = \Delta_{water} + \Delta s_{reservoir}$$
$$= 0.873 + (-0.788) = 0.085 \text{ kJ/kg water K.}$$

***Note*:** During this process the entropy of the universe has increased. Hence, the process is irreversible. In Chapter 7 on reversibility, it had been qualitatively stated that heat transfer with finite temperature difference makes a process irreversible. Here, the same is shown to be true quantitatively with the use of entropy.

(b) Water is heated first from 30°C to 65°C by heat transfer from the reservoir at 65°C and then from 65°C to 100°C by heat transfer from the reservoir at 100°C.

Following the same sequence of calculations as in (a), for the first stage of heating

$$\Delta s_{water,\text{ I}} = c \ln\left(\frac{T_2}{T_1}\right) = 4.2 \ln \frac{(273+65)}{(273+30)}$$
$$= 0.459 \text{ kJ/kg K}$$

$$\text{Heat transfer to water } = Q_1 = c(t_2 - t_1) = 4.2(65 - 30) = 147 \text{ kJ/kg}$$

$$\Delta s_{res,\text{ I}} = \int \frac{\delta Q_{rev}}{T} = \frac{Q}{T} = \frac{-147}{(273+65)}$$
$$= -0.435 \text{ kJ/kg water K}$$

$$\text{Similarly, } \Delta s_{water,\text{ II}} = c \ln\left(\frac{T_2}{T_1}\right) = 4.2 \ln \frac{(273+100)}{(273+65)}$$
$$= 0.414 \text{ kJ/kg K}$$

$$\text{Heat transfer to water } = Q_{II} = c(t_2 - t_1) = 4.2(1000 - 65)$$
$$= 147 \text{ kJ/kg}$$

$$\Delta s_{res,\text{ II}} = \int \frac{\delta Q_{rev}}{T} = \frac{Q}{T} = \frac{-147}{(273+100)}$$
$$= -0.394 \text{ kJ/kg water K.}$$

$$\Delta s_{universe} = \Delta s_{water,\text{ I}} + \Delta s_{reservoir,\text{ I}} + \Delta s_{water,\text{ II}} + \Delta s_{reservoir,\text{ II}}$$
$$= 0.459 + (-0.435) + 0.414 + (-0.394)$$
$$= 0.044 \text{ kJ/kg water K.}$$

***Note*:** Entropy increase of the universe in (b) is less than that in (a). This is because the irreversibility in (b) is less than that in (a). Although the heating of water in both the cases is from an initial temperature of 30°C to the final temperature of 100°C, step heating in (b) decreases the finite temperature difference causing heat transfer in each stage. This reduces the irreversibility.

8.6 A certain gas with a constant c_v of 0.7 kJ/kg K is expanded reversibly and adiabatically from 0.0625 m^3/kg, 500 K to 0.1875 m^3/kg. During this process, the temperature drops by 178 degrees. In another irreversible but adiabatic process of the same gas, from the same initial state to the same final specific volume, the drop in temperature is given to be only 38 degrees. Find the change in entropy in the second process.

▶ Solution:

Shown on T–S coordinates, in Figure 8.17, are the reversible adiabatic process 1–2 and the irreversible adiabatic process 1–3, both starting at the same initial state and ending at the same final specific volume.

Entropy change in process 1–2 is zero as it is reversible and adiabatic. Process 1–3 is irreversible but adiabatic and hence $s_3 - s_1 > 0$. The entropy change $(s_3 - s_1)$ can be calculated using the $T ds$ relations (8.14), provided the required data at the two end states are available. But the data is not sufficient to adapt this method. Hence, an indirect method has to be used. It can be seen from Figure 8.17 that,

$$(s_3 - s_1) = (s_3 - s_2) + (s_2 - s_1) = (s_3 - s_2) + 0$$

Therefore, $(s_3 - s_1) = (s_3 - s_2)$

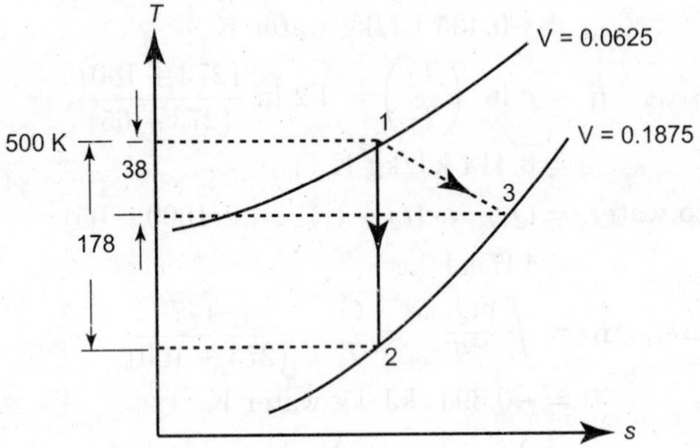

Fig. 8.17 Diagram for Illustrative Problem 8.6.

2–3 is a constant-volume process at $v = 0.1875$ m^3/kg.

From Tds relation, $\quad ds = \dfrac{du}{T} + \dfrac{pdv}{T}.$

For the process 2–3, $\quad dv = 0$ and hence $ds = \dfrac{du}{T} = c_v \dfrac{dT}{T}$

or

$$(s_3 - s_2) = \int ds = c_v \ln\left(\frac{T_3}{T_2}\right)$$

$$= 0.7 \ln \frac{(500 - 38)}{(500 - 178)} = 0.2527 \text{ kJ/kg K.}$$

Therefore, $\quad (s_3 - s_1) = (s_3 - s_2) = 0.2527$ kJ/kg K.

8.7 As shown in Figure 8.18, a reversible engine is operating having heat interactions with three reservoirs at temperatures of $T_1 = 500$ K, $T_2 = 300$ K and $T_3 = 250$ K. When Q_1, the heat transfer to the heat engine from the reservoir at T_1, is 500 kJ the work output of the engine is given to be 175 kJ. Find the magnitudes and directions of Q_2 and Q_3.

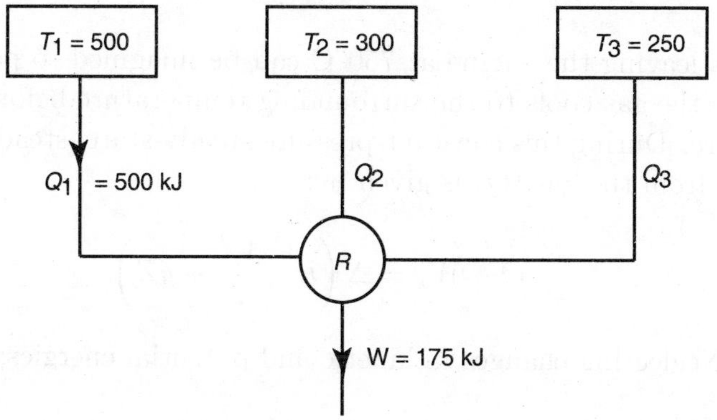

Fig. 8.18 Diagram for Illustrative Problem 8.7.

► **Solution:**

For the heat engine operating in a thermodynamic cycle, from the first law,

$$\Sigma Q = \Sigma W,$$

that is, $Q_1 + Q_2 + Q_3 = \Sigma W,$

where, from data, $Q_1 = +500$ kJ and $\Sigma W = +175$ kJ.

Therefore, $500 + Q_2 + Q_3 = 175$ or $Q_2 + Q_3 = -325.$ (i)

As the heat engine (cycle) is reversible, according to the Clausius inequality, $\oint \frac{\delta Q}{T} = 0$ and hence,

$$\frac{Q_1}{T_1} + \frac{Q_2}{T_2} + \frac{Q_3}{T_3} = 0$$

Therefore,

$$\frac{500}{500} + \frac{Q_2}{300} + \frac{Q_3}{250} = 0 \text{ or}$$

$$\frac{Q_2}{300} + \frac{Q_3}{250} = -1 \quad \text{(ii)}$$

Solving for Q_2 and Q_3 between (i) and (ii)

$Q_2 = -450$ kJ and $Q_3 = 125$ kJ that is, while 450 kJ is rejected by the engine to the reservoir at 300 K, 125 kJ, is transferred to the engine from the reservoir at 250 K.

8.8 Exhaust gases leave an internal combustion engine at 750°C and 1 atm, after having done 450 kJ work per kilogram gas in the engine cylinder. Assume that the enthalpy of the gas is a function of temperature only and that $c_p = 1.1$ kJ/kg K. Assume the temperature of the surroundings to be 27°C. Calculate (a) the available and unavailable parts of the energy in every kilogram gas discharged to the surroundings and (b) the ratio of the available energy lost to the engine work.

▶ **Solution:**

The exhaust gas leaving the engine at 750°C can be imagined to pass through a heat exchanger where the gas cools to the surrounding temperature before being discharged to the atmosphere. During this constant-pressure steady-state, steady-flow process, the heat transferred from the gas (Q) is given by:

$$Q - W_x = \Delta\left(h + \frac{V^2}{2} + gZ\right)$$

Here, $W_x = 0$. Neglecting changes in kinetic and potential energies,

$$Q = \Delta h$$

For the gases, as $h = h(T)$ only and c_p is a constant,

$$Q = \Delta h = c_p(t_2 - t_1) = 1.1 \times (27 - 750) = -795.3 \text{ kJ/kg}$$

The negative sign indicates that the heat transfer is from the gas to the surroundings.

This constant-pressure cooling process of the gas is shown by the path $1_g - 2_g$ in Figure 8.19*a*. The area below the path is equal to Q. The change in entropy during this process, Δs, is calculable using the $T ds$ relation.

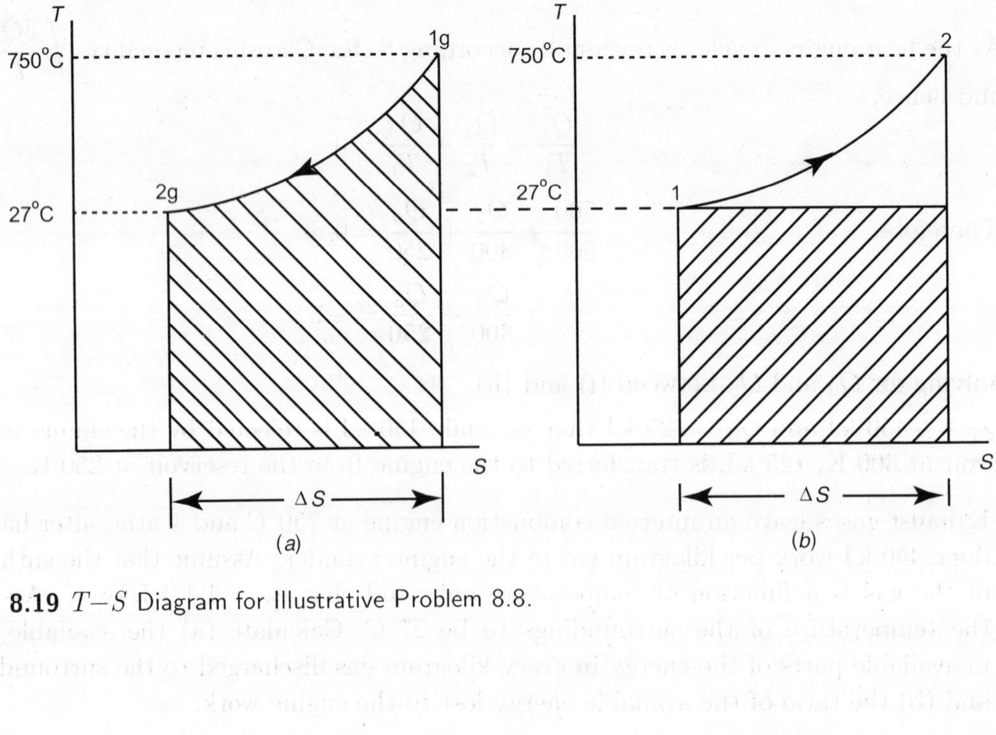

Fig. 8.19 T–S Diagram for Illustrative Problem 8.8.

$$ds = \frac{dh}{T} - \left(\frac{v}{T}\right) dp,$$

here $\quad dp = 0$ and $dh = c_p\, dT$

and hence, $\quad ds = c_p \dfrac{dT}{T}$

and therefore $\quad \Delta s = \displaystyle\int c_p \frac{dT}{T}.$

$$\Delta s = c_p \ln\left(\frac{T_2}{T_1}\right) = 1.1 \ln\left[\frac{(273+27)}{(273+750)}\right]$$

$$= -1.349 \text{ kJ/kg K}$$

If the heat transfer from the gas could be used in an externally reversible heat engine, the path of the heat addition process has to exactly match the cooling process in Figure 8.19 a, but the two process paths will be in opposite directions. This is shown by the process 1–2 in Figure 8.19*b*. During this process, Δs is equal to the value of Δs during the process 1_g–2_g. When this engine rejects heat to the surroundings at 27°C the magnitude of heat rejected (hatched area) is the unavailable part of the energy in the gases.

Unavailable energy $= T_0\Delta s = (273+27) \times 1.349 = 404.7$ kJ/kg gas.

Available energy $= Q - T_0\Delta s = 795.3 - 404.7 = 390.6$ kJ.

(b) The required ratio $= \dfrac{390.6}{450} = 0.868.$

Note: The useful part of energy in the exhaust gases is 86.8% of the work developed by the engine per every kilogram gas. This analysis shows that it is worth attempting to recover this energy which otherwise is thrown as waste to the atmosphere.

8.9 In a certain power station, a binary vapour cycle employing mercury and water as the working substance is used for power generation. In this cycle, the condensing mercury vapour at 272°C is used to vapourize water at 252°C. Assuming the surroundings to be at 27°C, calculate the increase in unavailable energy due to irreversible heat transfer between the mercury and water. Obtain the results based on 1 kg water evaporated. The heat transfer required for vapourizing 1 kg water = 1700 kJ.

► Solution:

Refer to Figure 8.20, in which 1_{Hg}–2_{Hg} represents the process of mercury condensation and 1_w–2_w represents the evaporation of water.

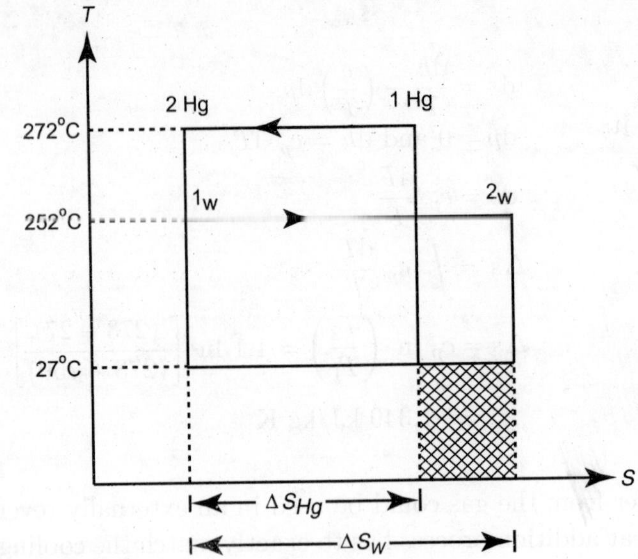

Fig. 8.20 Diagram for Problem 8.9.

Area below $1_w - 2_w$ is equal to heat transfer to water and this, from data, is equal to 1700 kJ/kg water. This amount of heat transfer to the water must have come from the condensing mercury and, hence, area below $1_{Hg} - 2_{Hg}$ is also equal to 1700 kJ/kg water evaporated.

$$\Delta S_{water} = \left(\frac{Q}{T}\right)_{water} = \frac{1700}{(273+252)} = 3.24 \text{ kJ/kg K.}$$

$$\Delta S_{Hg} = \left(\frac{Q}{T}\right)_{Hg} = \frac{-1700}{(273+272)} = -3.12 \text{ kJ/kg water K}$$

$$\begin{aligned}\text{Increase in unavailable energy} &= T_0 \Delta S_{net}\\ &= \text{double-hatched area}\\ &= T_0(\Delta s_{water} + \Delta s_{Hg})\\ &= (273+27)(3.24 - 3.12) = 36 \text{ kJ/kg water.}\end{aligned}$$

Problems

8.1 A system undergoes a reversible process during which heat added is given by $Q = aT^2 + bT$, where $a = 3.18$ J/K^2 and $b = 200$ J/K. Calculate the entropy change of the system as its temperature increases from 30 to 200°C.

8.2 A system undergoes a reversible process such that the path of the process is a straight line on a T–S diagram. Some properties at the initial and final states are as follows: $t_1 = 60°\text{C}$, $t_2 = 170°\text{C}$; $U_1 = 170$ kJ, $U_2 = 190$ kJ; $H_1 = 220$ kJ, $H_2 = 246$ kJ; $S_1 = 0.23$ kJ/K, $S_2 = 0.3$ kJ/K. Calculate the work during the process.

8.3 Prove that the minimum work input to a refrigerator that cools a rigid body from atmospheric temperature of $T_0(K)$ to a lower temperature, T_L, is equal to $T_0(S_0 - S_L) + (U_L - U_0)$, where S and U are the entropy and internal energy of the body being cooled.

8.4 An open insulated vessel is divided into two parts by a vertical non-conducting partition. On one side of the partition is 6 kg water at a temperature of 32°C while on the other side is 12 kg of water at a temperature of 77°C. When the partition is removed the two masses of water mix and after a time the condition is uniform throughout the vessel. Evaluate the increase in entropy of the system comprising 18 kg water. Assume for water $c = 4.2$ kJ/kg K and that there are no other heat transfers.

8.5 A steel tool of mass 0.5 kg, at a temperature of 400°C is plunged suddenly into an insulated vessel containing 10 kg oil at 30°C. After some time, the temperature of the contents of the vessel becomes uniform, during which period no oil evaporates. Calculate the increase in entropy of the final contents of the vessel. Assume for oil and steel the specific heats are, respectively, 1.9 kJ/kg K and 0.5 kJ/kg K.

8.6 A mass of m kg of a liquid at a temperature of T_1 K is mixed with an equal mass of the same liquid at a temperature of T_2 K, adiabatically. Show that the entropy increase of the universe is given by $2mc_p \ln\left[\dfrac{(T_1+T_2)/2}{(T_1T_2)^{0.5}}\right]$.

8.7 An electric current of 10 amp is maintained for 1 s in a resistor of 25 ohms, which is kept immersed in a large pool of water. The temperatures of the resistor and water remain constant at 37°C and 27°C, respectively. Calculate the entropy change for (i) the resistor, (ii) the water and (iii) the universe. Is the process reversible or irreversible?

8.8 Sketch the following processes on $T-S$ diagram and calculate the change in entropy in each case.

(a) One kilogram of water is heated at constant pressure from 25°C to 95°C. Assume for water $c = 4.2$ kJ/kg K.

(b) A certain 0.2 kg gas contained in an insulated rigid vessel is stirred, resulting in its temperature increasing from 25°C to 35°C. Assume for the gas $u = u(T)$ only and that $c_v = 700$ J/kg K.

(c) 1 kilogram of steam at 1 atm pressure and 100°C temperature is condensed completely to water at the same pressure and temperature. The specific enthalpy of steam decreases by 2257 kJ/kg.

8.9 One kilogram of water is heated at constant pressure from 25°C to 95°C. Assume for water $c = 4.2$ kJ/kg K. If the heat interaction is reversible, show the processes for the heat source and 1 kg water on T–S diagram and calculate the available and unavailable energy of the heat interaction. Also represent them by areas on T–S diagram. Assume the surrounding atmosphere to be at 1 atm and 27°C.

8.10 One kilogram of water is heated at constant pressure from 25°C to 95°C. Assume for water $c = 4.2$ kJ/kg K. The source for heating is steam condensing at 1 atm and 100°C. Sketch the processes for the steam and the water on T–S diagram and calculate the entropy increase of the universe per kilogram water. Show the increase in unavailable energy on the diagram. Also, calculate the same and express it as a percentage of the energy received by 1 kilogram water. Assume that there are no other heat losses and that the atmosphere is at 27°C and 1 atm.

8.11 One kilogram of water is heated at constant pressure from 25°C to 95°C. Assume for water $c = 4.2$ kJ/kg K. Suppose the heating is in a counter-flow heat exchanger with a stream of water cooling from 100°C to 30°C. Show the processes for the two streams of water on T–S diagram. Calculate the entropy increase of the universe assuming the heat exchanger to be adiabatic. Calculate the increase in unavailable energy due to irreversible heat transfer and express it as a percentage of the heat received by 1 kg water being heated. Compare the results with those of the previous problem and comment on the choice of a source to heat a fluid isobarically from t_1 to t_2. Assume the surroundings are at the same condition as in problem (8.10).

8.12 (a) A fluid system at a temperature of 70°C and a pressure of 1 bar undergoes a reversible process during which the temperature of the system remains constant. Given that heat transfer to the system is 120 kJ, evaluate the increase in entropy of the system.

(b) The system in (a) has a mass of 2.35 kg. Evaluate the increase in the specific entropy of the system.

(c) A second fluid, identical to that in (a) undergoes an irreversible isothermal process from the same initial state to the same final state as in (a). The heat transfer to the system in this process is 80 kJ. Evaluate the increase in entropy of the system.

8.13 A refrigerator patterned after a reversed Carnot cycle operates between −25°C and 35°C; 15 000 kJ is to be removed from the cold room at −25°C. The only difference in the cycle, compared to the reversed Carnot cycle, is that the reversible adiabatic expansion is replaced by an irreversible adiabatic process during which the entropy of the system increases by 10% of the entropy change of the system during the reversible isothermal compression process. Sketch the cycle on T–S diagram and show the area that corresponds to the heat absorbed at −25°C. Compute the power required.

8.14 An engine operates on a modified Carnot cycle between two reservoirs at 250°C and 30°C. During the isothermal expansion, the entropy of the system undergoing cycle changes by 1.3 kJ/K for 2 kg of the working substance. Fluid friction causes an increase in entropy of 0.04 kJ/K during the adiabatic expansion process. Sketch the cycle on T–S diagram. Compute heat added, heat rejected, work output and efficiency of the cycle. Compare the values with those obtained for the actual Carnot cycle under identical conditions.

8.15 Two identical solid bodies of mass m kg each are initially at a temperature of T_1 K. Suppose a refrigerator is made to operate between these two bodies until one of them is cooled to a lower temperature of T_2 K. Show that the theoretical minimum work required by the refrigerator is given by:

$W = mc\left[\left(\dfrac{T_1^2}{T_2}\right) + T_2 - 2T_1\right]$, where c is the specific heat of the solid.

8.16 A certain heat engine is designed to have four heat interactions in a cycle. The heat transfers and the temperatures are as follows: (i) +9 kJ at 1200°C; (ii) +18 kJ at 900°C; (iii) –3 kJ at 100°C and (iv) –2.5 kJ at 27°C. The heat engine is claimed to deliver a work of 5300 J. Determine whether or not the heat engine is feasible, and if feasible, classify it as either reversible or irreversible.

8.17 Consider two identical solid bodies A and B, each of mass m kg, at temperatures of T_A and T_B, respectively, where $T_A > T_B$. Suppose a heat engine is operated between these bodies until such a time that the two bodies reach a common temperature. Find the theoretical maximum work that can be developed in such a case and also the common temperature of these two bodies.

8.18 A reversible heat engine operates between three reservoirs as shown in Figure 8.21. If its thermal efficiency is 40% and $W = 2100$ kJ, calculate the values of Q_2 and t_1.

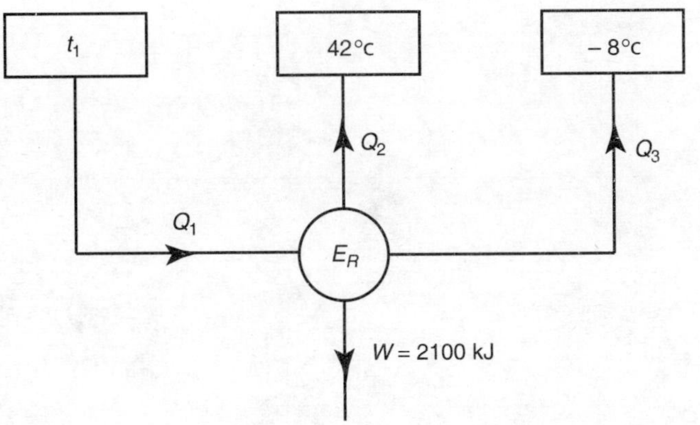

Fig. 8.21 Diagram for Problem 8.18.

8.19 In a certain power station, the evaporation of water at 250°C is done in a boiler by heat transfer from a furnace. Find the increase in entropy of the universe and the increase in unavailable energy due to irreversible heat transfer between the cooling gases and evaporating water. The gases are cooled at constant pressure from 1100°C to 550°C and all the heat transfer from the gases go to water. Assume for the gases $h = h(T)$ only and that $c_p = 0.95$ kJ/kg K. Obtain the results based on 1 kg water assuming 1620 kJ heat is required to evaporate 1 kg water.

8.20 Three heat engines, A, B and C, are operating separately between two reservoirs at 327°C and 27°C. The heat transfer to each engine is 1000 kJ. It is claimed that engine A develops 500 kJ work, B develops 600 kJ and C develops 400 kJ. Identify the engines as reversible, irreversible or impossible using: (a) Carnot's statement, (b) the Clausius inequality and (c) principle of increase of entropy.

IDEAL GAS AND IDEAL GAS MIXTURES

9

9.1 Introduction

Earlier chapters have focussed on the fundamentals of thermodynamics, evaluation of heat and work interactions in various processes, the relationship among these interactions and the changes in properties of the system in question. No attempt has been made in these chapters to explain how the various thermodynamic properties of a system at different states could be obtained. Properties like pressure, volume and temperature of a system at any given instant are measurable. However, in most thermodynamic analyses, besides p, V and T, values of various other properties, such as energy, specific heat, enthalpy, entropy etc ., may be required. These properties are not measurable in nature. They can only be calculated using the data on measurable properties. Before embarking on thermodynamic analyses of various processes, one must have information on how the required properties of the given working substance could be obtained either through empirical relations or through property tables. Among several commonly used working substances in engineering practice, the simplest are gases, gas mixtures, wet and dry vapours, mixtures of gases and vapours. This chapter is devoted to the study of a particular type of gas, namely, the *ideal gas* and the *mixtures of ideal gases.*

9.2 Ideal Gas

Any equation that represents the relation among pressure, volume and temperature, the three measurable properties, is called the equation of state of the substance in reference.

An ideal gas is one that follows the equation of state,

$$pv = \mathrm{R}T, \text{ or } pV = m\mathrm{R}T \tag{9.1}$$

at all ranges of pressure and temperature. In this equation, p is the pressure in N/m^2; v, the specific volume in m^3/kg; V, the total volume in m^3; T, the temperature in K; m, the mass in kg and R, the particular gas constant in J/kg K.

No real gas follows the equation $pV = m\mathrm{R}T$ at all ranges of pressure and temperature. The $p-V-T$ relations for most real gases are quite complex and they have been established only for a very few widely used gases. However, most real gases, especially those with low molar mass, tend to behave as ideal gases at high temperatures and/or low pressures. Thus, an ideal gas is only a mathematical model used to simplify the thermodynamic analyses in many situations.

When m kg of an ideal gas undergoes a change from state 1 to 2, irrespective of the nature of the process, the properties at these two states are related by

$$p_1V_1 = m\mathrm{R}T_1, \text{ or } \frac{p_1V_1}{T_1} = m\mathrm{R} \tag{i}$$

and

$$p_2V_2 = m\mathrm{R}T_2, \text{ or } \frac{p_2V_2}{T_2} = m\mathrm{R} \tag{ii}$$

As right-hand sides of both (i) and (ii) are equal,

$$\frac{p_1V_1}{T_1} = \frac{p_2V_2}{T_2} \tag{9.2}$$

9.2.1 Universal Gas Constant

Consider 1 kmol (M kg, where M is the molar mass or molecular weight of the gas) of any gas at one standard atmospheric pressure, p_o, equal to 1.01325 bar and a standard temperature, T_0, equal to 273.15 K. According to Avogadro's principle, all gases in this state occupy the same volume, V_0, equal to 22.416 m^3. Substitution of these numerical values in the ideal gas equation gives the following result

$$\frac{p_0V_0}{T_0} = M\mathrm{R} = 1.01325 \times 10^5 \times \frac{22.416}{273.15} \approx 8314.3 \text{ J/kmol K}$$

It may be observed that the value of $M\mathrm{R}$ $\left(= \frac{p_0V_0}{T_0}\right)$, is the same for all gases and is a constant. It is for this reason that $M\mathrm{R}$ is called the *universal gas constant* and is denoted by $\mathcal{R}$ (or $\bar{\mathrm{R}}$). But, the value of the *particular gas constant*, R, for each gas is different and the two are obviously related by

$$\mathrm{R} = \frac{\mathcal{R}}{M} = \frac{8314.3}{M} \text{ J/kg K} \tag{9.3}$$

where M is the molar mass of the gas, in kg/kmol.

The ideal gas equation can be written in any one of the following forms:

(i) $pv = \mathrm{R}T$

(ii) $pV = m\mathrm{R}T$

(iii) $pV = m\left(\frac{\mathcal{R}}{M}\right)T = \left(\frac{m}{M}\right)\mathcal{R}T$

(iv) $pV = n\mathcal{R}T$, where $n = \frac{m}{M}$ = number of kilomoles.

(v) $p\overline{v} = \mathcal{R}T$, where $\overline{v}$ = molar specific volume in m^3/kmol

Example 9.1

What is the volume occupied by 6 kg nitrogen (M = 28 kg/kmol) at 1 bar and 27°C.

▶ **Solution:**

As $$pV = m\text{R}T,\; V = m\left(\frac{\mathcal{R}}{M}\right)\frac{T}{p}$$

$$= 6\times\left(\frac{8314.3}{28}\right)\times\frac{(273+27)}{10^5} = 5.345 \text{ m}^3.$$

Example 9.2

Air confined in a piston–cylinder mechanism expands from 10 bar, 227°C, 0.01 m^3 to 1 bar, 27°C. Find the final volume.

▶ **Solution:**

$$\frac{p_1V_1}{T_1} = \frac{p_2V_2}{T_2}$$

where suffixes 1 and 2 refer to initial and final states, respectively. Therefore,

$$V_2 = \left(\frac{p_1}{p_2}\right)\left(\frac{T_2}{T_1}\right)(V_1)$$

$$= \left(\frac{10}{1}\right)\left[\frac{(273+27)}{(273+227)}\right](0.01) = 0.06 \text{ m}^3.$$

9.2.2 Internal Energy and Enthalpy of an Ideal Gas

The internal energy and enthalpy of an ideal gas are functions of temperature only. This fact is elaborated below:

Consider an infinitesimal reversible cycle 1–2–3–4–1 undergone by an ideal gas. This cycle, comprising two constant temperature processes (1–2 and 3–4) and two constant volume processes (2–3 and 4–1) is shown plotted in Figure 9.1 on both T–s and p–v coordinates.

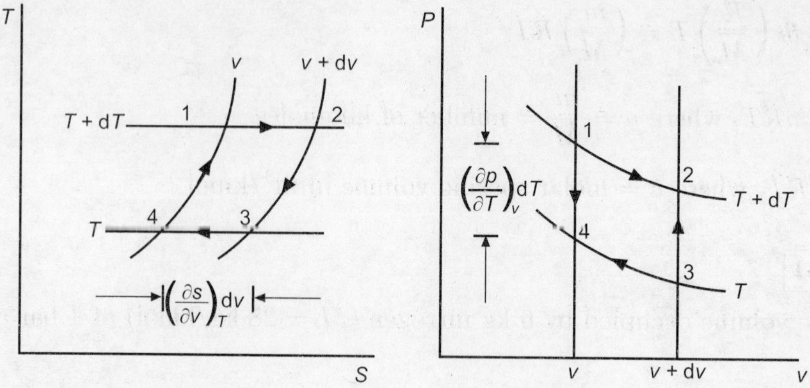

Fig. 9.1 Reversible cycle on $T-s$ and $p-v$ diagrams.

Area bound by the cycle on a $p-v$ plane is equal to the net work interaction during the cycle and the area bound by the cycle on a $T-s$ plane is equal to the net heat interaction during the cycle. According to the first law of thermodynamics, net work is equal to net heat in a cycle and hence the two areas must be equal.

Thus, in the limit, area of the cycle 1–2–3–4–1 on $T-s$ diagram is equal to the area of the same cycle on $p-v$ diagram.

That is,
$$\left[\frac{\partial s}{\partial v}\right]_T \mathrm{d}v \times \mathrm{d}T = \left[\frac{\partial p}{\partial T}\right]_v \mathrm{d}T \times \mathrm{d}v$$

Or,
$$\left[\frac{\partial s}{\partial v}\right]_T = \left[\frac{\partial p}{\partial T}\right]_v \tag{i}$$

An ideal gas is a pure substance and, hence, $\mathrm{d}u = T\mathrm{d}s - p\mathrm{d}v$. Differentiating this equation with respect to v at constant T,

$$\left[\frac{\partial u}{\partial v}\right]_T = T\left[\frac{\partial s}{\partial v}\right]_T - p \tag{ii}$$

Eliminating $\left[\frac{\partial s}{\partial v}\right]_T$ between (i) and (ii),

$$\left[\frac{\partial u}{\partial v}\right]_T = T\left[\frac{\partial p}{\partial T}\right]_v - p \tag{iii}$$

For an ideal gas, $pv = \mathrm{R}T$, or $p = \frac{T\mathrm{R}}{v}$. Differentiating with respect to T at constant volume,

$$\left[\frac{\partial p}{\partial T}\right]_v = \frac{\mathrm{R}}{v} \tag{iv}$$

Substituting for $\left[\dfrac{\partial p}{\partial T}\right]_v$ in (iii) from (iv),

$$\left[\frac{\partial u}{\partial v}\right]_T = \mathrm{T}\left(\frac{\mathrm{R}}{v}\right) - p = p - p = 0.$$

As $\left[\dfrac{\partial u}{\partial v}\right]_T = 0$, the internal energy of an ideal gas, at a given temperature, is independent of v. In other words, internal energy depends on temperature only.

Now, for a pure substance, $h = u + pv$ and for an ideal gas $pv = \mathrm{R}T$. Therefore, for an ideal gas, $h = u + \mathrm{R}T$. Here, u is a function of temperature only. Therefore, h also has to be independent of v and dependent on temperature only.

The properties v and T are independent for gases. For every different value of v at a given temperature the internal energy is the same. Each set of values of v and T gives a state point which has a particular pressure. Thus, when u is independent of v it must also be independent of p. So will h be. Therefore, u and h are functions of temperature only. This is mathematically represented as: $u = u(T)$ and $h = h(T)$.

Example 9.3

(a) It is given that the enthalpy of air at 1 bar and 50°C is 324.6 kJ/kg. Assume air behaves as an ideal gas and that its molar mass is 28.97 kg/kmol. Find its internal energy. (b) find the enthalpy and internal energy of air at (i) 10 bar, 50°C and (ii) 0.085 $\mathrm{m^3/kg}$ and 50°C.

► **Solution:**

$$h = u + pv \text{ and for an ideal gas } pv = \mathrm{R}T \text{ and hence}$$

$$h = u + \mathrm{R}T = u + \left(\frac{\mathcal{R}}{M}\right)T$$

or

$$u = h - \left(\frac{\mathcal{R}}{M}\right)T$$

$$= 324.6 - \left(\frac{8314.3 \times 10^{-3}}{28.97}\right) \times (273 + 50) = 231.9 \text{ kJ/kg}$$

(b)

(i) h and u are functions of temperature only. Temperature in this case is also 50°C, the same as in (a). Hence, the internal energy and enthalpy have the same values as in (a) irrespective of pressure.

(ii) For the same reason, in this case also u and h have the same values as in (a).

9.2.3 Specific Heats of an Ideal Gas

By definition, for a pure substance, the specific heats at constant volume (c_v) and constant pressure (c_p) are given by: $c_v = \left[\dfrac{\partial u}{\partial T}\right]_v$ and $(c_p) = \left[\dfrac{\partial h}{\partial T}\right]_p$. An ideal gas is a pure substance and its u and h are independent of v and p and are functions of temperature only. Therefore, the above two expressions for the two specific heats can be written as,

$$c_v = \frac{\mathrm{d}u}{\mathrm{d}T} \text{ and } c_p = \frac{\mathrm{d}h}{\mathrm{d}T} \tag{9.4}$$

or,
$$\mathrm{d}u = c_v \mathrm{d}T \text{ and } \mathrm{d}h = c_p \mathrm{d}T \tag{9.5}$$

Therefore,
$$\Delta u = u_2 - u_1 = \int c_v \mathrm{d}T \tag{9.6}$$

and
$$\Delta h = h_2 - h_1 = \int c_p \mathrm{d}T \tag{9.7}$$

9.2.4 Relation among Specific Heats and Gas Constant

From (9.4), $c_p = \dfrac{\mathrm{d}h}{\mathrm{d}T}$.

Also, for an ideal gas, $h = u + \mathrm{R}T$. Therefore,

$$c_p = \frac{\mathrm{d}(u + \mathrm{R}T)}{\mathrm{d}T} = \frac{\mathrm{d}u}{\mathrm{d}T} + \mathrm{R}$$
$$= c_v + \mathrm{R}.$$

That is,
$$c_p - c_v = \mathrm{R} \tag{9.8}$$

For an ideal gas, the difference between the two specific heats is always equal to its gas constant.

Example 9.4

Calculate the value of c_p for air ($M = 28.97$ kg/kmol) given that its $c_v = 718$ J/kgK.

▶ Solution:

For an ideal gas, $c_p - c_v = \mathrm{R}$, and, $\mathrm{R} = \dfrac{\mathcal{R}}{M}$ Therefore,

$$c_p - c_v = \frac{\mathcal{R}}{M}$$

Hence, $c_p - 718 = \dfrac{8314.3}{28.97}$ or $c_p = 1005$ J/kg K

9.2.5 Perfect Gas and Semi-Perfect Gas

The internal energy of an ideal gas is a function of temperature only. The relation between u and T may either be linear or non-linear.

(i) Suppose the relation is linear and is of the form,

$$u = \mathrm{A} + \mathrm{B}T, \text{ where A and B are constants.}$$

Then, $$c_v = \frac{\mathrm{d}u}{\mathrm{d}T} = \mathrm{B} \text{ which happens to be a constant.}$$

Also, $$h = u + \mathrm{R}T = (\mathrm{A} + \mathrm{B}T) + \mathrm{R}T$$

Therefore, $$c_p = \frac{\mathrm{d}h}{\mathrm{d}T} = (\mathrm{B} + \mathrm{R}), \text{ which is again a constant.}$$

(ii) Suppose the relation between u and T is non-linear and is, say, of the form,

$$u = \mathrm{A} + \mathrm{B}T + \mathrm{C}T^2 + \mathrm{D}T^3 + \cdots$$

Then, $$c_v = \frac{\mathrm{d}u}{\mathrm{d}T} = \mathrm{B} + 2\mathrm{C}T + 3\mathrm{D}T^2 + \cdots$$

Hence, c_v will also be a function of temperature.

Again, $$h = u + \mathrm{R}T = (\mathrm{A} + \mathrm{B}T + \mathrm{C}T^2 + \mathrm{D}T^3 + \ldots) + \mathrm{R}T$$

$$c_p = \frac{\mathrm{d}h}{\mathrm{d}T} = (\mathrm{B} + 2\mathrm{C}T + 3\mathrm{D}T^2 + \ldots) + \mathrm{R}$$

Thus, c_p is also a function of temperature. Hence, for an ideal gas, the specific heats could either be constants or functions of temperature. These two possibilities lead to the classification of an ideal gas as either a perfect gas or a semi-perfect gas.

A *perfect gas* is an ideal gas whose specific heats are constants and a *semi-perfect gas* is an ideal gas whose specific heats are functions of temperature.

From (9.6) and (9.7),

$$\Delta u = u_2 - u_1 = \int_1^2 c_v \mathrm{d}T$$

and $$\Delta h = h_2 - h_1 = \int_1^2 c_p \mathrm{d}T.$$

As a sequel to the classification of an ideal gas as either a perfect gas or a semi-perfect gas, to calculate the values of Δu and Δh, while c_p and c_v can be taken out of the integral sign for a perfect gas it cannot be done so for a semi-perfect gas.

Therefore, for a perfect gas,

$$\Delta u = u_2 - u_1 = c_v \int_1^2 \mathrm{d}T = c_v(T_2 - T_1) \tag{9.9}$$

and $$\Delta h = h_2 - h_1 = c_p \int_1^2 \mathrm{d}T = c_p(T_2 - T_1) \tag{9.10}$$

and for a *semi-perfect gas*,

$$\Delta u = u_2 - u_1 = \int_1^2 c_v \mathrm{d}T \neq c_v \int \mathrm{d}T \neq c_v(T_2 - T_1) \text{ and}$$

$$\Delta h = h_2 - h_1 = \int_1^2 c_p \mathrm{d}T \neq c_p \int \mathrm{d}T \neq c_p(T_2 - T_1).$$

Example 9.5

(a) Gaseous nitrogen is compressed in a piston–cylinder device from 1 bar, 27°C to 10 bar, 327°C. Calculate the increase in its specific enthalpy, assuming its c_p is given by $c_p = 1.4 - 18.3\left(\frac{T}{100}\right)^{-1.5} + 38.3\left(\frac{T}{100}\right)^{-2} - 29.3\left(\frac{T}{100}\right)^{-3}$, where c_p is in kJ/kgK and T is in kelvin. (b) Calculate the change in enthalpy, considering nitrogen as a perfect gas with $c_p = 1.042$ kJ/kg K.

▶ Solution:

As c_p is a function of temperature, the gas is a semi-perfect gas. Therefore,

$$\Delta h = h_2 - h_1 = \int_1^2 c_p \mathrm{d}T$$

$$= \int \left[1.4 - 18.3\left(\frac{T}{100}\right)^{-1.5} + 38.3\left(\frac{T}{100}\right)^{-2} - 29.3\left(\frac{T}{100}\right)^{-3}\right] \mathrm{d}T$$

$$= 1.4(T_2 - T_1) - 18.3\frac{(T_2^{-0.5} - T_1^{-0.5})}{(-0.5)100^{-1.5}}$$

$$+ 38.3 \times \frac{(T_2^{-1} - T_1^{-1})}{(-1)100^{-2}} - \frac{29.3(T_2^{-2} - T_1^{-2})}{(-2)100^{-3}}$$

Substituting for $T_1 = (273 + 27) = 300$ K and $T_2 = (273 + 327) = 600$ K

$h = 317.34$ kJ/kg

(b) $\Delta h = h_2 - h_1 = \int c_p \mathrm{d}T$

$$= c_p(T_2 - T_1) = 1.042 \times (600 - 300) = 312.6 \text{ kJ/kg}$$

9.3 Analysis of Some Reversible Processes

The procedure to be used in evaluating heat and work interactions and change in properties of a perfect gas (an ideal gas with constant specific heats) in some common reversible (quasi-static) processes is discussed below.

9.3.1 Reversible Constant Pressure Process

A constant pressure process is shown in Figure 9.2 on $p-V$ and $T-S$ coordinates. In such a process, the pressure remains the same all through the processes, starting from its initial state 1 to its final state 2. Hence, $p_1 = p = p_2 =$ a constant. An example of such a process is the expansion or compression of an ideal gas in a vertical frictionless piston–cylinder mechanism in which the piston carries a constant weight. In such a case, the work interaction is only the displacement work W which is given by $\int p dV$.

$$\begin{aligned} W &= \int p\mathrm{d}V = p\int \mathrm{d}V \\ &= p(V_2 - V_1) = (pV_2 - pV_1) \\ &= (p_2V_2 - p_1V_1) \end{aligned} \tag{9.11}$$

Also,
$$\begin{aligned} W &= (m\mathrm{R}T_2 - m\mathrm{R}T_1) \\ &= m\mathrm{R}(T_2 - T_1) \end{aligned} \tag{9.12}$$

For a perfect gas, $\Delta u = c_v(T_2 - T_1)$ (9.9)

Therefore, $\Delta U = m\Delta u = mc_v(T_2 - T_1)$

From the first law, $Q - W = \Delta U.$

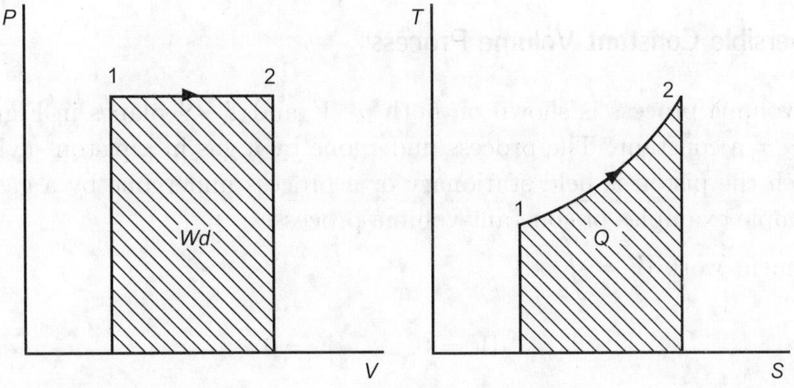

Fig. 9.2 Constant pressure process on $p-V$ and $T-S$ diagrams.

Therefore, $Q - m\mathrm{R}(T_2 - T_1) = mc_v(T_2 - T_1)$, or

$$\begin{aligned} Q &= m\mathrm{R}(T_2 - T_1) + mc_v(T_2 - T_1) \\ &= m(\mathrm{R} + c_v)(T_2 - T_1) \\ &= mc_p(T_2 - T_1) \end{aligned} \tag{9.13}$$

Also, for a perfect gas, $\Delta h = c_p(T_2 - T_1)$ (9.10)

Therefore, $\Delta H = m\Delta h = mc_p(T_2 - T_1)$

Substituting for $mc_p(T_2 - T_1)$ in (9.13)

$$Q = \Delta H \tag{9.14}$$

That is, heat transfer in a constant pressure process is always equal to the change in enthalpy.

Reversible constant pressure process:

$$W = p(V_2 - V_1) = m\mathrm{R}(T_2 - T_1)$$
$$Q = mc_p(T_2 - T_1) = \Delta H$$
$$\Delta U = mc_v(T_2 - T_1)$$

Also, graphically, while the area bound by the path 1–2 and the extreme ordinates on the p–V diagram gives the displacement work interaction, the area on the T–S diagram gives the heat interaction. The process shown here is a process of expansion as $V_2 > V_1$. The analysis is valid even if $V_2 < V_1$, in which case, we would be dealing with a compression process and hence the directions of heat and work interactions would be different.

9.3.2 Reversible Constant Volume Process

A constant volume process is shown on both p–V and T–S planes in Figure 9.3. Here $V_1 = V = V_2 =$ a constant. The process undergone by a gas in a piston–cylinder mechanism in which the piston is held stationary or a process undergone by a gas in any rigid vessel are simple examples of constant-volume processes.

Displacement work $W = \int p\mathrm{d}V$

As $\mathrm{d}V = 0, \quad W = 0$ (9.15)

For a perfect gas $\Delta U = m\Delta u = mc_{\bar{v}}(T_2 - T_1)$

From the first law, $Q - W = \Delta U$.

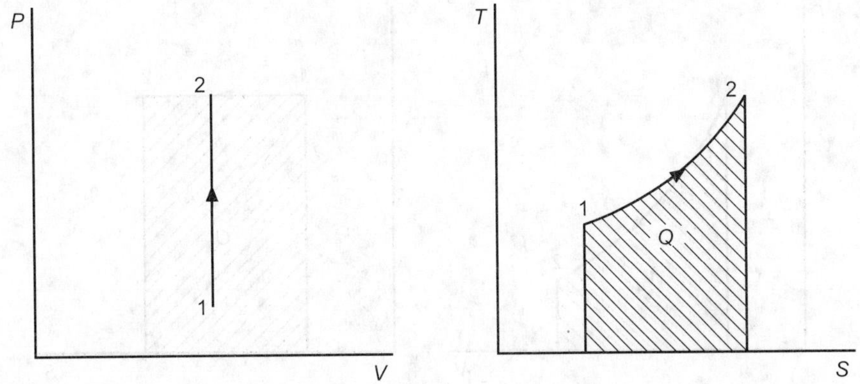

Fig. 9.3 Constant volume process on $p\text{–}V$ and $T\text{–}S$ diagrams.

Therefore, $Q - 0 = mc_v(T_2 - T_1)$ or,

$$Q = mc_v(T_2 - T_1) = \Delta U \tag{9.16}$$

> **Reversible constant volume process:**
> $W = 0$ and
> $Q = mc_v(T_2 - T_1) = \Delta U$

9.3.3 Reversible Hyperbolic Process

This process, which follows the path pV = constant, is shown in Figure 9.4 on $p\text{–}V$ and $T\text{–}S$ diagrams. For an ideal gas, $pv = \mathrm{R}T$. Hence if pv = constant, T must also be a constant. Thus, for an ideal gas, a hyperbolic process is also a constant-temperature or isothermal process.

Let $p_1V_1 = pV = p_2V_2 = \mathrm{C}$, a constant.

That is, $m\mathrm{R}T_1 = m\mathrm{R}T = m\mathrm{R}T_2 = \mathrm{C}$

Therefore $T_1 = T = T_2 = \dfrac{\mathrm{C}}{m\mathrm{R}}$ = another constant.

Displacement work $W = \int p\mathrm{d}V$

Multiplying and dividing the right-hand side of the above relation by V,

$$W = \int \left(\frac{pV}{V}\right) \mathrm{d}V, \text{ and therefore}$$

$$W = \int \left(\frac{\mathrm{C}}{V}\right) \mathrm{d}V = \int \mathrm{C}\left(\frac{\mathrm{d}V}{V}\right) = \mathrm{C}\ \ln\left(\frac{V_2}{V_1}\right), \quad \text{or}$$

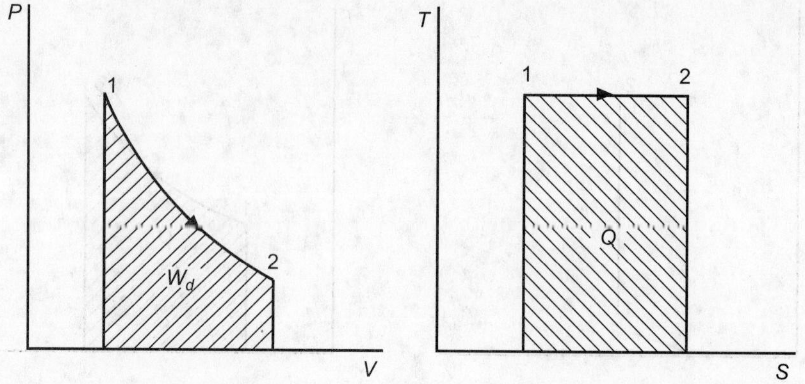

Fig. 9.4 $p-V\, T-S$

$$W = p_1V_1 \ln\left(\frac{V_2}{V_1}\right) = p_2V_2 \ln\left(\frac{V_2}{V_1}\right) \tag{9.17}$$

Also, as p is inversely proportional to V, $\left(\frac{V_2}{V_1}\right) = \left(\frac{p_1}{p_2}\right)$

Therefore
$$W = p_1V_1 \ln\left(\frac{p_1}{p_2}\right) = p_2V_2 \ln\left(\frac{p_1}{p_2}\right) \tag{9.18}$$

and
$$\Delta U = m\Delta u = mc_v(T_2 - T_1) = 0 \text{ as } T_2 = T_1.$$

From the first law,
$$Q - W = \Delta U. \text{ Therefore, as } \Delta U = 0, Q = W. \tag{9.19}$$

Reversible hyperbolic process pV = Constant
(Also, an isothermal process for an ideal gas)

$$Q = W = p_1V_1 \ln\left(\frac{V_2}{V_1}\right) = p_2V_2 \ln\left(\frac{V_2}{V_1}\right)$$

$$Q = W = p_1V_1 \ln\left(\frac{p_1}{p_2}\right) = p_2V_2 \ln\left(\frac{p_1}{p_2}\right) \text{ and}$$

$$\Delta U = 0$$

9.3.4 Reversible Polytropic Process

A reversible polytropic process is one which follows the path pV^n = constant, where n is the exponent of expansion or compression. This process is shown in Figure 9.5 on both $p-V$ and $T-S$ planes.

Here, $p_1V_1^n = pV^n = p_2V_2^n$ = constant, say = C.

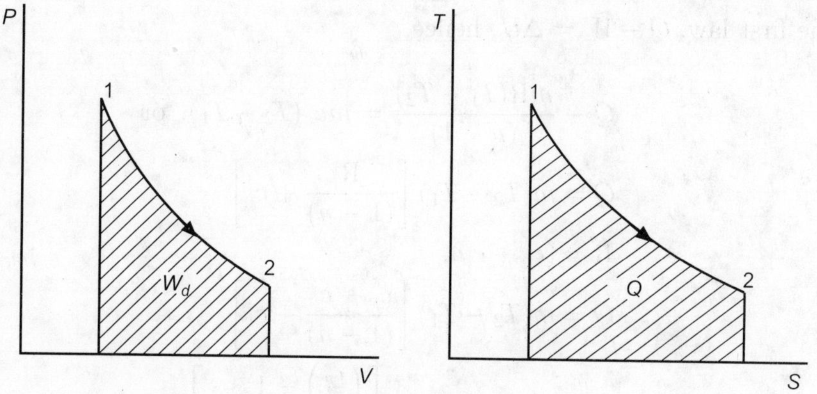

Fig. 9.5 Polytropic process on $p-V$ and $T-S$ diagrams.

Displacement work $W = \int p\mathrm{d}V$

Dividing and multiplying the right-hand side by V^n

$$\begin{aligned}
W &= \int \left(\frac{pV^n}{V^n}\right)\mathrm{d}V \\
&= \int \left(\frac{C\mathrm{d}V}{V^n}\right) = C\int V^{-n}\mathrm{d}V \\
&= C\left[\frac{(V_2^{-n+1} - V_1^{-n+1})}{(-n+1)}\right] \\
&= \frac{CV_2^{-n+1} - CV_1^{-n+1}}{(-n+1)} \\
&= \frac{(p_2V_2^nV_2^{-n+1} - p_1V_1^nV_1^{-n+1})}{(-n+1)} \\
&= \frac{(p_2V_2 - p_1V_1)}{(-n+1)} \\
&= \frac{(p_1V_1 - p_2V_2)}{(n-1)} \qquad (9.20)
\end{aligned}$$

From ideal gas equation $p_1V_1 = m\mathrm{R}T_1$ and $p_2V_2 = m\mathrm{R}T_2$

Substituting for p_1V_1 and p_2V_2 in (9.20)

$$W = \frac{(m\mathrm{R}T_1 - m\mathrm{R}T_2)}{(n-1)} = \frac{m\mathrm{R}(T_1 - T_2)}{(n-1)} \qquad (9.21)$$

Also, $\Delta U = m\Delta u = mc_v(T_2 - T_1)$

From the first law, $Q - W = \Delta U$, hence,

$$Q - \frac{m\text{R}(T_1 - T_2)}{(n-1)} = mc_v(T_2 - T_1), \text{ or}$$

$$Q = m(T_2 - T_1)\left[\frac{\text{R}}{(1-n)} + c_v\right]$$

As

$$\text{R} = (c_p - c_v),$$

$$Q = m(T_2 - T_1)\left[\frac{c_p - c_v}{(1-n)} + c_v\right]$$

$$= mc_v(T_2 - T_1)\left[\frac{\left(\frac{c_p}{c_v}\right) - 1}{(1-n)} + 1\right]$$

The ratio of specific heats, $\left(\dfrac{c_p}{c_v}\right)$, is denoted by γ. Therefore,

$$Q = mc_v(T_2 - T_1)\left[\frac{(\gamma - 1)}{(1-n)} + 1\right], \text{ or}$$

$$Q = mc_v(T_2 - T_1)\left[\frac{(\gamma - n)}{(1-n)}\right] \tag{9.22}$$

As

$$\Delta U = m\Delta u = mc_v(T_2 - T_1)$$

$$Q = \Delta U\left[\frac{(\gamma - n)}{(1-n)}\right] \tag{9.23}$$

Reversible polytropic process pV^n = Constant

$$W = \frac{(p_1V_1 - p_2V_2)}{(n-1)} = \frac{mR(T_1 - T_2)}{(n-1)}$$

$$Q = mc_v(T_2 - T_1)\left[\frac{(\gamma - n)}{(1-n)}\right] = \Delta U\left[\frac{(\gamma - n)}{(1-n)}\right]$$

$$\Delta U = mc_v(T_2 - T_1)$$

9.3.4.1 To show that pv^γ = Constant for a Reversible Adiabatic Process

As shown above, for a reversible polytropic process, pv^n = constant. Heat transfer during this process is given by:

$$Q = \Delta U\left[\frac{(\gamma - n)}{(1-n)}\right] \tag{9.23}$$

Q is equal to zero for an adiabatic process, and therefore

$$\Delta U \left[\frac{(\gamma - n)}{(1-n)}\right] = 0$$

that is, $$(\gamma - n) = 0 \text{ or } \gamma = n$$

Hence, $n = \gamma$ for a reversible adiabatic process. In other words, a reversible adiabatic process is one which follows the path,

$$pv^{\gamma} = \text{constant} \tag{9.24}$$

9.3.4.2 General Nature of the Process pv^n = Constant

Equations (9.20–9.23) give the work and heat interactions in a reversible polytropic process. These expressions are general in nature and can be used to evaluate the work and heat interactions in any reversible process using an appropriate value for n.

For example, if $n = 0, pv^0 = \text{constant}$, which means $p = \text{constant}$. Thus $pv^n = \text{constant}$ represents a reversible constant pressure process when $n = 0$. The expressions for heat and work interactions are obtained by substituting $n = 0$ in (9.20–9.23) which are the same as those derived for such a process in an orthodox manner in section 9.3.1.

Similarly if $n = \infty, pv^{\infty} = \text{constant}$, which means $v = \text{constant}$. Thus $pv^n = \text{constant}$ represents a reversible constant volume process when $n = \infty$. The expressions for heat and work interactions we get by substituting $n = 0$ in (9.20–9.23) are the same as those derived for such a process in an orthodox manner in section 9.3.2.

Further if $n = 1$, $pv^1 = \text{constant}$, which means $pv = \text{constant}$. Thus, $pv^n = \text{constant}$ represents a reversible hyperbolic process (also a reversible isothermal process for an ideal gas) when $n = 1$. However, the expressions for heat and work interactions cannot be obtained by substituting $n = 1$ in (9.20–9.23) as they turn out to be indeterminate. Thus, to find the work and heat interactions, only the orthodox method explained in section 9.3.3 has to be used.

Lastly, as explained under section 9.3.4.1, if $n = \gamma$, $pv^{\gamma} = \text{constant}$. Thus $pv^n = \text{constant}$ represents a reversible adiabatic process when $n = \gamma$. (This is proved in section 9.3.4.1).

9.3.5 Diagrammatic Representation of Processes

All the processes discussed hitherto have been shown together on a single $p-V$ diagram, in Figure 9.6. In this diagram, $a-b$ represents the constant-pressure process for which $n = 0$. Similarly, $c-d$ is the reversible isothermal process for which $n = 1$; $e-f$ is the reversible adiabatic process for which $n = \gamma$; $g-h$ is a general reversible polytropic process for which $n > \gamma$ and $i-j$ is the constant-volume process for which $n = \infty$. It is worth noting here that the slopes of the paths of various processes change gradually with an increase in the value of n. Also, if it is assumed that each process starts from the same initial state 'o', the processes proceeding towards the right of 'o', such as $o-a$, $o-c$, $o-e$, $o-g$ and $o-i$, represent the expansion processes during which there is generally a decrease in pressure and an increase in volume. Similarly, the processes proceeding towards the left of 'o' represent the compression processes during which generally there is an increase in pressure and a decrease in volume.

The above processes are also shown together on $T-s$ plane in Figure 9.7. However, no such order as observed in the $p-v$ diagram of Figure 9.6 is seen here.

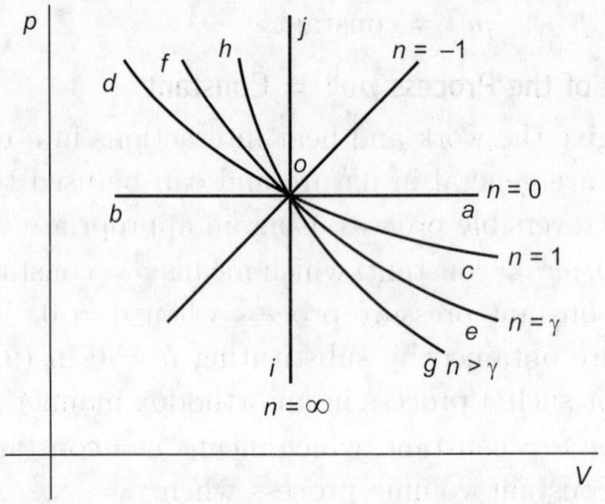

Fig. 9.6 Various processes on $p-V$ diagram.

9.3.6 Relationship among p, V and T in a Reversible Polytropic Process

Discussed below are the relations between p and V, T and V and T and p for a reversible polytropic process.

A reversible polytropic process is represented by $pV^n = C$. (i)

For an ideal gas, $$p = \frac{m\mathrm{R}T}{V} \qquad \text{(ii)}$$

Replacing p in (i) by $\frac{m\mathrm{R}T}{V}$ from (ii),

$$\left(\frac{m\mathrm{R}T}{V}\right) V^n = \mathrm{C}, \text{ or } TV^{n-1} = \frac{C}{m\mathrm{R}} = \text{another constant.}$$

Therefore, the relation between T and V is

$$TV^{n-1} = \text{constant} \qquad \text{(iii)}$$

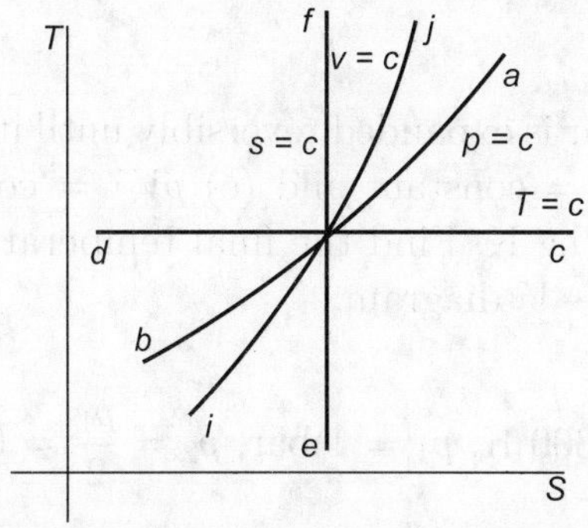

Fig. 9.7 Various processes on $T\text{–}S$ diagram.

From the ideal gas equation of state, $V = \dfrac{m\mathrm{R}T}{p}$.

Substituting for V in (i) by $\dfrac{m\mathrm{R}T}{p}$,

$p\left[\dfrac{m\mathrm{R}T}{p}\right]^n = \mathrm{C}$, or $T^n p^{1-n} = \dfrac{C}{(m\mathrm{R})^n}$ = another constant.

Therefore, the relation between T and p is

$$T^n p^{1-n} = \text{constant} \tag{iv}$$

For a reversible polytropic process, occurring between an initial state 1 and a final state 2, the relations in (i), (iii) and (iv) may be summarized as given below:

(i) $pV^n =$ constant, hence $p_1 V_1^n = p_2 V_2^n$, or

$$\left[\frac{p_2}{p_1}\right] = \left[\frac{V_1}{V_2}\right]^n \tag{9.25}$$

(ii) $TV^{n-1} =$ constant, hence $T_1 V_1^{n-1} = T_2 V_2^{n-1}$, or

$$\left[\frac{T_2}{T_1}\right] = \left[\frac{V_1}{V_2}\right]^{n-1} \tag{9.26}$$

(iii) $T^n p^{1-n} =$ constant, hence $T_1^n p_1^{1-n} = T_2^n p_2^{1-n}$, or

$$\left[\frac{T_2}{T_1}\right] = \left[\frac{p_2}{p_1}\right]^{(n-1)/n} \tag{9.27}$$

Example 9.6

One kilogram air at 27°C, 1 bar is expanded reversibly until its pressure is halved according to (a) V = constant, (b) pV = constant and (c) pV^γ = constant. Assume for air c_p = 1005 J/kg K and c_v = 718 J/kg K. Find the final temperature of air in each case. Show the processes on a common $p-V$ diagram.

▶ **Data:**

$$m = 1 \text{ kg}, T_1 = (27+273) = 300 \text{ K}, p_1 = 1 \text{ bar}, p_2 = \frac{p_1}{2} = 0.5 \text{ bar}$$

▶ **Solution:**

(a) It is a constant-volume process. From ideal gas rule

$$V_1 = \frac{mRT_1}{p_1}, \text{ where R} = c_p - c_v = 1005 - 718 = 287 \text{ J/kgK.}$$

Therefore, $V_1 = 1 \times 287 \times \dfrac{300}{1 \times 10^5} = 0.861 \text{ m}^3.$

Finally, $V_2 = V_1 = 0.861 \text{ m}^3$ and $p_2 = 0.5$ bar.

From the ideal gas rule, $T_2 = \dfrac{p_2V_2}{mR} = 0.5 \times 10^5 \times \dfrac{0.861}{1 \times 287} = 150 \text{ K}$

(b) As the process follows pV = constant, for an ideal gas it must also be a constant-temperature process.

Therefore, $T_2 = T_1 = 300$ K.

Also, as $p_1V_1 = p_2V_2, V_2 = V_1\left(\dfrac{p_1}{p_2}\right) = 0.861\left(\dfrac{1}{0.5}\right) = 1.722 \text{ m}^3.$

(c) The process is, $p_1V_1^\gamma = p_2V_2^\gamma$, where $\gamma = \dfrac{c_p}{c_v} = \dfrac{1005}{718} = 1.4$

Therefore, $V_2 = V_1\left(\dfrac{p_1}{p_2}\right)^{1/\gamma} = 0.861\left(\dfrac{1}{0.5}\right)^{1/1.4} = 1.4126 \text{ m}^3.$

From the ideal gas rule, $T_2 = \dfrac{p_2V_2}{mR} = 0.5 \times 10^5 \times \dfrac{1.4126}{1 \times 287} = 246 \text{ K}.$

Alternatively, $\left[\dfrac{T_2}{T_1}\right] = \left[\dfrac{p_2}{p_1}\right]^{\frac{(\gamma-1)}{\gamma}} = \left(\dfrac{0.5}{1}\right)^{\frac{0.4}{1.4}} = 0.8203$

Therefore, $T_2 = T_1 \times 0.8203 = 300 \times 0.8203 = 246 \text{ K}.$

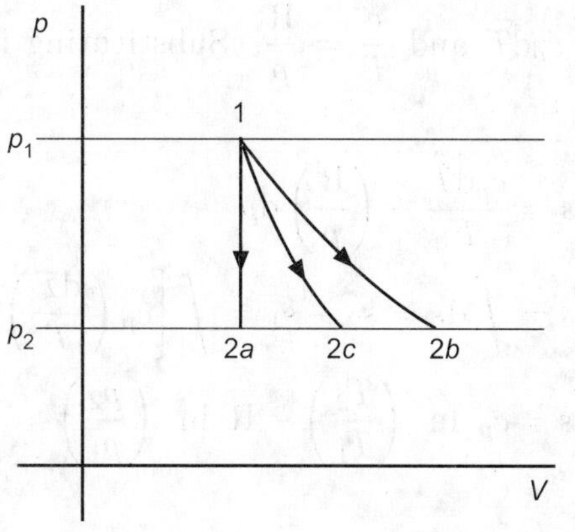

Fig. 9.8 p– V diagram sought in Example 9.6.

9.3.7 Calculation of Entropy

It has been explained in Chapter 8, that entropy of a pure substance can be calculated using the two $T\mathrm{d}s$ relations,

$$T\mathrm{d}s = \mathrm{d}u + p\mathrm{d}v \quad (8.11)$$

and

$$T\mathrm{d}s = \mathrm{d}h - v\mathrm{d}p. \quad (8.12)$$

The relation (8.11) can be written as,

$$\mathrm{d}s = \frac{\mathrm{d}u}{T} + \left(\frac{p}{T}\right)\mathrm{d}v$$

For an ideal gas $\mathrm{d}u = c_v\mathrm{d}T$ and $\dfrac{p}{T} = \dfrac{\mathrm{R}}{v}$. Substituting for du and $\dfrac{p}{T}$ in the above equation,

$$\mathrm{d}s = \frac{c_v\mathrm{d}T}{T} + \left(\frac{\mathrm{R}}{v}\right)\mathrm{d}v$$

Therefore, $$\Delta s = \int \mathrm{d}s = s_2 - s_1 = \int \left[c_v\left(\frac{\mathrm{d}T}{T}\right) + \mathrm{R}\left(\frac{\mathrm{d}v}{v}\right)\right]$$

or, $$\Delta s = s_2 - s_1 = c_v \ \ln\left(\frac{T_2}{T_1}\right) + \mathrm{R}\ \ln\left(\frac{v_2}{v_1}\right) \quad (9.28)$$

From the relation (8.12),

$$\mathrm{d}s = \frac{\mathrm{d}h}{T} - \left(\frac{v}{T}\right)\mathrm{d}p$$

For an ideal gas, $\mathrm{d}h = c_p \mathrm{d}T$ and $\frac{v}{T} = \frac{\mathrm{R}}{p}$. Substituting for d$h$ and $\frac{v}{T}$ in the above equation,

$$\mathrm{d}s = \frac{c_p \mathrm{d}T}{T} - \left(\frac{\mathrm{R}}{p}\right) \mathrm{d}p$$

Therefore,

$$\Delta s = \int \mathrm{d}s = s_2 - s_1 = \int \left[c_p \left(\frac{\mathrm{d}T}{T}\right) - \left(\frac{\mathrm{R}\mathrm{d}p}{p}\right)\right]$$

or,

$$\Delta s = c_p \ \ln \left(\frac{T_2}{T_1}\right) - \mathrm{R} \ \ln \left(\frac{p_2}{p_1}\right) \tag{9.29}$$

Example 9.7

(a) Nitrogen is to be stored at 10 MPa, 27°C in a bottle of volume 50 dm^3. How many kilograms of nitrogen is stored in the bottle? (b) It is desired to limit the pressure in the bottle to 15 MPa whenever there is an undesirable heat transfer from the surroundings. This is achieved by providing a spring-loaded safety-valve on the bottle which would open when the pressure in the bottle reaches the maximum permitted pressure. (i) Find the temperature of nitrogen when the pressure in the bottle reaches the maximum permitted value. Also find (ii) the increase in internal energy, (iii) increase in enthalpy and (iv) the increase in entropy of nitrogen during the process of rise in pressure from 10 MPa to 15 MPa. Assume that nitrogen behaves as an ideal gas with $M = 28$ and $\gamma = 1.4$.

► **Data:**

$p_1 = 10 \times 10^6$ N/m^2, $T_1 = (273 + 27) = 300$ K, $V_1 = 50 \times 10^{-3}$ m^3, M = 28, $\gamma = 1.4$

► **Solution:**

(a)

$$m = \frac{p_1 V_1}{\mathrm{R} T_1}, \text{ where R} = \frac{\mathcal{R}}{M}$$

Therefore,

$$\mathrm{R} = \frac{8314.3}{28} = 296.94 \text{ J/kg K}$$

And,

$$m = \frac{10 \times 10^6 \times 50 \times 10^{-3}}{(296.94 \times 300)} = 5.613 \text{ kg}$$

(b)

(i)

$$T_2 = \frac{p_2 V_2}{m\mathrm{R}}, \text{ where } V_2 = V_1 = 50 \times 10^{-3} \text{ m}^3 \text{ and } p_2 = 15 \times 10^6 \text{ N/m}^2$$

Therefore,

$$T_2 = \frac{15 \times 10^6 \times 50 \times 10^{-3}}{5.613 \times 296.94} = 450 \text{ K} = 177°\text{C}.$$

(ii)

$$\Delta U = m(u_2 - u_1) = mc_v(T_2 - T_1).$$

From data,

$$\gamma = \frac{c_p}{c_v} = 1.4 \text{ and R} = c_p - c_v = 296.94 \text{ J/kgK}$$

Solving for c_v and c_p, $c_v = 742.35$ J/kgK and $c_p = 1039.3$ J/kgK.

Hence, $\Delta U = mc_v(T_2 - T_1) = 5.613 \times 742.35 \times (450 - 300)$

$= 625{,}000 \text{ J} = 625 \text{ kJ}$

(iii) $\Delta H = mc_p(T_2 - T_1) = 5.613 \times 1039.3 \times (450 - 300)$

$= 875{,}000 \text{ J} = 875 \text{ kJ}$

(iv)
$$\Delta S = m(s_2 - s_1)$$
$$= m\left[c_v \ln\left(\frac{T_2}{T_1}\right) + \text{R} \ln\left(\frac{v_2}{v_1}\right)\right] \tag{9.28}$$

During the process, the volume does not change. Hence, $V_2 = V_1$, or $\ln\left(\frac{V_2}{V_1}\right) = 0$.

Therefore,
$$\Delta S = mc_v \ln\left(\frac{T_2}{T_1}\right) = 5.613 \times 742.35 \ln\left(\frac{450}{300}\right)$$
$$= 1689.5 \text{ J/K} = 1.6895 \text{ kJ/K}$$

Alternatively,
$$\Delta S = m(s_2 - s_1)$$
$$= m\left[c_p \ln\left(\frac{T_2}{T_1}\right) - \text{R} \ln\left(\frac{p_2}{p_1}\right)\right] \tag{9.29}$$
$$= 5.613\left[1039.3 \ln\left(\frac{450}{300}\right) - 296.94 \ln\left(\frac{15}{10}\right)\right]$$
$$= 1689.5 \text{ J/K} = 1.6895 \text{ kJ/K}$$

Example 9.8

Air is expanded irreversibly and adiabatically in a turbine from 800 K, 1.5 MPa to 540 K. If the process had been reversible, the final temperature would have been 485 K for the same discharge pressure. Calculate the work per kilogram of air and the change in entropy. Assume for air $c_v = 718$ J/kg K and $\gamma = 1.4$.

▶ Data:

Shown on the T–S diagram in Figure 9.9 are both the reversible and irreversible adiabatic processes. While 1–2s is the reversible adiabatic process, 1–2 is the irreversible adiabatic process.

(i) $T_1 = 800$ K, $p_1 = 1.5$ MPa, $T_2 = 540$ K, process is irreversible and $Q = 0$.

(ii) If the process were to be reversible and adiabatic $T_{2s} = 485$ K and p_2 in both cases is the same. $c_v = 0.718$ kJ/kgK and $\gamma = 1.4$.

The final pressure cannot be calculated using the data on irreversible process, as the path is not known, and work cannot be evaluated using $\int p dV$ for the same reason.

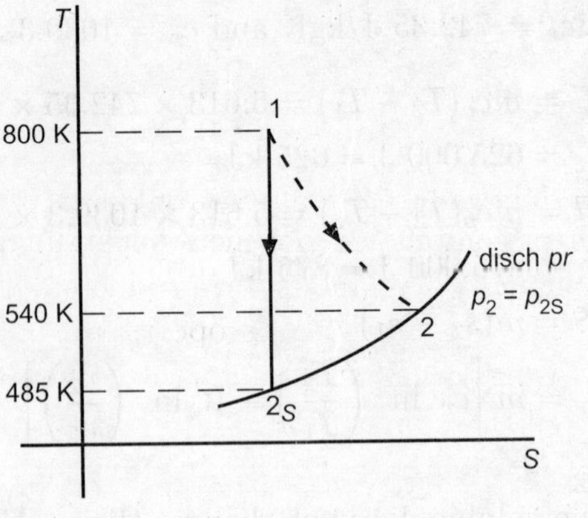

Fig. 9.9 T–S Diagram for Example 9.8.

A reversible adiabatic process is one for which pV^{γ} = constant and, hence,

$$\frac{p_{2s}}{p_1} = \left(\frac{T_{2s}}{T_1}\right)^{\frac{\gamma}{(\gamma-1)}}$$

and hence, $$p_{2s} = p_1 \left(\frac{T_{2s}}{T_1}\right)^{\frac{\gamma}{(\gamma-1)}}$$

that is, $$p_{2s} = 1.5 \left(\frac{485}{800}\right)^{\frac{1.4}{(1.4-1)}} = 0.26 \text{ MPa}$$

From data, $$p_2 = p_{2s} = 0.26 \text{ MPa}$$

Now, for air which has undergone an irreversible process,

$$\Delta u = c_v(T_2 - T_1) = 0.718 \times (540 - 800) = -186.7 \text{ kJ/kg}$$

The process is adiabatic and hence $Q = 0$. From the first law,

$$W = Q - \Delta u = 0 - (-186.7) = 186.7 \text{ kJ/kg}$$

Now, $$\Delta s = c_p \ \ln \left(\frac{T_2}{T_1}\right) - \text{R} \ \ln \left(\frac{p_2}{p_1}\right) \tag{9.29}$$

Here, $$c_p = (\gamma c_v) = 1.4 \times 0.718 = 1.005 \text{ kJ/kgK and}$$

$$\text{R} = c_p - c_v = 1.005 - 0.718 = 0.287 \text{ kJ/kgK}$$

Therefore, $$\Delta s = 1.005 \times \ \ln \left(\frac{540}{800}\right) - 0.287 \times \ \ln \left(\frac{0.26}{1.5}\right)$$

$$= -0.395 - (-0.503) = +0.108 \text{ kJ/kgK}$$

9.4 Ideal Gas Mixture

9.4.1 Introduction

Many times, the working substance encountered happens to be a mixture of gases. A particular example is that of combustion products expanding in the turbine of an open-cycle gas turbine power plant. The combustion products could be a mixture of several gases such as CO_2, CO, O_2, N_2 etc, in varying proportions. To carry out a thermodynamic analysis of systems comprising gas mixtures, it is necessary to know how the required thermodynamic properties of these mixtures could be obtained. Explained in this chapter is the procedure adopted to evaluate the thermodynamic properties of ideal gas mixtures in terms of the properties of the constituent ideal gases.

The procedure is valid only under the following assumptions: (1) The constituent gases are ideal gases; (2) the mixture behaves as an ideal gas itself and (3) the mixture is homogeneous and non-reactive. Before proceeding further, some terminology, frequently used in this chapter, are defined and explained.

9.5 Analysis of Gas Mixtures for their Composition

Common sense tells us that the properties of a gas mixture depends not only on the properties of the constituent gases, but also on what they are and to what extent each gas exists either by mass or by volume.

9.5.1 Gravimetric Analysis

The analysis of a gas mixture, based on mass, that estimates the extent to which each individual gas exists in the mixture is called *Gravimetric Analysis.*

Consider m kg of a gas mixture. Let the constituent individual gases in the mixture be A, B, C $\cdots$ etc, and their masses in kg, respectively, be $m_A, m_B, m_C \cdots$ etc. Then,

$$m = m_A + m_B + m_C + \cdots \tag{i}$$

Dividing the right-hand side and left-hand side of (i) by m,

$$1 = \frac{m_A}{m} + \frac{m_B}{m} + \frac{m_C}{m} + \cdots \tag{ii}$$

Each term on the right-hand side is called the *mass fraction*. Thus, $\frac{m_A}{m}$ = mass fraction of gas A in the mixture, $\frac{m_B}{m}$ = mass fraction of gas B and $\frac{m_C}{m}$ = mass fraction of C, etc. Mass fraction is denoted by y with the name of the gas as its suffix.

Therefore, $\dfrac{m_A}{m} = y_A, \dfrac{m_B}{m} = y_B, \dfrac{m_C}{m} = y_C$, etc. Equation (ii) can hence be rewritten as,

$$y_A + y_B + y_C + \cdots = 1$$

or, $$\Sigma\, y_i = 1 \tag{9.30}$$

where $i =$ A, B, C,$\cdots$ etc.

That is, the sum of the mass fractions of the individual constituents in a mixture is equal to one.

9.5.2 Volumetric or Molar Analysis

The analysis of a gas mixture, based on volume, that estimates the extent to which each individual gas exists in the mixture is called *volumetric* or *molar analysis.*

Let the number of molecules of the individual gases A, B, C $\cdots$. etc that make up N molecules of the mixture be $N_A, N_B, N_C \cdots$ etc. Then,

$$N = N_A + N_B + N_C + \cdots \tag{iii}$$

Dividing the left-hand side and right-hand side by the Avogadro number, N_0,

$$\frac{N}{N_0} = \frac{N_A}{N_0} + \frac{N_B}{N_0} + \frac{N_C}{N_0} + \cdots \tag{iv}$$

In (iv), while N_A, N_B, N_C and N are proportional to corresponding masses, N_0 is proportional to the corresponding molar masses M. Therefore,

$$\frac{m}{M} = \frac{m_A}{M_A} + \frac{m_B}{M_B} + \frac{m_C}{M_C} + \cdots \tag{v}$$

Each term in (v) is the number of moles of the corresponding gas, n. Therefore,

$$n = n_A + n_B + n_C + \cdots \tag{vi}$$

Dividing the left-hand side and right-hand side of (vi) by n and rearranging,

$$\frac{n_A}{n} + \frac{n_B}{n} + \frac{n_C}{n} + \cdots = 1 \tag{vii}$$

Each term on the left-hand side of (vii) is called the mole fraction, x, of the constituent gases. Therefore,

$$x_A + x_B + x_C + \cdots = 1, \text{ or}$$

$$\Sigma\, x_i = 1 \tag{9.31}$$

that is, the sum of the mole fractions of the individual constituents in a mixture is equal to one.

Note: Hereafter, while all properties denoted with a suffix correspond to the individual constituents, the properties denoted without any suffix correspond to the mixture. For example, while $m_A, t_A, V_A \cdots$ correspond to mass, temperature, volume etc. of the constituent A in the mixture $m, t, V \cdots$ correspond to mass, temperature, volume etc. of the mixture.

9.5.3 Dalton's Law of Additive Pressures

Dalton's law of additive pressures states that the pressure of a mixture of ideal gases is equal to the sum of the partial pressures of the constituent gases; and the partial pressure of an individual constituent gas is the pressure it exerts if it were to exist alone at the mixture temperature and volume.

The Dalton's law can also be alternately stated as follows:

In a mixture of ideal gases each constituent gas behaves as though it existed alone at the mixture temperature and volume.

As a mixture of ideal gases behaves like an ideal gas,

$$p = \frac{n\mathcal{R}T}{V}$$

Here,

$$n = n_A + n_B + n_C + \cdots, \text{ and hence,}$$

$$p = (n_A + n_B + n_C + \cdots)\frac{\mathcal{R}T}{V}, \text{ or}$$

$$p = \frac{n_A\mathcal{R}T}{V} + \frac{n_B\mathcal{R}T}{V} + \frac{n_C\mathcal{R}T}{V} + \cdots$$

Each term on the right-hand side equals the pressure the individual gases A, B, C, etc. would exert if each gas, at the mixture temperature, had alone occupied the mixture volume. These pressures are called the partial pressures and are represented by p_A, p_B, p_C, etc. In other words, $p_A = \dfrac{n_A\mathcal{R}T}{V}$, $p_B = \dfrac{n_B\mathcal{R}T}{V}$, $p_C = \dfrac{n_C\mathcal{R}T}{V}, \cdots$. Thus, as defined earlier, the partial pressure of a constituent gas is equal to the pressure it would exert if it alone existed at mixture temperature and volume. Hence,

$$p = p_A + p_B + p_C \cdots \text{ or } p = \Sigma p_i \tag{9.32}$$

Equation (9.32) embodies the Dalton's law.

By definition, $p_i = \dfrac{n_i\mathcal{R}T}{V}$. Dividing and multiplying the right-hand side by n the number of moles of the mixture,

$$p_i = n_i\left(\frac{n}{n}\right)\frac{\mathcal{R}T}{V} = \left(\frac{n_i}{n}\right) \times \left(\frac{n\mathcal{R}T}{V}\right) \tag{i}$$

The first term in parentheses on the right-hand side of (i) is equal to the mole fraction of the ith constituent and the terms in the second set of parentheses is equal to the mixture pressure. Hence,

$$p_i = x_i p \tag{9.33}$$

The partial pressure of an individual constituent in a mixture of ideal gases is equal to the product of the mole fraction of the constituent and the pressure of the mixture.

9.5.4 Amagat's Law of Additive Volumes

Amagat's law of additive volumes states that

The volume of an ideal gas mixture is equal to the sum of the partial volumes of the constituent gases; and the partial volume of an individual constituent is defined as the volume it alone occupies at the mixture temperature and pressure.

For an ideal gas mixture,

$$V = \frac{nRT}{p}$$

Here, n, the number of moles of the mixture is given by

$$n = n_A + n_B + n_C + \cdots \text{and hence,}$$
$$V = (n_A + n_B + n_C + \cdots)\frac{RT}{p}, \text{ or}$$
$$V = \frac{n_A RT}{p} + \frac{n_B RT}{p} + \frac{n_C RT}{p} + \cdots$$

Each term on the right-hand side equals the volume of the individual gases A, B, C, etc, if each gas, at the mixture temperature, had alone existed at the mixture pressure. These volumes are called the partial volumes and are represented by V_A, V_B, V_C, etc.

In other words, $V_A = \dfrac{n_A RT}{p}, V_B = \dfrac{n_B RT}{p}, V_C = \dfrac{n_C RT}{p}, \cdots$

Thus, the partial volume of a constituent gas, as defined earlier, is equal to the volume it occupies if it existed alone at the mixture temperature and pressure. Hence,

$$V = V_A + V_B + V_B \cdots$$

or

$$V = \Sigma V_i \tag{9.34}$$

Equation (9.34) embodies Amagat's Law.

By definition, $V_i = \dfrac{n_i RT}{p}$. Dividing and multiplying the right-hand side of this equation by n,

$$V_i = \left(\frac{n_i}{n}\right) \times \frac{nRT}{p} = x_i \times V.$$

Therefore, the partial volume of a constituent gas is equal to the product of its mole fraction and the volume of the mixture. That is,

$$V_i = x_i V \tag{9.35}$$

9.5.5 Mole-Fractions and Volume-Fractions

More often than not, the analysis of a gas mixture is based on volume. For example, when the analysis of exhaust gas from an engine is carried out using a simple Orsat's apparatus, the percentages of various constituent gases are obtained on volume basis. Suppose a_i is the volume fraction of the ith constituent, then a_i is the ratio of the volume of the individual gas, V_i, to the sample volume V of the mixture taken for analysis. The experiment is conducted such that both V_i and V are measured at the same pressure and temperature. Thus $\dfrac{V_i}{V} = a_i$, is also equal to the mole fraction x_i according to (9.35). Therefore, $a_i = x_i$.

It can, hence, be generalized that mole fractions are also equal to volume fractions. In other words, the mole fractions represent the volume fractions of the individual constituents in an ideal gas mixture.

9.5.6 Relation between Mass-Fraction and Mole-Fraction

By definition, the mass fraction and mole fraction of any constituent are given by

$$y_i = \frac{m_i}{m}, \text{ and } x_i = \frac{n_i}{n}$$

Also, $n_i = \dfrac{m_i}{M_i}$ and $n = \dfrac{m}{M}$, where M_i and M are, respectively, the relative molar masses (molecular weights) of the ith constituent and the mixture.

Therefore, $m_i = n_i M_i$ and $m = nM$

or, $$y_i = \frac{n_i M_i}{nM} = \left(\frac{n_i}{n}\right) \times \left(\frac{M_i}{M}\right) = x_i \left(\frac{M_i}{M}\right)$$

Thus, the relation between mass fraction, y, and the mole fraction, x, of any individual constituent is given by

$$y_i = x_i \left(\frac{M_i}{M}\right) \tag{9.36}$$

9.6 Properties of a Gaseous Mixture in terms of Properties of Individual Constituents

Explained below are a few important properties of a gaseous mixture:

9.6.1 Apparent Molar Mass (M) and Gas Constant (R)

(i) As, $m_A + m_B + m_C + \cdots = m$, and by definition, $m_i = n_i M_i$, and $m = nM$,

$$n_A M_A + n_B M_B + n_C M_C + \cdots = nM,$$
$$M = \left(\frac{n_A}{n}\right) M_A + \left(\frac{n_B}{n}\right) M_B + \left(\frac{n_C}{n}\right) M_C + \cdots$$
$$M = x_A M_A + x_B M_B + x_C M_C + \cdots$$
$$M = \Sigma\, x_i M_i \tag{9.37}$$

M, the apparent molar mass of a mixture is equal to the sum of the products of the mole fractions and the corresponding molar masses of the constituents.

(ii) The gas constant, R, for the mixture is given by

$$\mathrm{R} = \frac{\mathcal{R}}{M}, \text{ where, for a mixture, from (9.37), } M = \Sigma\, x_i M_i$$

Therefore,
$$\mathrm{R} = \frac{\mathcal{R}}{\Sigma\, x_i M_i} \tag{9.38}$$

The gas constant of a mixture is equal to universal gas constant divided by the sum of the products of the mole fractions and corresponding molar masses of the constituents.

(iii) R and M can also be evaluated using the mass fractions of the constituents, as explained below:

$$pV = n\mathcal{R}T = n_A\mathcal{R}T + n_B\mathcal{R}T + n_C\mathcal{R}T + \cdots$$
$$= n_A M_A \mathrm{R}_A T + n_B M_B \mathrm{R}_B T + n_C M_C \mathrm{R}_C T + \cdots$$
$$(\text{as } \mathcal{R} = M_i \mathrm{R}_i)$$

that is,
$$m\mathrm{R}T = m_A \mathrm{R}_A T + m_A \mathrm{R}_A T + m_A \mathrm{R}_A T + \cdots$$
$$(\text{as } n_i M_i = m_i \text{ and } pV = m\mathrm{R}T)$$

Dividing throughout by mT,

$$\mathrm{R} = \left(\frac{m_A}{m}\right)\mathrm{R}_A + \left(\frac{m_B}{m}\right)\mathrm{R}_B + \left(\frac{m_C}{m}\right)\mathrm{R}_C + \cdots$$
$$\mathrm{R} = y_A \mathrm{R}_A + y_B \mathrm{R}_B + y_C \mathrm{R}_C + \cdots, \text{ or}$$
$$\mathrm{R} = \Sigma\, y_i \mathrm{R}_i \tag{9.39}$$

Gas constant of a mixture is equal to the sum of the products of the mass fractions and gas constants of the corresponding constituents.

(iv) $$R = \frac{\mathcal{R}}{M} = y_A R_A + y_B R_B + y_C R_C + \cdots$$

Therefore, $$\frac{\mathcal{R}}{M} = \frac{y_A \mathcal{R}}{M_A} + \frac{y_B \mathcal{R}}{M_B} + \frac{y_C \mathcal{R}}{M_C} + \cdots \left(\text{as} \quad R_i = \frac{\mathcal{R}}{M_i} \right)$$

Dividing throughout by $\mathcal{R}$,

$$\frac{1}{M} = \frac{y_A}{M_A} + \frac{y_B}{M_B} + \frac{y_C}{M_C} + \cdots = \frac{\Sigma\, y_i}{M_i}$$

$$M = \frac{1}{\Sigma \frac{y_i}{M_i}} \tag{9.40}$$

Apparent molar mass of a gas mixture is equal to the reciprocal of the sum of the ratios of mass fractions and corresponding molar masses of the constituents. (A gas mixture is not a chemical compound and hence strictly speaking it cannot have a molar mass. It is for this reason that M for a mixture is called 'apparent' molar mass).

Example 9.9

Orsat's analysis of combustion products from an engine gave the following volume fractions: 10% CO_2, 9% O_2, 1% CO and the rest N_2. Find the molar mass and the gas constant of the mixture. Also find the mass fractions of the constituents. Assume molar masses of CO_2, O_2, CO and N_2 to be, respectively, 44, 32, 28 and 28.

▶ Data:

Volume fractions are equal to mole fractions and hence, $x_{CO_2} = 0.1, x_{O_2} = 0.09$, $x_{CO} = 0.01$ and $x_{N_2} = 0.8$.

▶ Solution:

$$M = \Sigma\, x_i M_i = (0.1 \times 44 + 0.09 \times 32 + 0.01 \times 28 + 0.8 \times 28)$$
$$= 29.96 \text{ kg/kmol}$$

$$\text{R} = \frac{\mathcal{R}}{M} = \frac{8314.3}{29.96} = 277.5 \text{ J/kgK}$$

$$y_i = x_i \left(\frac{M_i}{M} \right)$$ and therefore the various mass fractions are

$$y_{CO_2} = 0.10 \times \left(\frac{44}{29.96} \right) = 0.1469$$

$$y_{O_2} = 0.09 \times \left(\frac{32}{29.96}\right) = 0.0961$$

$$y_{CO} = 0.01 \times \left(\frac{28}{29.96}\right) = 0.0094$$

$$y_{N_2} = 1 - (y_{CO_2} + y_{O_2} + y_{CO})$$

$$= 1 - (0.1467 + 0.0961 + 0.0093) = 0.7477$$

Example 9.10

1.9 kg of an ideal gas mixture at 20°C and 150 kPa consists of 10% oxygen, 75% nitrogen, 12% carbon dioxide and 3% carbon monoxide by mass. Find the (a) molar mass and gas the constant for the mixture and (b) mole fractions, partial pressures and partial volumes of the constituents. Assume the following: $M_{O_2} = 32$, $M_{N_2} = 28$, $M_{CO_2} = 44$ and $M_{CO} = 28$.

▶ **Data:**

$m = 1.9$ kg, $T = 273 + 20 = 293$ K, $p = 150$ kPa, $y_{O_2} = 0.1$, $y_{N_2} = 0.75$, $y_{CO_2} = 0.12$ and $y_{CO} = 0.03$.

▶ **Solution:**

(a)

$$M = \frac{1}{\left(\frac{y_i}{M_i}\right)} = \frac{1}{\left[\left(\frac{0.1}{32}\right) + \left(\frac{0.75}{28}\right) + \left(\frac{0.12}{44}\right) + \left(\frac{0.03}{28}\right)\right]}$$

$$= 29.67 \text{ kg/kmol}$$

$$\text{R} = \frac{\mathcal{R}}{M} = \frac{8314.3}{29.67}$$

$$= 280.27 \text{ J/kgK}$$

$$V = \frac{m\text{R}T}{p} = \frac{1.9 \times 280.27 \times 293}{(150 \times 10^3)}$$

$$= 1.04 \text{ m}^3$$

(b) The mole fractions, partial pressures and partial volumes of the constituents are computed and presented in Table 9.1.

Table 9.1 Characteristics evaluated for Example 9.10(b).

Constituent gas	Mole fraction $x_i = y_i \left(\frac{M}{M_i}\right)$	Partial pressure, $p_i = x_i p$, in kPa,	Partial volumes, $V_i = x_i V$, in m ^{3},
Oxygen	$0.1\left(\frac{29.67}{32}\right) = 0.093$	$0.093 \times 150 = 13.95$	$0.093 \times 1.04 = 0.0967$
Nitrogen	$0.75\left(\frac{29.67}{28}\right) = 0.795$	$0.795 \times 150 = 119.25$	$0.795 \times 1.04 = 0.8268$
Carbondioxide	$0.12\left(\frac{29.67}{44}\right) = 0.081$	$0.081 \times 150 = 12.15$	$0.081 \times 1.04 = 0.0842$
Carbon Monoxide	$1 - (x_{O_2} + x_{N_2} + x_{CO_2}) = 0.031$	$(150 - 13.95 - 119.25 - 12.15) = 4.76$	$1.04 - 0.0967 - 0.8268 - 0.0842 = 0.0323$

9.6.2 Specific Properties

Suppose Z is an extensive property of a mixture (it could be volume, V, or internal energy, U, or enthalpy, H, or entropy, S). For the mixture, Z, is equal to the sum of Z_i for the individual constituents.

$$Z = Z_A + Z_B + Z_C + \cdots$$

If z (in lower case) were to denote the corresponding specific property, then

$$mz = m_A z_A + m_B z_B + m_B z_B + \cdots$$

Dividing throughout by m,

$$z = y_A z_A + y_B z_B + y_B z_B + \cdots \text{ or}$$

$$z = \Sigma\, y_i z_i \tag{9.41}$$

That is,

$$v = \Sigma\, y_i v_i$$

$$u = \Sigma\, y_i u_i,$$

$$h = \Sigma\, y_i h_i \text{ and}$$

$$s = \Sigma\, y_i s_i$$

Similarly, it can also be shown that

$$c_p = \Sigma\, y_i c_{p,i} \quad \text{and}$$

$$c_v = \Sigma\, y_i c_{v,i}$$

The specific properties of a mixture are equal to the sum of the products of the mass fractions and the corresponding specific properties of the constituent gases.

From (9.36), $y_i = x_i \left(\frac{M_i}{M}\right)$

Substituting for y_i in (9.41),

$$z = \Sigma\, x_i \left(\frac{M_i}{M}\right) z_i$$

and therefore,

$$Mz = \Sigma\, x_i (M_i z_i)$$

Mz is the molar specific property and is usually denoted by $\bar{z}$. Hence,

$$\bar{z} = \Sigma\, x_i \bar{z}_i \tag{9.42}$$

Molar specific properties of a gas mixture are equal to the sum of the products of the mole fractions and the corresponding molar specific properties of the constituent gases.

Table 9.2 lists the properties of some of the commonly encountered gases.

Table 9.2 Ideal gas properties.

Gas	Chemical formula	M in kg/kmol (A)	M in kg/kmol (B)*	R kJ/kg K	c_p kJ/kg K	c_v kJ/kg K
Air	–	28.97	28.97	0.287	1.005	0.718
Argon	Ar	39.948	40.0	0.208	0.520	0.312
Butane	C_4H_{10}	58.124	58.0	0.143	1.716	1.573
Carbon dioxide	CO_2	44.010	44.0	0.189	0.842	0.653
Carbon monoxide	CO	28.010	28.0	0.297	1.041	0.744
Ethane	C_2H_6	30.070	30.0	0.276	1.766	1.490
Ethylene	C_2H_4	28.054	28.0	0.296	1.548	1.252
Helium	He	4.003	4.0	2.077	5.193	3.116
Hydrogen	H_2	2.016	2.0	4.135	14.21	10.085
Methane	CH_4	16.040	16.0	0.518	2.254	1.736
Neon	Ne	20.183	20.0	0.421	1.039	0.618
Nitrogen	N_2	28.013	28.0	0.297	1.042	0.745
Oxygen	O_2	31.999	32.0	0.260	0.922	0.662
Propane	C_3H_8	44.097	44.0	0.188	1.679	1.491

(* Approximate values of M used in all the problems in this book.)

Example 9.11

It is given that the gas constant and the specific heat at constant pressure for a mixture of nitrogen and hydrogen are, respectively, 1828 J/kgK and 6310 J/kgK. Evaluate the mass fractions and volume fractions of nitrogen and hydrogen in the mixture. Assume that the values of M and c_p for N_2 are, respectively, 28 and 1042 J/kgK, and the corresponding values for H_2 are, respectively, 2 and 14 210 J/kgK.

▶ **Data:**

R = 1828 J/kg K, c_p = 6310 kJ/kg K.

Using N and H as suffixes for the properties of nitrogen and hydrogen, respectively, $M_\text{N} = 28$, $M_\text{H} = 2$, $c_{p\text{N}} = 1042$ J/kg K and $c_{p\text{H}} = 1\,4210$ J/kg K.

▶ **Solution:**

$$\text{R} = \Sigma\, y_i \text{R}_i = \Sigma\, y_i \left(\frac{\mathcal{R}}{M_i}\right)$$

Therefore,
$$\text{R} = \mathcal{R}\left(\frac{y_\text{N}}{M_\text{N}} + \frac{y_\text{H}}{M_\text{H}}\right)$$

$$1828 = 8314.3\left[\frac{y_\text{N}}{28} + \frac{y_\text{H}}{2}\right]$$

that is,
$$0.0357 y_\text{N} + 0.5 y_\text{H} = 0.2199 \qquad \text{(i)}$$

Also,
$$c_p = (y_\text{N} c_{p\text{N}} + y_\text{H} c_{p\text{H}})$$

$$6310 = [y_\text{N} \times 1042 + y_\text{H} \times 14\,210]$$

that is,
$$y_\text{N} + 13.637 y_\text{H} = 6.056 \qquad \text{(ii)}$$

Solving for y_N and y_H between (i) and (ii),

$$y_\text{H} = 0.4 \text{ and } y_\text{N} = 0.6.$$

$$\text{volume fraction} = x_i = y_i\left(\frac{M}{M_i}\right), \text{ where } M = \frac{\mathcal{R}}{\text{R}} = \frac{8314.3}{1828} = 4.548.$$

Therefore,
$$x_\text{N} = 0.6 \times \left(\frac{4.548}{28}\right) = 0.0975$$

and, $x_\text{H} = 1 - x_\text{N} = 0.9025$. The answers are given in Table 9.3.

Table 9.3 Answer to Problem 9.11.

Gas	Mass fraction	Mole fraction
Nitrogen	0.6	0.0975
Hydrogen	0.4	0.9025

Example 9.12

Assuming internal energies and entropies of each constituent in Example 9.10 to be zero at the datum state of 100 kPa and 0°C determine the following for the mixture: (a) specific heats, (b) internal energy, (c) enthalpy and (d) entropy. Assume c_p for O_2, N_2, CO_2 and CO to be, respectively, 0.922, 1.042, 0.842 and 1.041 kJ/kg K.

▶ **Data:**

m = 1.9 kg, $T = 273 + 20 = 293$ K, $p = 150$ kPa, p_{dat} =100 kPa, $T_{dat} = 273$ K, $u_{dat,i} = 0$ and $s_{dat,i} = 0$. $y_{O_2} = 0.1$, $y_{N_2} = 0.75$, $y_{CO_2} = 0.12$ and $y_{CO} = 0.03$.

▶ **Solution:**

(a) The mole fractions, R and M, as evaluated in Example 9.10 are: $x_{O_2} = 0.093$, $x_{N_2} = 0.795$, $x_{CO_2} = 0.081$ and $x_{CO} = 0.031$. R = 280.22 J/kg K and M = 29.67 kg/kmol.

$$c_p = \Sigma\, y_i c_{pi}$$

Therefore,

$$c_p = [0.1 \times 0.922 + 0.75 \times 1.042 + 0.12 \times 0.842 + 0.03 \times 1.041]$$
$$= 1.006 \text{ kJ/kgK}$$
$$c_v = c_p - R$$
$$= 1.006 - 0.28022 = 0.7257 \text{ kJ/kgK}$$

(b) The internal energy of each constituent, at any given state, can be calculated from the datum state value by

$$u_i - u_{dat,i} = c_{vi}(t_i - t_{dat,i}) \text{ where } u_{dat,i} = 0 \text{ at } t_{dat,i} = 0$$

Therefore, $$u_i = c_{vi} t_i$$

Also for a mixture, $u = \Sigma\, y_i u_i$

and hence, $$u = \Sigma\, y_i c_{vi} t_i = t_i \Sigma\, y_i c_{vi} = 20 \times c_v = 20 \times 0.7257 = 14.514 \text{ kJ/kg}$$

Therefore, $$U = mu = 1.9 \times 14.513 = 27.58 \text{ kJ}$$

(c) $H = U + pV = U + m\text{R}T = 27.58 + 1.9 \times 0.28022 \times 293 = 183.6$ kJ

(d) Entropy of each constituent is calculated as though its state changes from the datum state to its partial pressure and mixture temperature.

$$s_i - s_{dat,i} = c_{p_i} \ln\left(\frac{T_i}{T_{dat}}\right) - \text{R}_i \ln\left(\frac{p_i}{p_{dat}}\right)$$

where p_i = partial pressure = $x_i p$ and p_{dat} = 100 kPa, $s_{dat,i} = 0$, $T_i = 293$ K and $T_{dat} = 273$ K.

Therefore, $$s_i = c_{pi} \ln\left(\frac{293}{273}\right) - \text{R}_i \ln\left(\frac{x_i p}{p_{dat}}\right)$$
$$= c_{pi} \ln\left(\frac{293}{273}\right) - \text{R}_i \ln\left[x_i\left(\frac{p}{p_{dat}}\right)\right]$$
$$= c_{pi} \ln\left(\frac{293}{273}\right) - \text{R}_i \ln\left(x_i\frac{150}{100}\right)$$
$$= 0.0707c_{pi} - \left(\frac{\mathcal{R}}{M_i}\right) \ln(1.5x_i)$$

Substituting the values of c_{pi} and M_i,

$$s_{O_2} = 0.0707 \times 0.922 - \left(\frac{8.3143}{32}\right) \times \ln(0.093 \times 1.5) = 0.5773 \text{ kJ/kgK}$$
$$s_{N_2} = 0.0707 \times 1.042 - \left(\frac{8.3143}{28}\right) \times \ln(0.795 \times 1.5) = 0.0214 \text{ kJ/kgK}$$
$$s_{CO_2} = 0.0707 \times 0.842 - \left(\frac{8.3143}{44}\right) \times \ln(0.081 \times 1.5) = 0.4579 \text{ kJ/kgK}$$
$$s_{CO} = 0.0707 \times 1.041 - \left(\frac{8.3143}{28}\right) \times \ln(0.0.031 \times 1.5) = 0.9849 \text{ kJ/kgK}$$

Also, $$s = \Sigma\, y_i s_i$$

Therefore, $$s = 0.1 \times 0.5773 + 0.75 \times 0.0214 + 0.12 \times 0.4579 + 0.03 \times 0.9849$$
$$= 0.1583 \text{ kJ/kg K}$$
$$S = ms = 1.9 \times 0.1583 = 0.3008 \text{ kJ/K}$$

9.7 Mixing of Ideal Gases at Different Pressures and Temperatures properties of Resulting Mixtures

What has been discussed hitherto is only the methodology adopted to evaluate the properties of a homogeneous mixture of ideal gases in terms of the properties of the individual constituents. Together with some additional relations the methodology can also be extended to find the properties of a gas mixture resulting by mixing several ideal gases at different pressures and temperatures, adiabatically or otherwise.

9.7.1 Adiabatic Mixing of Gases Initially Existing in Separate Containers

As an example, let us consider three ideal gases A, B and C in three different rigid but insulated vessels, as shown in Figure 9.10. Let these three vessels be interconnected by a small tube and valves. When the valves are opened, the gases get mixed, and after a long time each vessel contains an uniform mixture at the same pressure and temperature.

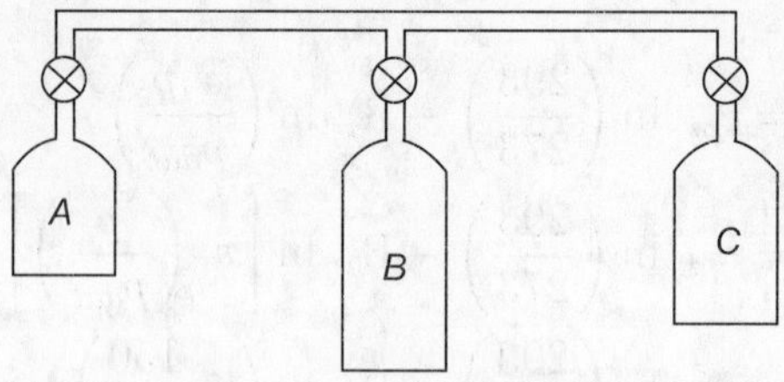

Fig. 9.10 Mixing of gases initially present in different states.

Applying the first law of thermodynamics to the mixing process,

$$Q - W = \Delta U$$

During the process of mixing $Q = 0$ and $W = 0$ and hence,

$$\Delta U = 0$$

that is, $\quad U - (U_A + U_B + U_C) = 0$

where U is the internal energy of the mixture and U_A, U_B and U_C are the internal energies of the constituents before mixing.

Therefore, $\quad U = (U_A + U_B + U_C)$

In general, the final temperature of the mixture is different from the temperatures of A, B or C. Thus, the internal energy of each gas in the mixture has to be different from its initial value before mixing, subject to the condition that their sum is still equal to U. This is possible if and only if the sum of the changes in the internal energy of each gas is equal to zero. Hence,

$$\Delta U_A + \Delta U_B + \Delta U_C = 0$$

Assuming the specific heats are constant and that the mixture temperature is T,
$m_A c_{vA}(T - T_A) + m_B c_{vB}(T - T_B) + m_C c_{vC}(T - T_C) = 0$, or

$$T = \frac{(m_A c_{vA} T_A + m_B c_{vB} T_B + m_C c_{vC} T_C)}{(m_A c_{vA} + m_B c_{vB} + m_C c_{vC})},$$

or, in general,
$$T = \frac{\Sigma\, m_i c_{vi} T_i}{\Sigma\, m_i c_{vi}} \tag{9.43}$$

Once the mixture temperature is determined, from the ideal gas rule, the pressure is given by, $p = \dfrac{m\mathrm{R}T}{V}$, where $m = \Sigma\, m_i, V = \Sigma\, V_i$ and $\mathrm{R} = \Sigma\, y_i \mathrm{R}_i$.

The mixing process is irreversible. Therefore, from the principle of increase of entropy,

$$(\Delta S_{\text{system}} + \Delta S_{\text{surr}}) > 0$$

Here, the mixing is assumed to occur adiabatically. Therefore, during the process of mixing, the gases have no heat (insulated vessels) or work (rigid vessels) interactions, which makes $\Delta S_{\text{surr}} = 0$. Thus, during the mixing process,

$$\Delta S > 0$$

that is,

$$\Delta S_A + \Delta S_B + \Delta S_C > 0 \tag{9.44}$$

The change in entropy of each constituent can be calculated using Dalton's law that each gas in a mixture behaves as though it existed alone at the mixture temperature and volume. In other words, each gas changes its state from its initial state before mixing to its final state identified by its partial pressure and mixture temperature. that is,

$$\Delta S_i = m_i \left[c_{pi} \ln\left(\frac{T}{T_i}\right) - \mathrm{R}_i \ln\left(\frac{x_i p}{p_i}\right) \right] \tag{9.45}$$

where T is the mixture temperature, T_i is the temperature of the constituent before mixing, p_i is the pressure before mixing and $x_i p$ (x_i is the mole fraction of the constituent and p is the mixture pressure) is the partial pressure of the constituent in the mixture.

9.7.2 Non-Adiabatic Mixing

The properties of the mixture can be evaluated similarly even if the mixing process were not to be adiabatic. The procedure to be adopted in such a case is outlined below. Besides, the methodology to be used to find the temperature after mixing, or the heat transfer during the mixing process, depending on the data given, is also explained.

Consider an example as in section (9.7.1), with the only difference being that the vessels are not insulated. During the process of mixing, the gases can have heat interaction with the surroundings. The direction of heat transfer could be either way, to the gases or from the gases.

(i) Let us now see how to analyse the problem if the final temperature were to be given. Let this given final temperature be T_f.

The mass, volume and other properties for the final temperature, T_f, can be found using the relations:

$$m = \Sigma\, m_i,$$

$$V = \Sigma\, V_i$$

$$\mathrm{R} = \Sigma\, y_i \mathrm{R}_i,$$

and

$$p = \frac{m\mathrm{R}T_f}{V}$$

Let it now be assumed that the gases have mixed adiabatically. Then, T_{ad} is calculated as in section (9.7.1).

$$T_{ad} = \frac{\Sigma\, m_i c_{vi} T_i}{\Sigma\, m_i c_{vi}} \tag{9.43}$$

Let the mixture be then brought to T_{final} by heat transfer Q. This heat transfer is given by

$$Q = mc_v \times (T_{final} - T_{ad})$$

(ii) Let us now see how to analyse the problem if the magnitude and direction of heat transfer during the mixing process is given. Let the heat transfer be Q.
Even in this case, let us first assume that the mixing process is adiabatic and then evaluate T_{ad} from,

$$T_{ad} = \frac{(m_i c_{vi} T_i)}{(m_i c_{vi})} \tag{9.43}$$

Let T_f be the actual final temperature. Then,

$$Q = mc_v(T_{final} - T_{ad})$$

Once the final temperature is known, the other properties can be evaluated as in Section 9.7.1.

Example 9.13

At 30°C, 0.3 m^3 of helium at 20 bar is mixed with 0.7 m^3 of oxygen at 6 bar and 2°C by opening the valve between the two tanks. The final temperature of the mixture is 27°C. Calculate (a) the magnitude and direction of heat transfer and (b) the pressure of the mixture. Assume for helium, R = 2077 J/kg K, c_v = 3116 J/kg K and for oxygen, R = 260 J/kg K and c_v = 662 J/kg K.

▶ **Data:**

$V_{\text{He}} = 0.3$ m^3, $p_{\text{He}} = 20\ bar$, $T_{\text{He}} = (273 + 30) = 303$ K, $V_{\text{O}_2} = 0.7$ m^3, $p_{\text{O}_2} = 6$ bar, $T_{\text{O}_2} = (273 + 2) = 275$ K.

▶ **Solution:**

$$m_{\text{He}} = \left[\frac{pV}{\text{RT}}\right]_{\text{He}} = \frac{(20 \times 10^5 \times 0.3)}{(2077 \times 303)} = 0.9534 \text{ kg}$$

$$m_{\text{O}_2} = \left[\frac{pV}{\text{RT}}\right]_{\text{O}_2} = \frac{(6 \times 10^5 \times 0.7)}{(260 \times 275)} = 5.874 \text{ kg}$$

Assuming mixing to be adiabatic,

$$\text{Temperature after mixing} = T_{ad} = \frac{\Sigma\, m_i c_{vi} T_i}{\Sigma\, m_i c_{vi}} \tag{9.43}$$

$$= \frac{(m_{\text{He}} c_{v\text{He}} T_{\text{He}} + m_{\text{O}_2} c_{v\text{O}_2} T_{\text{O}_2})}{(m_{\text{He}} c_{v\text{He}} + m_{\text{O}_2} c_{v\text{O}_2})}$$

$$= \frac{(0.9534 \times 3116 \times 303 + 5.874 \times 662 \times 275)}{(0.9534 \times 3116 + 5.874 \times 662)}$$

$$= 287.12 \text{ K}$$

But, it is given that the final temperature is 300 K. Hence, there must have been heat transfer during the mixing process whose magnitude would be that required to heat the mixture from 287.12 K to 300 K.

$$Q = mc_v(T_{final} - T_{ad}) = mc_v \times (300 - 287.12)$$

where for the mixture $\quad c_v = \Sigma\, y_i c_{vi}$

Here, $\quad m = m_{\text{He}} + m_{O_2} = 0.9534 + 5.874 = 6.8274 \text{ kg}$

$$y_{\text{He}} = \frac{m_{\text{He}}}{m} = \frac{0.9534}{6.8274} = 0.1396$$

$$y_{O_2} = 1 - y_{\text{He}} = 1 - 0.1396 = 0.8604$$

Therefore, $\quad c_v = 0.1396 \times 3.116 + 0.8604 \times 0.662 = 1.0046 \text{ kJ/kg K}$

$$Q = 6.8274 \times 1.0046 \times (300 - 287.12) = 88.34 \text{ kJ}$$

(b) p_f can be found using $p_f = \frac{mRT_f}{V_f}$, where $R = \Sigma y_i R_i = \Sigma y_i \frac{\text{R}}{M_i}$

9.7.3 Mixing of Gases in Steady Flow

Properties of a gas mixture resulting from adiabatic mixing in steady flow, can also be analysed in the manner described above.

In this case, with the assumption that changes in kinetic and potential energies are negligible and that the specific heats are constant, the mixture temperature can be shown to be given by

$$T = \frac{\Sigma\, \dot{m}_i c_{pi} T_i}{\Sigma\, \dot{m}_i c_{pi}} \tag{9.46}$$

where m_i is the flow rate of the ith constituent.

Example 9.14

Ethane (C_2H_6) at 20°C, 200 kPa and methane (CH_4) at 45°C, 200 kPa mix adiabatically in steady flow as shown in Figure 9.11. The mass-flow rate of ethane is 9 kg/s, which is twice that of methane. Determine (a) the mixture temperature and (b) the rate of entropy generation during the process of mixing. Assume for ethane $M = 30$, $c_p = 1766$ J/kg K and for methane $M = 16$ and $c_p = 2254$ J/kg K.

▶ Data:

Using E and M as suffixes for the properties of ethane and methane, respectively, $T_E = (273 + 20) = 293$ K, $p_E = 200$ kPa, $T_M = (273 + 45) = 318$ K, $p_M = 200$ kPa, $\dot{m}_E = 9$ kg/s and $\dot{m}_M = 4.5$ kg/s

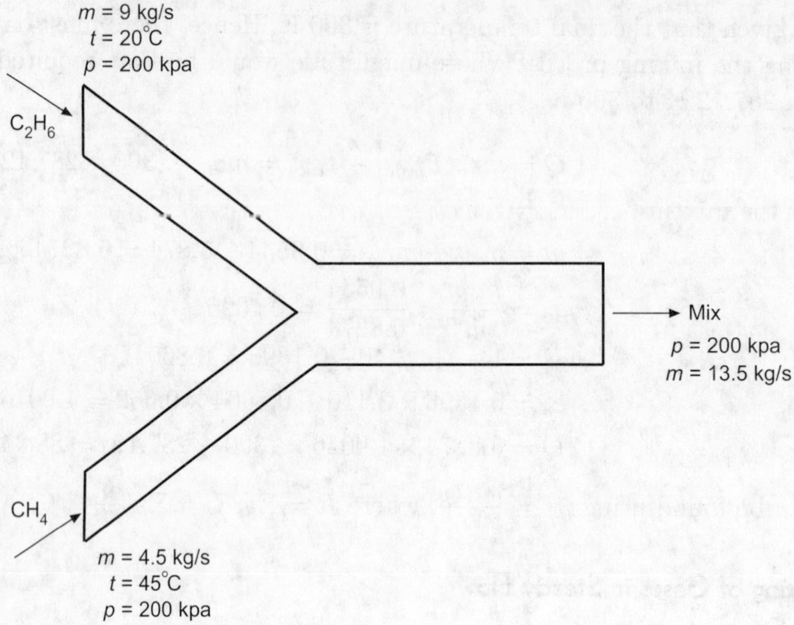

Fig. 9.11 Mixing of gases in steady flow.

▶ **Solution:**

(a)
$$T = \frac{\Sigma\, \dot{m}_i c_{pi} T_i}{\Sigma\, \dot{m}_i c_{pi}} \tag{9.46}$$
$$= \frac{(\dot{m}_E c_{pE} T_E + \dot{m}_M c_{pM} T_M)}{(\dot{m}_E c_{pE} + \dot{m}_M c_{pM})}$$
$$= \frac{(9 \times 1766 \times 293 + 4.5 \times 2254 \times 318)}{(9 \times 1766 + 4.5 \times 2254)}$$
$$= 302.7 \text{ K.}$$

(b)
$$\Delta S = \Delta S_E + \Delta S_M = m_E \Delta s_E + m_M \Delta s_M$$
$$\Delta s_E = c_{pE} \;\ln\left(\frac{T}{T_E}\right) - R_E \;\ln\left(\frac{p_{Ep}}{p_E}\right)$$

where T is the mixture temperature, T_E = temperature of ethane before mixing, P_{Ep} = partial pressure of ethane in the mixture (= $x_E p$, where p is the mixture pressure) and p_E = ethane pressure before mixing. As the pressure of ethane and methane streams are both 200 kPa, the final mixture pressure must also be equal to 200 kPa. Therefore, $p_{Ep}/p_E = x_E p/p_E = x_E$.

$$\Delta s_E = c_{pE} \;\ln\left(\frac{T}{T_E}\right) - \text{R}_E \;\ln(x_E) \tag{i}$$

$$\text{R} = \Sigma\, y_i \text{R}_i = y_E \text{R}_E + y_M \text{R}_M$$

Here, $$y_E = \frac{\dot{m}_E}{(\dot{m}_E + \dot{m}_M)} = \frac{9}{(9 + 4.5)} = 0.666,$$

and $$y_M = (1 - y_E) = 0.334$$

$$\text{R}_E = \frac{8314.3}{30} = 277 \text{ J/kg K} \text{ and } \text{R}_M = \frac{8314.3}{16} = 0.520 \text{ J/kgK}$$

$$\text{R} = 0.666 \times 277 + 0.334 \times 520 = 358 \text{ J/kgK}$$

$$M = \frac{\mathcal{R}}{\text{R}} = \frac{8314.3}{358} = 23.22 \text{ kg/kmol}$$

$$x_E = y_E \times \left(\frac{M}{M_E}\right) = 0.666 \times \left(\frac{23.22}{30}\right) = 0.516$$

$x_M = 1 - x_E = 1 - 0.516 = 0.484$. Sub. for c_{pE} and x_E in(i)

$$\Delta s_E = c_{pE} \ln\left(\frac{T}{T_E}\right) - \text{R}_E \ln(x_E)$$

$$= 1766 \times \ln\left(\frac{302.7}{293}\right) - 277 \times \ln(0.516)$$

$$= 240.8 \text{ J/kgK} = 0.2408 \text{ kJ/kg K}$$

Similarly, $$\Delta s_M = c_{pM} \ln\left(\frac{T}{T_M}\right) - \text{R}_M \ln(x_M)$$

$$= 2254 \times \ln\left(\frac{302.7}{318}\right) - 520 \times \ln(0.484)$$

$$= 266.2 \text{ J/kgK} = 0.2662 \text{ kJ/kgK}$$

The rate of entropy production $= \dot{m}_E x \Delta s_E + \dot{m}_M x \Delta s_E$

$$= 9 \times 0.2397 + 4.5 \times 0.2662 = 3.3534 \text{ kJ/s K}$$

Illustrative Problems

9.1 Oxygen (molar mass = 32) is contained in a rigid tank of volume 0.5 m^3 at a pressure of 10 MPa and a temperature of 37°C. Assume that oxygen behaves as an ideal gas and calculate its mass, in kg, and also the two specific heats, in kJ/kg K, given that their ratio is 1.4.

▶ **Data:**

$$V = 0.5 \text{ m}^3, p = 10 \times 106 \text{ N/m}^2, T = 273 + 37 = 310 \text{ K and } \frac{c_p}{c_v} = 1.4$$

► **Solution:**

For an ideal gas, $pV = mRT$, where $\mathrm{R} = \dfrac{\mathcal{R}}{M}$ and M for oxygen is 32.

Hence, $$m = \frac{pV}{\mathrm{R}T} = \frac{10 \times 10^6 \times 0.5}{[(\frac{8314.3}{32}) \times 310]} = 62 \text{ kg}$$

For an ideal gas $$c_p - c_v = \mathrm{R} = \frac{\mathcal{R}}{M}, \text{ and hence for. oxygen}$$

$$c_p - c_v = \mathrm{R} = \frac{\mathcal{R}}{M} = \frac{8314.3}{32} = 259.82 \text{ J/kgK} \quad \text{(i)}$$

From data, $$\frac{c_p}{c_v} = 1.4, \text{ or } \quad c_p = 1.4c_v \quad \text{(ii)}$$

Substituting for c_p in (i) from (ii), $1.4\, c_v - c_v = 259.82$, and therefore $c_v = 649.5$ J/kg K $= 0.6495$ kJ/kg K

Substituting for c_v in (ii), $c_p = 1.4 \times 0.6495 = 0.909$ kJ/kg K

9.2 0.1 m^3 of air at 5 MPa, 350°C contained in a cylinder expands reversibly and isothermally to 0.25 MPa. Calculate for air (a) the work transfer, (b) heat transfer and (c) the change in entropy assuming that air behaves as an ideal gas with R = 287 J/kg K

► **Data:**

$V_1 = 0.1$ m^3, $p_1 = 5 \times 10^6$ N/m^2, $T_1 = (273 + 350) = 623$ K, $p_2 = 0.25 \times 10^6$ N/m^2, R = 287 J/kg K

► **Solution:**

For an ideal gas, an isothermal process is also a reversible hyperbolic process and hence the process is pV = constant.

(a) $$W = p_1 V_1 \ln \frac{p_1}{p_2} \quad (9.18)$$

$$= 5 \times 10^6 \times 0.1 \ln \left(\frac{5}{0.25}\right)$$

$$= 1.4979 \times 10^6 \text{ J} = 1497.9 \text{ kJ}$$

(b) For an isothermal process, $Q = W$ (9.19).

Therefore, $Q = 1497.9$ kJ

(c)
$$\Delta s = c_p \ \ln \left(\frac{T_2}{T_1}\right) - \text{R} \ \ln \left(\frac{p_2}{p_1}\right)$$
$$= -\text{R} \ \ln \left(\frac{p_2}{p_1}\right), \text{ because } T_2 = T_1$$
$$\Delta s = -287 \ \ln \left(\frac{0.25}{5}\right) = 859.8 \text{ J/kg K} = 0.8598 \text{ kJ/kg K}$$

But, $\Delta S = m\Delta s$, where m is the mass of air. From the ideal gas rule,
$$m = \frac{p_1 V_1}{\text{R}T_1} = 5 \times 10^6 \times \frac{0.1}{287 \times 623} = 2.796 \text{ kg}$$
Therefore,
$$\Delta S = m\Delta s = 2.796 \times 0.8598 = 2.40 \text{ kJ/K}$$

9.3 0.5 kg of nitrogen is cooled in a rigid tank from 227°C to 27°C. The initial pressure is 15 bar. Calculate the final pressure and the changes in internal energy, enthalpy and entropy. Assume that nitrogen behaves as an ideal gas with $c_p = 1.042$ kJ/kg K and $c_v = 0.745$ kJ/kg K.

► **Data:**
$m = 0.5$ kg, the process is a constant-volume process as it occurs in a rigid vessel and therefore $V_2 = V_1, T_1 = (273 + 227) = 500$ K, $T_2 = (273 + 27) = 300$ K, $p_1 = 15$ bar, $c_p = 1.042$ kJ/kg K and $c_v = 0.745$ kJ/kg K.

► **Solution:**
As nitrogen is given to behave as an ideal gas,

$$\frac{p_1 V_1}{T_1} = \frac{p_2 V_2}{T_2} \tag{9.2}$$

As, $V_2 = V_1, \dfrac{p_1}{T_1} = \dfrac{p_2}{T_2}$, or $p_2 = \left(\dfrac{T_2}{T_1}\right) p_1$

Therefore,
$$p_2 = \left(\frac{300}{500}\right) \times 15 = 9 \text{ bar}$$
$$\Delta U = mc_v \Delta T = 0.5 \times 0.745 \times (300 - 500) = -745 \text{ kJ}$$
$$\Delta H = mc_p \Delta T = 0.5 \times 1.042 \times (300 - 500) = -1042 \text{ kJ}$$
$$\Delta S = m \left[c_v \ \ln \left(\frac{T_2}{T_1}\right) + \text{R} \ \ln \left(\frac{v_2}{v_1}\right)\right]$$

As $(v_2 = v_1)$,
$$\Delta S = mc_v \ \ln \left(\frac{T_2}{T_1}\right)$$
$$= 0.5 \times 0.745 \times \ \ln \left(\frac{300}{500}\right) = -0.190 \text{ kJ/K}$$

9.4 A certain mass of air ($c_p = 1.005$ kJ/kg K, $c_v = 0.718$ kJ/kg K) initially at 150 kPa, 0.13 m^3 is compressed in a reversible polytropic process to 700 kPa. During this process, heat transfer to air is 20 kJ and the internal energy of air increases by 50 kJ. Find the final volume of air.

▶ **Data:**

$p_1 = 150$ kPa, $V_1 = 0.13$ m^3, $p_2 = 700$ kPa, the process is described by $pV^n =$ constant, $Q = +\,20$ kJ, $\Delta U = 50$ kJ.

▶ **Solution:**

As,
$$p_1V_1^n = p_2V_2^n,$$
$$V_2 = \left(\frac{p_1}{p_2}\right)^{1/n} V_1 \tag{i}$$

The values of all quantities in the right-hand side of the above equation are given except that of n.

Hence, the value of n has to be first found to evaluate the value of final volume. From the first law of thermodynamics,

$$Q - W = \Delta U, \text{ on substituting for } Q \text{ and } \Delta U,$$
$$W = Q - \Delta U = 20 - 50 = -30 \text{ kJ}$$

As the process is reversible polytropic,

$$W = \frac{(p_1V_1 - p_2V_2)}{(n-1)} \qquad 9.20$$

$$= \frac{m\mathrm{R}\,(T_1 - T_2)}{(n-1)} \tag{ii}$$
$$= -30 \text{ kJ}$$

Also,
$$\Delta U = mc_v(T_2 - T_1) = 50 \text{ kJ, and therefore,}$$
$$m(T_1 - T_2) = \frac{-50}{c_v} = \frac{-50}{0.718} = -69.64 \text{ kg K} \tag{iii}$$

Substituting for $m(T_1 - T_2)$ from (iii) in (ii),

$$\frac{\mathrm{R} \times (-69.4)}{(n-1)} = -30, \text{ where } \mathrm{R} = c_p - c_v = 0.287 \text{ kJ/kg K}$$

Therefore, $0.287 \times (-69.4) = (n-1) \times (-30)$, and hence, $n = 1.664$.

Substituting for n in (i),

$$V_2 = \left(\frac{p_1}{p_2}\right)^{\frac{1}{n}} V_1 = \left(\frac{150}{700}\right)^{\frac{1}{1.664}} \times 0.13 = 0.0515 \text{ m}^3$$

9.5 Consider a closed cylinder divided initially to two compartments of equal volume of 25 litres each by an insulated frictionless piston. All the walls of the cylinder are insulated except for its left-hand face which is a good conductor of heat. Initially, the cylinder contains air and hydrogen, as shown in Figure 9.12, each at a pressure of 150 kPa and 27°C. Air is now heated slowly until its pressure increases to 300 kPa. Calculate the final temperatures of air and hydrogen and also the heat transferred to air. Assume that air and hydrogen behave as ideal gases and that $\gamma_{\text{hydrogen}} = 1.4$, $c_{v,\text{air}} = 0.718$ kJ/kg K and $R_{\text{air}} = 0.287$ kJ/kg K.

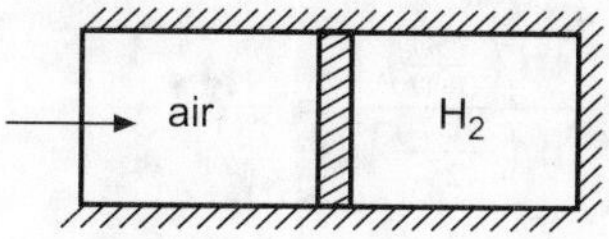

Fig. 9.12 Diagram for Illustrative Problem 9.5.

▶ **Solution:**

The pressure and temperature of air increases during the period it is being heated. Also the piston travels slowly to the right thereby causing the air to expand and compress the hydrogen reversibly (frictionless piston) and adiabatically (insulated piston and walls). At any instant during the process, $p_{\text{air}} = p_{\text{hydrogen}}$ and hence at the end of the process both are equal to 300 kPa.

Considering the hydrogen in the right-hand compartment

$$p_1 = 150 \text{ kPa},$$
$$T_1 = 273 + 27 = 300 \text{ K}$$
$$V_1 = 25 \times 10^{-3} \text{ m}^3 \text{ and}$$
$$p_2 = 300 \text{ kPa}.$$

Hydrogen is subjected to a reversible adiabatic process and hence,

$$p_1 V_1^\gamma = p_2 V_2^\gamma \text{ or}$$

$$V_2 = \left(\frac{p_1}{p_2}\right)^{\frac{1}{\gamma}} V_1 = \left(\frac{150}{300}\right)^{\frac{1}{1.4}} \times 25 = 15.24 \text{ l},$$

and also,

$$T_2 = \left(\frac{V_1}{V_2}\right)^{\gamma-1} T_1 = \left(\frac{25}{15.24}\right)^{0.4} \times 300$$
$$= 365.7 \text{ K} = 92.7°C$$

Therefore, the final temperature of hydrogen = 365.7 K = 92.7°C

Considering now air in the left-hand side compartment, air expands as it is heated. The process can be idealized as a reversible polytropic process pV^n = constant. For air, p_1 = 150 kPa, $T_1 = 273 + 27 = 300$ K, $V_1 = 25 \times 10^{-3}$ m^3 and $p_2 = 300$ kPa.

Also, $V_2 = V_1$ + decrease in volume of hydrogen

$$= 25 + (25 - V_{2,\text{hydrogen}}) = 25 + (25 - 15.24) = 34.76 \text{ l}$$

As, $p_1 V_1^n = p_2 V_2^n$

$$n = \frac{\ln\left(\frac{p_1}{p_2}\right)}{\ln\left(\frac{V_2}{V_1}\right)} = \frac{\ln\left(\frac{150}{300}\right)}{\ln\left(\frac{34.76}{25}\right)} = -2.1$$

Also, for air undergoing a reversible polytropic process, with $n = -2.1$,

$$T_2 = T_1 \left[\frac{V_1}{V_2}\right]^{n-1} = 300 \times \left[\frac{25}{34.76}\right]^{-2.1-1} = 833.4 \text{ K} = 560.4°\text{C}.$$

From the ideal gas rule, $m_{\text{air}} = \left[\frac{p_1 V_1}{RT_1}\right]_{\text{air}}$

$$= \frac{150 \times 10^3 \times 25 \times 10^{-3}}{287 \times 300} = 0.04355 \text{ kg}$$

$$\Delta U_{\text{air}} = mc_v(T_2 - T_1) = 0.04355 \times 0.718 \times (833.4 - 300) = 16.68 \text{ kJ}$$

$$W_{\text{air}} = \frac{(p_1 V_1 - p_2 V_2)}{(n-1)} = \frac{m\text{R}(T_1 - T_2)}{(n-1)} = \frac{0.04355 \times 0.287 \times (300 - 833.4)}{(-2.1 - 1)} = 2.15 \text{ kJ}$$

Applying the first law to air, $Q = W + \Delta U = 2.15 + 16.68 = 18.83$ kJ

9.6 0.5 kg air, initially at 90 kPa, 70°C, in a piston–cylinder device, is compressed reversibly and adiabatically to 0.4 MPa. It is then expanded reversibly at constant pressure to its initial volume. The air is then cooled to its initial state. Sketch the cycle on $p-V$ and $T-S$ diagrams. Calculate the work and heat interactions in each process and the cycle. Also find the mep and the cycle efficiency. Assume that air behaves as an ideal gas with R = 287 J/kg K and $\gamma = 1.4$.

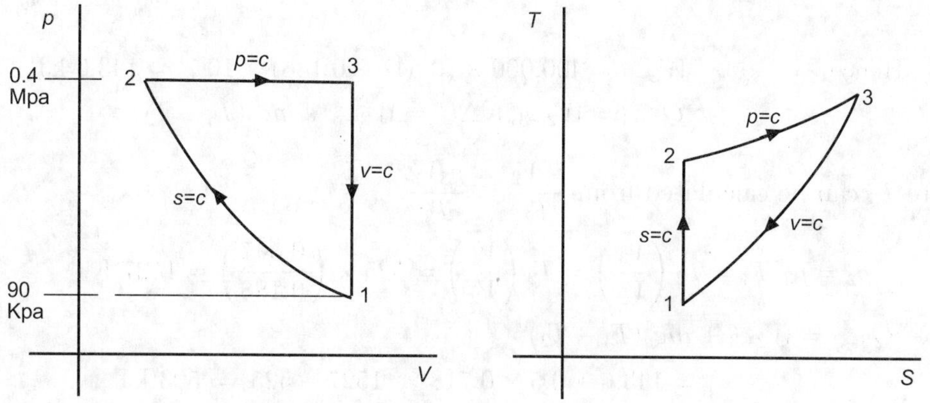

Fig. 9.13 $p-V$ and $T-S$ diagrams to solve Problem 9.6.

▶ **Data:**

$m = 0.5$ kg, $p_1 = 90$ kPa $= 90{,}000$ N/m^2, $T_1 = 273 + 70 = 343$ K, $p_3 = p_2 =$ 0.4 MPa $= 400{,}000$ N/m^2, $V_3 = V_1$

▶ **Solution:**

At the initial state, $V_1 = \dfrac{mRT_1}{p_1} = \dfrac{0.5 \times 287 \times 343}{90{,}000} = 0.547$ m^3. The process 1–2 is reversible and adiabatic and hence follows the path $pv^\gamma =$ constant.

Therefore,
$$\left[\frac{T_2}{T_1}\right] = \left[\frac{p_2}{p_1}\right]^{\frac{(\gamma-1)}{\gamma}}$$

$$T_2 = T_1\left[\frac{p_2}{p_1}\right]^{\frac{(\gamma-1)}{\gamma}} = 343 \times \left[\frac{0.4 \times 10^6}{90 \times 10^3}\right]^{\frac{0.4}{1.4}} = 525 \text{ K}$$

$$W_{1-2} = \frac{(p_1V_1 - p_2V_2)}{(\gamma-1)} = \frac{mR(T_1 - T_2)}{(\gamma-1)}$$

$$= \frac{0.5 \times (287 \times 10^{-3}) \times (343 - 525)}{0.4} = -65.3 \text{ kJ}$$

$$Q_{1-2} = 0$$

Process 2–3 is a constant-pressure process. Therefore,

$$W_{2-3} = p_2(V_3 - V_2)$$
$$= p_2(V_1 - V_2), \quad (\text{since } V_3 = V_1)$$

where V_2 can be calculated from process 1–2 as,

$$V_2 = V_1\left(\frac{p_1}{p_2}\right)^{\frac{1}{\gamma}} = 0.547 \times \left(\frac{9}{40}\right)^{\frac{1}{1.4}} = 0.188 \text{ m}^3.$$

Hence, $W_{2-3} = 400{,}000 \times (0.547 - 0.188) \times 10^{-3} = 143.6 \text{ kJ}$

$$Q_{2-3} = W_{2-3} + \Delta U = W_{2-3} + mc_v(T_3 - T_2)$$

where T_3 can be calculated from $\dfrac{p_2V_2}{T_2} = \dfrac{p_3V_3}{T_3}$.

As, $p_2 = p_3, T_3 = T_2\left(\dfrac{V_3}{V_2}\right) = T_2\left(\dfrac{V_1}{V_2}\right) = 525 \times \left(\dfrac{0.547}{0.188}\right) = 1527 \text{ K}.$

$$Q_{2-3} = W_{2-3} + mc_v(T_3 - T_2)$$

$$= 143.6 + 0.5 \times 0.718 \times (1527 - 525) = 503 \text{ kJ}.$$

For process 3–1, $W_{3-1} = 0$ as it is at a constant volume.

For the cycle 1–2–3–1, from the first law, $\Sigma Q = \Sigma W$.

Therefore, $Q_{1-2} + Q_{2-3} + Q_{3-1} = W_{1-2} + W_{2-3} + W_{3-1}$

that is, $0 + 503 + Q_{3-1} = -65.3 + 143.6 + 0$

Therefore, $Q_{3-1} = -424.7 \text{ kJ} \approx -425 \text{ kJ}$

Alternatively, $Q_{3-1} = mc_v(T_1 - T_3) = 0.5 \times 0.718 \times (343 - 1527) = -425 \text{ kJ}$

$$W_{\text{cycle}} = W_{1-2} + W_{2-3} + W_{3-1} = -65.3 + 143.6 + 0 = 78.3 \text{ kJ}$$

and $Q_{\text{cycle}} = W_{\text{cycle}} = 78.3 \text{ kJ}.$

$$\text{mep} = \frac{W_{\text{net}}}{\text{swept volume}} = \frac{78.3}{(V_1 - V_2)} = \frac{78.3}{(0.547 - 0.188)} = 218 \text{ kPa}$$

$$\text{cycle efficiency} = \frac{\text{network}}{\text{heat input}}$$

$$= \frac{78.3}{Q_{2-3}} = \frac{78.3}{503} = 0.156$$

9.7 One kilogram of air (R = 287 J/kg K and c_p = 1005 J/kg K) undergoes a cyclic process comprising three process 1–2, 2–3 and 3–1. At state 1, the pressure and temperature are, respectively, 1 MPa and 27°C. 1–2 is an irreversible adiabatic constant-pressure process, 2–3 is a reversible adiabatic process and 3–1 is a reversible isothermal process. At state 3, p = 100 kPa. (a) Sketch the cycle on p–v and T–S coordinates, (b) Find the heat and work interactions in each of the three processes and the net work per cycle and (c) analyse quantitatively whether the cycle is reversible or irreversible.

▶ **Data:**

$p_1 = 10^6 \text{ N/m}^2, T_1 = 273 + 27 = 300 \text{ K}, p_3 = 10^5 \text{ N/m}^2,$

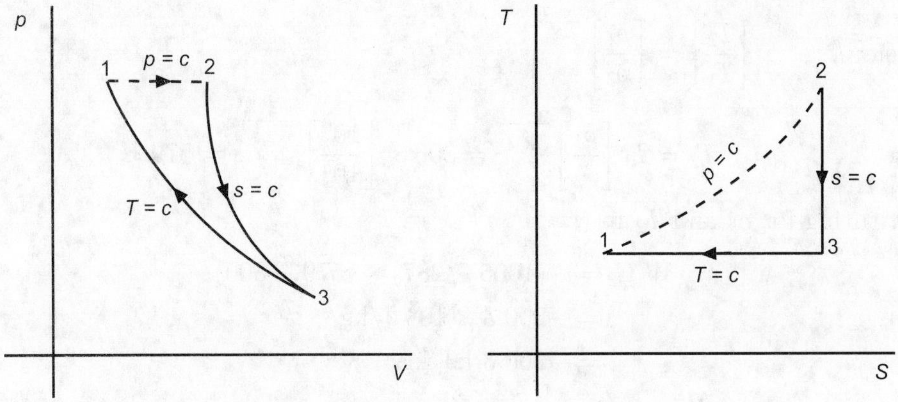

Fig. 9.14 $p-V$ and $T-S$ diagrams.

During process 1–2, $p_1 = p_2, Q = 0$

During process 2–3, which is reversible and adiabatic, $p_2V_2^\gamma = p_3V_3^\gamma$

During process 3–1, which is reversible isothermal, because air is an ideal gas, $T_3 = T_1$ and $p_3V_3 = p_1V_1$

▶ **Solution:**

(a) The $p-V$ and $T-S$ diagrams are shown in Figure 9.14

(b) For the constant pressure process 1–2

Work cannot be evaluated using $W = p(V_2 - V_1) = m\mathrm{R}(T_2 - T_1)$, as the process 1–2 is irreversible. The above expression for work is derived using pdv relation which is valid only for a reversible process.

However, applying the first law to process 1–2,

$$Q - W = m\Delta u = mc_v(T_2 - T_1)$$

as $$m = 1, Q - W = c_v(T_2 - T_1)$$

Here, $Q = 0$ and therefore,

$$W_{1-2} = -c_v(T_2 - T_1) \tag{i}$$

To find out T_2,

For process 2–3,

$$\left[\frac{T_2}{T_3}\right] = \left[\frac{p_2}{p_3}\right]^{\frac{(\gamma-1)}{\gamma}}$$ and from data $T_3 = T_1$ and $p_2 = p_1 = 10^6\ \mathrm{N/m^2}$.

Also, $$\gamma = \frac{c_p}{c_v} = \frac{c_p}{(c_p - \mathrm{R})} = \frac{1005}{(1005 - 287)} = 1.4.$$

Therefore, $$\left[\frac{T_2}{T_1}\right] = \left[\frac{p_1}{p_3}\right]^{\frac{(\gamma-1)}{\gamma}}$$

Thus, $$T_2 = T_1\left[\frac{p_1}{p_3}\right]^{\frac{(\gamma-1)}{\gamma}} = 300 \times \left[\frac{10^6}{10^5}\right]^{\frac{(1.4-1)}{1.4}} = 579 \text{ K}$$

Substituting for c_v and T_2 in (i)

$$\begin{aligned} W_{1-2} &= -(1005 - 287) \times (579 - 300) \\ &= -200.3 \times 10^3 \text{ J/kg} \\ &= -200.3 \text{ kJ/kg} \end{aligned}$$

For process 2–3,

$Q = 0$, as the process is reversible and adiabatic. Also,

$$\begin{aligned} W_{2-3} &= \frac{(p_2v_2 - p_3v_3)}{(\gamma - 1)} \\ &= \frac{\text{R}(T_2 - T_3)}{(\gamma - 1)} \quad \text{where} \quad T_3 = T_1 \end{aligned}$$

Hence, $$W_{2-3} = \frac{(287 \times 10^{-3}) \times (579 - 300)}{0.4} = 200.2 \text{ kJ/kg.}$$

For process 3–1,

It is a reversible isothermal process, for which $pv =$ Constant,

$$\begin{aligned} Q = W_{3-1} &= p_3v_3 \ln\left(\frac{p_3}{p_1}\right) \\ &= \text{R}T_3 \ln\left(\frac{p_3}{p_1}\right) = \text{R}T_1 \ln\left(\frac{p_3}{p_1}\right) \\ &= (287 \times 10^{-3}) \times 300 \ln\left(\frac{10^5}{10^6}\right) = -198.2 \text{ kJ/kg} \end{aligned}$$

$$\begin{aligned} \text{Net work during the cycle} &= W_{1-2} + W_{2-3} + W_{3-1} \\ &= -200.3 + 200.2 - 198.2 \\ &= -198.3 \text{ kJ/kg} \end{aligned}$$

(c) The cycle is irreversible because of process 1–2. However, the cycle can be assessed for its reversibility or otherwise quantitatively, using the Clausius inequality statement, $\oint \frac{\delta Q}{T} \leq 0$. Here,

$$\oint \frac{\delta Q}{T} = \int_1^2 \frac{\delta Q}{T} + \int_2^3 \frac{\delta Q}{T} + \int_3^1 \frac{\delta Q}{T}.$$

While $Q = 0$ for processes 1–2 and 2–3, Q_{3-1} occurs at a constant temperature of $T_3 = T_1$. Therefore,

$$\oint \frac{\delta Q}{T} = 0 + 0 + \frac{Q_{3-1}}{T_1} = + \left(\frac{-198.2}{300}\right) = -0.66 \text{ kJ/kgK}$$

As, $\oint \frac{\delta Q}{T} < 0$, the cycle is irreversible.

9.8 (a) Calculate the power required by a compressor if air flowing at a rate of 1.5 kg/s enters at 100 kPa, 27°C with a velocity of 50 m/s and leaves at 200 kPa, 100°C with a velocity of 100 m/s. Heat transfer from air to the cooling water circulating around the compressor casing is 1.35 kJ/s. (b) Had the air at the same inlet conditions, as in (a), been compressed in a reversible adiabatic steady-flow device to the same discharge pressure and velocity as in (a) find the exit temperature and power required. Assume air is an ideal gas with $c_p = 1.005$ kJ/kg K and $\gamma = 1.4$.

▶ **Data:**

(a) $\dot{m} = 1.5$ kg/s, $p_1 = 100$ kPa, $T_1 = 273 + 27 = 300$ K, $V_1 = 50$ m/s, $p_2 = 200$ kPa, $T_2 = 373$ K, $V_2 = 100$ m/s. $\dot{Q}_{\text{air}} = -1.35$ kJ/s.

▶ **Solution:**

From the steady-state, steady-flow energy equation,

$$\dot{Q} - \dot{W} = \dot{m}\Delta\left(h + \frac{V^2}{2} + gZ\right)$$

$$\dot{Q} - \dot{W} = \dot{m}\left[(h_2 - h_1) + \frac{(V_2^2 - V_1^2)}{2} + g(Z_2 - Z_1)\right]$$

Neglecting changes in potential energy,

$$\dot{Q} - \dot{W} = \dot{m}\left[c_p(T_2 - T_1) + \frac{(V_2^2 - V_1^2)}{2}\right]$$

$$-1.35 - \dot{W} = 1.5\left[1.005(373 - 300) + \frac{(100^2 - 50^2)}{(2 \times 1000)}\right]$$

Therefore, $\dot{W} = -117 \text{ kJ/s} = -117 \text{ kW}$

(b) Had the compression been reversible and adiabatic to the same discharge pressure,

$$\left[\frac{T_2}{T_1}\right] = \left[\frac{p_2}{p_1}\right]^{\frac{(\gamma-1)}{\gamma}}$$

Therefore, $$T_2 = T_1\left[\frac{p_2}{p_1}\right]^{\frac{(\gamma-1)}{\gamma}} = 300 \times \left[\frac{200}{100}\right]^{\frac{0.4}{1.4}} = 365.7 \text{ K}$$

From the SS, SFEE,

$$\dot{Q} - \dot{W} = \dot{m}\left[(h_2 - h_1) + \frac{(V_2^2 - V_1^2)}{2} + g(Z_2 - Z_1)\right]$$

$$\dot{Q} - \dot{W} = \dot{m}\left[c_p(T_2 - T_1) + \frac{(V_2^2 - V_1^2)}{2}\right]$$

$$0 - \dot{W} = 1.5 \times \left[1.005(365.7 - 300) + \frac{(100^2 - 50^2)}{(2 \times 1000)}\right]$$

Therefore, $\dot{W} = -104.67$ kW

9.9 At a given state, show that, for an ideal gas, the constant volume line is steeper than the constant pressure line on a T–S diagram.

▶ **Solution:**

On a T–S diagram, at a given temperature, T, while the slope of a constant-volume line is $\left[\frac{\partial T}{\partial s}\right]_v$, the slope of a constant-pressure line is $\left[\frac{\partial T}{\partial s}\right]_p$.
For a pure substance, $T\mathrm{d}s = \mathrm{d}u + p\mathrm{d}v$, and an ideal gas is a pure substance. Also, for an ideal gas $\mathrm{d}u = c_v\mathrm{d}T$ and hence,

$$T\mathrm{d}s = c_v\mathrm{d}T + p\mathrm{d}v$$

For a constant volume line $dv = 0$, and thus

$$\left[\frac{\partial T}{\partial s}\right]_v = \frac{T}{c_v} \tag{i}$$

Similarly, for an ideal gas, $T\mathrm{d}s = \mathrm{d}h - v\mathrm{d}p$, and $\mathrm{d}h = c_p\mathrm{d}T$.

Hence, $T\mathrm{d}s = c_p\mathrm{d}T - v\mathrm{d}p$

For a constant-pressure line, $\mathrm{d}p = 0$ and thus

$$\left[\frac{\partial T}{\partial s}\right]_p = \frac{T}{c_p} \tag{ii}$$

For a gas, $c_p > c_v$, therefore $\frac{T}{c_v} > \frac{T}{c_p}$ and hence the right-hand side of (i) is greater than the right-hand side of (ii).

Therefore, $\left[\frac{\partial T}{\partial s}\right]_v > \left[\frac{\partial T}{\partial s}\right]_p$, which means, on a T–S diagram, the slope of a constant-volume line at a given state is greater than the slope of a constant-pressure line at the same state.

9.10 Starting from the Tds relation show that the reversible adiabatic process of an ideal gas with constant specific heats follows the path pv^γ = constant, where γ is the ratio of specific heats.

▶ **Solution:**

From the Tds relation

$$Tds = du + pdv \tag{i}$$

A reversible adiabatic process is also an isentropic process and hence $ds = 0$. Therefore, from (i),

$$du + pdv = 0 \tag{ii}$$

For an ideal gas, $du = c_v dT$, and hence from (ii)

$$c_v dT + pdv = 0 \tag{iii}$$

The differential form of the ideal gas equation $pv = RT$ is

$$pdv + vdp = RdT \tag{iv}$$

Eliminating dT between (iii) and (iv)

$$\frac{c_v(pdv + vdp)}{R} + pdv = 0$$

That is, $$c_v pdv + c_v vdp + Rpdv = 0$$

as $R = (c_p - c_v)$, $$c_v pdv + c_v vdp + (c_p - c_v)pdv = 0$$

That is, $$c_v vdp + c_p pdv = 0$$

Dividing by $c_v pv$, $$\frac{dp}{p} + \gamma\left(\frac{dv}{v}\right) = 0$$

Integration of this relation gives,

$$pv^\gamma = \text{Constant}$$

9.11 An insulated rigid tank of volume 0.5 m^3 initially contains air at 250°C and 8 MPa. A valve mounted on the tank is slowly leaking. Because of this, the pressure in the tank is slowly decreasing. After a certain time, the pressure in the tank is observed to be 500 kPa. Calculate the mass of air that leaked out of the tank. Assume that air behaves as an ideal gas with c_p = 1005 J/kg K and c_v = 718 J/kg K.

▶ **Solution:**

Mass leaked out $= (m_1 - m_2)$ where m_1 is the initial mass in the tank and m_2 is the final, mass in kilograms.

$$m_1 = \left(\frac{p_1 V_1}{\mathrm{R}T_1}\right) \quad \text{where } \mathrm{R} = c_p - c_v = 287 \text{ J/kgK}$$

$$m_1 = \frac{(8 \times 10^6 \times 0.5)}{(287 \times 523)} = 26.65\ kg$$

Considering the tank as the control volume, with respect to time, there is a decrease in energy of the control volume as air is leaking out of it.
The general energy equation for the control volume is

$$\delta Q - \delta W = \mathrm{d}E_c + \mathrm{d}m_{\text{out}}\left[h + \frac{V_2}{2} + gZ\right]_{\text{out}} - \mathrm{d}m_{\text{in}}\left[h + \frac{V_2}{2} + gZ\right]_{\text{in}}, \tag{5.19}$$

where, $\mathrm{d}E_c$ is the change in energy of the control volume. $\mathrm{d}E_c$ at any instant is equal to $\mathrm{d}(mu)$ where m and u are, respectively, the mass and internal energy of air in the tank at the considered instant. Hence.

$$\mathrm{d}E_c = \mathrm{d}mu + m\mathrm{d}u$$

Also $\delta Q = 0$ and $\delta W = 0$ and $\mathrm{dm_{in}} = 0$. Neglecting changes in kinetic and potential energies, the general energy equation reduces to,

$$\mathrm{d}mu + m\mathrm{d}u + \mathrm{d}m_{\text{out}}h_{\text{out}} = 0$$

At any instant, mass leaving the tank is equal to $\mathrm{d}m$ the mass depleting in the tank. Therefore, $\mathrm{d}m_{\text{out}} = -\mathrm{d}m$. Also, at any instant $h_{\text{out}} = h =$ enthalpy of air in the tank. Hence,

$$\mathrm{d}mu + m\mathrm{d}u - \mathrm{d}mh = 0$$

For an ideal gas $h = u + \mathrm{R}T$ and thus,

$$\mathrm{d}mu + m\mathrm{d}u - \mathrm{d}m(u + \mathrm{R}T) = 0$$

Therefore

$$m\mathrm{d}u - \mathrm{d}m\mathrm{R}T = 0$$

Also, $\mathrm{d}u = c_v\mathrm{d}T$, and therefore,

$$mc_v\mathrm{d}T - \mathrm{d}m\mathrm{R}T = 0, \quad \text{or}$$

$$\frac{\mathrm{d}m}{m} = \left(\frac{c_v}{\mathrm{R}}\right) \times \left(\frac{\mathrm{d}T}{T}\right)$$

Integrating the above equation,

$$\frac{m_2}{m_1} = \left[\frac{T_2}{T_1}\right]^{\frac{c_v}{R}} \tag{i}$$

But $$\frac{c_v}{R} = \frac{c_v}{(c_p - c_v)} = \frac{1}{(\gamma - 1)}$$

Therefore, $$\frac{m_2}{m_1} = [T_2]^{\frac{1}{(\gamma-1)}} \times [T_1]^{\frac{-1}{(\gamma-1)}}$$

Here, $m_2 = \left(\frac{p_2V_2}{RT_2}\right)$, where $p_2 = 500$ kPa, $V_2 = V_1 = 0.5$ m^3 (rigid tank) and T_2 is not known.

$$T_2 = \left(\frac{p_2V_2}{Rm_2}\right)$$

Therefore, $$\frac{m_2}{m_1} = \left[\frac{p_2V_2}{Rm_2}\right]^{\frac{1}{(\gamma-1)}} \times [T_1]^{\frac{1}{(1-\gamma)}}$$

$$m_2^{\frac{\gamma}{(\gamma-1)}} = m_1\left[\frac{p_2V_2}{R}\right]^{\frac{1}{(\gamma-1)}} \times (T_1)^{\frac{1}{(1-\gamma)}}$$

Substituting the values for m_1, p_2, $V_2(= V_1)$, T_1 and γ

$$m_2^{3.5} = 26.65\left[\frac{500 \times 10^3 \times 0.5}{287}\right]^{\frac{1}{0.4}} \times (523)^{\frac{-1}{0.4}}$$

$$m_2 = 3.678 \text{ kg}$$

Mass of air leaked out of the tank $= m_1 - m_2 = 26.65 - 3.678 = 22.972$ kg

▶ Alternative Method 1:

The same steps to be followed upto (i),

$$\frac{m_2}{m_1} = \left[\frac{T_2}{T_1}\right]^{\frac{1}{(\gamma-1)}}$$

As, $$p_2V_2 = m_2RT_2 \text{ and } p_1V_1 = m_1RT_1, \quad \frac{T_2}{T_1} = \frac{p_2V_2m_1}{p_1V_1m_2}$$

Therefore, $$\frac{m_2}{m_1} = \left[\frac{p_2V_2m_1}{p_1V_1m_2}\right]^{\frac{1}{(\gamma-1)}}. \text{ Here, } \quad V_2 = V_1, \quad \text{and hence,}$$

$$\frac{m_2}{m_1} = \left[\frac{p_2}{p_1}\right]^{\frac{1}{(\gamma-1)}} \times \left[\frac{m_1}{m_2}\right]^{\frac{1}{(\gamma-1)}}, \text{ or}$$

$$\frac{m_2}{m_1} = \left[\frac{p_2}{p_1}\right]^{\frac{1}{\gamma}}$$

$$m_2 = m_1 \times \left[\frac{p_2}{p_1}\right]^{\frac{1}{\gamma}} = 26.65 \times \left[\frac{500 \times 10^3}{8 \times 10^6}\right]^{\frac{1}{1.4}} = 3.678 \text{ kg}$$

▶ **Alternative Method 2:**

m_2 can be calculated by a simpler method using an unique but appropriate description. The mass of air finally remaining in the tank (m_2) at p_2, T_2 can be imagined to be that mass which initially could have existed at p_1, T_1, occupying a smaller volume. The change from the imagined initial state to the final state should occur adiabatically as the tank is insulated. Also, during the process, there are no irreversibilities. Hence, the assumed process would be reversible and adiabatic. Therefore,

$$\frac{T_2}{T_1} = \left[\frac{p_2}{p_1}\right]^{\frac{(\gamma-1)}{\gamma}} \cdot \text{ Substituting the values for } T_1,\ p_2,\ p_1 \text{ and } \gamma$$

$$T_2 = \left[\frac{500 \times 10^3}{8 \times 10^6}\right]^{\frac{0.4}{1.4}} = 236.85 \text{ K}$$

$$m_2 = \left(\frac{p_2 V_2}{\text{R}T_2}\right) = 500 \times 10^3 \times \frac{0.5}{287 \times 236.85} = 3.678 \text{ kg}$$

9.12 A piston–cylinder device contains a mixture of 0.2 kg H_2 and 1.6 kg N_2 at 100 kPa and 300 K. Heat is transferred to the mixture at constant pressure until the volume is doubled. Determine (a) the heat transfer and (b) the entropy change. Assume for H_2, R = 4.135 kJ/kg K, c_v = 10.085 kJ/kg K and for N_2, R = 0.297 kJ/kg K, c_v = 0.745 kJ/kg K.

▶ **Data:**

$m_{\text{H}} = 0.2$ kg, $m\text{N} = 1.6$ kg, $m = m_{\text{H}} + m_{\text{N}} = 1.8$ kg, $p_1 = 100$ kPa, $T_1 = 300$ K, $V_2 = 2V_1$, $p_1 = p_2$.

▶ **Solution:**

$$y_{\text{H}} = \frac{m_{\text{H}}}{m} = \frac{0.2}{1.8} = 0.111. \text{ Therefore, } y_{\text{N}} = 1 - y_{\text{H}} = 0.889$$

$$\text{R} = \Sigma\, y_i \text{R}_i = 0.111 \times 4.135 + 0.889 \times 0.297 = 0.723 \text{ kJ/kgK}$$

$$c_v = \Sigma\, y_i c_{vi} = 0.111 \times 10.085 + 0.889 \times 0.745 = 1.7817 \text{ kJ/kgK}$$

During the heating process, considering the mixture to be an ideal gas,

$$\frac{p_1 V_1}{T_1} = \frac{p_2 V_2}{T_2}. \text{ From data, } V_2 = 2V_1 \text{ and } p_1 = p_2.$$

Hence, $T_2 = 2T_1 = 600$ K

Work done $= \int p\mathrm{d}V = p(V_2 - V_1)$, where $p = p_1 = p_2$

$$= p_2V_2 - p_1V_1 = m\mathrm{R}(T_2 - T_1)$$
$$= 1.8 \times 0.723 \times (600 - 300) = 390.42 \text{ kJ}$$
$$\Delta U = mc_v(T_2 - T_1) = 1.8 \times 1.7817 \times 300 = 962 \text{ kJ}$$

Applying the first law,

$$Q = W + \Delta U = 390.42 + 962 = 1352.42 \text{ kJ}$$
$$\Delta s = c_p \ \ln\left(\frac{T_2}{T_1}\right) - \mathrm{R} \ \ln\left(\frac{p_2}{p_1}\right)$$

Because $p_2 = p_1, \Delta s = c_p \ \ln\left(\frac{T_2}{T_1}\right)$

$$= (R + c_v) \ \ln\left(\frac{600}{300}\right)$$
$$= (0.723 + 1.7817) \times \ \ln 2 = 1.736 \text{ kJ/kgK}$$
$$\Delta S = m\Delta s = 1.8 \times 1.736 = 3.125 \text{ kJ/K}$$

9.13 A mixture by mass of 80% N_2 and 20% CO_2 enters the nozzle of a turbo-jet engine at 600 kPa, 1000 K with negligible velocity and expands reversibly and adiabatically to 80 kPa. Determine (a) the exit temperature and (b) the exit velocity. Assume for N_2, $c_p = 1042$ J/kg K, $c_v = 745$ J/kg K and for CO_2, $c_p = 842$ kJ/kg K, $c_v = 653$ J/kg K.

▶ **Data:**

Using suffixes N and C for N_2 and CO_2, respectively, $y_{\mathrm{N}} = 0.8$, $y_{\mathrm{C}} = 0.2$, $p_1 = 600$ kPa, $T_1 = 1000$ K, $p_2 = 80$ kPa.

▶ **Solution:**

$$c_p = \Sigma\, y_i c_{pi} = y_N c_{pN} + y_c c_{pc}$$
$$= 0.8 \times 1.042 + 0.2 \times 0.842 = 1.002 \text{ kJ/kgK}$$

Similarly, $c_v = 0.8 \times 0.745 + 0.2 \times 0.653 = 0.726$ kJ/kgK.

Therefore, $\gamma = \dfrac{c_p}{c_v} = \dfrac{1.002}{0.726} = 1.38$

The process is reversible and adiabatic, and hence

$$\frac{T_2}{T_1} = \left(\frac{p_2}{p_1}\right)^{\frac{(\gamma-1)}{\gamma}} = \left(\frac{80}{600}\right)^{\frac{0.38}{1.38}} = 0.574$$
$$T_2 = 0.574T_1 = 574 \text{ K}$$

Applying SS, SFEE,

$$Q - W_x = \Delta\left(h + \frac{V_2}{2} + gZ\right)$$

Here, $Q = 0, W_x = 0$ and neglecting the change in potential energy,

$$\Delta\left(h + \frac{V_2}{2}\right) = 0$$

that is,
$$\frac{(V_2^2 - V_1^2)}{2} = h_1 - h_2 = c_p(T_1 - T_2)$$

As
$$V_1 = 0, V_2 = [2c_p(T_1 - T_2)]^{0.5}$$
$$= [2 \times 1.002 \times 10^3(1000 - 574)]^{0.5} = 924 \text{ m/s}$$

9.14 Consider two or more ideal gases in as many closed containers as available. The pressure and temperature in each container is the same. When these gases are mixed adiabatically, show that the entropy change depends only on the mole fractions of the gases and not on what they are and what their properties are.

▶ **Solution:**

Let us consider two ideal gases, A and B, initially in separate containers, but both at the same pressure, p, and temperature, T. Let the number of kmols of gas A be n_A and that of B be n_B. Let the two gases mix adiabatically. As there is no heat or work interactions, the final temperature and pressure of the mixture shall be the same as those of the individual gases before mixing. The mole fractions of A and B in the mixture would be,

$$x_A = \frac{n_A}{(n_A + n_B)} \text{ and } x_B = \frac{n_B}{(n_A + n_B)}$$

$$\Delta S_A = n_A\left[c_{p,A} \ln\left(\frac{T_{\text{A,after mix}}}{T_{\text{A,before mix}}}\right) - \mathcal{R} \ln\left(\frac{x_A p}{p}\right)\right]$$

where $x_A p$ is the partial pressure of A in the mixture and p is the mixture pressure.

As
$$T_{\text{A,after mix}} = T_{\text{A,before mix}}$$
$$\Delta S_A = n_A\left[-\mathcal{R} \ln\left(\frac{x_A p}{p}\right)\right] = -\mathcal{R} n_A \ln(x_A)$$

Similarly,
$$\Delta S_B = -\mathcal{R} n_B \ln(x_B)$$

Entropy of mixing
$$= \Delta S_A + \Delta S_B = -\mathcal{R}[n_A \ln(x_A) + n_B \ln(x_B)]$$
$$= -\mathcal{R}\Sigma n_i \ln(x_i)$$

Thus, the entropy change depends upon only n and x (i.e., the mole fractions of the constituents) and not on any other properties of the gases.

9.15 A well insulated rigid vessel of volume 0.05 m^3 containing air at 100 kPa and 20°C is connected through a valve to an insulated pipeline in which helium is flowing at 350 kPa, 50°C with negligible velocity. The valve is now opened and the temperature and pressure of the resulting mixture in the vessel is continuously recorded. When the pressure reaches 200 kPa, the valve is closed. At this instant the temperature is recorded to be 87°C. Assuming no air flows out of the vessel, find the mass of helium that has entered the vessel. Assume c_p of helium and air to be 5.193 kJ/kg K and 1.005 kJ/kg K, respectively.

► **Solution:**

The data is incorporated in Figure 9.15
Using suffixes h and a for helium and air, respectively, and suffixes i and f for initial state of air and final state of mixture in the vessel and applying the first law for the control volume, shown in Figure 9.15,

$$\delta Q - \delta W = \mathrm{d}E_{cv} + \mathrm{d}m_{\text{out}}\left[h + \frac{V^2}{2} + gZ\right] - \mathrm{d}m_{\text{in}}\left[h + \frac{V^2}{2} + gZ\right]$$

Here, $\delta Q = 0, \delta W = 0$ and $\mathrm{d}m_{\text{out}} = 0$. Also $E = U$ as the material in the CV at any instant possesses only internal energy. In addition, neglecting the changes in kinetic and potential energies,

$$\mathrm{d}U_{cv} - \mathrm{d}m_{\text{in}}[h] = 0$$

On integration,

$$(U_f - U_i) - m_{\text{in}}h_{\text{in}} = 0$$

Here, m_{in} = mass of helium in and $h_{\text{in}} = h_{h1}$, the enthalpy of helium in the pipeline. Therefore,

$$(U_f - U_i) - m_h h_{h1} = 0$$

Now,

$$U_f = (m_h + m_a)h_f - p_f V_f, \text{ and}$$

$$U_i = m_a h_{ai} - p_i V_i$$

also,

$$h_f = \left[\frac{m_h}{(m_h + m_a)}\right] h_{h,f} + \left[\frac{m_a}{(m_h + m_a)}\right] h_{a,f}$$

Therefore,

$$[m_h h_{h,f} + m_a h_{a,f} - p_f V_f] - [m_a h_{a,i} - p_i V_i] = m_h h_{h,1}$$

$$m_h(h_{h,f} - h_{h,1}) + m_a(h_{a,f} - h_{a,i}) = V(p_f - p_i), \text{ or}$$

$$m_h c_{p,h}(T_{h,f} - T_{h,i}) + m_a(T_{a,f} - T_{a,i}) = V(p_f - p_i)$$

From data,

$$m_a = \frac{pV}{RT} = \frac{100 \times 10^3 \times 0.05}{(287 \times 293)} = 0.06 \text{ kg}$$

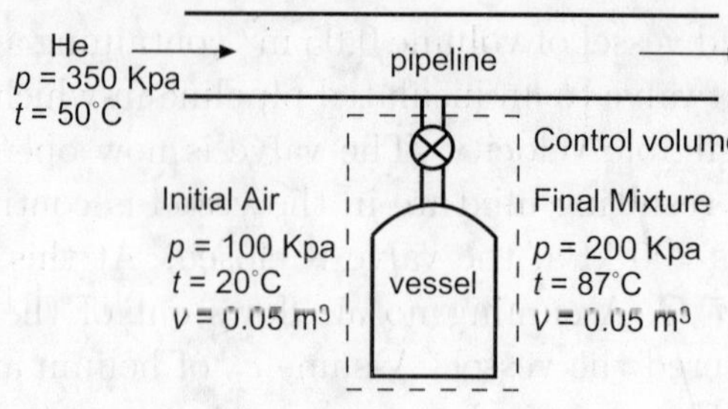

Fig. 9.15 Schematic for Illustrative Problem 9.15.

$$\text{Hence,}\quad m_h \times 5.193(87 - 50) + 0.06 \times 1.005(87 - 20) = \frac{0.05(200 - 100) \times 10^3}{10^3}$$

$$\text{that is,}\qquad m_h = 5 \times 10^{-3}\ \text{kg} = 5\ \text{g}$$

Problems

9.1 A certain gas has specific heats as follows: $c_p = 2260$ J/kg K and $c_v = 1735$ J/kg K. Find the molar mass and gas constant.

9.2 Tanks of 5 m^3 are to be used for storing different gases at 17.5 bar and 30°C. The gases are oxygen, nitrogen, helium, hydrogen and carbon monoxide. Find the mass of each gas when the corresponding tanks are full. Assume R for O_2, N_2, He, H_2 and CO to be, respectively, 0.260, 0.297, 2.077, 4.135 and 0.297 kJ/kg K.

9.3 Gaseous hydrogen is stored in a cylinder of volume 6 m^3 at 1 atm and 27°C. The cylinder is being evacuated using a vacuum pump. Find the mass of hydrogen pumped out when the cylinder pressures drops to 60 mm Hg vacuum given that the universal gas constant is 8314.3 J/kmol K.

9.4 A certain ideal gas has $c_p = 1.97$ kJ/kg K and $c_v = 1.51$ kJ/kg K. Determine its M and R. A rigid chamber of volume 0.3 m^3 contains 2 kg of this gas at 5°C. Heat is transferred to the gas until the temperature increases to 100°C. Find $W, Q, \Delta U, \Delta H$, and ΔS.

9.5 A small steel cylinder of volume 250 cm^3 contains air at 15 MPa, 20°C. This cylinder is placed inside an evacuated and insulated chamber of volume 0.25 m^3. By some suitable means, the small cylinder is allowed to discharge its contents into the large chamber. Calculate the final air pressure in the chamber when equilibrium conditions prevail. Assume for air $c_p = 1005$ J/kg K and $c_v = 718$ J/kg K.

9.6 One kilogram of nitrogen at a temperature of 150°C occupies a volume of 0.2 m^3. The nitrogen undergoes a fully resisted constant pressure expansion process to a final volume of 0.36 m^3. Evaluate the final pressure and temperature, work done, increase in internal energy and heat transfer. Assume for nitrogen R = 297 J/kg K and c_v =745 J/kg K.

9.7 A volume, 0.1 m^3, of air at 1 bar, 27°C undergoes a resisted *isobaric process* until the volume gets doubled. Find the work transfer, heat transfer, change in internal energy and entropy for the air during the given process. Assume for air c_p = 1005 J/kg K and c_v = 718 J/kg K.

9.8 A volume, 0.1 m^3, of air at 1 bar, 27°C undergoes a resisted *isothermal process* until the volume gets halved. Find the work transfer, heat transfer, change in internal energy and entropy during the given process. Assume for air c_p = 1005 J/kg K and c_v = 718 J/kg K.

9.9 A mass, 0.1 kg, of a perfect gas of relative molar mass 28 is compressed reversibly and polytropically according to $pV^{1.2}$ = Constant. Given that p_1 = 100 kPa, t_1 = 90°C, p_2 = 500 kPa and c_p for the gas = 1040 J/kg K, determine the following: (a) gas constant, (b) ΔU and ΔH, (c) W and Q and (d) ΔS.

9.10 One kilogram of air, initially at 30°C is heated reversibly at constant pressure until the volume is doubled; and then heated reversibly at constant volume until the pressure is doubled. Find for the total process the work transfer, heat transfer and the change in entropy. Assume for air c_p = 1005 J/kg K and c_v = 718 J/kg K.

9.11 0.01 kg of air in a piston–cylinder device undergoes a resisted expansion process from 10 bar, 227°C to 2 bar 27°C. The path of the process on a p–V diagram is a straight line. Sketch the p–V diagram and find the magnitudes and directions of work and heat interactions during the process. Assume for air c_p = 1005 J/kg K and c_v = 718 J/kg K.

9.12 A certain mass of air expands reversibly according to $pV^{1.3}$ = constant, from 200 kPa, 0.2 m^3 and 150°C to 0.6 m^3. Assume for air c_p = 1005 J/kg K and c_v = 718 J/kgK. (a) Find the final pressure and temperature of air. (b) Show the process path on both p–V and T–S planes. (c) Find the heat and work interactions and (d) the change in entropy.

9.13 A mass of 0.5 kg of argon gas undergoes a resisted process from 6 atm, 300 K until the volume gets trebled. During the process, internal energy does not change. Find the final pressure and temperature and also the change in entropy during the process. Assume for argon, c_p = 0.520 kJ/kg K and c_v = 0.312 kJ/kg K.

9.14 An ideal gas with constant specific heats undergoes a resisted polytropic process, pV^n = constant, from state 1 to state 2. Show that the specific entropy change during the process is given by: $s_2 - s_1 = \left[\frac{(n-\gamma)}{n(n-\gamma)}\right] R \ln \left(\frac{p_2}{p_1}\right)$.

9.15 An ideal gas with constant specific heats passes steadily through a control volume reversibly and polytropically. Show that the specific entropy change of the gas is given by $s_2 - s_1 = \left[\frac{(n-\gamma)}{(1-n)}\right] \ln \left(\frac{T_2}{T_1}\right)$, where suffixes 1 and 2 correspond, respectively, to inlet and exit sections of the control volume.

9.16 The specific heats of a certain ideal gas are given by: $c_p = a + kT$ and $c_v = b + kT$, where a, b, and k are constants and T the temperature is in kelvin. Show that when this gas expands reversible and adiabatically the path of the process is given by: $T^b v^{(a-b)} e^{kT}$ = constant, where v is the specific volume.

9.17 An ideal gas reversibly undergoes a change of state from state 1 to 2, the path of which on a $p-V$ plane is a straight line. The gas is then cooled at constant pressure to State 3. Finally, the gas is brought back to the initial state 1 along an isothermal process. Given that, $p_1 = 5$ bar, $V_1 = 0.005$ m^3, $p_2 = 1.5$ bar, $V_2 = 0.02$ m^3, find the work done in each process and the net heat for the cycle.

9.18 For a certain ideal gas, the specific heat at constant pressure, on molar basis, in the temperature range of 273 K and 1770 K is given by: $c_p = 28106 + 1.9665\, T + 4.802 \times 10^3\, T^{-2} - 1.966 \times 10^6\, T^{-3}$ J/kmol K. Evaluate the following quantities when the gas is heated at constant pressure from 300 to 1300 K: (i) the increase in specific enthalpy, (ii) the increase in specific internal energy and (iii) the increase in specific entropy.

9.19 Oxygen, initially at 22°C, 4 bar and 0.35 m^3 is reversibly and isothermally compressed to a final volume of 0.05 m^3. Calculate the mass, the final pressure, the increase in internal energy, the work done, the heat transfer and the change in entropy. Assume for oxygen R = 260 J/kg K.

9.20 One kilogram of air is reversibly and adiabatically compressed from 80 kPa, 60°C to 0.4 MPa, and is then expanded at constant pressure to the original volume. Sketch the processes on $p-V$ and $T-S$ planes. Compute the heat transfer and work transfer for the entire process. Assume for air $c_p = 1005$ J/kg K and $c_v =$ 718 J/kg K.

9.21 Air, initially at 350 kPa and 50°C in a piston–cylinder mechanism, reversibly and isothermally expands to 175 kPa. At the end of this process the air expands reversibly and adiabatically to 100 kPa. Sketch the processes on $p-V$ and $T-S$ planes. For the entire process, calculate the net heat transfer, change in specific internal energy and change in specific entropy. Assume for air $c_p = 1005$ J/kg K and $\gamma = 1.4$.

9.22 One kilogram air is heated at constant volume from 60°C, 3.5 bar to 280°C, compressed reversibly and adiabatically to 13.5 bar, and then cooled under constant volume conditions to 100°C. Determine the change in entropy of air for each process and for the entire process. Assume for air $c_p = 1005$ J/kg K and $c_v = 718$ J/kg K.

9.23 A reversible constant-pressure process begins with 1 kg air at 20°C occupying a volume of 0.3 m^3. Heat is transferred to air until its temperature is 200°C. Find the heat transfer, work transfer, changes in internal energy, enthalpy and entropy. Assume for air $c_p = 1005$ J/kg K and $c_v = 718$ J/kg K.

9.24 1.5 kmol, of a gas of molar mass 25 is contained in a rigid vessel of volume 4 m^3 at a temperature of 100°C. Evaluate the mass, pressure, specific volume and the gas constant of the gas. (b) Evaluate the values of c_p and c_v given that their ratio is 1.38. (c) If the gas were to cool to atmospheric temperature of 15°C find the final pressure. (d) Assuming that c_p and c_v are constants, calculate the increase in internal energy, enthalpy and the magnitude and direction of heat transfer.

9.25 A rigid chamber of volume 0.3 m^3 contains 2 kg of a certain gas, whose $c_p = 1.97$ kJ/kg K and $c_v = 1.51$ kJ/kg K, at 15°C. The gas is heated until its temperature rises to 100°C. Determine the following: (a) increase in specific internal energy, (b) increase in specific enthalpy, (c) increase in specific entropy and (d) magnitudes and directions of heat and work transfers.

9.26 Air expands reversibly and adiabatically in a piston cylinder device from 14 bar to 2 bar, 27°C. What is the initial temperature of air? Also calculate the change in specific enthalpy and the magnitude and direction of heat and work interactions per kilogram of air. Assume for air $c_p = 1005$ J/kg K and $\gamma = 1.4$.

9.27 (a) One kilogram of air at 1 MPa, 100°C undergoes a reversible polytropic process according to $pv^{1.1}$ = constant. The final pressure is 2 bar. (i) Find the final specific volume, final temperature and increase in entropy. (ii) Evaluate the work done and the heat transfer. (b) Redo this problem for an irreversible adiabatic process between the same end states as in (a). Assume for air $c_p = 1005$ J/kg K and $c_v =$ 718 J/kg K.

9.28 A mass of air is initially at 260°C, 700 kPa occupies 0.03 m^3. The air is made to expand at constant pressure to 0.1 m^3. A polytropic process ($n = 1.5$) is then carried out, followed by a constant-temperature process, which brings the air to its initial state. All the processes are reversible. (a) Sketch the cycle on the $p-V$ and $T-S$ planes. (b) Find the net heat transfer to and from the air and (c) Calculate the efficiency of the cycle. Assume for air $c_p = 1005$ J/kg K and $c_v = 718$ J/kg K.

9.29 Air is contained in a vertical cylinder fitted with a frictionless piston and a set of stops as shown in Figure 9.16. The cross-sectional area of the piston is 0.1 m^2 and the air is initially at 250 kPa, 400°C. The air is then cooled. (a) What is the temperature of the air inside the cylinder as the piston just touches the stops? (b) If the cooling is continued until the temperature reaches 30°C, what is the pressure in the cylinder? Assume for air $c_p = 1005$ J/kg K and $c_v = 718$ J/kg K.

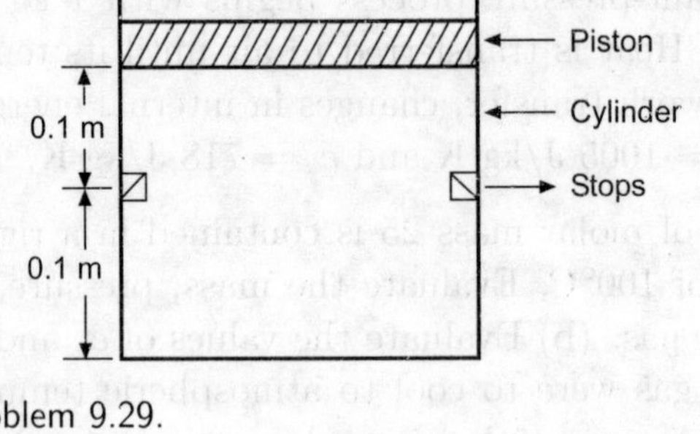

Fig. 9.16 Schematic for Problem 9.29.

9.30 Consider the system shown in Figure 9.17. The insulated rigid tank B has a volume of 0.3 m^3 and is initially evacuated. The uninsulated cylinder A has an initial volume of 0.15 m^3 and contains air at 20°C, 3.5 Mpa. The spring constant is 40 kN/m and the piston area is 0.03 m^2. It may be assumed that the spring force varies linearly with distance. The valve connecting A and B is opened and air travels from A to B until $p_B = 1.5$ MPa. During this process, heat is transferred to A from the surroundings so as to maintain t_A constant. Assume for air $h = c_pT$ and $u = c_vT$, where T is in K. The atmospheric pressure is 1 bar. Determine (a) the volume in cylinder A and (c) magnitude of heat transfer to A.

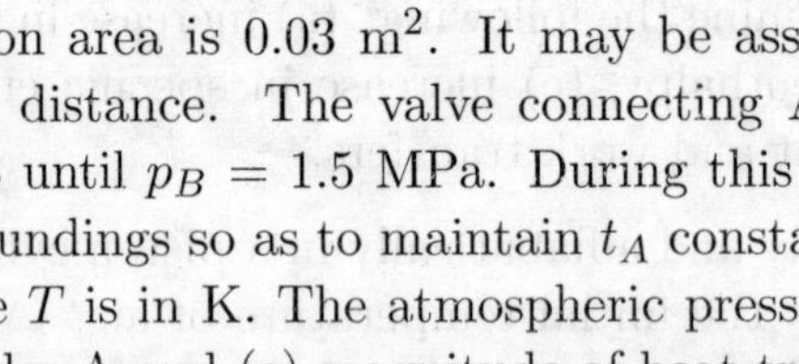

Fig. 9.17 Schematic for Problem 9.30.

9.31 Air flows through a nozzle of circular cross-section at a rate of 600 kg/h. The properties of air at the entry to the nozzle are 2 MPa, 127°C and 300 m/s. The exit pressure is 0.5 MPa. Find the exit velocity of air and also the inlet and exit diameters of the nozzle. Assume for air $c_p = 1005$ J/kg K and $c_v = 718$ J/kg K.

9.32 (a) In the compression of air in a certain cylinder cooled by a water jacket, the indicator diagram shows that pressure and volume are related by $pv^{1.3}$ = constant. The process begins at 27°C, 1 bar and the pressure rises to 7 bar. Find the final temperature, work and heat transfers per kilogram of air. (b) If the air at the same initial conditions as in (a) is compressed in a reversible adiabatic process in steady flow and the final pressure is same as before, find the final temperature and shaft work done per kilogram of air. Neglect changes in elevation and velocities. Assume for helium c_p = 5193 J/kg K and c_v = 3116 J/kg K.

9.33 Helium is contained in an insulated tank at 7 MPa and 70°C. A valve on the tank is opened and 40% of the helium is allowed to escape. The valve is then closed. The volume of the tank is 5 m^3. Calculate the final pressure in the tank. Assume for helium c_p = 5193 J/kg K and c_v = 3116 J/kg K.

9.34 Pressurized air stored in a 12 000 m^3 cave at 500 kPa and 400 K is to be used to drive a turbine at times of high demand for electric power. If the turbine exit conditions are 100 kPa and 300 K, determine the amount of work delivered by the turbine when the air pressure in the cave drops to 300 kPa. Assume both the cave and the turbine to be adiabatic. Assume for air c_p = 1005 J/kg K and c_v = 718 J/kg K.

9.35 A hair-drier is basically a duct in which a few layers of electric resistors are placed. A small fan pulls the air in and forces it through the resistors where it is heated. Air enters a 1200 W hair-drier at 100 kPa and 22°C and leaves at 47°C. The cross-sectional area of the hair-drier at the exit is 60 cm^2. Neglecting the power consumed by the fan and the heat losses through the walls of the hair-drier, determine (a) the volume flow rate of air at the inlet and (b) the velocity of the air at the exit. Assume for air c_p = 1005 J/kg K and c_v = 718 J/kg K.

9.36 Nitrogen expands in a cylinder from 690 kPa and 260°C to 210 kPa and 40°C. Calculate the work done on the face of the piston per kilogram of nitrogen. Assume the process is resisted and follows pV^n = constant.

9.37 A 2 m^3 rigid tank initially contains air at 100 kPa and 22°C. The tank is connected to a supply line through a valve. Air is flowing in the supply line at 600 kPa and 22°C. The valve is opened, and air is allowed to enter the tank until the pressure in the tank reaches the line pressure, at which point the valve is closed. A thermometer placed in the tank indicates that the air temperature at the final state is 77°C. Determine (a) the mass of air that has entered the tank and (b) the amount of heat transfer. Assume for air c_p = 1005 J/kg K and c_v = 718 J/kg K.

9.38 An insulated 0.08 m^3 tank contains helium at 2 MPa and 80°C. A valve is now opened, allowing some helium to escape. The valve is closed when one-half of the initial mass has escaped. Determine the final temperature and pressure in the tank. Assume for air $c_p = 1005$ J/kg K and $c_v = 718$ J/kg K.

9.39 A 1 m^3 tank containing air at 25°C and 500 kPa is connected through a valve to another tank containing 5 kg of air at 35°C and 200 kPa. Now the valve is opened, and the entire system is allowed to reach thermal equilibrium with the surroundings which is at 20°C. Determine the volume of the second tank and the final equilibrium pressure of air. Assume for air $c_p = 1005$ J/kg K and $c_v = 718$ J/kg K.

9.40 An insulated 5 m^3 rigid tank contains air at 500 kPa and 52°C. A valve connected to the tank is now opened, and air is allowed to escape until the pressure inside drops to 200 kPa. The air temperature during this process is maintained constant by an electric resistance heater placed in the tank. Determine the electrical work done during this process. Assume for air $c_p = 1005$ J/kg K and $c_v = 718$ J/kg K.

9.41 In a chemical plant, carbon dioxide flows steadily into a turbine at a pressure of 8 bar, 240°C; there it expands reversibly and adiabatically to a final pressure of 1 bar. Changes in velocity and elevation are negligible. Calculate the final specific volume, the final temperature and the increase in entropy. Given that the mass flow rate is 10 kg/s evaluate the heat transfer rate from the gas and the power delivered by the turbine. For carbon monoxide assume that $c_v = 0.653$ kJ/kmol K.

9.42 A balloon initially contains 50 m^3 of hydrogen at a pressure of 100 kPa and 25°C. The balloon is connected through a valve to a large reservoir that supplies hydrogen at 150 kPa and 30°C. When the valve is opened, hydrogen flows into the balloon until pressure equilibrium is reached with the reservoir. The material of the balloon is such that its volume increases linearly with pressure. Assume the process is adiabatic. Determine the final temperature in the balloon. Assume for H_2, $c_p = 14.21$ kJ/kg K and $c_v = 10.085$ kJ/kg K.

9.43 Propane at 700 kPa, 35°C, flows steadily through a pipeline. A tank of 0.06 m^3 volume that contains air at 101.3 kPa, 15°C, is connected to the pipeline and propane is allowed to flow into the tank until the total pressure in the tank is 350 kPa. Assume that no air leaves the tank and determine the heat transfer that is required to maintain the contents of the tank at 15°C. Assume the following if required: For propane, $c_p = 1679$ J/kg K and $M = 44$; For air $c_p = 1005$ kJ/kg K and $M = 28.97$.

Note: In the following problems, any missing data for the constituents may be suitably taken from Table 9.2

9.44 A gas mixture at 400 K and 150 kPa consists of 1 kg CO_2 and 3 kg CH_4. Determine the partial pressure of each gas and the apparent molar mass of the mixture.

9.45 A 0.3 m^3 rigid tank contains 0.6 kg N_2 and 0.4 kg O_2 at 300 K. Determine the partial pressure of each gas and the pressure of the mixture.

9.46 A gaseous mixture of 5 kg helium and 2 kg oxygen at 100 kPa and 27°C, is compressed reversibly and adiabatically to 800 kPa. Calculate: (i) Initial and final partial pressures of helium and oxygen, (ii) final mixture temperature, (iii) change in specific enthalpy and internal energy and (iv) entropy change of helium and oxygen.

9.47 Two kilomoles of carbon dioxide in a cylinder at 1.5 bar, 70°C are mixed adiabatically with three kilomoles of nitrogen in another cylinder at 2.5 bar, 50°C. Find the pressure and temperature of the mixture.

9.48 The volumetric analysis of a certain combustion product is as follows: 15% CO_2, 2% O_2, 1.5% CO and the rest N_2. Find the mass fractions of the constituents and the molar mass and the gas constant of the mixture.

9.49 A container has a mixture of hydrogen and oxygen in the ratio of 2 to 1 by volume. Find the mass of hydrogen and the volume of the container per kg oxygen given that the mixture is at 330 K and 100 kPa.

9.50 A rigid tank which contains 2 kg N_2 at 25°C and 200 kPa is connected to another rigid tank which contains 3 kg O_2 at 25°C and 500 kPa. The valve connecting the two tanks is opened, and the two gases mix adiabatically. Determine the volume of each tank and the final mixture pressure. Determine also the entropy change of the universe given that the surroundings are at 25°C.

9.51 Three gas cylinders interconnected by short pipes and valves contain, respectively, 5 kg nitrogen at 600 kPa, 27°C, 4 kg argon at 400 kPa, 37°C and 2 kg oxygen at 200 kPa, 47°C. When the valves are open, the gases mix adiabatically. Compute (a) the temperature, (b) apparent molar mass, (c) specific heats, (d) gas constant and (e) the pressure for the mixture. Find also the partial pressures of each gas in the mixture and the entropy change during mixing.

9.52 Three gas cylinders interconnected by short pipes and valves contain, respectively, 5 kg nitrogen at 600 kPa, 27°C, 4 kg argon at 400 kPa, 37°C and 2 kg oxygen at 200 kPa, 47°C. When the valves are open, the gases mix. If the final temperature of the mixture is 30°C find the direction and magnitude of heat transfer and the final pressure of mixture.

9.53 A stream of oxygen at 105 kPa, 30°C is mixed adiabatically in steady flow with a stream of hydrogen at 105 kPa, 30°C. The mass ratio of oxygen to hydrogen is 8 to 1. Neglect the changes in kinetic and potential energies find the change in specific enthalpy and specific entropy for the mixture.

9.54 A gas mixture at 150 kPa, 27°C consists of 1 kg helium and 2.5 kg nitrogen. If the mixture is compressed reversibly and adiabatically in a piston-cylinder device to a temperature of 327°C, compute the following: (a) ratio of specific heats, (b) initial and final partial pressures, (c) change in specific enthalpy and energy, (d) work interaction.

9.55 One kilogram of an ideal gas mixture is compressed reversibly and adiabatically from 27°C, 100 kPa to 500 kPa. The mixture consists of N_2 and another gas X. The mass fractions of N_2 and X are respectively 0.4 and 0.6. If the decrease in entropy of gas X during compression process is 72.4 J/K, find the final temperature of the mixture.

9.56 Two kilomoles of CO at 1.5 bar, 70°C and 3 kilomoles of N_2 at 2.5 bar, 50°C, in two different vessels, are mixed by opening the valve between them. When equilibrium conditions prevail the mixture temperature was equal to the surrounding temperature of 27°C. Find the entropy change of the universe during the above process.

9.57 A rigid vessel of 0.01 m^3 volume contains an ideal gas mixture of 0.04 kg N_2 and some CO_2 at 20°C. The mixture is then heated to 41.4°C by an heat transfer of 10 kJ/kg. Find the following: (a) Mole fractions of the constituents, (b) mass fractions of the constituents, (c) mass of the mixture in kg, (d) initial and final mixture pressure in kPa and (e) change in entropy in kJ/K.

10 PROPERTIES OF PURE SUBSTANCE—WATER

10.1 Introduction

Depending upon its pressure, temperature and volume, a pure substance may exist in distinctly different physical forms (phases) such as solid, liquid, or vapour or as a homogeneous mixture of all these or any two phases. Qualitatively, almost all pure substances behave identically in this respect. The properties of such pure substances and the methodology adopted to identify them depend on the phase of the substance. This chapter aims to explain the behaviour of pure substances over wide ranges of pressure and temperature and to develop methods to find their thermodynamic properties as a function of any two independent properties. This is done considering water as the example of a pure substance, because water in solid, liquid or, especially, in vapour form is extensively used in several industrial and domestic applications.

The knowledge obtained through the study of properties of water can be generalized and the methodology can be extended to any other pure substance as well. Listed below are a few examples of practical applications of water in vapour form (steam):

- As a working substance in turbines to develop shaft power which drive the generators in a steam power plant or other machineries elsewhere.
- For large-scale heating such as in community cooking systems.
- To sterilize bottles and cans, in dairies.
- For bleaching, in paper and textile industry.
- For different processes in sugar, chemical and pharmaceutical industries.
- To produce lower temperatures using steam jet refrigeration systems.
- To heat a room, using radiators, in winter periods.

In view of the fact that steam is extensively used in industrial applications, it is therefore essential for an engineer to know everything about the formation and properties of steam.

10.2 Constant-Pressure Heating and Formation of Steam

Consider a vertical piston–cylinder device, as shown in Figure 10.1, containing initially a certain mass of water at a low temperature whose physical condition, for any pressure, is in the solid phase (*ice*). The cylinder is provided with means to measure pressure, temperature and volume of the contents of the cylinder at any instant.

The piston may be loaded with a predetermined weight that subjects the water in the cylinder to a desired pressure. For example: in experiment (i), let the weight W be so selected that it subjects the water to a pressure of 0.001 bar. Let the water, starting from a very low temperature, be now heated steadily at this constant pressure. The measurements made would reveal that the volume and temperature of water (in the ice phase) keep increasing steadily until the temperature reaches a certain value below 0°C. Further heating will not increase the temperature but the volume keeps increasing. If the contents of the cylinder were to be seen through a transparent window, it would be noticed that the ice is getting converted to water vapour (steam) in this period. As the heating continues, the quantity of ice keeps decreasing and the quantity of steam keeps increasing. Eventually, a time will come when all the ice is converted to steam. From this instant onwards, further heating will increase both the temperature and volume of steam for ever. The variation in temperature with heating is shown by the curve (i) on $t - \tau$ diagram of Figure 10.2. In this diagram, time is plotted on the x axis. As the heating is assumed to be steady the time marked on this axis is also proportional to heat transfer.

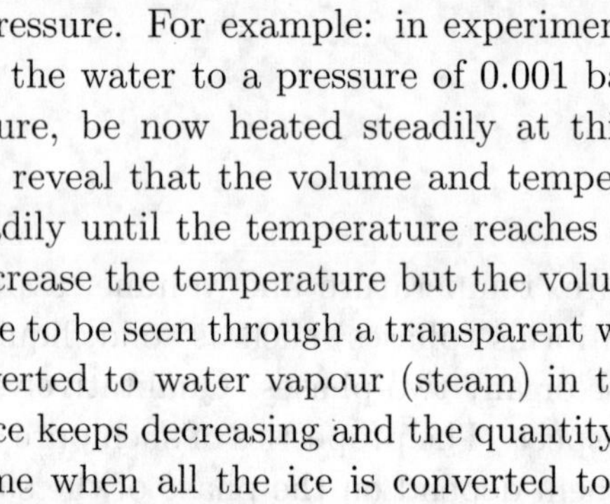

Fig. 10.1 Constant-pressure heating of water.

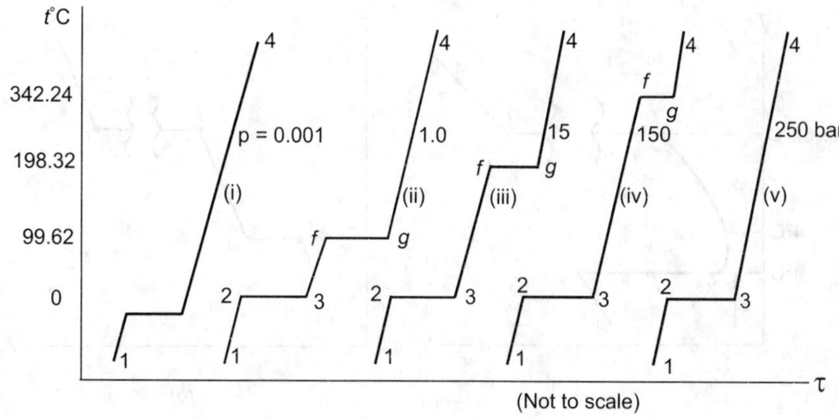

Fig. 10.2 Temperature–time diagram for five arbitrary constant-pressure processes.

Such experiments can be conducted at various pressures starting from a very low value to as high a pressure as practically feasible. Let, in each experiment, the pressure, temperature and volume be measured at regular intervals and recorded. Let us consider five experiments, four more besides the one already explained above, conducted at arbitrarily selected pressures of 0.001, 1, 15, 150 and 250 bar. These five experiments cover fairly a wide range of pressures and hence the results recorded may be generalized.

Consider experiment (ii) conducted at a higher pressure of 1 bar. The path of this constant pressure heating process is shown by curve (ii) in Figure 10.2. Initially, as heating progresses, the temperature and volume of solid keep increasing. When the temperature reaches 0°C further heating will not increase the temperature, but the volume keeps decreasing at the same temperature as ice starts melting. As done in experiment (i), a look into the cylinder will reveal the fact that ice is getting converted to water at 0°C. Until the last particle of ice melts, the temperature will not rise. (Actually, during the melting process and a little beyond, the decrease in volume is unique to water. Shown in Figure 10.3*a*, on $T-v$ coordinates, is the variation in volume of water when heated at a constant pressure of 1 bar. It can be seen that volume decreases between 2 and 3 as ice melts and this decrease continues till water reaches 4°C. The variation in volume during the process of melting of other pure substances, in general, is as shown in Figure 10.3*b*. The volume keeps increasing continuously all through the constant pressure heating process).

When all the ice melts, further heating increases both the temperature and volume of water. Surprisingly, once again the temperature ceases to rise beyond 99.62°C although the heating is steady and continuous. A look into the cylinder reveals that water has started to boil at this temperature and it is getting converted to steam. There will be a large change in volume during this process of vapourization. Until the last drop of water evaporates the temperature does not rise with heating. After the last drop of water evaporates further heating increases the temperature of steam forever.

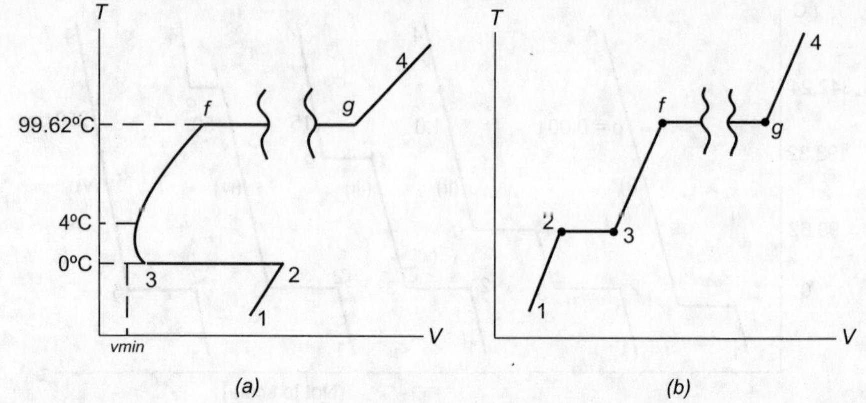

Fig. 10.3*a* Change in volume of water during constant-pressure heating; *b* change in volume of a pure substance, other than water, during constant-pressure heating.

Now, consider a similar experiment, (iii), conducted at a pressure of 15 bar. A perusal of path (iii) on the $t-\tau$ diagram reveals that similar changes take place here as in experiment (ii), except that the second horizontal step, during which water boils, occurs at a higher temperature, 198.32°C. Also, the width of this higher horizontal step is less than that in (ii). This means vapourization is faster at higher pressures. Hence, the magnitude of heat transfer required to change the phase from liquid to vapour is small at high pressure.

Experiment (iv), conducted at 150 bar, reveals the same information as in (ii) and (iii), except that the second horizontal step is very small in width and that it occurs at an even high temperature of 342.24°C.

The last one, experiment (v), conducted at a pressure of 250 bar, reveals that the second horizontal step has totally disappeared.

The following observations may be made on the results of these five constant-pressure heating experiments.

- Each experimental result shows horizontal steps on a $t-\tau$ diagram indicating that the temperature ceases to rise for sometime although the heating is continuous.
- Excepting the first and the last curves on the $t-\tau$ diagram, the others show two horizontal steps each. While, during the first horizontal step the phase changes from solid to liquid, during the second step liquid changes to vapour. It can, hence, be generally stated that during phase changes, at any given pressure, the temperature does not change.
- With an increase in pressure the second horizontal step occurs at higher temperatures. This means the boiling temperature increases with an increase in pressure.
- The higher steps decrease in width with an increase in pressure. The width is proportional to duration of heating, which in turn is proportional to the magnitude

of heat transfer. This means the heat required to bring about a change of phase from liquid to vapour decreases with an increase in pressure.

Among the five examples cited above, barring the first and the last, identical features are observed. For these experiments, it may be summarized that:

- During the process 1–2, the temperature and volume of solid (ice) increases.
- At state 2, ice starts melting at 0°C.
- During the process 2–3, ice continues to melt at a constant temperature of 0°C. There is a decrease in volume. The melting is complete at state 3.
- During the process 3–f, the temperature and volume of water keeps increasing.
- At state f, boiling or evaporation commences.
- During the process f–g, water continues to evaporate at a constant temperature. This temperature depends upon the pressure; higher the pressure higher is the temperature. There is a substantial increase in volume.
- At state g, the evaporation is complete.
- During the process g–4, the temperature and volume of steam keep increasing for ever.

Now, let the temperatures at which the various horizontal steps occurred on the t–τ diagrams of Figure 10.2, be plotted against the corresponding pressures on a t–p diagram as shown in Figure 10.4. This figure is only representative in nature and is not drawn for any particular scale. We get two curves: a steeply rising curve called the *liquid–vapour saturation line* which is the locus of boiling temperatures, and the other, more or less, a straight line, called the *solid–liquid saturation line* is the locus of melting temperatures. The point of intersection of these two curves is labelled TP and the point where the top curve suddenly vanishes in labelled CP. The extended portion of the *liquid–vapour saturation line* below TP is called the *solid–vapour saturation line.*

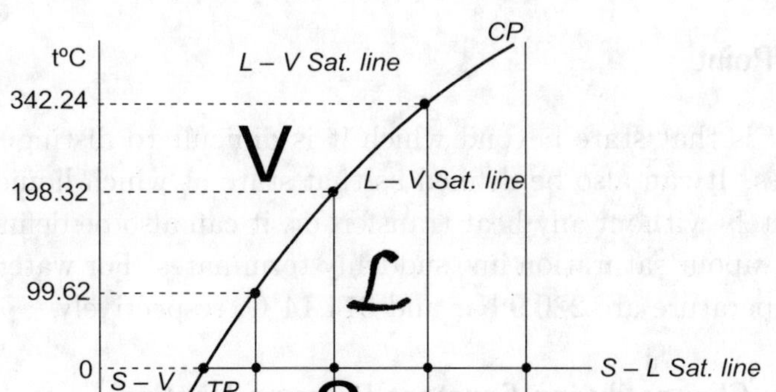

Fig. 10.4 t–p diagram for water.

A careful study of this figure reveals the following peculiarities:

- The saturation lines separate the various phases of water. While the S–L saturation line separates solid and liquid regions, the L–V saturation line separates liquid and vapour regions and the S–V saturation line separates the solid and vapour regions.
- Heating at pressures between the pressures corresponding to TP and CP makes the water change in an orthodox manner gradually from solid (ice) to liquid (water) to vapour (steam). The occurrence of these phase changes can be seen through a transparent window provided in the cylinder. The change of phase from solid to liquid is called *melting*. The melting temperature is more or less independent of pressure. The change of phase from liquid to vapour is called *boiling* or *vapourization*. Boiling temperature increases with an increase in pressure.
- At pressures greater than the pressure at CP it is not possible to evaporate the liquid as the line separating the two phases is extinct.
- Heating at pressures less than the pressure corresponding to TP causes the solid to directly become vapour, skipping the liquid region. This change of phase is called *sublimation*. Sublimation temperature increases with an increase in pressure.
- At TP all the three phases coexist in equilibrium.
- The L–V saturation curve suddenly ends at CP.

10.2.1 Triple Point

The point TP on the t–p diagram is called the *triple point*. It is that state at which all the three phases of a pure substance coexists in equilibrium. For water, the triple point pressure and temperature are, respectively, 0.006113 bar and 0.01°C.

10.2.2 Critical Point

The *critical point* is that state beyond which it is difficult to distinguish between liquid and vapour phases. It can also be defined as that state at which liquid can be converted to vapour completely without any heat transfer, or, it can also be defined as that state at which the liquid–vapour saturation line suddenly terminates. For water, the critical point pressure and temperature are 220.9 bar and 374.14°C, respectively.

10.2.3 Enthalpy Change during Constant-Pressure Heating

When water is heated at constant pressure, from the first law of thermodynamics,

$$Q - W = U_{\text{final}} - U_{\text{initial}} \tag{i}$$

where Q and W are the heat and work transfers during the heating period. The initial and the final states could be selected arbitrarily. The work involved here is only displacement work at constant pressure and, hence,

$$W = \int p\mathrm{d}V = p\int \mathrm{d}V = p\,(V_{\text{final}} - V_{\text{initial}}) \tag{ii}$$

Substituting for W from (ii) in (i),

$$\begin{aligned} Q &= p\,(V_{\text{final}} - V_{\text{initial}}) + (U_{\text{final}} - U_{\text{initial}}) \\ &= (pV_{\text{final}} + U_{\text{final}}) - (pV_{\text{initial}} + U_{\text{initial}}) \end{aligned}$$

For a constant pressure process $p = p_{\text{final}} = p_{\text{initial}}$ and, hence,

$$Q = (p_{\text{final}}V_{\text{final}} + U_{\text{final}}) - (p_{\text{initial}}V_{\text{initial}} + U)$$

By definition, enthalpy $H = U + pV$,

and therefore, $Q = H_{\text{final}} - H_{\text{initial}} = \Delta H$ (iii)

Between any two state points, the magnitude of heat transfer during a constant-pressure process is equal to the change in enthalpy of the substance being heated. And, per kilogram of substance, it is equal to the change in specific enthalpy, h. Therefore,

$$q = \Delta h = h_{\text{final}} - h_{\text{initial}} \tag{iv}$$

10.3 Frequently Used Terms and their Definitions

Consider any one constant-pressure heating path on the t–τ diagram of Figure 10.2. Let it be path (ii). This path, when transferred on to a t–h diagram, would retain the same shape as on a t–τ diagram because, in both diagrams, temperature is represented on the y axis and τ and h, which are proportional to each other and are indicators of heat transfer, are represented on the x axis.

During process 1–2, on the t–h diagram of Figure 10.5, the temperature of solid (ice) is increasing at constant pressure. The heat transfer, q, per kilogram of ice during this period is given by

$$q = (h_2 - h_1) = c_{ice}\,(t_2 - t_1) \tag{10.1}$$

where c_{ice} is the specific heat of ice. c_{ice} is taken to be 2.04 kJ/kg K, unless specified otherwise .

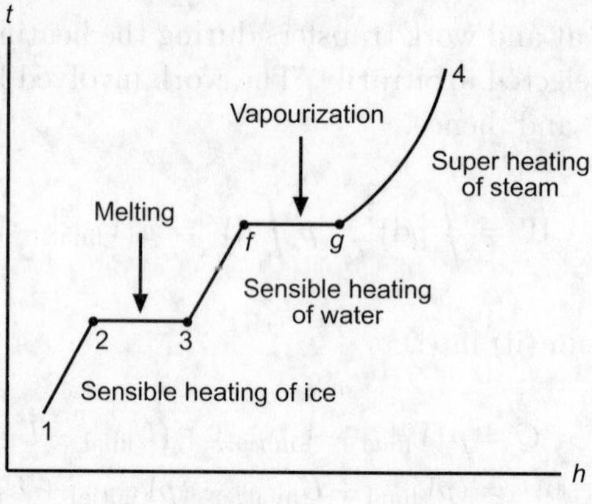

Fig. 10.5 t–h diagram for water.

At state 2 ice starts melting and at state 3 the melting is just complete. During the heating process 2–3, there is no change in temperature, but a phase change from solid to liquid and a change in volume occur. At state 2 the substance is said to be in a saturated state. It is called the *saturated solid state.* A saturated solid state is that state at which a phase change from solid to liquid can be brought about by heating without change in either temperature or pressure. Similarly, state 3 is also a saturated state and it is called the *saturated liquid state* with respect to solidification. It is so because when water is cooled it starts to solidify at state 3. During solidification, there would be no change in either pressure or temperature.

Heat transfer per kilogram of water, between states 2 and 3, that brings about complete change of phase from solid to liquid is equal to $(h_3 - h_2)$ and is called the *enthalpy of melting* at the given pressure. The other name that is being used for this enthalpy change, though not strictly correct, is *latent heat of melting.* The negative value of this latent heat, applicable for the process 3–2, is called the *enthalpy of fusion* or latent heat of fusion.

Further heating from state 3 will increase the temperature of liquid water steadily up to the state f. Beyond state f, the temperature again resists to increase for sometime although heating is continuous. At f, boiling or vapourization starts. At f, water is said to be in *saturated liquid state.* A saturated liquid state is one at which during a constant pressure heating the phase starts changing from liquid to vapour without any change in temperature. Heating of water from state 3 up to its boiling point is called *sensible heating.* Maximum sensible heat is the heat transfer required to heat the water from 0°C to its boiling point which in turn is equal to enthalpy change. Maximum sensible heat $= (h_f - h_3)$, where h_f is the enthalpy of the saturated liquid and h_3 is the enthalpy of water at 0°C. (As an approximation, h_3 is taken to be zero. In reality, the internal energy of liquid water at its triple point of 0.01°C is taken to be zero. This makes the value of enthalpy of liquid water equal to 0.01 kJ/kg at its triple point. The enthalpy at 0°C has to be evaluated with respect to the above reference, or datum. However, it approximates to zero.)

$$\text{Thus,}\quad (h_f - h_3) = h_f = c_{\text{wat}}\,(t_f - t_3) = 4.189\,(t_{\text{sat}} - 0) = 4.189 t_{\text{sat}}\ \text{kJ/kg}, \qquad (10.2)$$

where, t_{sat} is that temperature at which the phase change occurs from liquid to vapour and is called the *saturation temperature*. Each pressure has a unique saturation temperature and it increases with an increase in pressure.

Instead of using (10.2), h_f corresponding to a given pressure can also be directly read from the saturation steam tables, which will be explained later.

The change of phase is complete at state g. Between state f and g vapourization is in progress. The heat transfer required to completely convert one kilogram of liquid at state f to vapour at g, which is equal to $(h_g - h_f)$, and abbreviated as 'h_{fg}' is called as *enthalpy of vapourization* or *latent heat of vapourization*. This is equal to the heat transfer during the second horizontal step in curves of Figure 10.2. As the width of this step decreases with an increase in pressure, h_{fg} decreases with an increase in pressure and at CP, $h_{fg} = 0$.

At state g water is in a vapour state. It is just born vapour. If this vapour at state g is cooled, it starts condensing without change in pressure or temperature. Thus, at this state g, water is said to be in a *saturated vapour state*. A saturated vapour state is that state at which when cooled the vapour starts condensing without change in either pressure or temperature.

Heating further from state g will increase the temperature beyond t_{sat} and increases the volume of vapour beyond v_g. Thus, process $g - 4$ is said to be the process of *superheating*. Superheating is the process of increasing the temperature of vapour beyond its saturation temperature. The difference between the temperatures of superheated steam and saturated steam, at a given pressure, is called the *degree of superheating*. Thus, degree of superheating $= t_{\text{sup}} - t_{\text{sat}}$.

10.4 Phase Change Diagrams

10.4.1 p–V Diagram

Let us consider again the constant-pressure heating processes described in Section 10.2 and the t–τ diagrams in Figure 10.2. For each pressure, suppose the volumes at various marked states, such as 1, 2, 3, f, g and 4, are measured and plotted on a p–V diagram. Such a diagram is shown in Figure 10.6.

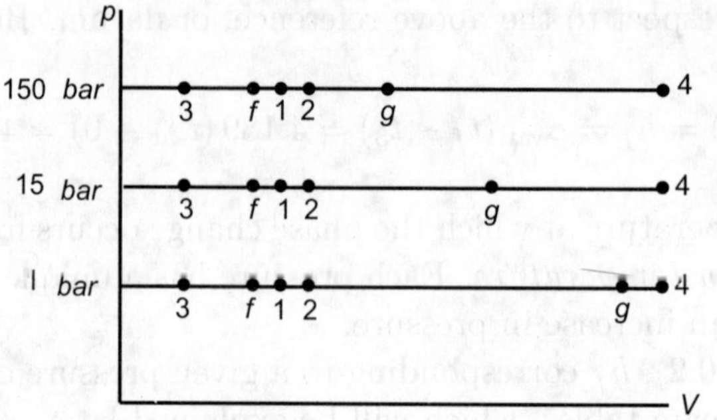

Fig. 10.6 $p-V$ diagram for water.

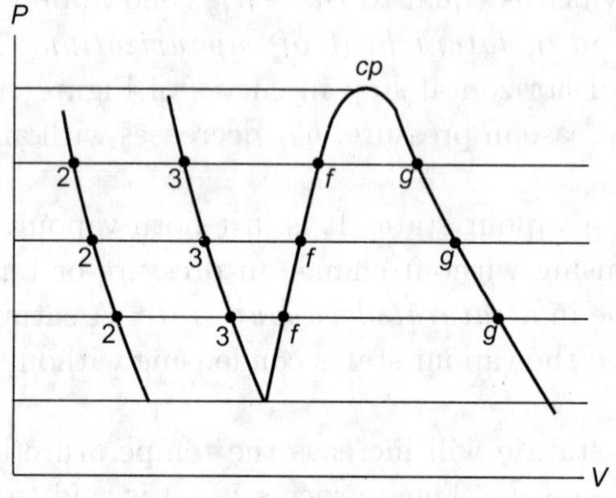

Fig. 10.7 A generalized $p-V$ diagram for a pure substance.

Reduction in volume during melting (i.e., between states 2 and 3 for any pressure), as already mentioned and shown in Figure 10.3*a*, and represented again in Figure 10.6, is an unique phenomenon for water. For most other pure substances, volume continuously increases, as shown in Figure 10.3*b*. In such a case, the saturated states of a pure substance on a $p-V$ diagram would be as shown in Figure 10.7.

The lines joining the corresponding saturation states are called *saturated lines.* Thus, in Figure 10.7, the line joining points 2 at various pressures is called the *saturated solid line.* Similarly, the line joining points 3 is called the *saturated liquid line* (*with respect to solidification*). The line joining points f is also called the *saturated liquid line* (*with respect to evaporation*) and the line joining points g is called the *saturated vapour line.* It is only across these saturation lines that a change in phase can begin or end during a constant pressure heating process. The phase change diagram on $p-V$ coordinates in Figure 10.8, is the same as that in Figure 10.7, except for the fact that in the former the saturation lines and various phase regions are marked explicitly over a wide range of pressures.

The saturated liquid and vapour lines join at the critical point. The triple point cannot be shown on the p– V plane, as volume at this point can have various values depending upon the proportions of solid, liquid and vapour in equilibrium. Thus, only a triple point line can be shown.

When a pure substance is heated at a constant pressure, between p_{TP} and p_{CP}, in conformity with the earlier explanations, there is an orthodox change of phase from solid to liquid and then from liquid to vapour. This is confirmed by the phase regions marked in Figure 10.8. At pressures less than p_{TP}, solid sublimes directly to vapour and at pressures beyond p_{CP}, the distinction between liquid and vapour phases ceases to exist.

Figure 10.9 is essentially the same as Figure 10.8. The additional feature of this figure is the marking of constant-temperature lines. Four representative constant-temperature lines running through various phase regions are shown. Here, $t_{cp} > t_3 > t_2 > t_1$. It can be seen that the temperature keeps increasing when heated at constant pressure, except during the change of phases.

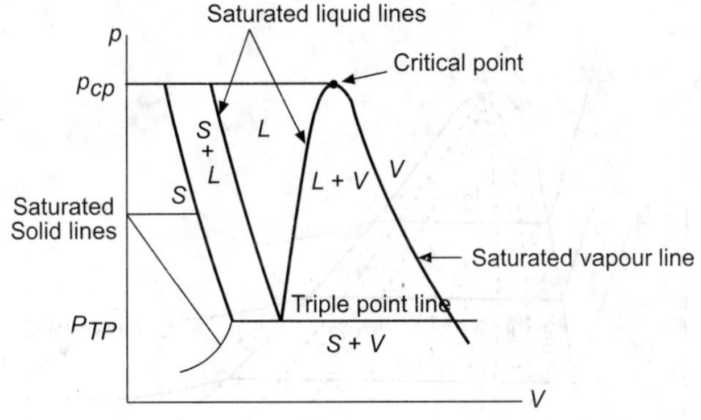

Fig. 10.8 Phase diagram on p– V coordinates.

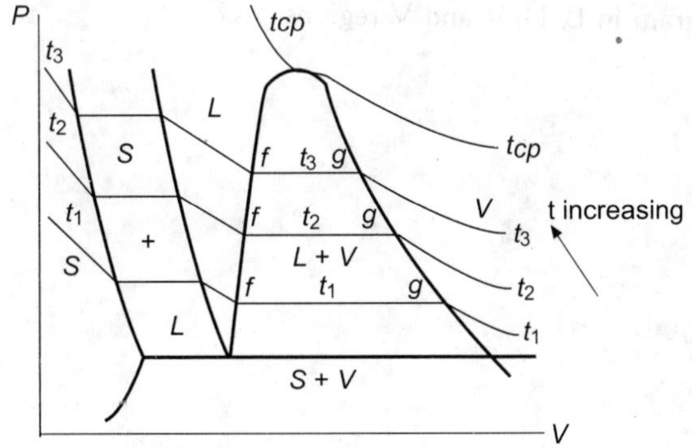

Fig. 10.9 Isotherms on a phase diagram.

Because of the fact that all pure substances, including water, behave alike qualitatively in L, L + V and V regions and that these are the most important and frequently encountered regions of pure substances in practical applications, a typical phase change diagram confining to these regions on $p-V$ coordinates is shown in Figure 10.10. The region marked L, to the left of saturated liquid line is called the *compressed liquid region*. In this region, for any pressure, the temperature of the liquid is less than the boiling or the saturation temperature. The region between the saturated liquid line and the saturated vapour line is called the *mixture region* or the region of *wet vapour*. From f to g the vapour proportion is increasing from 0% to 100%. The region marked V, to the right of saturated vapour line is the *superheated vapour region*. The temperature of the vapour for any pressure in this region is higher than the saturation temperature.

Shown in Figures 10.11, 10.12 are the $T-v$ and $T-S$ diagrams, respectively, in L, L + V and V regions. Specifically shown in these diagrams are the constant-pressure lines in the three regions.

In Figures 10.10 to 10.13, 'x' denotes the dryness fraction explained later in Section 10.4.3

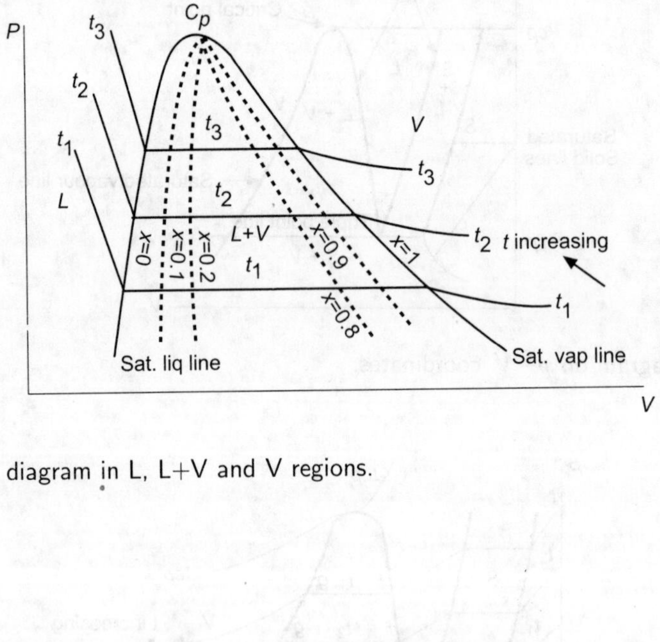

Fig. 10.10 A $p-V$ diagram in L, L+V and V regions.

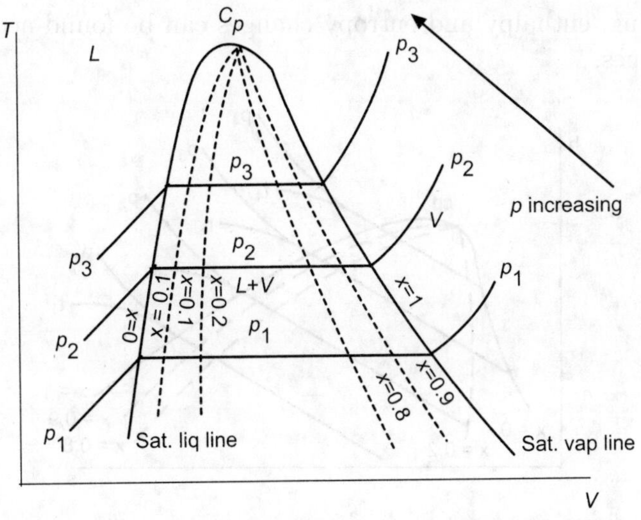

Fig. 10.11 A $T{-}v$ diagram in L, L+V and V regions.

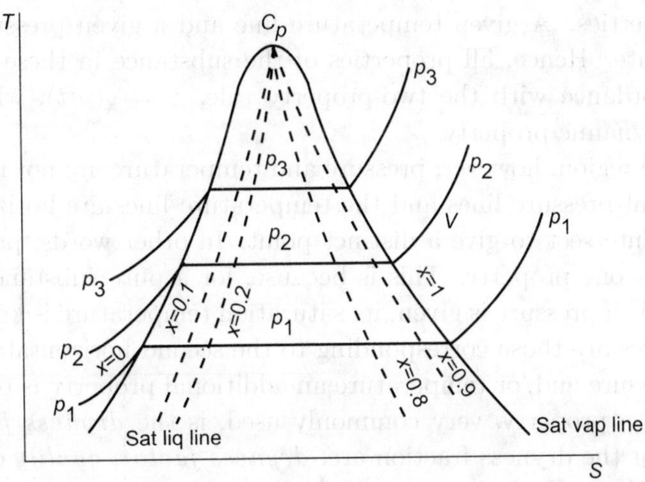

Fig. 10.12 A $T{-}S$ diagram in L, L+V and V regions.

10.4.2 Mollier Diagram

A Mollier diagram is generally the phase diagram on $h{-}s$ coordinates. Shown in Figure 10.13 is the $h{-}s$ diagram for water. It consists of constant-pressure lines in the mixture and superheated regions, constant dryness lines in the mixture region and constant-temperature lines in the superheated region. Also marked are the constant-volume lines in both regions. This diagram is generally made use of to quickly find the enthalpy and/or entropy change in constant-pressure, constant-temperature and isentropic processes. Though it is time-consuming, enthalpy and entropy changes can be found accurately using tabulated property values.

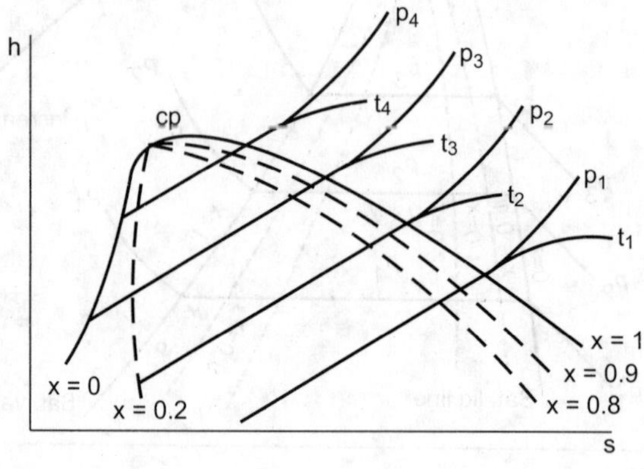

Fig. 10.13 A Mollier diagram.

10.4.3 Dryness Fraction

In the compressed liquid and superheated vapour regions, pressure and temperature are independent properties. A given temperature line and a given pressure line intersect to give a distinct state. Hence, all properties of the substance in these two regions can be evaluated in accordance with the two-property rule, $z = z(p, t)$, where z is any other required thermodynamic property.

In the mixture region, however, pressure and temperature are not independent properties. Both constant-pressure lines and the temperature lines are horizontal in this region and they do not intersect to give a distinct point. In other words, pressure and temperature serve as just one property. This is because, for a pure substance in general, or for water in particular, if pressure is given, its saturation temperature is known and vice versa. These temperatures are those corresponding to the second horizontal step in Figure 10.2. Thus, besides pressure and/or temperature an additional property is to be given to fix the state point. Such a property, very commonly used, is the *dryness fraction.* The other names in vogue for the dryness fraction are, *dryness factor*, *quality* or simply *dryness.*

Suppose a rigid vessel contains m kg of wet steam at a pressure p. This wet steam, also called a mixture, consists of both liquid water (wet part) and steam (dry vapour). The two phases are in equilibrium with each other. In other words, both phases are at the same pressure and temperature. Let m_f be the mass of liquid water and m_g the mass of steam. Then,

$$m = m_f + m_g \tag{10.3}$$

The dryness fraction of this wet steam, or mixture, usually denoted by x, is defined as the ratio of the mass of the steam (dry part) to the total mass. Thus,

$$x = \frac{m_g}{m} = \frac{m_g}{(m_f + m_g)} \tag{10.4}$$

At the saturated liquid state f, there is no steam, and hence $m_g = 0$. Therefore, $x_f = 0$.

At the saturated vapour state g, there is no liquid and hence $m_f = 0$. Therefore, $x_g = 1$.

Thus the dryness fraction can vary only between 0 and 1. As dryness fraction is defined only in the mixture region where the liquid phase and the vapour phase exist in equilibrium, x is irrelevant in the compressed liquid region and the superheated vapour region.

Suppose we consider 1 kg mixture, then $m = (m_f + m_g) = 1$. Therefore, $x = \dfrac{m_g}{m} = m_g$ and $(1 - x) = (1 - m_g) = m_f$.

Generally, the lines of constant dryness fraction are shown in all phase diagrams. These are shown as broken lines in Figures 10.10, 10.11 and 10.12. These lines start at the critical point. While the value of x is zero on the saturated liquid line it is equal to 1 on the saturated vapour line.

Example 10.1

A rigid vessel contains 1.5 kg wet steam at a certain pressure. If 8% of this mass is liquid find the dryness fraction of the mixture.

► **Data:**

$m = 1.5$ kg, $m_f = 0.08\ m = 0.08 \times 1.5 = 0.12$ kg.

► **Solution:**

As $\quad m = m_f + m_g,\ m_g = 1 - m_f = 1.5 - 0.12 = 1.38$ kg.

$$x = \frac{m_g}{m} = \frac{1.38}{1.5} = 0.92 = 92\%$$

Example 10.2

A boiler is generating steam at a pressure of 10 bar. The quality of steam is 0.85. Find the ratio of the mass of saturated liquid to saturated steam in the steam supplied by the boiler.

► **Data:**

$x = 0.85$

► **Solution:**

Every kilogram of mixture contains x kg of saturated steam and $(1 - x)$ kg saturated liquid. Therefore,

$$m_g \text{ per kg} = x = 0.85 \text{ kg/kg and } m_f = (1 - x) = (1 - 0.85) = 0.15 \text{ kg/kg}$$

Hence, $\quad \dfrac{m_f}{m_g} = \dfrac{0.15}{0.85} = 0.17647.$

10.4.4 Evaluation of other Properties in the Mixture Region

Let Z be any extensive thermodynamic property of the wet steam. Then, Z will be equal to sum of the Z values corresponding to the liquid part and the vapour part. Hence,

$$Z = Z_f + Z_g$$

Suppose z, z_f and z_g are the corresponding specific properties, and the mass of the mixture considered is m kg, with m_f kg saturated liquid and m_g kg saturated vapour, then

$$mz = m_f z_f + m_g z_g$$

that is, $$z = \left(\frac{m_f}{m}\right) z_f + \left(\frac{m_g}{m}\right) z_g$$

If $m = 1$ kg, we know that $m_f = (1 - x)$ and $m_g = x$.

Therefore, $$z = (1 - x) z_f + x z_g \tag{10.5}$$

Normally, $(z_g - z_f)$ is denoted by z_{fg}. This is equal to the change in the corresponding specific property between the saturated vapour and saturated liquid states.

From (10.5), $$z = z_f + x z_g - x z_f$$

Therefore, $$z = z_f + x z_{fg} \tag{10.6}$$

Equation (10.6) can be used when z_{fg} values are readily available in property tables. z can be specific volume (v), or specific internal energy (u), or specific enthalpy (h), or specific entropy (s). Hence, in the mixture region, combining Eqs.(10.5) and (10.6):

$$\begin{aligned} v &= (1 - x) v_f + x v_g = v_f + x v_{fg} \\ u &= (1 - x) u_f + x u_g = u_f + x u_{fg} \\ h &= (1 - x) h_f + x h_g = h_f + x h_{fg} \\ s &= (1 - x) s_f + x s_g = s_f + x s_{fg} \end{aligned} \tag{10.7}$$

10.4.5 Table of Properties for Water

Properties of most commonly encountered pure substances are made available in thermodynamic data books in the form of tables. For each substance, the properties are provided in four different tables, namely (i) pressure-based saturation property table, (ii) temperature-based saturation property table, (iii) superheated property table and (iv) compressed-liquid property table. Here, we concentrate on tables that give the properties of water. They are widely known as *steam tables.* They provide various properties such as specific volume, specific internal energy, specific enthalpy and specific entropy.

In preparing the tables, the internal energy of liquid water at triple point ($t = 0.01°$C) is taken to be zero. Since $h = u + pv$, the enthalpy of saturated water at 0.01°C has a negligible value because the value of the term pv is exceedingly small. The entropy of saturated water at its triple point is also chosen to be zero.

10.4.5.1 Saturated Steam (or Water) Tables

It is being once again repeated here that during the change of phase from liquid water to vapour at a constant pressure the temperature remains constant. Each pressure has a fixed temperature of change of phase. This is the saturation temperature. During the change of phase the liquid water and steam are in equilibrium. Thus, in the mixture region, pressure and temperature are not independent properties. If the pressure is given, the temperature is fixed and vice versa. The saturated property table lists the properties at two distinct states in the mixture region, namely, the saturated liquid state and the saturated vapour state for various pressures as in pressure-based table, or for various saturation temperatures as in temperature-based table. The properties provided in these tables are normally the specific volumes (v_f and v_g), specific internal energies (u_f and u_g), specific enthalpies (h_f and h_g) and specific entropies (s_f and s_g).

Table 10.1 Saturated Water and Steam Properties Based on Pressure

Pressure	Tem.	Specific Volume		Specific Enthalpy		Specific Entropy	
p	t	m^3/kg		kJ/kg		kJ/kg K	
kPa	°C	v_f	v_g	h_f	h_g	s_f	s_g
0.6113	0.01	0.001000	206.132	0.00	2501.3	0	9.1562
1.0	6.98	0.001000	129.208	29.29	2514.2	0.1059	8.9756
1.5	13.03	0.001001	87.98	54.7	2525.3	0.1956	8.8278
2.0	17.50	0.001001	67.004	73.47	2533.5	0.2607	8.7236
2.5	21.08	0.001002	54.254	88.47	2540.0	0.3120	8.6431
3.0	24.08	0.001003	45.665	101.03	2545.5	0.3545	8.5775
4.0	28.96	0.001004	34.80	121.44	2554.4	0.4226	8.4746
5.0	32.88	0.001005	28.193	137.79	2561.4	0.4763	8.3950
7.5	40.29	0.001008	19.238	168.77	2574.8	0.5763	8.2514
10.0	45.81	0.001010	14.674	191.81	2584.6	0.6492	8.1501
15.0	53.97	0.001014	10.022	225.91	2599.1	0.7548	8.0084
20.0	60.06	0.001017	7.649	251.38	2609.7	0.8319	7.9085
25.0	64.97	0.001020	6.204	271.90	2618.2	0.8930	7.8313
30.0	69.10	0.001022	5.229	289.21	2626.3	0.9439	7.7686
40.0	75.87	0.001026	3.993	317.55	2636.7	1.0258	7.6700
50.0	81.33	0.001030	3.240	340.47	2645.9	1.0910	7.5939
75.0	91.77	0.001037	2.217	384.36	2663.0	1.2129	7.4563
100.0	99.62	0.001043	1.6940	417.44	2675.5	1.3025	7.3593

(*Table 10.1 Continued*)

(*Table 10.1 Continued*)

Pressure p kPa	Tem. t °C	Specific Volume m^3/kg		Specific Enthalpy kJ/kg		Specific Entropy kJ/kg K	
		v_f	v_g	h_f	h_g	s_f	s_g
101.325	100.0	0.001043	1.6735	419.07	2675.7	1.3069	7.3545
125.0	105.99	0.001048	1.3749	443.30	2685.3	1.3739	7.2843
150.0	111.37	0.001053	1.1593	467.08	2693.5	1.4335	7.2232
175.0	116.06	0.001057	1.0036	486.97	2700.5	1.4848	7.1717
200.0	120.23	0.001061	0.8857	504.68	2706.6	1.5300	7.1271
225.0	124.00	0.001064	0.7933	520.69	2712.0	1.5705	7.0878
250.0	127.43	0.001067	0.7187	535.34	2716.9	1.6072	7.0526
275.0	130.60	0.001070	0.6573	548.87	2721.3	1.6407	7.0208
300.0	133.55	0.001073	0.6058	561.45	2725.3	1.6717	6.9918
325.0	136.30	0.001076	0.5620	573.23	2729.0	1.7005	6.9651
350.0	138.88	0.001079	0.5243	584.31	2732.4	1.7274	6.9404
375.0	141.32	0.001081	0.4914	594.79	2735.6	1.7527	6.9174
400.0	143.63	0.001084	0.4625	604.73	2738.5	1.7766	6.8958
450.0	147.93	0.001088	0.4140	623.24	2743.9	1.8206	6.8565
500.0	151.86	0.001093	0.3749	640.21	2748.7	1.8606	6.8212
550.0	155.48	0.001097	0.3427	655.91	2752.9	1.8972	6.7892
600.0	158.85	0.001101	0.3157	670.54	2756.8	1.9311	6.7600
650.0	162.01	0.001104	0.2927	684.26	2760.3	1.9627	6.7330
700.0	164.97	0.001108	0.2729	697.20	2763.5	1.9922	6.7080
750.0	167.77	0.001111	0.2556	709.45	2766.4	2.0199	6.6846
800.0	170.43	0.001115	0.2404	721.10	2769.1	2.0461	6.6627
850.0	172.96	0.001118	0.2270	732.20	2771.6	2.0709	6.6421
900.0	175.38	0.001121	0.2150	742.82	2773.9	2.0946	6.6225
950.0	177.69	0.001124	0.2042	753.00	2776.1	2.1171	6.6040
1000.0	179.91	0.001127	0.1944	762.79	2778.1	2.1386	6.5864

(*Table 10.1 Continued*)

(*Table 10.1 Continued*)

Pressure p MPa	Tem. t °C	Specific Volume m³/kg v_f	v_g	Specific Enthalpy kJ/kg h_f	h_g	Specific Entropy kJ/kg K s_f	s_g
1.10	184.09	0.001133	0.17753	781.32	2781.7	2.1791	6.5535
1.20	187.99	0.001139	0.1633	798.64	2784.8	2.2165	6.5233
1.30	191.64	0.001144	0.15125	814.91	2787.6	2.2514	6.4953
1.40	195.07	0.001149	0.14084	830.29	2790.0	2.2842	6.4692
1.50	198.32	0.001154	0.13177	844.87	2792.1	2.3150	6.4448
1.75	205.76	0.001166	0.11349	878.48	2796.4	2.3851	6.3895
2.00	212.42	0.001177	0.09963	908.77	2799.5	2.4473	6.3408
2.25	218.45	0.001187	0.08875	936.48	2801.7	2.5034	6.2971
2.50	223.99	0.001197	0.07998	962.09	2803.1	2.5546	6.2574
2.75	229.12	0.001207	0.07275	985.97	2803.9	2.6018	6.2208
3.00	233.90	0.001216	0.06668	1008.41	2804.1	2.6456	6.1869
3.25	238.38	0.001226	0.06152	1029.60	2804.0	2.6866	6.1551
3.50	242.60	0.001235	0.05707	1049.73	2803.4	2.7252	6.1252
4.00	250.40	0.001252	0.04978	1087.29	2801.4	2.7963	6.0700
5.00	263.99	0.001286	0.03944	1154.21	2794.3	2.9201	5.9733
6.00	275.64	0.001319	0.03244	1213.32	2784.3	3.0266	5.8891
7.00	285.88	0.001351	0.02737	1266.97	2772.1	3.1210	5.8132
8.00	295.06	0.001384	0.02352	1316.61	2757.9	3.2067	5.7431
9.00	303.40	0.001418	0.02048	1363.23	2742.1	3.2857	5.6771
10.00	311.06	0.001452	0.01803	1407.53	2724.7	3.3595	5.6140
11.00	318.15	0.001489	0.01599	1450.05	2705.6	3.4294	5.5527
12.00	324.75	0.001527	0.01426	1491.24	2684.8	3.4961	5.4923
13.00	330.93	0.001567	0.01278	1531.46	2662.2	3.5604	5.4323
14.00	336.75	0.001611	0.01149	1571.08	2637.5	3.6231	5.3716
15.00	342.24	0.001658	0.01034	1610.45	2610.5	3.6847	5.3097
16.00	347.43	0.001711	0.00931	1650.00	2580.6	3.7460	5.2454
17.00	352.37	0.001770	0.00836	1690.25	2547.2	3.8078	5.1776
18.00	357.06	0.001840	0.00749	1731.97	2509.1	3.8713	5.1044
19.00	361.54	0.001924	0.00666	1776.43	2464.5	3.9387	5.0227
20.00	365.81	0.002035	0.00583	1826.18	2409.7	4.0137	4.9269
21.00	369.89	0.002206	0.00495	1888.30	2334.7	4.1073	4.8015
22.00	373.80	0.002808	0.00353	2034.92	2159.0	4.3307	4.5224
22.09	374.14	0.003155	0.00315	0.00315	2099.3	4.4297	4.4297

Table 10.2 Saturated Water and Steam Properties Based on Temperature

Tem. t °C	Pressure p kPa	Specific Volume m^3/kg v_f	v_g	Specific Enthalpy kJ/kg h_f	h_g	Specific Entropy kJ/kg K s_f	s_g
0.01	0.6113	0.001000	206.132	0.00	2501.3	0.0000	9.1562
5	0.8721	0.001000	106.377	41.99	2519.7	0.1510	8.9007
10	1.2276	0.001000	77.925	62.98	2528.9	0.2245	8.7813
15	1.7051	0.001001	77.925	62.98	2528.9	0.2245	8.7813
20	2.3385	0.001002	57.790	83.94	2538.1	0.2966	8.6671
25	3.1691	0.001003	43.359	104.87	2547.2	0.3673	8.5579
30	4.2461	0.001004	32.893	125.77	2556.2	0.4369	8.4533
35	5.6280	0.001006	25.216	146.66	2565.3	0.5052	8.3530
40	7.3837	0.001008	19.523	167.54	2574.3	0.5724	8.2569
45	9.5934	0.001010	15.258	188.42	2583.2	0.6386	8.1647
50	12.35	0.001012	12.032	209.31	2592.1	0.7037	8.0762
55	15.758	0.001015	9.568	230.20	2600.9	0.7679	7.9912
60	19.941	0.001017	7.671	251.11	2609.6	0.8311	7.9095
65	25.033	0.001020	6.197	272.03	2618.2	0.8934	7.8309
70	31.188	0.001023	5.042	292.96	2626.8	0.9548	7.7552
75	38.578	0.001026	4.131	313.91	2635.3	1.0154	7.6824
80	47.390	0.001029	3.407	334.88	2643.7	1.0752	7.6121
85	57.834	0.001032	2.828	355.88	2651.9	1.1342	7.5444
90	70.139	0.001036	2.361	376.90	2660.1	1.1924	7.4790
95	84.554	0.001040	1.982	397.94	2668.1	1.2500	7.4158
100	101.35	0.001044	1.673	419.00	2676.0	1.3068	7.3548
105	120.82	0.001047	1.4194	440.13	2683.8	1.3629	7.2958
110	143.28	0.001052	1.2102	461.27	2691.5	1.4184	7.2386
115	169.06	0.001056	1.0366	482.46	2699.0	1.4733	7.1832
120	198.53	0.001060	0.8919	503.69	2706.3	1.5275	7.1295
125	232.10	0.001065	0.7706	524.96	2713.5	1.5812	7.0774
130	270.10	0.001070	0.6685	546.29	2720.5	1.6343	7.0269
135	313.00	0.001075	0.5822	567.67	2727.3	1.6869	6.9777
140	361.30	0.001080	0.5089	589.11	2733.9	1.7390	6.9298
145	415.40	0.001085	0.4463	610.61	2740.3	1.7906	6.8832
150	475.90	0.001090	0.3928	632.18	2746.4	1.8417	6.8378
155	543.10	0.001096	0.3468	653.82	2752.4	1.8924	6.7934

(Table 10.2 Continued)

(Table 10.2 Continued)

Tem. t °C	Pressure p kPa	Specific Volume m^3/kg v_f	v_g	Specific Enthalpy kJ/kg h_f	h_g	Specific Entropy kJ/kg K s_f	s_g
160	617.80	0.001102	0.3071	675.53	2758.1	1.9426	6.7501
165	700.50	0.001108	0.2727	697.32	2763.5	1.9924	6.7078
170	791.70	0.001114	0.2428	719.20	2768.7	2.0418	6.6663
175	892.00	0.001121	0.2168	741.16	2773.6	2.0909	6.6256
180	1002.2	0.001127	0.1941	763.21	2778.2	2.1395	6.5857
185	1122.7	0.001134	0.1741	785.36	2782.4	2.1878	6.5464
190	1254.4	0.001141	0.1565	807.61	2786.4	2.2358	6.5078
195	1397.8	0.001149	0.1411	829.96	2790.0	2.2835	6.4697
200	1553.8	0.001156	0.1274	852.43	2793.2	2.3308	6.4322

Tem. t °C	Pressure p MPa	Specific Volume m^3/kg v_f	v_g	Specific Enthalpy kJ/kg h_f	h_g	Specific Entropy kJ/kg K s_f	s_g
205	1.7230	0.001164	0.1152	875.03	2796.0	2.3779	6.3951
210	1.9063	0.001173	0.1044	897.75	2798.5	2.4247	6.3584
215	2.1042	0.001181	0.0948	920.61	2800.5	2.4713	6.3221
220	2.3178	0.001190	0.0862	943.61	2802.1	2.5177	6.2860
225	2.5477	0.001199	0.0785	966.77	2803.3	2.5639	6.2502
230	2.7949	0.001209	0.0716	990.10	2803.9	2.6099	6.2146
235	3.0601	0.001219	0.0654	1013.61	2804.1	2.6557	6.1791
240	3.3442	0.001229	0.0598	1037.31	2803.8	2.7015	6.1436
245	3.6482	0.001240	0.0547	1061.21	2802.9	2.7471	6.1083
250	3.9730	0.001251	0.0501	1085.34	2801.5	2.7927	6.0729
255	4.3195	0.001263	0.0460	1109.72	2799.5	2.8382	6.0374
260	4.6886	0.001276	0.0422	1134.35	2796.9	2.8837	6.0018
265	5.0813	0.001289	0.0388	1159.27	2793.6	2.9293	5.9661
270	5.4987	0.001302	0.0356	1184.49	2789.7	2.9750	5.9301
275	5.9418	0.001317	0.0328	1210.05	2785.0	3.0208	5.8937
280	6.4117	0.001332	0.0302	1235.97	2779.5	3.0667	5.8570
285	6.9094	0.001348	0.0278	1262.29	2773.3	3.1129	5.8198
290	7.4360	0.001366	0.0256	1289.04	2766.1	3.1593	5.7821
295	7.9928	0.001384	0.0235	1316.27	2758.0	3.2061	5.7436

(Table 10.2 Continued)

(Table 10.2 Continued)

Tem. t °C	Pressure p MPa	Specific Volume m³/kg v_f	v_g	Specific Enthalpy kJ/kg h_f	h_g	Specific Entropy kJ/kg K s_f	s_g
300	8.5810	0.001404	0.0217	1344.01	2748.9	3.2533	5.7044
305	9.2018	0.001425	0.0199	1372.33	2738.7	3.3009	5.6642
310	9.8566	0.001447	0.0184	1401.29	2727.3	3.3492	5.6229
315	10.547	0.001472	0.0169	1430.97	2714.4	3.3981	5.5803
320	11.274	0.001499	0.0155	1461.45	2700.1	3.4479	5.5361
330	12.845	0.001561	0.0130	1525.29	2665.8	3.5506	5.4416
340	14.586	0.001638	0.0108	1594.15	2622.0	3.6593	5.3356
350	16.514	0.001740	0.0088	1670.54	2563.9	3.7776	5.2111
360	18.651	0.001892	0.0069	1760.48	2481.0	3.9146	5.0525
370	21.028	0.002213	0.0049	1890.37	2332.1	4.1104	4.7972
374.14	22.089	0.003155	0.00315	2099.26	2099.3	4.4297	4.4297

10.4.5.2 Superheated Steam (or Water) Table

In the superheated region, for a given pressure, steam can exist at any temperature above the saturation temperature. Each temperature corresponds to a different state. In the superheated table, properties such as specific volume, enthalpy and entropy are listed for various temperatures for a given pressure. As innumerable values of temperatures are possible, the data can be very voluminous, making it difficult to use it. Hence to make the table compact, the temperatures are generally selected at intervals of 50 degrees. For a temperature which is not available in the table, the properties can be evaluated through linear interpolation.

Table 10.3 Superheated Steam Properties

$p = 10$ kPa				$p = 50$ kPa			
t	v	h	s	t	v	h	s
45.81	**14.674**	**2584.6**	**8.1501**				
50	14.869	2592.6	8.1749	**81.33**	**3.240**	**2645.9**	**7.5939**
100	17.196	2687.5	8.4479	100	3.418	2682.5	7.6947
150	19.513	2783.0	8.6881	150	3.889	2780.1	7.9400
200	21.825	2879.5	8.9037	200	4.356	2877.6	8.1579
250	24.136	2977.3	9.1002	250	4.821	2976.0	8.3555
300	26.445	3076.5	9.2812	300	5.284	3075.5	8.5372
400	31.063	3279.5	9.6076	400	6.209	3278.9	8.8641
500	35.679	3489.0	9.8977	500	7.134	3488.6	9.1545
600	40.295	3705.4	10.1608	600	8.058	3705.1	9.4177
700	44.911	3928.7	10.4028	700	8.981	3928.5	9.6599
800	49.526	4159.1	10.6281	800	9.904	4158.9	9.8852
900	54.141	4396.4	10.8395	900	10.828	4396.3	10.0967
1000	58.757	4640.6	11.0392	1000	11.751	4640.5	10.2964
1100	63.372	4891.2	11.2287	1100	12.674	4891.1	10.4858
1200	67.987	5147.8	11.4090	1200	13.597	5147.7	10.6662
1300	72.603	5409.7	11.5810	1300	14.521	5409.6	10.8382

$p = 100$ kPa				$p = 200$ kPa			
t	v	h	s	t'	v	h	s
99.62	**1.6940**	**2675.5**	**7.3593**				
100	1.6958	2676.2	7.3614	**120.23**	**0.88573**	**2706.6**	**7.1271**
150	1.9364	2776.4	7.6133	150	0.95964	2768.8	7.2795
200	2.1723	2875.3	7.8342	200	1.08034	2870.5	7.5066
250	2.4060	2974.3	8.0332	250	1.19880	2971.0	7.7085
300	2.6388	3074.3	8.2157	300	1.31616	3071.8	7.8926
400	3.1026	3278.1	8.5434	400	1.54930	3276.5	8.2217
500	3.5655	3488.1	8.8341	500	1.78139	3487.0	8.5132
600	4.0278	3704.7	0.0975	600	2.01297	3704.0	8.7769
700	4.4899	3928.2	9.3398	700	2.24426	3927.7	9.0194
800	4.9517	4158.7	9.5652	800	2.47539	4158.3	9.2450
900	5.4135	4396.1	9.7767	900	2.70643	4395.8	9.4565
1000	5.8753	4640.3	9.9764	1000	2.93740	4640.0	9.6563
1100	6.3370	4890.9	10.1658	1100	3.16834	4890.7	9.8458
1200	6.7986	5147.6	10.3462	1200	3.39927	5147.3	10.0262
1300	7.2603	5409.5	10.5182	1300	3.63018	5409.3	10.1982

$p = 300$ kPa				$p = 400$ kPa			
t	v	h	s	t	v	h	s
133.55	**0.60582**	**2725.3**	**6.9918**	**143.63**	**0.46246**	**2738.5**	**6.8958**
150	0.63388	2761.0	7.0778	150	0.47084	2752.8	6.9299
200	0.71629	2865.5	7.3115	200	0.53422	2860.5	7.1706
250	0.79636	2967.6	7.5165	250	0.59512	2964.2	7.3788
300	0.87529	3069.3	7.7022	300	0.65484	3066.7	7.5661
400	1.03151	3275.0	8.0329	400	0.77262	3273.4	7.8984
500	1.18669	3486.0	8.3250	500	0.88934	3484.9	8.1912
600	1.34136	3703.2	8.5892	600	1.00555	3702.4	8.4557
700	1.49573	3927.1	8.8319	700	1.12147	3926.5	8.6987
800	1.64994	4157.8	9.0575	800	1.23722	4157.4	8.9244
900	1.80406	4395.4	9.2691	900	1.35288	4395.1	9.1361
1000	1.95812	4639.7	9.4689	1000	1.46847	4639.4	9.3360
1100	2.11214	4890.4	9.6585	1100	1.58404	4890.1	9.5255
1200	2.26614	5147.1	9.8389	1200	1.69958	5146.8	9.7059
1300	2.42013	5409.0	10.0109	1300	1.81511	5408.8	9.8780

$p = 500$ kPa				$p = 600$ kPa			
t	v	h	s	t	v	h	s
151.86	**0.37489**	**2748.7**	**6.8212**	**158.55**	**0.31567**	**2756.8**	**6.7600**
200	0.42492	2855.4	7.0592	200	0.35202	2850.1	6.9665
250	0.47436	2960.7	7.2708	250	0.39383	2957.2	7.1816
300	0.52256	3064.2	7.4598	300	0.43437	3061.6	7.3723
350	0.57012	3167.6	7.6328	350	0.47424	3165.7	7.5463
400	0.61728	3271.8	7.7937	400	0.51372	3270.2	7.7078
500	0.71093	3483.8	8.0872	500	0.59199	3482.7	8.0020
600	0.80406	3701.7	8.3521	600	0.66974	3700.9	8.2673
700	0.89691	3926.0	8.5952	700	0.74720	3925.4	8.5107
800	0.98959	4157.0	8.8211	800	0.82450	4156.5	8.7367
900	1.08217	4394.7	9.0329	900	0.90169	4394.4	8.9485
1000	1.17469	4639.1	9.2328	1000	0.97883	4638.8	9.1484
1100	1.26718	4889.9	9.4224	1100	1.05594	4889.6	9.3381
1200	1.35964	5146.6	9.6028	1200	1.13302	5146.3	9.5185
1300	1.45210	5408.6	9.7749	1300	1.21009	5408.3	9.6906

p = 800 kPa				p = 1.0 MPa			
t	v	h	s	t	v	h	s
170.43	**0.24043**	**2769.1**	**6.6627**	**179.91**	**0.19444**	**2778.1**	**6.5864**
200	0.26080	2839.2	6.8158	200	0.20596	2827.9	6.6939
250	0.29314	2950.0	7.0384	250	0.23268	2942.6	6.9246
300	0.32411	3056.4	7.2327	300	0.25794	3051.2	7.1228
350	0.35439	3161.7	7.4088	350	0.28247	3157.7	7.3010
400	0.38426	3267.1	7.5715	400	0.30659	3263.9	7.4650
500	0.44331	3480.6	7.8672	500	0.35411	3478.4	7.7621
600	0.50184	3699.4	8.1332	600	0.40109	3697.9	8.0289
700	0.56007	3924.3	8.3770	700	0.44779	3923.1	8.2731
800	0.61813	4155.7	8.6033	800	0.49432	4154.8	8.4996
900	0.67610	4393.6	8.8153	900	0.54075	4392.9	8.7118
1000	0.73401	4638.2	9.0153	1000	0.58712	4637.6	8.9119
1100	0.79188	4889.1	9.2049	1100	0.63345	4888.5	9.1016
1200	0.84974	5145.8	9.3854	1200	0.67977	5145.4	9.2821
1300	0.90758	5407.9	9.5575	1300	0.72608	5407.4	9.4542

p = 1.2 MPa				p = 1.4 MPa			
t	v	h	s	t	v	h	s
187.99	**0.16333**	**2784.8**	**6.5233**	**195.07**	**0.14084**	**2790.0**	**6.4692**
200	0.17930	2815.9	6.5898	200	0.14302	2803.3	6.4975
250	0.19235	2935.0	6.8293	250	0.16350	2927.2	6.7467
300	0.21382	3045.8	7.0316	300	0.18228	3040.4	6.9533
350	0.23452	3153.6	7.2120	350	0.20026	3149.5	7.1359
400	0.25480	3260.7	7.3773	400	0.21780	3257.4	7.3025
500	0.29463	3476.3	7.6758	500	0.25215	3474.1	7.6026
600	0.33393	3696.3	7.9434	600	0.28596	3694.8	7.8710
700	0.37294	3922.00	8.1881	700	0.31947	3920.9	8.1160
800	0.41177	4153.9	8.4149	800	0.35281	4153.0	8.3431
900	0.45051	4392.2	8.6272	900	0.38606	4391.5	8.5555
1000	0.48919	4637.0	8.8274	1000	0.41924	4636.4	8.7558
1100	0.52783	4888.0	9.0171	1100	0.45239	4887.5	8.9456
1200	0.56646	5144.9	9.1977	1200	0.48552	5144.4	9.1262
1300	0.60507	5406.9	9.3698	1300	0.51864	5406.5	9.2983

p = 1.6 MPa				p = 1.8 MPa			
t	*v*	*h*	*s*	*t*	*v*	*h*	*s*
201.4	**0.12380**	**2794.0**	**6.4217**	**207.15**	**0.11042**	**2797.1**	**6.3793**
225	0.13287	2857.2	6.5518	225	0.11673	2846.7	6.4807
250	0.14184	2919.2	6.6732	250	0.12497	2911.0	6.6066
300	0.15862	3034.8	6.8844	300	0.14021	3029.2	6.8226
350	0.17456	3145.4	7.0693	350	0.15457	3141.2	7.0099
400	0.19005	3254.2	7.2373	400	0.16847	3250.9	7.1793
500	0.22029	3471.9	7.5389	500	0.19550	3469.7	7.4824
600	0.24998	3693.2	7.8080	600	0.22199	3691.7	7.7523
700	0.27937	3919.7	8.0535	700	0.24818	3918.6	7.9983
800	0.30859	4152.1	8.2808	800	0.27420	4151.3	8.2258
900	0.33772	4390.8	8.4934	900	0.30012	4390.1	8.4386
1000	0.36678	4635.8	8.6938	1000	0.32598	4635.2	8.6390
1100	0.39581	4887.0	8.8837	1100	0.35180	4886.4	8.8290
1200	0.42482	5143.9	9.0642	1200	0.37761	5143.4	9.0096
1300	0.45382	5406.0	9.2364	1300	0.40340	5405.6	9.1817

p = 2.0 MPa				p = 2.5 MPa			
t	*v*	*h*	*s*	*t*	*v*	*h*	*s*
212.42	**0.09963**	**2799.5**	**6.3408**	**223.99**	**0.07998**	**2803.1**	**6.2574**
225	0.10377	2835.8	6.4146	225	0.08027	2806.3	6.2638
250	0.11144	2902.5	6.5452	250	0.08700	2880.1	6.4084
300	0.12547	3023.5	6.7663	300	0.09890	3008.8	6.6437
350	0.13587	3137.0	6.9562	350	0.10976	3126.2	6.8402
400	0.15120	3247.6	7.1270	400	0.12010	3239.3	7.0147
500	0.17568	3467.6	7.4316	500	0.13998	3462.0	7.3233
600	0.19960	3690.1	7.7023	600	0.15930	3686.2	7.5960
700	0.22323	3917.5	7.9487	700	0.17832	3914.6	7.8435
800	0.24668	4150.4	8.1766	800	0.19716	4148.2	8.0270
900	0.27004	4389.4	8.3895	900	0.21590	4387.6	8.2853
1000	0.29333	4634.6	8.5900	1000	0.23458	4633.1	8.4860
1100	0.31659	4885.9	8.7800	1100	0.25322	4884.6	8.6761
1200	0.33984	5142.9	8.9606	1200	0.27185	5141.7	8.8569
1300	0.36306	5405.1	9.1328	1300	0.29046	5404.0	9.0291

$p = 3.0$ MPa				$p = 3.5$ MPa			
t	v	h	s	t	v	h	s
233.90	**0.06668**	**2804.1**	**6.1869**	**242.60**	**0.05707**	**2803.4**	**6.1252**
250	0.07058	2855.8	6.2871	250	0.05873	2829.2	6.1748
300	0.08114	2993.5	6.5389	300	0.06842	2977.5	6.4460
350	0.09053	3115.3	6.7427	350	0.07678	3104.0	6.6578
400	0.09936	3230.8	6.9211	400	0.08453	3222.2	6.8404
500	0.11619	3456.5	7.2337	500	0.09918	3450.9	7.1571
600	0.13243	3682.3	7.5084	600	0.11324	3678.4	7.4338
700	0.14838	3911.7	7.7571	700	0.12699	3908.8	7.6837
800	0.16414	4146.0	7.9862	800	0.14056	4143.8	7.9135
900	0.17980	4385.9	8.1999	900	0.15402	4384.1	8.1275
1000	0.19541	4631.6	8.4009	1000	0.16743	4630.1	8.3288
1100	0.21098	4883.3	8.5911	1100	0.18080	4881.9	8.5191
1200	0.22652	5140.5	8.7719	1200	0.19415	5139.3	8.7000
1300	0.24206	5402.8	8.9442	1300	0.20749	5401.7	8.8723

$p = 4.0$ MPa				$p = 4.5$ MPa			
t	v	h	s	t	v	h	s
250.40	**0.04978**	**2801.4**	**6.0700**	**257.48**	**0.04406**	**2798.3**	**6.0198**
275	0.05457	2886.2	6.2284	275	0.04730	2863.1	6.1401
300	0.05884	2960.7	6.3614	300	0.05135	2943.1	6.2828
350	0.06645	3092.4	6.5820	350	0.05840	3080.6	6.5130
400	0.07341	3213.5	6.7689	400	0.06475	3204.7	6.7046
500	0.08643	3445.2	7.0900	500	0.07651	3439.5	7.0300
600	0.09885	3674.4	7.3688	600	0.08765	3670.5	7.3109
700	0.11095	3905.9	7.6198	700	0.09847	3903.0	7.5631
800	0.12287	4141.6	7.8502	800	0.10911	4139.4	7.7942
900	0.13469	4382.3	8.0647	900	0.11965	4380.6	8.0091
1000	0.14645	4628.7	8.2661	1000	0.13013	4627.2	8.2108
1100	0.15817	4880.6	8.4566	1100	0.14056	4879.3	8.4014
1200	0.16987	5138.1	8.6376	1200	0.15098	5136.9	8.5824
1300	0.18156	5400.5	8.8099	1300	0.16139	5399.4	8.7548

$p = 5.0$ MPa				$p = 6.0$ MPa			
t	v	h	s	t	v	h	s
263.99	**0.03944**	**2794.3**	**5.9733**	**275.64**	**0.03244**	**2784.3**	**5.8891**
275	0.04141	2338.3	6.0543	300	0.03616	2884.2	6.0673
300	0.04532	2924.5	6.2083	350	0.04223	3043.0	6.3334
350	0.05194	3068.4	6.4492	400	0.04739	3177.2	6.5407
400	0.05781	3195.6	6.6458	450	0.05214	3301.8	6.7192
500	0.06857	3433.8	6.9758	500	0.05665	3422.1	6.8802
600	0.07869	3666.5	7.2588	600	0.06525	3658.4	7.1676
700	0.08849	3900.1	7.5122	700	0.07352	3894.3	7.4234
800	0.09811	4137.2	7.7440	800	0.08160	4132.7	7.6566
900	0.10762	4378.8	7.9593	900	0.08958	4375.3	7.8727
1000	0.11707	4625.7	8.1612	1000	0.09749	4622.7	8.0751
1100	0.12648	4878.0	8.3519	1100	0.10536	4875.4	8.2661
1200	0.13587	5135.7	8.5330	1200	0.11321	5133.3	8.4473
1300	014526.	5398.2	8.7055	1300	0.12106	5396.0	8.6199

$p = 7.0$ MPa				$p = 8.0$ MPa			
t	v	h	s	t	v	h	s
285.88	**0.02737**	**2772.1**	**5.8132**	**295.06**	**0.02352**	**2757.9**	**5.7431**
300	0.02947	2838.4	5.9304	300	0.02426	2785.0	5.7905
350	0.03524	3016.0	6.2282	350	0.02995	2987.3	6.1300
400	0.03993	3158.1	6.4477	400	0.03432	3138.3	6.3633
450	0.04416	3287.0	6.6326	450	0.03817	3272.0	6.5550
500	0.04814	3410.3	6.7974	500	0.04175	3398.3	6.7239
600	0.05565	3650.3	7.0894	600	0.04845	3642.0	7.0205
700	0.06283	3888.4	7.3476	700	0.05481	3882.5	7.2812
800	0.06981	4128.3	7.5822	800	0.06097	4123.8	7.5173
900	0.07669	4371.8	7.7991	900	0.06702	4368.3	7.7350
1000	0.08350	4619.8	8.0020	1000	0.07301	4616.9	7.9384
1100	0.09027	4872.8	8.1933	1100	0.07896	4870.3	8.1299
1200	0.09703	5130.9	8.3747	1200	0.08489	5128.5	8.3115
1300	0.10377	5393.7	8.5472	1300	0.09080	5391.5	8.4842

$p = 9.0$ MPa				$p = 10.0$ MPa			
t	v	h	s	t	v	h	s
303.40	**0.02048**	**2742.1**	**5.6771**	**311.06**	**0.01803**	**2724.7**	**5.6140**
325	0.02327	2855.9	5.8711	325	0.01986	2809.0	5.7568
350	0.02580	2956.5	6.0361	350	0.02242	2923.4	5.9442
400	0.02993	3117.8	6.2853	400	0.02641	3096.5	6.2119
450	0.03350	3256.6	6.4843	450	0.02975	3240.8	6.4189
500	0.03677	3386.1	6.6575	500	0.03279	3373.6	6.5965
600	0.04285	3633.7	6.9588	600	0.03837	3625.3	6.9028
700	0.04857	3876.5	7.2221	700	0.04358	3870.5	7.1687
800	0.05409	4119.4	7.4597	800	0.04859	4114.9	7.4077
900	0.05950	4364.7	7.6782	900	0.05349	4361.2	7.6272
1000	0.06485	4613.9	7.8821	1000	0.05832	4611.0	7.8315
1100	0.07016	4867.7	8.0739	1100	0.06312	4865.1	8.0236
1200	0.07544	5126.2	8.2556	1200	0.06789	5123.8	8.2054
1300	0.08072	5389.2	8.4283	1300	0.07265	5387.0	8.3783

$p = 12.5$ MPa				$p = 15$ MPa			
t	v	h	s	t	v	h	s
327.89	**0.01350**	**2673.8**	**5.4623**	**342.24**	**0.010338**	**2610.5**	**5.3097**
350	0.01613	2826.2	5.7117	350	0.011470	2692.4	5.4420
400	0.02000	3039.3	6.0416	400	0.015649	2975.4	5.8810
450	0.02299	3199.8	6.2718	450	0.018446	3156.2	6.1403
500	0.02560	3341.7	6.4617	500	0.020800	3308.5	6.3442
550	0.02801	3475.1	6.6289	550	0.022927	3448.6	6.5198
600	0.03029	3604.0	6.7810	600	0.024911	3582.3	6.6775
700	0.03460	3855.4	7.0536	700	0.028612	3840.1	6.9572
800	0.03869	4103.7	7.2965	800	0.032096	4092.4	7.2040
900	0.04267	4352.5	7.5181	900	0.035457	4343.8	7.4279
1000	0.04658	4603.8	7.7237	1000	0.038748	4596.6	7.6347
1100	0.05045	4858.8	7.9165	1100	0.042001	4852.6	7.8282
1200	0.05430	5118.0	8.0987	1200	0.045233	5112.3	8.0108
1300	0.05813	5381.4	8.2717	1300	0.048455	5375.9	8.1839

p = 17.5 MPa				p = 20 MPa			
t	*v*	*h*	*s*	*t*	*v*	*h*	*s*
354.75	***0.0079204***	***2528.8***	***5.1418***	***365.81***	***0.0058342***	***2409.7***	***4.9269***
400	0.0124477	2902.8	5.7212	400	0.0099423	2818.1	5.5539
450	0.0151740	3109.7	6.0182	450	0.0126953	3060.1	5.9016
500	0.0173585	3274.0	6.2382	500	0.0147683	3238.2	6.1400
550	0.0192877	3421.4	6.4229	550	0.0165553	3393.5	6.3347
600	0.0210640	3560.1	6.5866	600	0.0181781	3537.6	6.5048
650	0.0227372	3693.9	6.7356	650	0.0196929	3675.3	6.6582
700	0.0243365	3824.7	6.8736	700	0.0211311	3809.1	6.7993
800	0.0273849	4081.1	7.1245	800	0.0238532	4069.8	7.0544
900	0.0303071	4335.1	7.3507	900	0.0264463	4326.4	7.2830
1000	0.0331580	4589.5	7.5588	1000	0.0289666	4582.5	7.4925
1100	0.0359695	4846.4	7.7530	1100	0.0314471	4840.2	7.6874
1200	0.0387605	5106.6	7.9359	1200	0.0339071	5101.0	7.8706
1300	0.0415417	5370.5	8.1093	1300	0.0363574	5365.1	8.0441

[Units: *t* in °C, *v* in m^3/kg, *h* in kJ/kg and *s* in kJ/kg K. The rows with bold italics correspond to saturated vapour properties]

TRIPLE POINT DATA

Temperature: 0.01°C.
Pressure: 0.006113 bar.

Phase	*v*(m^3/kg)	*u* (kJ/kg)	*h* (kJ/kg)	*s* (kJ/kgK)
Ice	0.0010905	−333.5	−333.5	−1.2209
Water	0.0010002	0	0.0006113	0
Steam	206.1629	2375.3	2501.3	9.1562

CRITICAL POINT DATA

p : 220.9 bar
t : 374.14°C
v : 0.003155 m^3/kg
h : 2099.3 kJ/kg
s : 4.4297 kJ/kg K

The other properties which are available in many tables are the change in volume during the change of phase from f to $g\,(v_{fg})$, change in internal energy (u_{fg}), change in enthalpy (h_{fg}) and change in entropy (s_{fg}). It is not always necessary to list these values as they can be quickly calculated from the listed properties at the saturation states. For example, $v_{fg} = (v_g - v_f)$, $h_{fg} = (h_g - h_f)$ etc. Also, it may not be necessary to list the values of u as it is easily calculable from other properties. For example, $u_f = (h_f - pv_f)$ and $u_g = (h_g - pv_g)$.

In the property tables provided above, while Table 10.1 is the saturated steam table based on pressure, Table 10.2 is the saturated steam table based on temperature. These tables do not list the values of u, v_{fg}, h_{fg} and s_{fg}. Table 10.3 is the superheated steam table.

10.4.5.3 Compressed-Water Table

In this region also, for each pressure, the state point depends upon the second property given. Let the other given property be temperature. Then for each value of temperature, there exists a unique state. Each of these states, fixed by the given pressure and temperature would have unique values of specific properties such as volume, internal energy, enthalpy and entropy. These are tabulated in the compressed-water property tables. Thus, in structure, the compressed-water property table looks similar to superheated steam table. Extensive compressed-water property tables are not generally available. This is because the properties of liquid water in this region depend strongly on temperature and not on pressure. Hence, if temperature is known, irrespective of the value of pressure, the other properties of the liquid can be taken to be equal to the corresponding values of properties of saturated liquid at the given temperature. Also, the properties in this region can be evaluated fairly accurately using various property relations. Both methods are illustrated in Examples 10.3 and 10.4.

10.4.6 How to Fix the States in Different Phase Regions and Evaluate Other Properties

Suppose the pressure or temperature and one other specific property, z, is given. The following procedure may be adopted to find the remaining properties using the property tables.

Step 1: If the pressure is given, always refer to saturated-property table based on pressure. If the temperature is given, refer to the saturated-property table based on temperature.

Step 2: For the given pressure or temperature read the values of z_f and z_g from the table. (Example: Suppose pressure and specific enthalpy h are given, read from the pressure-based table the values of h_f and h_g for the given pressure).

Step 3: The phase region where the state of water lies is decided by comparing z with z_f and z_g.

(a) If $z < z_f$, the water is in the compressed-liquid region.

(b) If $z = z_f$, the water is in saturated-liquid state.

(c) If $z_f < z < z_g$, the steam is in a mixture state.

(d) If $z = z_g$, the steam is in saturated-vapour state

(e) If $z > z_g$, the steam is in the superheated state.

Case (a): Examples 10.3 and 10.4 illustrate how to evaluate the properties when water is in the compressed-liquid region.

Example 10.3

A rigid vessel contains water at 2 bar and 100°C. Find its volume, enthalpy, internal energy and entropy per kilogram of water.

▶ Data:

$p_1 = 2$ bar $= 200$ kPa, $t_1 = 100°$C.

▶ Solution:

From the saturated-property table based on pressure, Table 10.1, at $p = 200$ kPa, $t_s = 120.23°$C. Hence,

$$t_1 = 100°\text{C} < t_s = 120.23°\text{C}$$

Therefore, water is in the compressed-liquid region. Its state is shown as 1 in the T–S diagram of Figure 10.14. All the properties corresponding to this state are approximately equal to the saturated water properties corresponding to $t_s = 100°$C. These properties can be directly read from the saturated property table based on temperature (Table 10.2). Thus,

$$v_1 = v_f|_{t=100} = 0.001044 \text{ m}^3/\text{kg}$$
$$h_1 = h_f|_{t=100} = 419 \text{ kJ/kg}$$
$$s_1 = s_f|_{t=100} = 1.3068 \text{ kJ/kg K}$$
$$u_1 = u_f|_{t=100} = h - pv$$
$$= 419 - 200 \times 0.001044 = 418.8 \text{ kJ/kg}$$

Example 10.4

Find the temperature and all other specific properties for water at $p = 5$ bar and $s = 1.3625$ kJ/kg K.

▶ Data:

$p_1 = 500$ kPa, $s_1 = 1.3625$ kJ/kg K

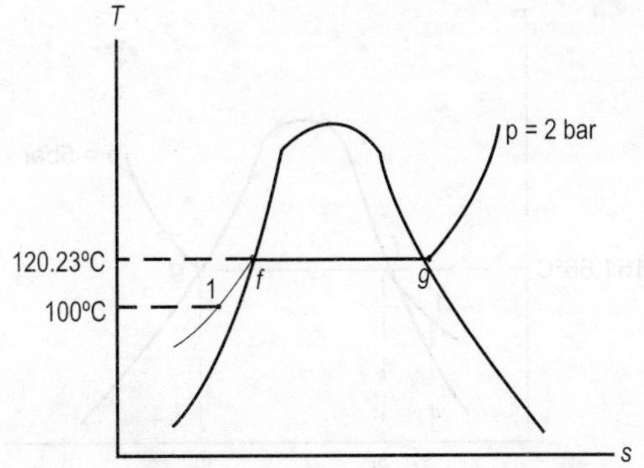

Fig. 10.14 $T\text{–}s$ diagram to solve Example 10.3.

▶ **Solution:**

Method 1: From the saturated-property table based on pressure (Table 10.1),

$$s_f|_{p=500\text{ kPa}} = 1.8606\text{ kJ/kg K},\quad \text{and}\quad s_g|_{p=500\text{ kPa}} = 6.8212\text{ kJ/kg K}.$$

As, $s_1 < s_f$, the given state must be in compressed liquid region. This state is shown as 1 on $T\text{–}S$ diagram in Figure 10.15.

All other properties are approximately equal to saturated-liquid properties corresponding to t_1. But t_1 is not directly given. t_1 must be equal to that saturation temperature at which $s_1 = s_f = 1.3625$ kJ/kg K. Scanning through the s_f column in both saturated-property tables, it is found from Table 10.2 that $s_f = 1.3629$ kJ/kg K (≈ 1.3625 kJ/kg K) at $t_s = 105°$C. Therefore,

$$\begin{aligned} t_1 &\approx 105°\text{C},\quad \text{and} \\ v_1 &= v_f|_{t=105} = 0.001047\text{ m}^3\text{/kg} \\ h_1 &= h_f|_{t=105} = 440.13\text{ kJ/kg} \\ u_1 &= u_f|_{t=105} = h - pv \\ &= 440.13 - 500 \times 0.001047 = 439.6\text{ kJ/kg K}. \end{aligned}$$

▶ **Alternatively** :

Method 2: It can be seen from Figure 10.15 that states 1 and f are both on 500 kPa pressure line. Hence state f can be obtained by heating the compressed liquid from 1 at a constant pressure of 500 kPa. Therefore,

$$s_f - s_1 = c_{wat}\ln\left(\frac{T_s|_{p=500\text{ kPa}}}{T_1}\right)$$

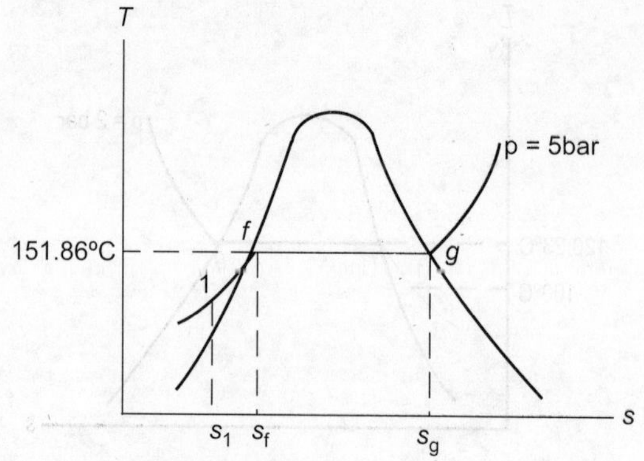

Fig. 10.15 T–s diagram to solve Example 10.4.

From Table 10.1, $T_s|_{p=500 \text{ kPa}} = (273 + 151.86) = 424.86$ K, and

$s_f|_{p=500 \text{ kPa}} = 1.8606$ kJ/kg K

Substituting for T_s, s_f and $c_{wat}(= 4.189$ kJ/kg K$)$,

$$1.8606 - 1.3625 = 4.189 \ln\left(\frac{424.86}{T_1}\right)$$

that is, $T_1 = 377.23$ K, or $t_1 = 104.23$°C.

$v_1 \approx 0.001$ m^3/kg (an approximate value that can be used for any state in the compressed liquid region).

Also, $(h_f - h_1)_{p=500 \text{ kPa}} = c_{wat}\,(t_f - t_1)$

From Table 10.1, $h_f = 640.21$ kJ/kg K

and therefore, $640.21 - h_1 = 4.189(151.86 - 104.23)$.

that is, $h_1 = 440.7$ kJ/kg.

$$u_1 = h_1 - p_1 v_1 = 440.7 - 500 \times 0.001 = 440.2 \text{ kJ/kg}$$

The various properties found from both methods are tabulated below to show that all the values agree fairly well.

Method	t (°C)	v (m^3/kg)	u (kJ/kg)	h (kJ/kg)
Method 1	105	0.001047	439.6	440.13
Method 2	104.23	0.001	440.2	440.7

Case (b): In this case, for the given pressure, as $z = z_f$, the water must be in the saturated-liquid state. Hence, all other saturated-water properties can be directly from the table for the given pressure.

Example 10.5

Water leaves the condenser of a steam power plant at a temperature of 50°C and an enthalpy of 209.31 kJ/kg. Find the other properties of water.

▶ **Data:**

$t = 50°\text{C}$, $h = 209.31$ kJ/kg.

▶ **Solution:**

From the saturated-property table based on temperature (Table 10.2) at $t = 50°\text{C}$,

$$h_f = 209.31 \text{ kJ/kg} \quad \text{and} \quad h_g = 2592.1 \text{ kJ/kg}$$

As $h = h_f$, the water is in the saturated-liquid condition.

Therefore, from the same table we can read the other required properties as,

$$p = p_s|_{t=50} = 12.35 \text{ kPa}$$

$$v = v_f|_{t=50} = 0.001012 \text{ m}^3/\text{kg}$$

$$u = h_f - pv_f = 209.31 - 12.35 \times 0.001012 = 209.3 \text{ kJ/kg}$$

and

$$s = s_f|_{t=50} = 0.7037 \text{ kJ/kg K}$$

Case (c): For the given pressure or temperature, if $z_f < z < z_g$, the steam is in a mixture state. In such a case, using the given value of z and the values of z_f and z_g found from the table, the value of the dryness fraction is determined first. Thereafter, the other required properties are calculated using this value of dryness.

Example 10.6

Steam leaves a boiler at a pressure of 12 bar and an enthalpy of 2705 kJ/kg. Find all the other properties of this steam.

▶ **Data:**

$p = 12$ bar, $h = 2705$ kJ/kg

▶ **Solution:**

From the pressure-based saturated-property table, at $p = 12$ bar,

$$h_f = 798.64 \text{ kJ/kg} \quad \text{and} \quad h_g = 2784.8 \text{ kJ/kg}.$$

As $h_f < h < h_g$, the steam must be in the mixture region. Dryness fraction of steam can now be found from,

$$h = (1-x)h_f + xh_g,$$
$$2705 = (1-x)798.64 + 2784.4x$$

that is, $x = 0.96$

Now, $v = |(1-x)v_f + xv_g|_{p=12 \text{ bar}}$

Picking the values of v_f and v_g from Table 10.1 at $p = 12$ bar,

$$v = (1-0.96)0.001139 + 0.96 \times 0.1633 = 0.1568 \text{ m}^3/\text{kg}.$$

Similarly, $s = (1-x)s_f + xs_g|_{p=12 \text{ bar}}$

$$= (1-0.96)2.2165 + 0.96 \times 6.5233 = 6.351 \text{ kJ/kg K}.$$

and, $u = h - pv = 2705 - 12 \times 10^5 \times 10^{-3} \times 0.1568 = 2517 \text{ kJ/kg}$

Case (d): For a given pressure or temperature if $z = z_g$, the steam exists as saturated vapour and hence other properties can be read directly from the saturated-property table.

Example 10.7

It is given that a certain mass of steam at 15 bar has a specific internal energy of 2594.5 kJ/kg. Find all the other thermodynamic properties.

▶ Data:

$p = 15$ bar, $u = 2594.5$ kJ/kg

▶ Solution:

The values of u are not available in the tables provided in this book. Hence, the values of u_f and u_g will have to be first evaluated using the listed values of h, p and v.

From the saturated-steam tables based on pressure, at $p = 15$ bar, $h_f = 844.87$ kJ/kg, $v_f = 0.001154$ m^3/kg, $h_g = 2792.1$ kJ/kg and $v_g = 0.13177$ m^3/kg.

Now, $u_f = |h_f - pv_f|_{p=15 \text{ bar}} = 844.87 - 15 \times 10^5 \times 10^{-3} \times 0.001154 = 843.14$ kJ/kg

and, $u_g = |h_g - pv_g|_{p=15 \text{ bar}} = 2792.1 - 15 \times 10^5 \times 10^{-3} \times 0.13177 = 2594.5$ kJ/kg.

As $u = u_g = 2594.5$ kJ/kg, the steam must be in the saturated-vapour state. Hence all other required properties can be directly read from the saturated-property table.

$$t = t_s|_{p=15 \text{ bar}} = 198.32^\circ\text{C}$$
$$v = v_g|_{p=15 \text{ bar}} = 0.13177 \text{ m}^3/\text{kg}$$
$$h = h_g|_{p=15 \text{ bar}} = 2792.1 \text{ kJ/kg} \quad \text{and}$$
$$s = s_g|_{p=15 \text{ bar}} = 6.4448 \text{ kJ/kg K}$$

Case (e): In this case, for the given pressure, as $z > z_g$, the steam must be in the superheated state. Hence, all other properties can be directly found from the superheated-property table for the given pressure.

Example 10.8

Find all the thermodynamic properties of steam at 10 MPa and 550°C.

▶ **Data:**

$p = 10$ MPa, $t = 550$°C

▶ **Solution:**

From the saturated-property table (Table 10.1) $t_s|_{p=10 \text{ Mpa}} = 311.06$°C.

As $t > t_s$, steam must be in superheated state.

It can be seen from the superheated-property table (Table 10.3) that at $p = 10$ MPa, the properties are not listed for $t = 550$°C. However, the properties are available at 500°C (a temperature immediately less than the given temperature of 550°C) and 600°C (a temperature immediately above it). Hence, the properties at 550°C may be evaluated by linear interpolation between the properties available in the table for 500° and 600°C.

From the superheated table,

at $\quad p = 10$ Mpa, $\quad v_{500°\text{C}} = 0.03279 \text{ m}^3/\text{kg}$ and

$$v_{600°\text{C}} = 0.03837 \text{ m}^3/\text{kg}$$

Therefore,

$$\frac{\Delta v}{\text{deg}} = \frac{(0.03837 - 0.03279)}{(600 - 500)} = 5.58 \times 10^{-5} \text{ m}^3/\text{kg/deg.}$$

$$v_{550°\text{C}} = v_{500°\text{C}} + \left(\frac{\Delta v}{\text{deg}}\right) \times (550 - 500)$$

$$= 0.03279 + 5.58 \times 10^{-5} \times 50 = 0.03558 \text{ m}^3/\text{kg}.$$

Similarly,

$$h_{550°\text{C}} = h_{500°\text{C}} + \left(\frac{\Delta h}{\text{deg}}\right) \times (550 - 500)$$

$$= 3373.6 + \left[\frac{(3625.3 - 3373.6)}{(600 - 500)}\right] \times (550 - 500)$$

$$= 3499.5 \text{ kJ/kg}$$

and,

$$s_{550°\text{C}} = s_{500°\text{C}} + \left(\frac{\Delta s}{\text{deg}}\right) \times (550 - 500)$$

$$= 6.5965 + \left[\frac{(6.9028 - 6.5965)}{(600 - 500)}\right] \times (550 - 500) = 6.7497 \text{ kJ/kg K}$$

Example 10.9

Find the other thermodynamic properties of steam at $t = 250$°C and $h = 2855.8$ kJ/kg

▶ **Data:**

$t = 250$°C and $h = 2855.8$ kJ/kg

▶ **Solution:**

From the saturated-property table based on temperature (Table 10.2),

at $t = 250°\text{C}$, $h_g = 2801.5$ kJ/kg

As $h > h_g$, the steam must be in the superheated state.

Now carefully scan through the superheated property table and find the pressure at which the given t and h are available together. Having found this pressure read all other properties corresponding to the this state.

$t = 250°\text{C}$ and $h = 2855.8$ kJ/kg are available together on the superheated table at the pressure of 3 MPa. Thus, $p = 3$ MPa. The other properties can be directly read from the table corresponding to $p = 3$ MPa and $t = 250°\text{C}$. Accordingly, $v = 0.07058$ m^3/kg and $s = 6.2871$ kJ/kg K.

And, $u = h - pv = 2855.8 - (3 \times 10^6 \times 10^{-3}) \times 0.07058 = 2644$ kJ/kg

10.4.7 Experimental Determination of Dryness Fraction of Steam

The dryness fraction of a sample of steam can be experimentally determined using a *throttling calorimeter* or *a combined separating and throttling calorimeter*. The former is used if the dryness fraction of the sample is fairly high. If the moisture content of the sample steam were to be high, the latter is used.

10.4.7.1 Throttling Calorimeter

Shown in Figure 10.16 is the schematic of a throttling calorimeter. A sample of wet steam is drawn in to the calorimeter through the sampling tube. The pressure of the sample as indicated by the pressure gauge mounted on the sampling tube is p_1. This sample is then expanded adiabatically through a throttle valve to a pressure p_2 such that the throttled steam is superheated by at least 5 degree. That is, $t_2 \geq \left(t_s|_{p=p_2} + 5 \right)$. Using the experimentally measured values of p_1, p_2 and t_2, the dryness of sample steam can be evaluated as follows.

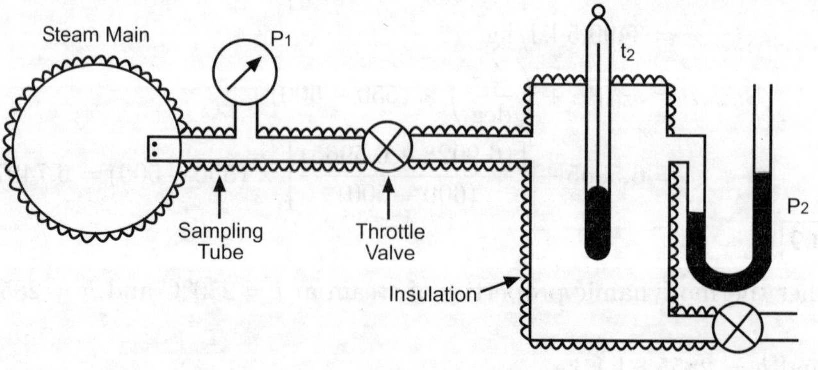

Fig. 10.16 Schematic of a throttling calorimeter.

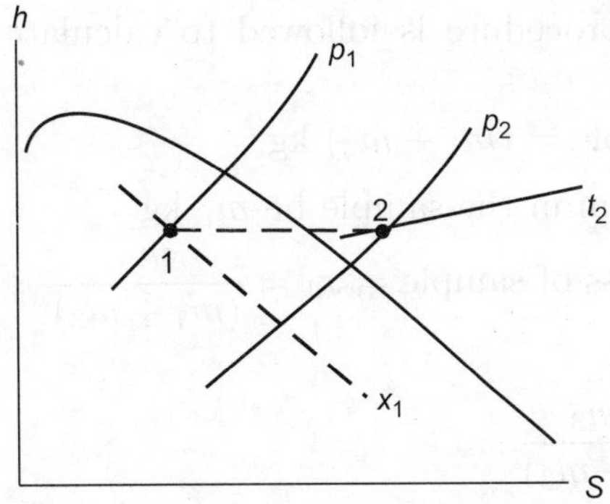

Fig. 10.17 h–s diagram to fix the dryness from the readings of a throttling calorimeter.

The adiabatic throttling process is a constant-enthalpy process. Therefore, $h_1 = h_2$. Here, h_2, corresponding to p_2 and t_2 can be readily found from the superheated-property tables. And, h_1 is given by,

$$h_1 = [(1 - x_1)\, h_f + x_1 h_g]_{p=p1}\,,$$

where h_f and h_g can be read from saturated-property table based on pressure and x_1 is the dryness to be determined. As $h_1 = h_2$,

$$h_2 = [(1 - x_1)\, h_f + x_1 h_g]_{p=p_1}$$

In the above equation, the only unknown is x_1.

Alternatively, x_1 can also be found out using the Mollier diagram for steam. As shown in Figure 10.17, state 2 can be fixed on the diagram for the measured values of p_2 and t_2. A horizontal line (a constant enthalpy line) is then drawn from state 2 to intersect the constant pressure line of p_1, at state 1. The dryness fraction line passing through state 1 gives the required value of x_1.

10.4.7.2 Combined Separating and Throttling Calorimeter

Steam from a main at a pressure p_1 and dryness x_1 is passed through the separating device of the calorimeter, as shown in Figure 10.18. In the separating device, because of the sudden change in the flow direction of steam, some moisture in the steam gets condensed and falls into the separator. This process occurs at a constant pressure of p_1. At a certain interval, let the mass of condensate collected in the separator be m_1 kg. The remaining steam, drier than the sample steam, enters the throttle valve at section 2. At section 2, while the pressure is still equal to p_1, the dryness fraction x_2 will be higher than x_1. Let the pressure and temperature of steam measured after throttling, which must be in the superheated region, be p_3 and t_3. The entire throttled steam is condensed through a water-cooled condenser, as shown in Figure 10.18. Let the mass of this condensed steam be m_3 kg. The following procedure is followed to calculate the dryness of the sample steam.

The mass of steam sample $= (m_1 + m_3)$ kg.

Let the mass of dry steam in the sample be m_v kg.

By definition, the dryness of sample $= x_1 = \dfrac{m_v}{(m_1 + m_3)}$

Also, $\quad m_v = x_2 m_3.$

Therefore, $\quad x_1 = \dfrac{x_2 m_3}{(m_1 + m_3)}$

Here, x_2 can be calculated using the relation,

$$h_3 = h_2 = [(1 - x_2)\, h_f + x_2 h_g]_{p=p_1}$$

where h_3 can be found from the superheated-property table corresponding to p_3 and t_3 and h_f and h_g can be found from the saturated-property table corresponding to the pressure of p_1.

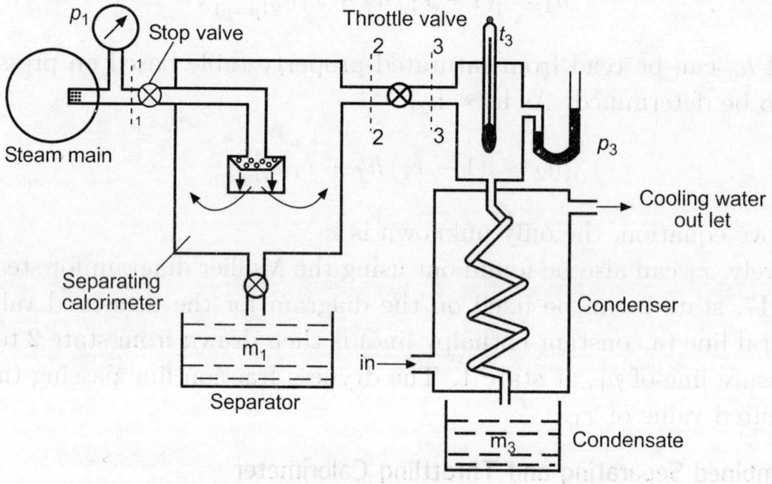

Fig. 10.18 Schematic of combined separating and throttling calorimeter.

Alternatively, x_1 can also be found out using the Mollier diagram. As shown in Figure 10.19, a horizontal line is drawn from state 3, which is fixed for measured values of p_3 and t_3, to intersect the p_1 line at state 2. The reading x_2 can now be read from the Mollier diagram. With the known value of x_2 and the measured values of m_1 and m_2, x_1 can be fixed from

$$x_1 = \frac{x_2 m_3}{(m_1 + m_3)}$$

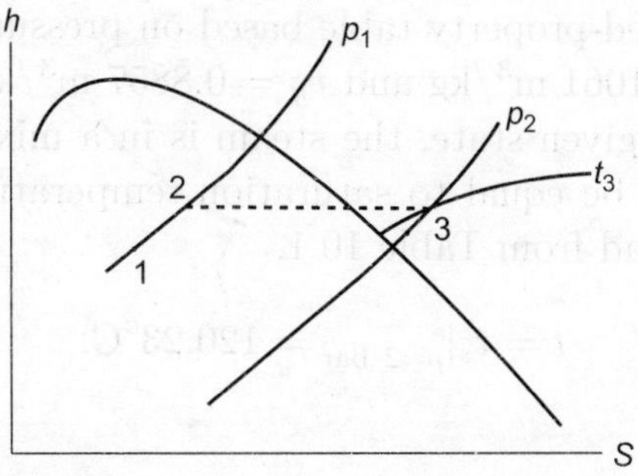

Fig. 10.19 A $h-s$ diagram to fix the dryness from the readings of a separating and throttling calorimeter.

10.5 Analyses of Some Processes Undergone by Steam

In actual practice, the steam that is confined either in a system boundary or flowing through a control surface may undergo several processes. These processes can be analysed with the help of the first and second laws of thermodynamics, as was done earlier for any pure substance including ideal gas.

Different thermodynamic properties of steam may be necessary for a certain analysis. These are generally obtained from the property tables corresponding to the given two independent properties.

Examples 10.10–10.14 illustrate how properties of steam can be used in the analyses of a few selected processes.

Example 10.10

A rigid vessel of volume 0.1 m^3 contains 0.16 kg steam at a pressure of 2 bar. Find the temperature of steam.

▶ Data:

$p = 2$ bar, $m = 0.16$ kg and $V = 0.1\ m^3$.

▶ Solution:

To find the temperature, the p–v–T relation can be used. If the vessel were to contain a pure substance, such as an ideal gas, the temperature could be calculated using the p–v–T relation, $T = \dfrac{pV}{m\text{R}}$. In the case of steam, the temperature has to be found using only the data available in property tables.

The two independent properties given to fix the initial state are:

$$p = 2 \text{ bar} \quad \text{and} \quad v = \frac{V}{m} = \frac{0.1}{0.16} = 0.625\ m^3/kg$$

Referring to the saturated-property table based on pressure (Table 10.1),

At $p = 2$ bar, $v_f = 0.001061 \text{ m}^3/\text{kg}$ and $v_g = 0.8857 \text{ m}^3/\text{kg}$

As $v_f < v < v_g$, at the given state, the steam is in a mixture condition. Hence, the temperature of steam must be equal to saturation temperature corresponding to 2 bar. This temperature can be read from Table 10.1.

$$t = t_s|_{p=2 \text{ bar}} = 120.23°\text{C}.$$

Example 10.11

The steam in Example 10.10 is cooled until its temperature drops to 100°C. Find the heat rejected from steam.

► **Data:**

From Example 10.10, $p_1 = 2$ bar, $v_1 = 0.625 \text{ m}^3/\text{kg}$; $t_1 = 120.23°\text{C}$.

► **Solution:**

Applying the first law for the cooling process, $Q - W = \Delta U$. The cooling occurs at constant volume and hence $W = 0$. Therefore,

$$Q = \Delta U = m(u_2 - u_1)$$

Let x_1 be the dryness fraction of steam at state 1. Then,

$$v_1 = |(1 - x_1) v_f + x_1 v_g|_{p=2 \text{ bar}}$$

where v_f and v_g can be obtained from saturation steam table based on pressure.

That is, $\quad 0.625 = (1 - x_1)\, 0.001061 + 0.8857x_1$

Therefore, $\quad x_1 = 0.705.$

Now, $\quad h_1 = |(1 - x_1) h_f + x_1 h_g|_{p=2 \text{ bar}}$

Picking the values of h_f and h_g from Table 10.1,

$$= (1 - 0.705)504.68 + 0.705 \times 2706.6 = 2057 \text{ kJ/kg}$$

Hence, $\quad u_1 = h_1 - p_1 v_1 = 2057 - (2 \times 10^5 \times 10^{-3} \times 0.625) = 1932 \text{ kJ/kg}$

As the cooling is at constant volume, $v_2 = v_1$ and also steam shall still be in the mixture region as some steam in the vessel condenses because of cooling. Consequently, there will be an increase in the moisture content or decrease in the dryness.

t_2 is given to be 100°C. Hence,

$$v_2 = v_1 = |(1 - x_2) v_f + x_2 v_g|_{t=100°\text{C}}$$

v_f and v_g can be read from saturated property table based on temperature (Table 10.2).

$$0.625 = (1 - x_2)\,0.001044 + 1.673x_2$$

Therefore $\quad x_2 = 0.373.$

Now, $\quad h_2 = |(1 - x_2)\,h_f + x_2 h_g|_{t=100°C}$

$$= (1 - 0.373)419 + 0.373 \times 2676 = 1260.86 \text{ kJ/kg}.$$

Also, $\quad p_2 = p_s|_{t=100°C} = 1.01325$ bar (from Table 10.2)

$$u_2 = h_2 - p_2 v_2 = 1260.86 - (1.01325 \times 10^5 \times 10^{-3}) \times 0.625$$

$$= 1197.5 \text{ kJ/kg}$$

Heat transfer $\quad Q = \Delta U = m\,(u_2 - u_1)$

$$= 0.16(1197.5 - 1932) = -117.5 \text{ kJ}$$

Example 10.12

0.1 kg saturated steam expands reversibly from 10 to 1 bar in a piston–cylinder device according to $pv^{1.3}$ = constant. Find the work and heat interactions during the expansion process.

► **Data:**

$p_1 = 10$ bar, $v_1 = v_g|_{p1=10\text{ bar}}$, $p_2 = 1$ bar, $pv^{1.3}$ = constant.

► **Solution:**

$$v_1 = v_g|_{p1=10\text{ bar}} = 0.1944 \text{ m}^3\text{/kg (from Table 10.1)}$$

As $\quad p_1 v_1^{1.3} = p_2 v_2^{1.3},\ v_2 = \left(\dfrac{p_1}{p_2}\right)^{\frac{1}{1.3}} \times v_1 = \left(\dfrac{10}{1}\right)^{\frac{1}{1.3}} \times 0.1944 = 1.1427 \text{ m}^3\text{/kg}$

$$W = \frac{m\,(p_1 v_1 - p_2 v_2)}{(n-1)} = \frac{0.1(10 \times 0.1944 - 1 \times 1.1427) \times 10^5 \times 10^{-3}}{(1.3-1)}$$

$$= 26.71 \text{ kJ}$$

Applying the first law to the process, $Q - W = m\,(u_2 - u_1)$,

where, $\quad u_1 = h_1 - p_1 v_1 = |h_g - p_1 v_g|_{p=10\text{ bar}}$

$$= (2778.1 - 10 \times 10^5 \times 10^{-3} \times 0.1944) = 2583.7 \text{ kJ/kg}$$

At state 2, $p = 1$ bar and $v_2 = 1.1427$ m^3/kg.

From Table 10.1, at $p = 1$ bar, $v_f = 0.001043$ m^3/kg and $v_g = 1.694$ m^3/kg.

As $v_f < v_2 < v_g$, the steam is in a mixture state. The dryness factor at state 2 can be found from,

$$v_2 = |(1 - x_2)\, v_f + x_2 v_g|_{p=1 \text{ bar}}$$
$$1.1427 = (1 - x_2)\, 0.001043 + 1.694 x_2$$

and hence, $x_2 = 0.674$

Now, $$h_2 = |(1 - x_2)\, h_f + x_2 h_g|_{p=1 \text{ bar}}$$
$$= (1 - x_2)\, 417.44 + 2675.5 x_2 = 1939.4 \text{ kJ/kg}$$

and $$u_2 = h_2 - p_2 v_2 = 1939.4 - 1 \times 10^5 \times 10^{-3} \times 1.1427 = 1825.1 \text{ kJ/kg}$$

Therefore, $$Q = W + m\,(u_2 - u_1) = 26.71 + 0.1(1825.1 - 2583.7) = -49.15 \text{ kJ}$$

Example 10.13

Steam enters the nozzle of an impulse turbine at 10 bar, 300°C with a velocity of 50 m/s. The exit pressure is 1 bar. Find the exit area of the nozzle, in cm^2, assuming the expansion to be reversible and adiabatic for a steam flow rate of 1.2 kg/s.

▶ **Data:**

$p_1 = 10$ bar, $t_1 = 300$°C, $V_1 = 50$ m/s, $p_2 = 1$ bar, $\dot{m} = 1.2$ kg/s.

▶ **Solution:**

The expansion process is reversible and adiabatic. The process may be represented by pv^γ = constant, where γ is the ratio of specific heats. But, steam does not have a unique value of γ over a wide range. Hence, this equation cannot be used here. However, for the given process $Q = 0$ and $\Delta s = 0$. Therefore, $s_1 = s_2$. This fact may be made use of to analyse the problem.

From the pressure-based saturated property table, at $p = 10$ bar, $t_s = 179.91$°C.
As $t_1 > t_s$, initially, steam is in superheated state.
From the superheated-property table, at $p_1 = 10$ bar, $t_1 = 300$°C,

$$h_1 = 3051.2 \text{ kJ/kg} \quad \text{and} \quad s_1 = 7.1228 \text{ kJ/kg K.}$$

Finally, $$p_2 = 1 \text{ bar, and } s_2 = s_1 = 7.1228 \text{ kJ/kg K}$$

From the pressure-based saturated-property table, at $p = 1$ bar,

$$s_f = 1.3025 \text{ kJ/kg K} \quad \text{and } s_g = 7.3593 \text{ kJ/kg K}$$

At the exit of the nozzle, as $s_f < s_2 < s_g$, steam is in a mixture state.

Hence, $$s_2 = |(1 - x_2)\, h_f + x_2 h_g|_{p=1 \text{ bar}}$$

Substituting for s_2, h_f and h_g,

$$7.1228 = (1 - x_2)\, 1.3025 + 7.3593 x_2, \quad \text{and hence, } x_2 = 0.96.$$

Now, $$h_2 = |(1 - x_2)\, h_f + x_2 h_g|_{p=1 \text{ bar}}$$
$$= (1 - 0.96)417.44 + 0.96 \times 2675.5 = 2585.2 \text{ kJ/kg}.$$

Also, $$v_2 = |(1 - x_2)\, v_f + x_2 v_g|_{p=1 \text{ bar}}$$
$$= (1 - 0.96)0.001043 + 0.96 \times 1.694 = 1.626 \text{ m}^3/\text{kg}.$$

Applying the first law (i.e., applying SS, SFEE) to the nozzle,

$$\dot{Q} - \dot{W} = \dot{m}\Delta\left[h + \frac{V^2}{2} + gZ\right]$$

Here, $\dot{Q} = 0$ and $\dot{W} = 0$. Besides, neglecting the change in potential energy,

$$0 - 0 = \dot{m}\left[(h_2 - h_1) + \frac{(V_2^2 - V_1^2)}{2}\right]$$

$$0 = 1.2\left[(2585.2 - 3051.2) \times 10^3 + \frac{(V_2^2 - 50^2)}{2}\right]$$

Therefore, $$V_2 = 966.7 \text{ m/s}.$$

From the continuity equation, $\dot{m} = \dfrac{A_2 V_2}{v_2}$ and hence

$$A_2 = \frac{\dot{m} v_2}{V_2} = 1.2 \times \frac{1.626}{966.7} = 2.02 \times 10^{-3} \text{ m}^2 = 20.2 \text{ cm}^2.$$

Example 10.14

An open feed water heater (OFWH) in a steam power plant operating on regenerative Rankine cycle is a steady-flow device in which a small amount of steam bled from a particular cross-section in the turbine is mixed adiabatically with sub-cooled feed water coming from the pump. Both streams enter the OFWH at the same pressure and leave it as a single stream of saturated liquid. Steam at 4 bar, 150°C and feed water at 4 bar, 50°C enter a certain OFWH steadily. The two streams mix adiabatically and leave the OFWH as saturated liquid at 4 bar. Given that the flow rate of steam in 0.2 kg/s, find the flow rate of feed water into the heater.

► **Data:**

Using suffixes 1, 2 and 3 for the entering steam, feed water and saturated liquid leaving the outlet, respectively, $\dot{m}_1 = 0.2$ kg/s, $h_1 = h|_{p=4 \text{ bar}, t=150°\text{C}}$, $h_2 = h\,|_{p=4 \text{ bar}, t=50°\text{C}}$, and $h_3 = h_f|_{p=4 \text{ bar}}$.

▶ Solution:

Applying SS, SFEE to the adiabatic feed water heater (neglecting changes in kinetic and potential energies)

$$\dot{m}_1 h_1 + \dot{m}_2 h_2 = (\dot{m}_1 + \dot{m}_2)\, h_3. \qquad \text{(i)}$$

Here, $h_1 = h|_{p=4 \text{ bar},\ t=150^\circ\text{C}}$.

From Table 10.1, at $p = 4$ bar, $t_s = 143.63°\text{C}$. As $t_1 > t_s$, the steam is superheated at entry. From the superheated-property table, $h_1 = 2752.8$ kJ/kg.

Feed water is also entering at the same pressure of 4 bar. As $t_2 < t_s$, the water must be in compressed or sub-cooled liquid state.

Hence, $h_2 = h|_{p=4 \text{ bar}, t=50^\circ\text{C}} \approx h_f|_{t=50^\circ\text{C}} = 209.31$ kJ/kg (from Table 10.2)

Also, $h_3 = h_f|_{p=4 \text{ bar}} = 604.73$ kJ/kg (from Table 10.1)

Substituting the above values in (i)

$$0.2 \times 2752.8 + \dot{m}_2 \times 209.31 = (0.2 + \dot{m}_2) \times 604.73$$

That is, $$\dot{m}_2 = 1.086 \text{ kg/s}$$

Illustrative Problems

10.1 For the states of water fixed by the given two properties in the following examples, using the steam tables, identify the condition of water as compressed or sub-cooled liquid, saturated liquid, mixture of saturated liquid and vapour, saturated vapour or superheated vapour?

(a) $p = 1$ bar, $t = 99.62°\text{C}$

(b) $p = 5$ bar, $h = 520$ kJ/kg

(c) $t = 50°\text{C}$, $s = 7.2345$ kJ/kg K

(d) $p = 100$ bar, $u = 1393$ kJ/kg

(e) $p = 125$ kPa, $v = 1.3749$ m^3/kg

(f) $p = 10000$ kPa, $t = 200°\text{C}$

(g) $p = 750$ kPa, $s = 6.8761$ kJ/kg K

▶ Data:

(a) $p = 1$ bar, $t = 99.62°\text{C}$

▶ **Solution:**

From Table 10.1 (saturated-steam tables based on pressure), $t_s|_{p\,=\,1\text{ bar}} = 99.62°\text{C}$. As $t = t_s$, the water must be in boiling condition at the given pressure and temperature. The given pressure and temperature are not independent properties. Thus, the state exists in the mixture region. One other additional property is required to exactly locate the state point.

▶ **Data:**

(b) $p = 5$ bar, $h = 520$ kJ/kg

▶ **Solution:**

From Table 10.1, at $p = 5$ bar, $h_f = 640.21$ kJ/kg and $h_g = 2748.7$ kJ/kg. As $h < h_f$, the state of water must be in the compressed liquid (sub-cooled) region.

▶ **Data:**

(c) $t = 50°\text{C}$, $s = 7.2345$ kJ/kg K

▶ **Solution:**

From Table 10.2 (saturated-steam tables based on temperature), at $t = 50°\text{C}$, $s_f = 0.7037$ kJ/kg K and $s_g = 8.0762$ kJ/kg K.

As $s_f < s < s_g$, the state of water must be in the mixture region.

▶ **Data:**

(d) $p = 100$ bar, $u = 1393$ kJ/kg.

▶ **Solution:**

From Table 10.1, at $p = 100$ bar, $h_f = 1407.53$ kJ/kg and $v_f = 0.001452\ \text{m}^3/\text{kg}$.

Therefore, $u_f = h_f - pv_f = 1407.53 - (100 \times 10^5) \times 10^{-3} \times 0.001452 = 1393$ kJ/kg.
As the given value of $u = u_f$, water is in saturated-liquid condition.

▶ **Data:**

(e) $p = 125$ kPa, $v = 1.3749\ \text{m}^3/\text{kg}$.

▶ **Solution:**

From Table 10.1, at $p = 125$ kPa, $v_f = 0.001048\ \text{m}^3/\text{kg}$ and $v_g = 1.3749\ \text{m}^3/\text{kg}$. As $v = v_g$, the water must be in saturated-vapour condition.

▶ **Data:**

(f) $p = 10\,000$ kPa, $t = 200°\text{C}$.

► **Solution:**

From Table 10.1, At $p = 10000$ kPa = 10 Mpa = 100 bar, $t_s = 311.06°$C. As $t < t_s$, the state of water must be in sub-cooled region.

► **Data:**

(g) $p = 750$ kPa, $s = 6.8761$ kJ/kg K.

► **Solution:**

From Table 10.1, at $p = 750$ kPa, $s_f = 2.0199$ kJ/kg K and $s_g = 6.6846$ kJ/kg K. It can be seen that $s > s_g$. Therefore, the state of water must be in the superheated region.

10.2 Given in Table 10.4 are six different states of water. Using the two properties given in each case find the other properties and fill in the blanks. Insert a dash if any property were to be irrelevant.

Table 10.4 Data for Illustrative Problem 10.2

No.	p kPa	t °C	v m³/kg	h kJ/kg	u kJ/kg	s kJ/kg	x	Deg of superheating
(i)	200	30	–	–	–	–	–	–
(ii)	250			2280				–
(iii)	300	–	0.7	–	–	–	–	–
(iv)	1000	350	–	–	–	–	–	–
(v)		200	–	–	–		0.9	–
(vi)		250	–	–	–	5.9825	–	–

► **Data:**

(i) $p = 200$ kPa, $t = 30°$C.

► **Solution:**

From Table 10.1, $t_s|_{p=200 \text{ kPa}} = 120.23°$C. As $t < t_s$, the water must be in sub-cooled state

***Method 1*:**

In the compressed-liquid region, the properties of water may be assumed to have the saturation-liquid properties corresponding to the given temperature of 30°C.

Hence, from the temperature-based saturated-property table (Table 10.2), at $t = 30°C$,

$$v = v_f = 0.001004 \text{ m}^3/\text{kg}$$
$$h = h_f = 125.77 \text{ kJ/kg}$$
$$u = h_f - pv_f$$
$$= 125.77 - 200 \times 0.001004$$
$$= 125.57 \text{ kJ/kg}$$
$$s = s_f|_{t\,=\,30°C} = 0.4369 \text{ kJ/kg K.}$$

***Method 2*:**

To find v: In the sub-cooled region, v can be assumed to be $0.001 \text{ m}^3/\text{kg}$.

To find h: Shown in the T–s diagram of Figure 10.20 are the constant-pressure line, initial state 1, the saturation-liquid and vapour lines, saturated-liquid and vapour states f and g. Theoretically, state f can be reached from state 1 through the constant-pressure process, $1 - f$, in the sub-cooled region. Therefore,

$$h_f - h_1 = c_{\text{wat}} \times (t_f - t_1) = 4.189 \times (t_s - t_1)$$

$$504.68 - h_1 = 4.189 \times (120.23 - 30) \text{ (from Table 10.1, } h_f|_{p=200 \text{ kPa}} = 504.68 \text{ kJ/kg)}$$

Therefore, $h_1 = 126.7$ kJ/kg.

To find 'u' : $u = h - pv = 126.7 - 200 \times 0.001 = 126.5$ kJ/kg.

To find s: Again, for the constant-pressure process $1 - f$,

$$\Delta s = s_f - s_1 = c_{\text{wat}} \ln\left(\frac{T_f}{T_1}\right) = 4.189 \ln\left[\frac{(273 + 120.23)}{(273 + 30)}\right]$$

From Table 10.1, $s_f|_{p=200 \text{ kPa}} = 1.5300$ kJ/kg K.

$$\text{Therefore,} \quad 1.5300 - s_1 = 4.189 \times \ln\left[\frac{(273 + 120.23)}{(273 + 30)}\right]$$

$$\text{Or,} \qquad s_1 = 0.4381 \text{ kJ/kg K.}$$

Here, dryness 'x' and degree of superheating are irrelevant.

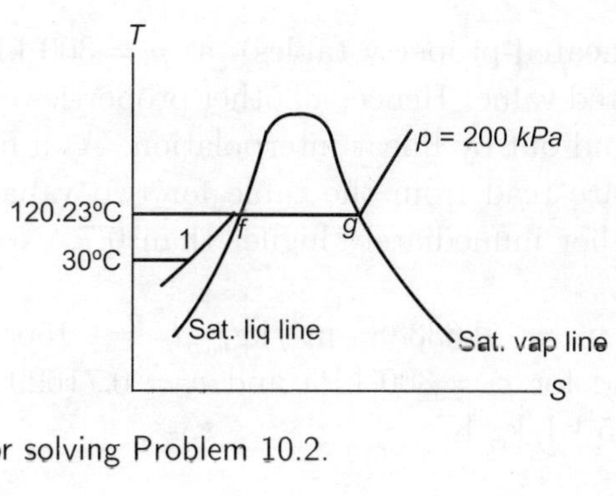

Fig. 10.20 T–S diagram for solving Problem 10.2.

Note: The values of different properties found from both the methods agree fairly well.

▶ **Data:**

(ii) $p = 250$ kPa, $h = 2280$ kJ/kg.

▶ **Solution:**

From Table 10.1, at $p = 250$ kPa, $h_f = 535.34$ kJ/kg and $h_g = 2716.9$ kJ/kg. As $h_f < h < h_g$, water must be in the mixture state. Therefore $t = t_s = 127.43$°C. Dryness can now be found using h values.

$$h = |(1-x)h_f + xh_g|_{p=250 \text{ kPa}}$$

That is, $2280 = (1-x)535.34 + 2716.9x$, or $x = 0.8$

$$v = |(1-x)v_f + xv_g|_{p=250 \text{ kPa}},$$

where v_f and v_g can be got from Table 10.1.

$$= 0.2 \times 0.001067 + 0.8 \times 0.7187 = 0.5752 \text{ m}^3/\text{kg}$$

$$u = h - pv = 2280 - 250 \times 0.5752 = 2136.2 \text{ kJ/kg}$$

And, $s = |(1-x)s_f + xs_g|_{p=250 \text{ kPa}}$,

where s_f and s_g can be got from Table 10.1

$$= 0.2 \times 1.6072 + 0.8 \times 7.0526 = 5.9635 \text{ kJ/kg K}$$

Degree of superheating is irrelevant in the mixture region.

▶ **Data:**

(iii) $p = 300$ kPa, $v = 0.7$ m^3/kg.

▶ **Solution:**

From Table 10.1, at $p = 300$ kPa, $v_f = 0.001073$ m^3/kg and $v_g = 0.6058$ m^3/kg. As $v > v_g$, water must be in superheated state.

From Table 10.3 (superheated-property tables), at $p = 300$ kPa, $v = 0.7$ m^3/kg is not directly available as a listed value. Hence, all other properties corresponding to the given state will have to be found out by linear interpolation. As a first step in this direction, all required properties are read from the table for two values of v, one immediately less than 0.7 and the other immediately higher than 0.7. Now, from the superheated tables, for
$p = 300$ kPa and $v = 0.63388$ m^3/kg, $t = 150$°C, $h = 2761$ kJ/kg, $s = 7.0778$ kJ/kg K and for $p = 300$ kPa and $v = 0.71629$ m^3/kg, $t = 200$°C, $h = 2865.5$ kJ/kg, $s = 7.3115$ kJ/kg K.

To find t: The above values of t and v read from the tables are shown plotted on a $t - v$ plane in Figure 10.21, assuming v varies linearly with t.

From similar triangles, $$\frac{cd}{ad} = \frac{be}{ae}$$

That is, $$\frac{(t - 150)}{(0.7 - 0.63388)} = \frac{(200 - 150)}{(0.71629 - 0.63388)}$$

Therefore $$t = 190°\text{C}.$$

To find h: The known values of t and h are plotted on a $t - h$ plane in Figure 10.22.

Here also, $$\frac{cd}{ad} = \frac{be}{ae}$$

That is, $$\frac{(190 - 150)}{(h - 2761)} = \frac{(200 - 150)}{(2865.5 - 2761)}$$

Therefore, $$h = 2844.6 \text{ kJ/kg}.$$

To find u: u $= h - pv$ = 2844.6-300 × 0.7 = 2634.6 kJ/kg.

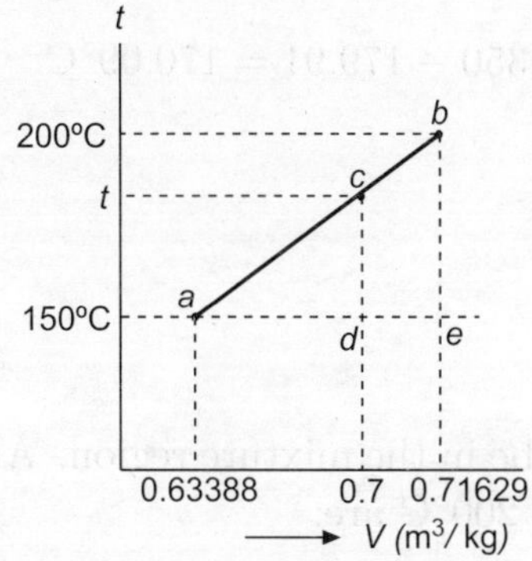

Fig. 10.21 The $t - v$ diagram for interpolation.

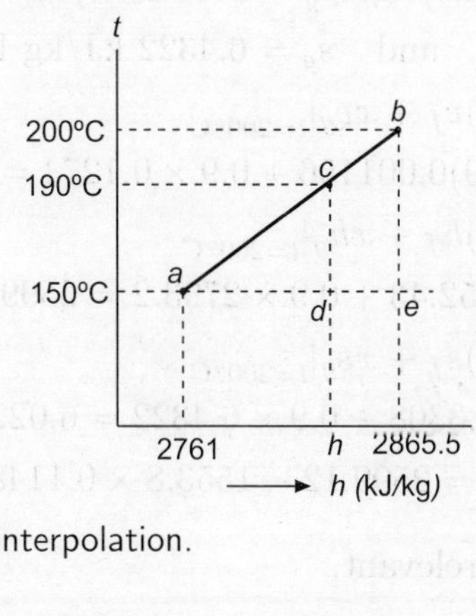

Fig. 10.22 The $t - h$ diagram for interpolation.

To find s: On the same lines as above,

$$\frac{(190 - 150)}{(s - 7.0778)} = \frac{(200 - 150)}{(7.3115 - 7.0778)}$$

Therefore, $s = 7.2648$ kJ/kg K.

Degree of superheating $t_{\text{sup}} - t_{\text{sat}} = 190 - 133.55 = 56.45$ deg.

▶ **Data:**

(iv) $p = 1000$ kPa, $t = 350°$C.

▶ **Solution:**

From Table 10.1, at $p = 1000$ kPa, $t_s = 179.91°$C. As $t > t_s$, the state of water must be in the superheated region. From Table 10.3, for $p = 1000$ kPa (= 1 MPa) and $t = 350°$C, $v = 0.28247$ m^3/kg, $h = 3157.7$ kJ/kg and $s = 7.3010$ kJ/kg K. Now, $u = h - pv = 3157.7 - 1000 \times 0.28247 = 2875.23$ kJ/kg. Dryness is irrelevant. *Degree of superheat* $= t_{\text{sup}} - t_{\text{sat}} = 350 - 179.91 = 170.09°$C.

▶ **Data:**

(v) $t = 200°$C, $x = 0.9$

▶ **Solution:**

As 0 < x < 1, the state must be in the mixture region. All required saturated properties read from Table 10.2, for $t = 200°$C are:

$$p = 1553.8 \text{ kPa}, \ v_f = 0.001156 \text{ m}^3/\text{kg}, \ v_g = 0.1274 \text{ m}^3/\text{kg},$$
$$h_f = 852.43 \text{ kJ/kg}, \ h_g = 2793.2 \text{ kJ/kg}, \ s_f = 2.3308 \text{ kJ/kg K}$$
$$\text{and} \quad s_g = 6.4322 \text{ kJ/kg K}.$$

Now,
$$v = |(1 - x)v_f + xv_g|_{t=200°\text{C}}$$
$$= (1 - 0.9)0.001156 + 0.9 \times 0.1274 = 0.1148 \text{ m}^3/\text{kg}$$

Similarly,
$$h = |(1 - x)h_f + xh_g|_{t=200°\text{C}}$$
$$= 0.1 \times 852.43 + 0.9 \times 2793.2 = 2599.12 \text{ kJ/kg}$$
$$s = |(1 - x)s_f + xs_g|_{t=200°\text{C}}$$
$$= 0.1 \times 2.3308 + 0.9 \times 6.4322 = 6.0221 \text{ kJ/kg K}$$

Also,
$$u = h - pv = 2599.12 - 1553.8 \times 0.1148 = 2420.74 \text{ kJ/kg}$$

Degree of superheating is irrelevant.

▶ **Data:**

(vi) $t = 250°C$, $s = 5.9825$ kJ/kg K.

▶ **Solution:**

From Table 10.2, for $t_s = 250°C$, $s_f = 2.7927$ kJ/kg K and $s_g = 6.0729$ kJ/kg K. As $s_f < s < s_g$, the state of water must be in mixture region. Now, the dryness can be found out from,

$$s = |(1-x)s_f + xs_g|_{t=250°C}$$

picking the values of s_f and s_g from tables,

$$5.9825 = (1-x)2.7927 + 6.0729x$$

Therefore, $x = 0.9724$

Now, $p = p_s$ for $t_s = 250°C$, and, hence, from tables $p = 3.973$ MPa.
Other properties can be evaluated using this value of x and the other listed saturated-state properties in Table 10.2 corresponding to $t = 250°C$.

$$v = |(1-x)v_f + xv_g|_{t=250°C}$$
$$= (1-0.9724) \times 0.001251 + 0.9724 \times 0.0501 = 0.04875 \text{ m}^3/\text{kg}$$

Similarly, $h = |(1-x)h_f + xh_g|_{t=250°C}$

$$= (1-0.9724) \times 1085.34 + 0.9724 \times 2801.5 = 2754.13 \text{ kJ/kg}$$

Now, $u = h - pv = 2754.13 - (3.973 \times 10^3) \times 0.04875 = 2560.4$ kJ/kg

Degree of superheating is irrelevant in the mixture region.
The properties found out in all the above six cases have been shown entered in Table 10.5 in bold characters.

Table 10.5 Answers for Illustrative Problem 10.2

No.	p kPa	t °C	v m^3/kg	h kJ/kg	u kJ/kg	s kJ/kg	x	Deg of superheat
(i)	200	30	**0.001**	**126.7**	**126.7**	**0.4381**	–	–
(ii)	250	**127.43**	**0.5752**	2280	**2136.2**	**5.9635**	**0.8**	–
(iii)	300	**190**	0.7	**2844.6**	**2634.6**	**7.2648**	–	**56.45**
(iv)	1000	350	**0.28247**	**3157.7**	**2875.23**	**7.3010**	–	**170.09**
(v)	**1553.8**	200	**0.1148**	**2599.12**	**2420.8**	**6.0221**	0.9	–
(vi)	**3973**	250	**0.04875**	**2754.13**	**2560.4**	5.9825	**0.9724**	–

10.3 A rigid vessel of volume 0.2 m^3 contains wet steam at a pressure of 1 bar. Given that the ratio of mass of liquid to vapour is $\frac{5}{95}$, find the mass of liquid, vapour and the dryness fraction of steam.

▶ **Data:**

$V = 0.2$ m^3, $p = 1$ bar, steam is wet, $\dfrac{m_f}{m_g} = \dfrac{5}{95}$.

▶ **Solution:**

From Table 10.1, at $p = 1$ bar, $v_f = 0.001043$ m^3/kg and $v_g = 1.694$ m^3/kg.

Now, $$V = V_f + V_g = m_f v_f + m_g v_g$$

Therefore, $$0.2 = 0.001043\ m_f + 1.694\ m_g \tag{a}$$

From data, $$m_f = \frac{5\ m_g}{95} \tag{b}$$

Eliminating m_f between (a) and (b),

$$0.2 = 0.001043\left(\frac{5\ m_g}{95}\right) + 1.694\ m_g$$

That is, $$m_g = 0.118 \text{ kg}$$

Substituting for m_g in (b), $m_f = 0.118 \times \dfrac{5}{95} = 6.2105 \times 10^{-3}$ kg.

Dryness $$x = \frac{m_g}{(m_f + m_g)} = \frac{0.118}{(0.118 + 6.2105 \times 10^{-3})} = 0.95$$

▶ **Alternatively 1:**

From data, $\dfrac{m_f}{m_g} = \dfrac{5}{95}$. Adding 1 to each side of the relation,

$$1 + \frac{m_f}{m_g} = 1 + \frac{5}{95}$$

That is, $$\frac{(m_g + m_f)}{m_g} = \frac{1}{x} = \frac{100}{95}, \text{ or } x = 0.95$$

▶ **Alternatively 2:**

Consider 100 kg wet steam. As $\dfrac{m_f}{m_g} = \dfrac{5}{95}$, every 100 kg wet steam would have 5 kg moisture and 95 kg dry saturated vapour. Therefore,

$$x = \frac{m_g}{(m_f + m_g)} = \frac{95}{(5 + 95)} = 0.95.$$

10.4 (a) A sealed rigid vessel contains 0.1 m^3 of wet steam at 2 bar. By volume, 95% is in the vapour form. Find the dryness fraction of steam. Also find the specific volume. (b) The steam in part (a) is heated until it becomes a saturated vapour. Find the final pressure and temperature; (c) also find the heat transfer.

▶ **Data:**

$$V = 0.1 \text{ m}^3, p = 2 \text{ bar}, V_g = 0.95 \text{ V, and thus } V_f = 0.05 \text{ V, or } \frac{V_g}{V_f} = 19$$

▶ **Solution:**

$$V = 0.1 = V_g + V_f = m_f v_f + m_g v_g$$

From Tables 10.1, at $p = 2$ bar, $v_f = 0.001061 \text{ m}^3/\text{kg}$ and $v_g = 0.8857 \text{ m}^3/\text{kg}$.

Therefore, $$0.1 = 0.001061 m_f + 0.8857 m_g \quad \text{(a)}$$

Also, from data, $$\frac{V_g}{V_f} = 19 = \frac{m_g v_g}{m_f v_f}$$

that is, $$\frac{m_g v_g}{m_f v_f} = \frac{0.8857 m_g}{0.001061 m_f} = 19$$

Therefore $$m_g = 0.02276\, m_f \quad \text{(b)}$$

Eliminating m_g between (a) and (b),

$$0.1 = 0.001061 m_f + 0.8857 \times 0.02276\, m_f, \text{ or } m_f = 4.7126 \text{ kg}$$

Substituting for m_f in (b), $m_g = 0.02276 \times 4.7126 = 0.1072$ kg.

Therefore, $x = \dfrac{m_g}{(m_f + m_g)} = \dfrac{0.1072}{(4.7126 + 0.1072)} = 0.02224$

Specific volume $= \dfrac{V}{(m_f + m_g)} = \dfrac{0.1}{(4.7126 + 0.1072)} = 0.0207 \text{ m}^3/\text{kg}$.

(b) The heating is at constant volume. When it is heated from its initial state 1, as shown in Figure 10.23, it becomes saturated vapour when the constant-volume line intersects the saturated-vapour line at 2. Now, $v_2 = v_1 = 0.0207 \text{ m}^3/\text{kg}$. Also $v_2 = v_g$. Therefore, $v_g = 0.0207 \text{ m}^3/\text{kg}$

A glance through Tables 10.1 and 10.2 reveals that a value of v_g close to $0.0207 \text{ m}^3/\text{kg}$ is available in Table 10.1. When $v_g = 0.02048 \approx 0.0207 \text{ m}^3/\text{kg}$, $p = 9.0$ MPa and $t = 303.4°\text{C}$.

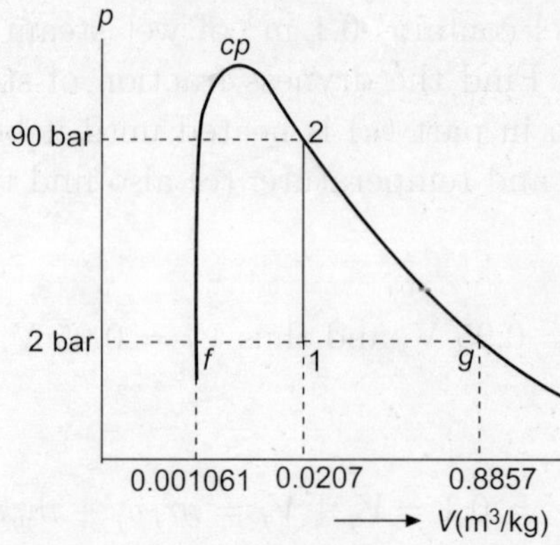

Fig. 10.23 The p–V diagram to solve Problem 10.4.

(c) 1–2 is a constant-volume process. Therefore,

$$Q = \Delta U = m(u_2 - u_1) \tag{i}$$

Now,
$$u_1 = h_1 - p_1 v_1 = [(1-x)h_f + xh_g]_{p=2\text{bar}} - p_1 v_1$$
$$= [(1 - 0.02224)504.68 + 0.02224 \times 2706.6] - 200 \times 0.0207$$
$$= 549.51 \text{ kJ/kg}$$
$$u_2 = h_2 - p_2 v_2 = h_g|_{p_2} - p_2 v$$
$$= 2742.1 - 9000 \times 0.0207 = 2555.8 \text{ kJ/kg}$$
$$m = \frac{V}{v} = \frac{0.1}{0.0207} = 4.83 \text{ kg}$$

Substituting the values for m, u_1 and u_2 in (i),

$$Q = 4.83(2555.8 - 549.51) = 9690 \text{ kJ.}$$

10.5 A vertical cylinder with a freely floating piston contains steam at a pressure of 100 kPa and a dryness of 0.9. The steam is heated until it becomes a saturated vapour. Find the final pressure and temperature and the heat transfer per kilogram steam.

▶ **Data:**

$p_1 = 100$ kPa $= 1$ bar, $x_1 = 0.9$. Heating is at constant pressure. Finally steam is in a saturated vapour state.

▶ **Solution:**

The heating process 1–2 is shown on the T–S diagram in Figure 10.24
There will be no change in temperature as the process happens to be evaporation at constant pressure.
From Table 10.1, at $p = 1$ bar, $t_s = 99.62°\text{C}$, $h_f = 417.44$ kJ/kg and $h_g = 2675.5$ kJ/kg.
Therefore, $t_1 = t_2 = t_s = 99.62°\text{C}$ and $p_1 = p_2 = 1$ bar.

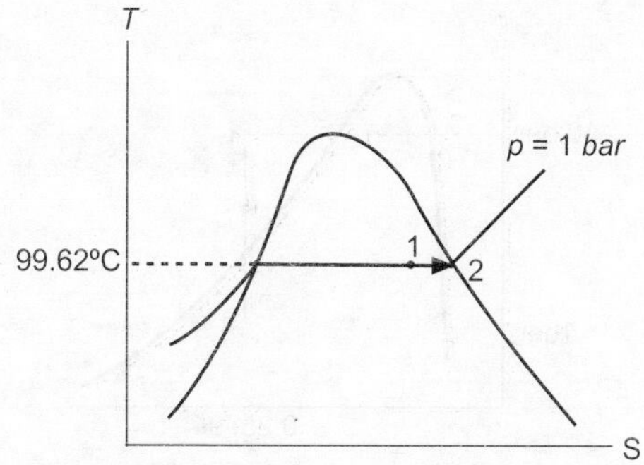

Fig. 10.24 The T–S diagram to solve Problem 10.5.

$$Q = \Delta h = h_2 - h_1$$

Now, $$h_1 = [(1 - x_1)h_f + x_1 h_g]_{p\,=\,1\text{ bar}}$$
$$= [(1 - 0.9) \times 417.44 + 0.9 \times 2675.5] = 2449.7 \text{ kJ/kg},$$

and. $$h_2 = h_g|_{p\,=\,1\text{ bar}} = 2675.5 \text{ kJ/kg}$$

Therefore, $$Q = (2675.5 - 2449.7) = 225.8 \text{ kJ/kg}$$

10.6 A rigid cylinder contains steam at 10 bar and a temperature of 300°C. The steam is cooled until its pressure decreases to 1 bar. Find the final state and the heat transfer per kilogram of steam.

▶ **Data:**

$p_1 = 10$ bar, $t_1 = 300$°C; cooling is at constant volume; $p_2 = 1$ bar.

▶ **Solution:**

From Table 10.1, at $p = 10$ bar, $t_s = 179.91$°C. As the given t is greater than t_s, at state 1, the steam must be in the superheated region as shown in T–S diagram of Figure 10.25. From Table 10.3 (superheated tables), at $p = 10$ bar and $t = 300$°C, $v = 0.25794$ m^3/kg and $h = 3051.2$ kJ/kg. Because the cooling occurs at constant volume, $v_2 = v_1 = 0.25794$ m^3/kg. It is given that $p_2 = 1$ bar.

From Table 10.1, at $p = 1$ bar, $v_f = 0.001043$ m^3/kg and $v_g = 1.694$ m^3/kg.

Hence, as $v_f < v_2 < v_g$, steam must be in a mixture state. The process of cooling is shown by the path 1–2 in Figure 10.25.

x_2 can now be evaluated using the relation,

$$v_2 = [(1 - x_2)v_f + x_2 v_g]_{p\,=\,1\text{ bar}}$$

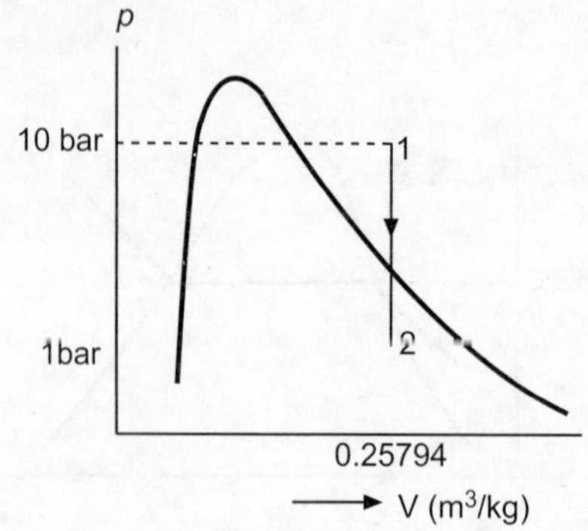

Fig. 10.25 The $p-V$ diagram to solve Example 10.6.

That is, $0.25794 = (1 - x_2)0.001043 + 1.694x_2$

Therefore, $x_2 = 0.1517$

Now, $Q = u_2 - u_1$ (i)

But, $u_1 = h_1 - p_1v_1 = 3051.2 - 1000 \times 0.25794 = 2793.26$ kJ/kg

And,
$$
\begin{aligned}
u_2 &= h_2 - p_2v_2 \\
&= [(1 - x_2)h_f + x_2h_g]_{p\,=\,1\text{ bar}} - p_2v_2 \\
&= [(1 - 0.1517) \times 417.44 + 0.1517 \times 2675.5] - 100 \times 0.25794 \\
&= 734.2 \text{ kJ/kg}
\end{aligned}
$$

Substituting for u_1 and u_2 in (i)

$$Q = 734.2 - 2793.26 = -2059.06 \text{ kJ/kg}$$

10.7 Saturated steam at 10 bar is expanded irreversibly in an adiabatic turbine to a pressure of 0.05 bar. The turbine develops 450 kJ of work per kilogram of steam. Find the temperature of steam at the exit of the turbine if it is superheated or find the dryness otherwise. Neglect changes in kinetic and potential energies.

▶ **Data:**

Initially, steam is in saturated vapour condition, $p_1 = 10$ bar, $p_2 = 0.05$ bar $= 5$ kPa, $W = 450$ kJ/kg and $Q = 0$.

▶ **Solution:**

From the first law for a control volume, neglecting changes in kinetic and potential energies

$$Q - W = \Delta h = h_2 - h_1$$

As $\quad Q = 0, W = h_1 - h_2$

Here, $W = 450$ kJ/kg and from Table 10.1, $h_1 = h_g|_{p=10\text{bar}} = 2778.1$ kJ/kg.

Therefore, $450 = 2778.1 - h_2$, or $h_2 = 2328.1$ kJ/kg.

At 2, $p_2 = 0.05$ bar, and $h_2 = 2328.1$ kJ/kg.

From Table 10.1, at $p = 0.05$ bar, $h_f = 137.79$ kJ/kg and $h_g = 2561.4$ kJ/kg.

As $h_f < h_2 < h_g$, at state 2, steam must be in the mixture region.

The initial and final states are labelled 1 and 2, respectively, on $p - v$ coordinates in Figure 10.26. The path of the process is not known as the process is given to be irreversible. Notionally it is shown by the broken line.

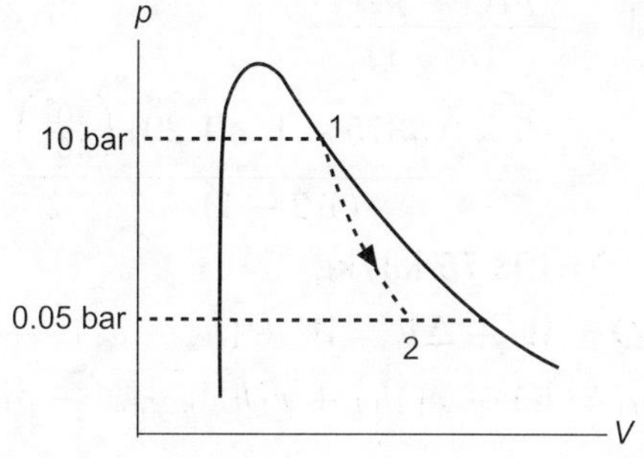

Fig. 10.26 The p– V diagram to solve Problem 10.7.

x_2 can be calculated from,

$$h_2 = [(1 - x_2)\, h_f + x_2 h_g]_{p=0.05 \text{ bar}}$$

That is, $$2328.1 = [(1 - x_2)\, 137.79 + 2561.4\, x_2]$$

Therefore, the dryness at exit $= x_2 = 0.904$. Also $t_2 = t_s|_{p\,=\,0.05\text{ bar}} = 32.88°\text{C}$.

10.8 0.1 kg steam at 5 bar, 0.9 dry expands to a pressure of 1bar in a resisted process that follows the path $pv^{1.2}$ = constant. Find work transfer, heat transfer and change in entropy.

▶ **Data:**

$m = 0.1$ kg, $p_1 = 5$ bar, $x_1 = 0.9$, $p_2 = 1$ bar and $pv^{1.2}$ = constant

▶ **Solution:**

From Table 10.1, at $p = 5$ bar, $v_f = 0.001093$ m^3/kg and $v_g = 0.3749$ m^3/kg.

Therefore, $v_1 = [(1 - x_1)v_f + x_1 v_g]_{p\,=\,5\text{ bar}}$

That is, $v_1 = [(1 - 0.9)0.001093 + 0.3749 \times 0.9] = 0.3375\ \text{m}^3/\text{kg}$

From data, $p_1 v_1^{1.2} = p_2 v_2^{1.2}$, and therefore, $v_2 = v_1 \left(\dfrac{p_1}{p_2}\right)^{\frac{1}{1.2}}$

That is, $v_2 = 0.3375 \left(\dfrac{5}{1}\right)^{\frac{1}{1.2}} = 1.29\ \text{m}^3/\text{kg}.$

Now, as it is a resisted process,

$$W = \frac{(p_1 v_1 - p_2 v_2)}{(n - 1)}$$

$$= \frac{(5 \times 0.3375 - 1 \times 1.29)\left(\frac{10^5}{10^3}\right)}{(1.2 - 1)}$$

$$= 198.75\ \text{kJ/kg}$$

From the first law, $Q = W + \Delta u = W + (u_2 - u_1)$ (i)

where $u_1 = [(1 - x_1)h_f + x_1 h_g]_{p=5\text{bar}} - p_1 v_1$

Taking the values of h_f and h_g from Table 10.1,

$$u_1 = [(1 - 0.9)640.21 + 0.9 \times 2748.7] - 5 \times 100 \times 0.3375$$

$$= 2369.1\ \text{kJ/kg}$$

At state 2, $p_2 = 1$ bar and $v_2 = 1.29\ \text{m}^3/\text{kg}$

From Table 10.1, at $p = 1$ bar, $v_f = 0.001043\ \text{m}^3/\text{kg}$ and $v_g = 1.694\ \text{m}^3/\text{kg}$.

As $v_f < v_2 < v_g$, the steam finally must be in mixture region. Therefore,

$$v_2 = [(1 - x_2)v_f + x_2 v_g]_{p=1\text{bar}}$$

That is, $1.29 = (1 - x_2) \times 0.001043 + x_2 \times 1.694$, or $x_2 = 0.761$.

The process of expansion can now be shown on the p–v diagram. From beginning to end the process occurs in the mixture region. The process is shown in Figure 10.27 on the p–v plane.

Now, $u_2 = h_2 - p_2 v_2 = [(1 - x_2)h_f + x_2 h_g]_{p\,=\,1\text{ bar}} - p_2 v_2$

$$= [(1 - 0.761) \times 417.44 + 0.761 \times 2675.5] - 1 \times 100 \times 1.29$$

$$= 2006.8\ \text{kJ/kg}$$

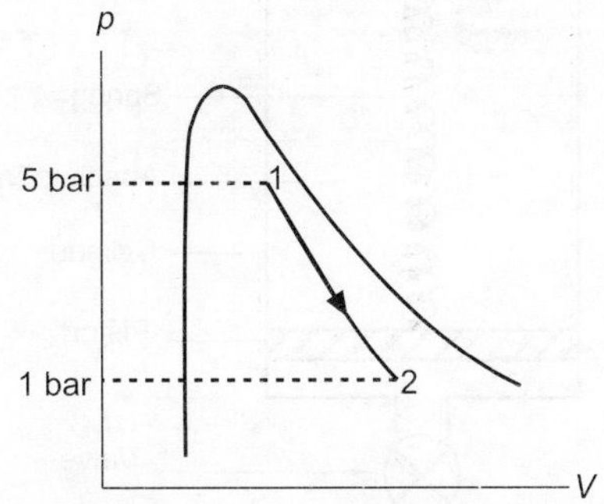

Fig. 10.27 Expansion process of steam on the p– V plane.

Substituting for W, u_1 and u_2 in (i),

$$Q = 198.75 + (2006.8 - 2369.1) = -163.55 \text{ kJ/kg}$$

As the mass of the system is 0.1 kg,

Net $\quad W = 0.1 \times 198.75 = 19.875$ kJ, and

Net $\quad Q = 0.1 \times -163.55 = -16.355$ kJ.

10.9 Steam is flowing in a pipe at 0.8 MPa, 250°C. An insulated cylinder fitted with a frictionless leak–proof piston and a spring as shown in Figure 10.28, is connected to the steam main through a valve. Initially, when the piston is at its bottom-most position, the spring is just touching the piston without exerting any force. The valve is now opened. Steam from the pipe enters the cylinder slowly rising the piston against the spring force. During this process the spring force varies linearly with the piston displacement. Find the temperature of steam inside the cylinder when the pressure reaches 0.8 MPa. Assume that at 0.8 MPa, the specific volume and specific internal energy of superheated steam are related to temperature by $v = 0.133 + 6.3 \times 10^{-4}t$ and $u = 2293 + 1.67t$, respectively, where v is in m^3/kg, u is in kJ/kg and t in °C.

► **Solution:**

When the valve is opened, steam from the main enters the cylinder with a pressure of 0.8 MPa and a temperature of 250°C. The entering steam raises the piston and at any instant the resisting spring force is proportional to the distance through which the piston moves up. The distance moved being proportional to volume, steam pressure inside the cylinder at any instant is proportional to volume as well.

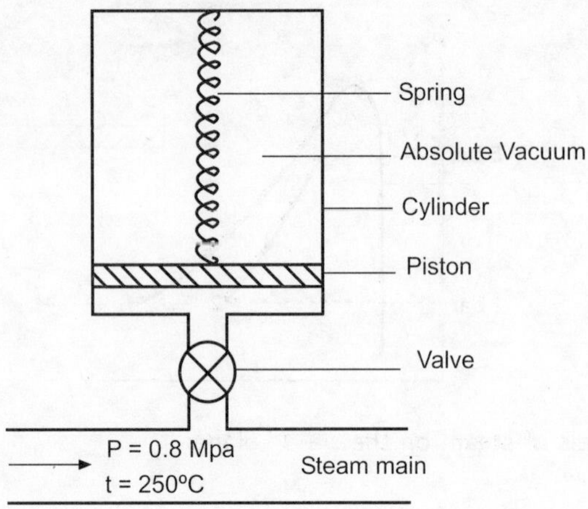

Fig. 10.28 Arrangement for Illustrative Problem 10.9.

Let the volume of the cylinder at the final state be V, and the mass of steam entering it be m kg. The problem can be easily analysed using the system concept. Let us imagine m kg steam which would ultimately enter the cylinder from the steam main as the system. The initial system boundary, s_1, is as shown in Figure 10.29a; all the m kg is in the pipe and nothing in the cylinder. At the end of the process, this m kg steam finds place inside the cylinder and nothing in the steam main. The final system boundary, s_2, will thus be as shown in Figure 10.29b. So, the process can be imagined to be one in which the volume of m kg steam in the main decreases at constant pressure from mv_m to zero, where v_m corresponds to the specific volume of steam in the mains at 0.8 MPa and 250°C. In the meanwhile, the volume of this m kg steam is increasing in the cylinder from zero to V and the pressure is increasing linearly.

Applying the first law to the system,

$$Q - W = \Delta U = m(u_2 - u_1),$$

or, as $$Q = 0, \; W = m(u_1 - u_2), \qquad \text{(i)}$$

where W is the net work interaction for the system, u_1 is the specific internal energy of steam in the main and u_2 is the specific internal energy of steam in the cylinder.

Now, $$W = W_m + W_c \qquad \text{(ii)}$$

where W_m is the work done on m kg steam in the main as its volume shrinks from mv_m to zero, and W_c is the work done by the steam as it enters the cylinder under linearly varying pressure to increase its volume from zero to V.

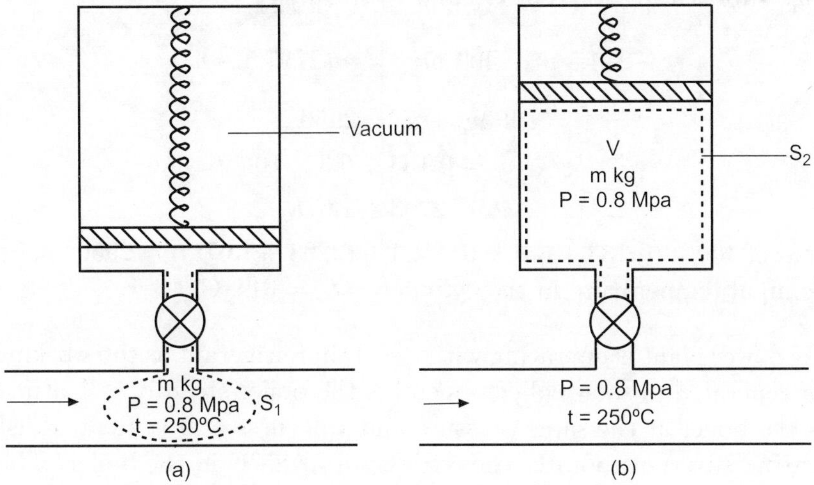

Fig. 10.29 Initial and final system boundaries for Problem 10.9.

Substituting for W in (i) from (ii),

$$(W_m + W_c) = m(u_1 - u_2) \quad \text{(iii)}$$

Now, $$W_m = \int p\mathrm{d}V = p_m \int \mathrm{d}V = p_m(0 - mv_m).$$

Here $p_m = 0.8$ MPa $= 800$ kPa, and $v_m = v|_{p=0.8\ \mathrm{MPa}}$ and $t = 250°\mathrm{C} = 0.29314\ \mathrm{m^3/kg}$,

Therefore, $$W_m = 800(-0.29314)m = (-234.5m)\ \mathrm{kJ}$$

Now again, $$W_c = \int p_c \mathrm{d}V = \int kV\mathrm{d}V \quad (\text{as } p_c \text{ is proportional to } V).$$

$$= k\left[\frac{V^2}{2}\right] \text{(because, final volume} = V \text{ and initial volume} = 0)$$

Finally, when volume in cylinder $= V$, $p = 800$ kPa, and hence, $k = \dfrac{p}{V} = \dfrac{800}{V}$. Substituting for k in the above equation,

$$W_c = \left(\frac{800}{V}\right)\left(\frac{V^2}{2}\right) = 400\ V = 400\ V_c = 400(mv_c)$$

where v_c is the specific volume of steam at 800 kPa and t_c, the final temperature of steam in the cylinder,.

Also, $u_1 = u|_{p\,=\,800\ \mathrm{kPa},\ t\,=\,250°\mathrm{C}} = (h - pv)_{p\,=\,800\ \mathrm{kPa},\ t\,=\,250°\mathrm{C}}$

$$= 2950.0 - 800 \times 0.29314 = 2715.5\ \mathrm{kJ/kg}$$

As $u_2 = u_c$, substituting for W_m, W_c and u_1 in (iii)

$$-234.5\ m + 400\ mv_c = m(2715.5 - u_c)$$

that is $$400v_c + u_c = 2950$$

From data, $$v_c = 0.133 + 6.3 \times 10^{-4}t_c,$$

$$u_c = 2293 + 1.67t_c$$

and therefore, $400 \times (0.133 + 6.3 \times 10^{-4}t_c) + (2293 + 1.67t_c) = 2950$

Therefore, final temperature in the cylinder $= t_c = 314°\text{C}$.

10.10 A solar power plant is operating with a certain refrigerant as the working substance. Steam generated in solar collectors enters the boiler steadily at 1 atm, 0.9 dry and leaves the boiler at the same pressure and a dryness of 0.1. Heat transfer from the condensing steam evaporates the refrigerant at 85°C in the boiler. The refrigerant vapours are then expanded in a turbine. The condenser temperature in the cycle is 35°C. The surroundings are at a temperature of 27°C. Assume the efficiency of the cycle is equal to the theoretical maximum value. Find the following per kilogram of steam: (a) Available and unavailable energies of heat from the steam, (b) Increase in unavailable energy because of irreversible heat transfer between the steam and the refrigerant, (c) Percentage of available energy in the condensing steam converted to work, (d) Increase in heat rejection over unavailable energy of steam, (e) Discuss whether the increase in unavailable energy and the increase in heat rejection are the same or not? (f) Show that the increase in unavailable energy is equal to $T_o\Delta S_{\text{net}}$ where T_o is the surroundings temperature and ΔS_{net} is the increase in entropy of the universe.

▶ **Data:**

Steam pressure = 1 atm = 101.325 kPa, $x_1 = 0.9$, $x_2 = 0.1$
Temperature of evaporation of the refrigerant = 85°C = 358 K
Surroundings temperature = T_o = 27°C = 300 K
Condensing temperature = heat rejection temp = 35°C = 308 K

▶ **Solution:**

(a) $t_s|_{p\,=\,1\text{ atm}\,=\,101.325\text{ kPa}} = 100°\text{C} = 373\text{ K}$. Hence, in the boiler, steam is condensing at 373 K. Refer to Figure 10.30; here, B–A is the condensation process of steam. Thus, the area below A–B must be equal to Q, the heat transfer from 1 kg steam.

Therefore, $Q = \text{area ABCDA} = (h_1 - h_2) = (x_1 - x_2)\,h_{fg} = (x_1 - x_2)\,(h_g - h_f)$
where, x_1 and x_2 are, respectively, the dryness of steam at entry and exit of the boiler and h_f and h_g are, respectively, the saturated-liquid and saturated-vapour enthalpies at 101.325 kPa.

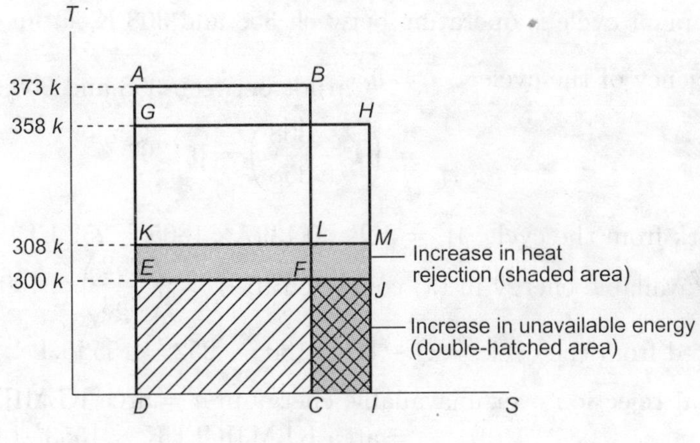

Fig. 10.30 The T–S diagram to solve Problem 10.10.

Picking the values of these enthalpies from Table 10.1 and substituting them in the above relation,

$$Q = (0.9 - 0.1)(2675.7 - 419.07) = 1805 \text{ kJ/kg}.$$

By definition, available energy of $Q = Q \times$ efficiency of the Carnot cycle operating between 373 K and 300 K.

$$\text{Therefore, available energy of } \quad Q = 1805 \times \left(1 - \frac{300}{373}\right) = 353 \text{ kJ}.$$

$$\text{Unavailable energy} = Q - \text{available energy} = \text{area EFCDE}$$
$$= 1805 - 353 = 1452 \text{ kJ}.$$

(b) Heat is transferred from the steam to the refrigerant at 358 K. There is a finite temperature difference. Hence, the heat transfer occurs irreversibly. However, the magnitude of heat transferred to refrigerant $= Q = 1805$ kJ/kg steam.
In Figure 10.30, G–H represents the heat transfer process to the refrigerant. Thus,

$$Q = \text{area GHIDG} = 1805 \text{ kJ}.$$

$$\text{Therefore, available energy of } Q = 1805 \times \left(1 - \frac{300}{358}\right) = 292 \text{ kJ}.$$

Unavailable energy $= Q -$ available energy $=$ area EJIDE $= 1805 - 292 = 1513$ kJ.

$$\text{Increase in unavailable energy} = \text{unavailable energy in (b)} - \text{unavailable energy in (a)}$$
$$= \text{area EJIDE} - \text{area EFCDE}$$
$$= 1513 - 1452 = 61 \text{ kJ/kg steam}$$
$$= \text{area FJICF}$$

(c) The power plant cycle is operating between 358 and 308 K. From data,

$$\text{efficiency of the cycle,} \quad \eta = \eta_{\text{Carnot between 358 and 308 K}}$$
$$= \left(1 - \frac{308}{358}\right) = 0.1397$$

Therefore, work from the cycle, $W = \eta Q = 0.1397 \times 1805 = 252$ kJ.

Percentage of available energy in (a) converted to $W = \left(\frac{252}{353}\right) \times 100 = 71.4\%$.

(d) Heat rejected from the cycle $= Q - W = 1805 - 252 = 1553$ kJ.

Increase in heat rejection over unavailable energy in a = area KLMIDK–EFCDE
= area KLMJICFEK = 1553–1452 = 101 kJ

(e) From (b), increase in unavailable energy = 61 kJ

From (d), increase in heat rejection = 101 kJ. These two are not the same.

(f)

$$\Delta S_{\text{net}} = \Delta S_{\text{steam}} + \Delta S_{\text{refrigerant}} + \Delta S_{\text{surroundings}}$$

While 1805 kJ is transferred from steam at 373 K, 1553 kJ is transferred to the surroundings at 308 K. There is no change in the entropy of the refrigerant as it undergoes a cycle.

Therefore, $$\Delta S_{\text{net}} = -\frac{1805}{373} + 0 + \frac{1553}{308} = 0.203 \text{ kJ/kg steam K.}$$

Hence, $$T_0 \Delta S_{\text{net}} = 300 \times 0.203 = 61 \text{ kJ/kg steam.}$$
$$= \text{Increase in unavailable energy in (b)}$$

10.11 A throttling calorimeter receives a sample of steam from a boiler at 1 MPa. After throttling the pressure and temperature are measured as 100 kPa and 115°C. Find the dryness of the boiler steam using (a) steam tables and (b) Mollier diagram.

▶ **Data:**

$p_1 = p_{\text{boiler}} = 1 \text{ MPa} = 1000 \text{ kPa}$, $p_2 = 100$ kPa, $t_2 = 115°\text{C}$

▶ **Solution:**

(a) After throttling, $p_2 = 100$ kPa, $t_2 = 115°\text{C}$

Now, from Table 10.1, $T_s|_{p=100 \text{ kPa}} = 99.62°\text{C}$.

As $t_2 > t_s$, the steam is in superheated condition after throttling.

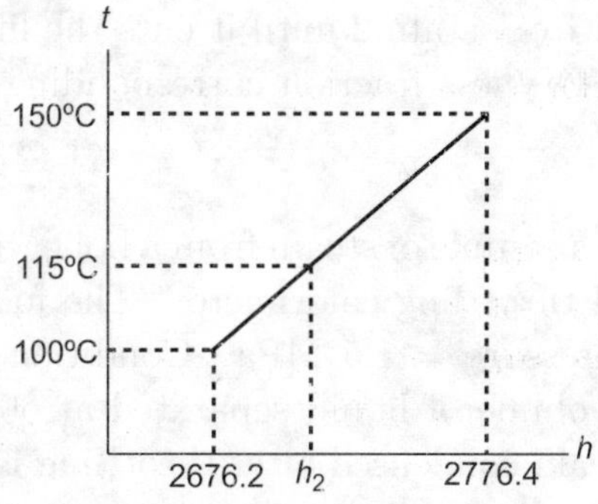

Fig. 10.31 The t–h diagram for linear interpolation.

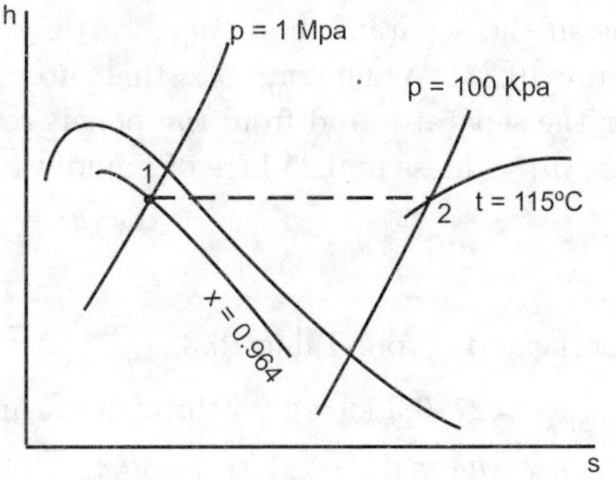

Fig. 10.32 Mollier diagram for dryness.

From superheated-property table at $p = 100$ kPa,

$$\text{when} \quad t = 100°\text{C}, \quad h = 2676.2 \text{ kJ/kg}, \quad \text{and}$$
$$\text{when} \quad t = 150°\text{C}, \quad h = 2776.4 \text{ kJ/kg}$$

Therefore h_2 at $t_2 = 115°$C can be found out by linear interpolation with the help of t–h diagram in Figure 10.31

$$\frac{(h_2 - 2676.2)}{(115 - 100)} = \frac{(2776.4 - 2676.2)}{(150 - 100)}$$

and hence, $h_2 = 2706.2$ kJ/kg.
Because steam is throttled from state 1 to state 2, $h_2 = h_1$. Let x_1 be the dryness of boiler steam. Then,

$$h_2 = \left[(1 - x_1)\, h_f + x_1 h_g\right]_{p=100 \text{ kPa}}$$
$$2706.2 = \left[(1 - x_1)\, 762.79 + 2778.1 x_1\right], \quad \text{or } x_1 = 0.964$$

(b) State 2 is fixed on the Mollier diagram as the point of intersection of $p_2 = 100$ kPa and $t_2 = 115°$C, as shown schematically in Figure 10.32.

A horizontal line is drawn from state 2 until it cuts the line of $p = 1$ MPa. The point of intersection is state 1. Dryness fraction corresponding to state 1 can now be read as $x_1 = 0.964$.

10.12 The dryness fraction of a sample of steam from a boiler drum is found experimentally using a separating and throttling calorimeter. The following are the experimental observations. Boiler pressure = 1.5 MPa. Condition of steam after throttling is 0.1 MPa and 120°C. Volume of liquid separated at 60°C during the experimental period is 0.25 litre. Steam condensed after throttling is 3.5 kg. Find the dryness of boiler steam.

▶ **Data:**

The processes occurring in the separator and the throttling device are shown plotted on h–s diagram in Figure 10.33. While state 3 is that after throttling, states 2 and 1 are those of steam after the separator and from the boiler, respectively. $p_3 = 100$ kPa, $t_3 = 120°\text{C}$, $m_3 = 3.5$ kg, m_1 = mass of 0.25 litre of liquid water at 60°C and $p_2 = p_1 =$ 1.5 MPa.

▶ **Solution:**

At state 3, steam is superheated. From Table 10.3,

$$h_3|_{p=100\text{ kPa},t=120°\text{C}} = 2716.3\text{ kJ/kg} \quad \text{(through linear interpolation)}$$
$$h_3 = h_2 = [(1 - x_2)\,h_f + x_2 h_g]_{p=1.5\text{ Mpa}}$$

Substituting the values of h_f and h_g from Table 10.1,

$$2716.3 = [(1 - x_2)\,844.87 + 2792.1x_2]$$

Therefore, the dryness of steam leaving the separator $= x_2 = 0.961$.

The dryness fraction of boiler steam is given by, $x_1 = \dfrac{x_2 m_3}{(m_1 + m_3)}$

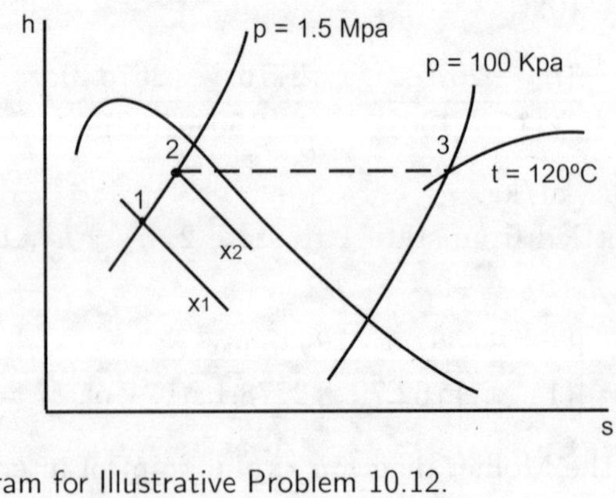

Fig. 10.33 The h–s diagram for Illustrative Problem 10.12.

where, $m_3 = 3.5$ kg and $m_1 = \dfrac{V}{(v)_{t_s} = 60°\text{C}} = \dfrac{(0.25 \times 10^{-3})}{0.001017} = 0.246$ kg.

Therefore, $x_1 = \dfrac{0.961 \times 3.5}{(0.246 + 3.5)} = 0.898.$

Problems

10.1 For each of the states of water given in Table 10.6, fill in the blanks in the table with property values. Insert a dash in the space if any property is to be irrelevant. Use the property tables given in this book.

Table 10.6 Data pertaining to Problem 10.1

No.	p kpa	t °C	v m³/kg	u kJ/kg	h kJ/kg	s kJ/kg K	x	Deg of superheat
1	100	25	—	—	—	—	—	—
2	250	—	—	—	480	—	—	—
3	300	—	—	—		1.5	—	—
4	1000	350	—	—	—	—	—	—
5		120	0.75	—	—	—	—	—
6		250	0.474	—	—	—	—	—
7	600		0.292	—	—	—	—	—
8	900	—	—	—	2773.9	—	—	—
9	700	—	—	—	2500	—	—	—
10	200	—	—	—	3300	—	—	—
11	500	—	—	—	—	1.5700		—
12	600	—	—	—	—	8.00	—	—
13	1200	—	—	2604	—	—	—	—
14		300	—	—	—	—	0.9	—
15	10^4	—	—	—	—	—	0.85	—
16	500	—	—	—	—	—	—	98.14

10.2 A vessel contains wet steam at a pressure of 2 bar, 12% of whose mass is liquid. Find the following: (a) temperature, (b) dryness fraction, (c) specific volume, (d) specific enthalpy and (e) specific internal energy.

10.3 A vessel of one cubic metre contains wet steam at a pressure of 2 bar, 12% of whose volume is liquid. Find the following: (a) temperature (b) dryness fraction (c) specific volume and (d) specific enthalpy.

10.4 Find the properties of steam in a vessel of volume one cubic metre given that the ratio of volumes of liquid to vapour is 10^{-3}.

10.5 A tank having a volume of 0.05 m^3 contains 80% saturated steam and 20% saturated water at 25°C. What is the quality of steam in the tank? Also find the masses of liquid and vapour.

10.6 It is found that wet steam at 150 kPa when heated in a rigid vessel of one cubic metre volume passes through the critical state. Find the mass of the steam and its initial dryness.

10.7 Ten kilogram of steam at 12 bar, 0.9 dryness is heated at constant pressure to 250°C. Find the heat transfer.

10.8 Water at 15 bar, 25°C is heated at constant pressure until it became wet steam with a dryness of 0.75. Find the heat transfer per kilogram.

10.9 Steam at 15 bar and 0.8 dryness is heated at constant pressure. Find the final state temperature, if the heat transfer is (a) 150 kJ/kg and (b) 500 kJ/kg.

10.10 A certain mass of steam at 1 bar, 0.273 dryness is heated in a rigid vessel until it becomes saturated vapour. Find the final temperature and heat transfer per kilogram steam.

10.11 A steam boiler has a total volume of 4 m^3. The boiler initially contains 3 m^3 of liquid water and 1 m^3 of vapour in equilibrium at 0.1 MPa. The boiler is fired up and heat is transferred to the water and steam in the boiler. The relief valve opens when the pressure reaches 5 MPa. How much heat was transferred to water and steam in the boiler before the relief valve just opened.

10.12 Consider the arrangement shown in Figure 10.34. When the piston is resting on the bottom stops the enclosed volume is 0.4 m^3. The enclosed volume is 0.6 m^3 when the frictionless piston encounters the top stops. Initially the cylinder is filled with water at 0.1 MPa and a dryness of 20%. The steam is heated until it becomes saturated. The mass of the piston is such that it exerts a pressure of 200 kPa. Atmospheric pressure is 1 bar. Determine (a) the final pressure in the cylinder and (b) heat and work interactions during the process.

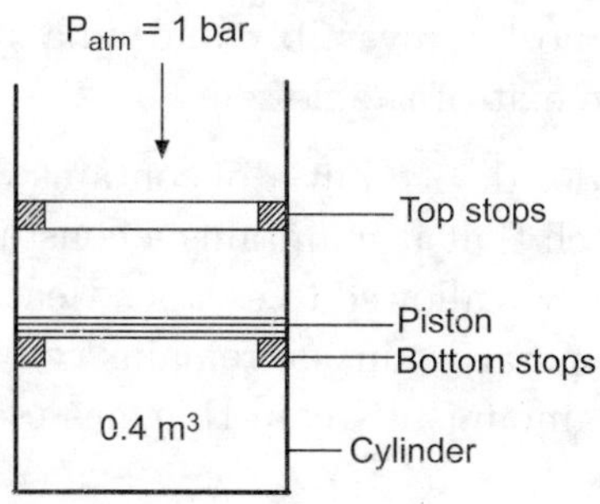

Fig. 10.34 Figure for Problem 10.12.

10.13 A piston–cylinder device contains 1 kg steam at 10 bar, 250°C. The steam is compressed by the motion of the piston inwards until the pressure increases to 20 bar. The work done on the steam is 610 kJ and the heat transfer from the steam is 890 kJ. Find the final temperature if the steam is superheated, otherwise find its dryness.

10.14 Saturated steam of mass 0.12 kg expands reversibly and polytropically in a cylinder from 10 to 1 bar, 0.17 m^3. Find the index of expansion and the heat transfer during the process. Sketch the process on a $T-s$ diagram. Find also the change in entropy of the steam.

10.15 Steam is compressed reversibly and adiabatically from 1 bar, 0.9 dryness to 10 bar. Determine (a) the final temperature (b) the increase in specific internal energy and (c) the increase in specific enthalpy.

10.16 Superheated steam at 10 bar, 400°C is expanded reversibly and adiabatically to 0.1 bar. Consider steam as the system and calculate the work per kilogram. If the process were to follow the relation pV^n = constant, find the value of n.

10.17 A number of years ago, a well-known architect designed a 1600 m tall building. Suppose that in such a building steam for the heating system enters a pipe at ground level as saturated vapour at 200 kPa. On the top floor of the building, the pressure in the pipe is 100 kPa, and the heat transfer from the steam as it flows up the pipe is 100 kJ/kg. What is the quality of the steam at the top floor pipe?

10.18 Steam enters a turbine with negligible velocity at 5000 kPa, 600°C, and leaves at 3 kPa, 5% moisture, with a velocity of 200 m/s. The flow rate is 25 000 kg/h and the power output is 7500 kW. The cross-sectional area of the exhaust opening is 1.51 m^2. Determine the heat loss (in kJ/kg) from the steam passing through the turbine.

10.19 (a) A pipe in which steam at 0.8 MPa, 250°C is flowing at a rate of 1 kg/s is connected to another pipe carrying steam at 0.8 MPa and 0.9 dryness. The two streams mix adiabatically and then flow through a common pipe of diameter 40 mm at a rate of 2.2 kg/s. Find the condition of steam after mixing. (b) The mixture is then expanded in a nozzle reversibly and adiabatically to 0.4 MPa. Find the velocity of steam at the exit of the nozzle.

10.20 A vertical piston–cylinder device initially contains 0.3 m^3 of steam at 250°C. The mass of the piston is such that it maintains a constant pressure of 300 kPa. Now a valve is opened and steam is allowed to escape. Heat transfer takes place during the process so that the temperature inside remains constant. If the final volume is 0.1 m^3, determine (a) the amount of steam that has escaped and (b) the mangnitude of heat transfer.

10.21 A frictionless piston, with no piston rod, made out of an insulating material divides a horizontal cylinder into two equal halves, each having a volume of 0.1 m^3. The left side of the cylinder contains some air at 40°C. The right side contains wet steam at 90°C and a dryness of 10%. The entire cylinder is insulated except the end plate on the right side. Steam is slowly heated through the conducting end plate until it becomes saturated at 150°C. Find the final temperature of air and also the magnitude of heat added to steam.

10.22 A cylinder of 250 mm diameter with a frictionless piston contains 0.02 kg steam at 6 bar, 200°C. As the piston moves out slowly a distance of 350 mm the steam undergoes a resisted expansion process according to pV^n = constant. The final pressure is 1 bar. Determine, (a) temperature of steam at the end of expansion, (b) the value of n, (c) work transfer and (d) heat transfer.

10.23 An insulated rigid vessel contains 2.2 kg water at 1 bar, 25°C. A mass of 0.2 kg steam at 1 bar, 0.8 dryness is added to the water. After some time the temperature of the contents of the vessel become uniform. Determine the final temperature and the increase in entropy.

10.24 Steam at a pressure of 20 bar and a temperature of 250°C enters steadily a turbine with negligible velocity. The steam leaves the turbine at a pressure of 0.15 bar with a velocity of 200 m/s. The rate of heat loss from the turbine casing is 160 kW. The power developed by the turbine is 3430 kW. The mass flow rate of steam is 6.1 kg/s. Determine the dryness of steam at the turbine exit and the exit cross–sectional area.

10.25 A 0.1 m^3 rigid tank initially contains saturated water vapor at 120°C. The tank is connected by a valve to a supply line that carries steam at 1 MPa and 300°C. The valve is opened, and steam is allowed to enter the tank. Heat transfer takes place with the surroundings such that the temperature in the tank remains constant at 120°C. The valve is closed when it is observed that one–half of the volume of the tank is occupied by liquid water. Determine (a) the final pressure in the tank, (b) the amount of steam that has entered the tank, and (c) the magnitude of heat transfer.

10.26 Air is heated steadily in a heat exchanger by condensing steam. Air at 1 atm and 10°C enters the heat exchanger at a flow rate of 50 m^3/min. Saturated steam enters the heat exchanger at 200 kPa and leaves as saturated liquid at the same pressure. The exit temperature of air is 30°C. Calculate the entropy change of the universe per kilogram steam.

10.27 Two insulated, interconnected vessels A and B, each one cubic metre in volume, initially contain saturated liquid at 120°C and steam at 0.2 MPa, 200°C, respectively. The valves separating the two vessels are now opened. The contents are allowed to mix and reach equilibrium condition. Find the final pressure. Also find the final quality or temperature depending upon whether the contents are in mixture condition or in superheated condition.

10.28 A throttling calorimeter sampling the steam generated by a boiler at a pressure of 800 kPa gives readings of 1 bar, 116°C. Evaluate the dryness of boiler steam.

10.29 The dryness fraction of steam from a boiler at 1.0 MPa, found experimentally using a separating and throttling calorimeter is 0.91. The pressure, temperature and mass of steam condensed after throttling are, respectively, 1 bar, 125°C and 4.0 kg. In the same duration, what would be the mass of liquid separated in the separator?

REAL GAS

11

11.1 Introduction

Chapter 9 deals with the ideal gas: this term refers to a gas that follows the equation of state $pV = mRT$ at all ranges of pressure and temperature. An ideal gas is only a mathematical model, and its equation of state is the simplest and the best-known relation for many gases. No real gas actually follows the ideal gas equation of state at all ranges of pressure and temperature. However, real gases, especially the gases of low molar mass, tend to behave as ideal gases either at low pressures or high temperatures. Thus, the applicability and accuracy of the ideal gas equation of state to real gases is limited to a judiciously selected range of pressures and temperatures. This range is not the same for all real gases. The ideal gas model fails to be true for real gases near the critical state and close to the saturation region. It can be conclusively proved from measured properties of real gases that most real gases behave as ideal gases, to an acceptable accuracy, at pressures very much lower than the critical pressure or at temperatures higher than the critical temperature. There is, thus, a need to develop, if possible, equations of state to describe the behaviour of real gases at all ranges and also to find various thermodynamic properties with the help of these newly developed equations of state and the two property rule. This chapter introduces briefly a few commonly used equations of state for a real gas.

11.2 Real Gas—Compressibility Factor

It is now clear that a real gas is one that does not follow the ideal gas equation of state. The initial attempts to find an alternative equation of state for a real gas concentrated on suitably modifying the ideal gas equation. The deviation from ideal gas behaviour at a given temperature and pressure can be accurately predicted by the introduction of a correction factor called the *compressibility factor*.

This factor denoted by Z, is defined as

$$Z = \frac{pv}{\mathrm{R}T} \tag{11.1}$$

The ideal gas equation of state is derived from the kinetic theory of gases which assumes that the intermolecular forces of attraction and repulsion are negligible and that the size of the molecules themselves are so small that they are neglected in comparison to the volume occupied by the gas. There will, thus, be a difference between the actual volume of a real gas and the ideal volume calculated based on the above assumptions. Equation (11.1) can be rearranged and written as $Z = \frac{v}{\left(\frac{\mathrm{R}T}{p}\right)}$. From the ideal gas equation, $\frac{\mathrm{R}T}{p} = v_{ideal}$. Therefore, Z can also be defined as

$$Z = \frac{v_{actual}}{v_{ideal}} \tag{11.2}$$

For real gases, Z can vary over a wide range from values less than one to greater than one. Illustrated in Figure 11.1 is a typical compressibility chart for a real gas. It shows the variation of Z with pressure for various temperatures. In this figure, each curve corresponds to one particular temperature. As temperature increases, the curve moves upwards or becomes flatter. Other features of this chart are:

- All isotherms up to a temperature of $2.5T_c$ approach $z = 1$ with negative slopes.
- The isotherm corresponding to $T = 2.5T_c$ approaches $z = 1$ with zero slope.
- All other isotherms corresponding to $T > 2.5T_c$ approach $z = 1$ with positive slopes.
- The isotherm $T = 5T_c$ has the maximum positive slope

Figure 11.1 shows that Z for any temperature tends to one as, the pressure tends to zero. Thus, at very low pressures, a real gas tends to behave as an ideal gas. Also, while at low pressures and moderate temperatures the value of Z is less than one, it is greater than one at high pressures. At low pressures, the molecules are strewn apart and, hence, the intermolecular forces of attraction are predominant so as to make the actual volume less than the ideal volume. Accordingly, in conformation with (11.2) the value of Z becomes less than one. At high pressures, as the molecules are closely packed, the intermolecular forces of repulsion are predominant and hence the actual volume tends to be greater than the ideal volume, leading to higher values of Z.

As can be seen from the compressibility chart, Z is a function of p and T. It is usually empirically determined, although principles of classical as well as statistical thermodynamics can be used in correlating Z with other properties. The value of Z at any state of a gas is a measure of its deviation from ideal gas behaviour. Also, it is a fact that at any given pressure and temperature the value of Z is different for different gases. Hence, to use Z as a correction factor, the compressibility chart must be made ready for each gas or gas mixture in question. It is not just that the generation of such data is very tedious but also such a data would be too voluminous and can never be exhaustive.

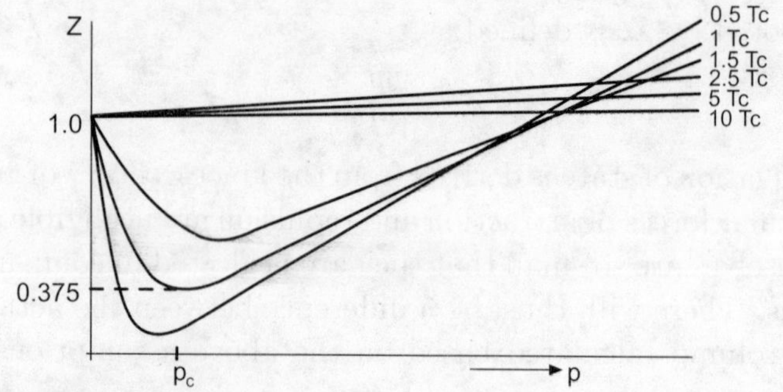

Fig. 11.1 A typical compressibility chart.

Example 11.1

Calculate the specific volume of propane (C_3H_8) at a pressure of 70 bar and a temperature of 150°C, given that the compressibility factor is equal to 0.55. Also, calculate the ideal gas specific volume.

▶ **Solution:**

As $pv = Z\mathrm{R}T, v = \dfrac{Z\mathrm{R}T}{p}$

$$\text{For propane,} \quad \mathrm{R} = \frac{\mathcal{R}}{M} = \frac{8314.3}{44} = 188.96 \text{ J/kg K}$$

$$\text{Therefore,} \quad v = 0.55 \times 188.96 \times \frac{(273+150)}{70 \times 10^5} = 6.28 \times 10^{-3} \text{ m}^3/\text{kg}$$

$$v_{idealgas} = \frac{\mathrm{R}T}{p} = 188.96 \times \frac{(273+150)}{70 \times 10^5} = 11.418 \times 10^{-3} \text{ m}^3/\text{kg}$$

11.3 Reduced Properties---Law of Corresponding States

The major disadvantage of any compressibility chart is that a separate chart is required for each gas. It would be very advantageous and convenient for computations if it is possible to construct a single compressibility chart that could be used for all gases. One method of doing this is based on the *law of corresponding states.* The statement of this law needs the knowledge of reduced properties or reduced coordinates.

The ratios of pressure, temperature and specific volume of a real gas to the corresponding critical values are called reduced properties. Thus,

$$p_{\mathrm{R}} = \frac{p}{p_c}, T_{\mathrm{R}} = \frac{T}{T_c} \text{ and } v_{\mathrm{R}} = \frac{v}{v_c}, \tag{11.3}$$

where, the subscript R denotes a *reduced* property and the suffix c denotes a property at the *critical state.* The reduced properties are dimensionless numbers. Values of p_c, T_c

Table 11.1 Gas constant and critical properties of some selected gases

Gas	Gas Constant R in kJ/kg K	Critical Properties T_c in K	p_c in MPa	V_c in m^3/kmol
Air	0.2870	132.5	3.77	0.0883
Ammonia	0.4882	405.5	11.28	0.0724
Argon	0.2081	151.0	4.86	0.0749
n-Butane	0.1430	425.2	3.80	0.2547
Carbon dioxide	0.1889	304.2	7.39	0.0943
Carbon monoxide	0.2968	133.0	3.50	0.0930
Chlorine	0.1173	417.0	7.71	0.1242
CCl_2F_2 (R–12)	0.06876	384.7	4.01	0.2179
$CHCl_2F$ (R–21)	0.08078	451.7	5.17	0.1973
Ethane	0.2765	305.5	4.48	0.1480
Ethylene	0.2964	282.4	5.12	0.1242
Helium	2.0769	5.3	0.23	0.0578
Hydrogen	4.1240	33.3	1.3	0.0649
Methane	0.5182	191.1	4.64	0.0993
Nitrogen	0.2968	126.2	3.39	0.0899
Oxygen	0.2598	154.8	5.08	0.0780
Propane	0.1885	370.0	4.26	0.1998
Propylene	0.1976	365.0	4.62	0.1810
Sulphur dioxide	0.1298	430.7	7.88	0.1217
R–134 a	0.08149	374.3	4.067	0.1847
CCl_3F (R–11)	0.06052	471.2	4.38	0.2478
Water	0.4615	647.3	22.09	0.0568

and v_c are available in thermodynamics data tables for several gases. Provided in Table 11.1 are the critical properties of some selected gases.

According to the law of corresponding states, the reduced volume is a function of reduced pressure and temperature.

That is,
$$v_R = f(p_R, T_R) \tag{11.4}$$

and that the function is the same for all gases. This means, any two gases having the same p_R and T_R will have the same reduced volumes. The law of corresponding states is only an approximation and is not adequately accurate over a wide range of engineering work. This fact is further elaborated below.

As $\quad v = \dfrac{ZRT}{p}$ and $v_c = \dfrac{ZRT_c}{p_c}$ equation (11.4) can be rewritten as

$$v_{\mathrm{R}} = \frac{v}{v_c} = \frac{\left(\frac{ZRT}{p}\right)}{\left(\frac{Z_cRT_c}{p_c}\right)} = f(p_{\mathrm{R}}, T_{\mathrm{R}})$$

$$= \left(\frac{Z}{Z_c}\right)\left(\frac{T}{T_c}\right)\left(\frac{p_c}{p}\right) = f(p_{\mathrm{R}}, T_{\mathrm{R}})$$

$$= \left(\frac{Z}{Z_c}\right)\left(\frac{T_{\mathrm{R}}}{p_{\mathrm{R}}}\right) = f(p_{\mathrm{R}}, T_{\mathrm{R}})$$

Therefore, $$\left(\frac{Z}{Z_c}\right) = \frac{f(p_{\mathrm{R}}, T_{\mathrm{R}})}{\left(\frac{T_{\mathrm{R}}}{p_{\mathrm{R}}}\right)} = \Phi(p_{\mathrm{R}}, T_{\mathrm{R}})$$

At low pressures, Z tends to 1 for all gases. Hence, according to the above relation, Z_c is only a function of p_R and T_R and thus Z_c must be the same for all gases. But experimental observations show it is not so. Hence, the law of corresponding states can only be an approximation.

The other interpretation of the law of corresponding states is that all gases at a given reduced state behave identically. In other words, the compressibility factor for all gases at a given p_{R} and T_{R} is the same.

That is, $$Z = f(p_{\mathrm{R}}, T_{\mathrm{R}})$$

Example 11.2

Given that $Z = 1.04$, find the reduced properties for air at 10 bar, 27°C.

▶ Solution:

From Table 11.1, for air, $p_c = 3.77$ MPa, $T_c = 132.5$ K and $v_c = 0.0883$ m^3/k mol.

Therefore, $$p_{\mathrm{R}} = \frac{p}{p_c} = \frac{10 \times 10^5}{3.77 \times 10^6} = 0.26525$$

$$T_{\mathrm{R}} = \frac{T}{T_c} = \frac{(273 + 27)}{132.5} = 2.26415$$

$$v = \frac{ZRT}{p} = 1.04 \times 287 \times \frac{(273 + 27)}{10 \times 10^6} = 0.0895 \text{ m}^3/\text{kg}$$

Hence, $$v_R = \frac{v}{v_c} = \frac{0.0895}{0.0883} = 1.014$$

Example 11.3

At what temperature and pressure would steam behave similar to air at 10 bar and 27°C.

▶ **Solution:**

For air, as calculated in Example 11.2, $p_R = 0.26525$, and $T_R = 2.26415$.

According to the law of corresponding states, gases behave alike at the same reduced states. Therefore, the reduced properties of steam must be equal to those of air.

From Table 11.1, for water, $p_c = 22.09$ bar and $T_c = 647.3$ K.

Therefore,

$$p = p_R p_c = 0.26525 \times 22.09 = 5.85 \text{ bar}$$

$$T = T_R T_C = 2.26415 \times 647.3 = 1465 \text{ K} = 1192°\text{C}$$

11.4 Generalized Compressibility Chart

The law of corresponding states permits to prepare a compressibility chart based on reduced properties. This can be prepared using the measured values on any one convenient and simple gas. Such a chart is valid for any gas according to the law of corresponding states. Figure 11.2 shows the features of such a generalized compressibility chart. While p_R is represented on the x axis, Z is represented on the y axis. Each curve corresponds to one particular T_R. It is drawn to scale approximately. For calculations it is advised to use the charts available in data books on thermodynamics for various ranges of p_R and T_R.

For a given p and T, with the knowledge of p_c and T_c for the given gas, the value of Z can be obtained from the chart. This Z is then used in the real gas relation $pv = ZRT$ to find the value of v.

Instead of p and T, if p and v or T and v are given, it is still possible to evaluate the third property using the generalized compressibility chart. However, this requires a process of some trial and error and thus the method happens to be quite tedious. It is for this reason that constant-v_R lines are also marked on the chart. The v_R marked in the generalized compressibility chart is called the *pseudo-reduced specific volume* and is defined as

$$v_R = \frac{v}{\left(\frac{RT_c}{p_c}\right)}$$

This v_R helps in findings out either T_R or p_R, as the case may be, without resorting to time consuming iterations.

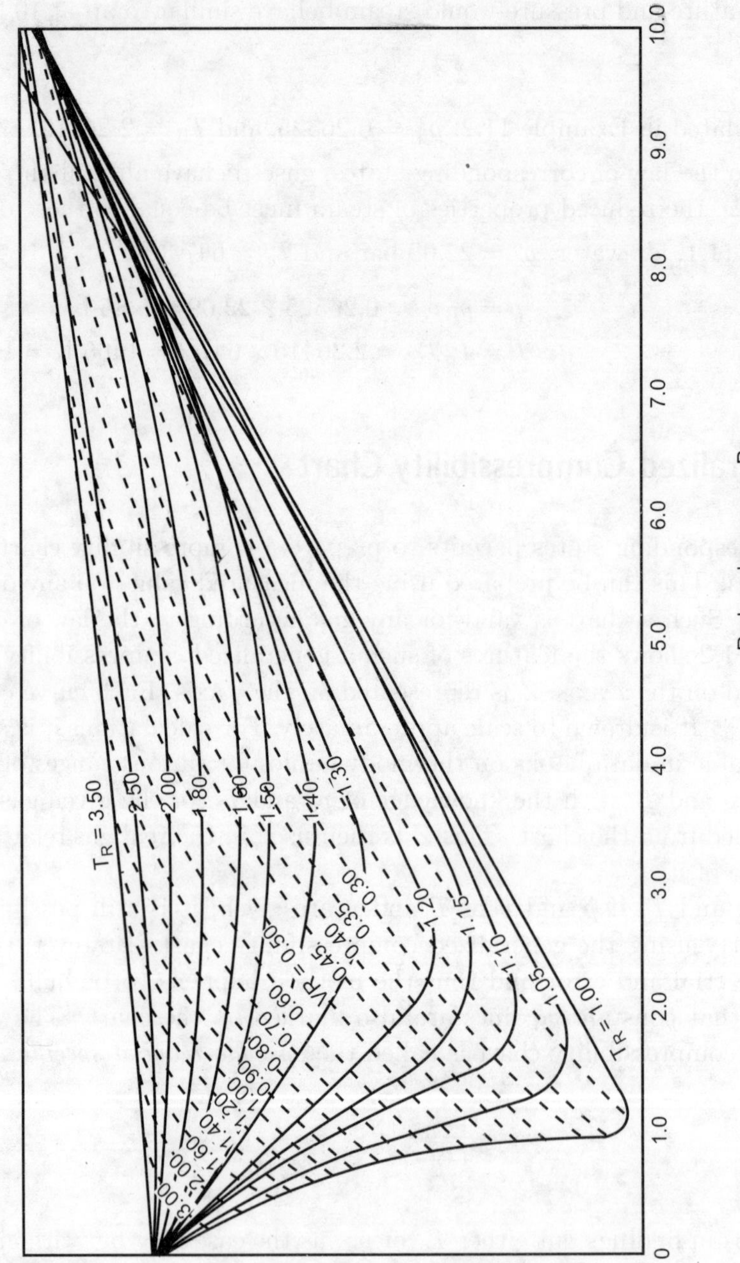

Fig. 11.2 Generalized compressibility chart.

Example 11.4

Using the generalized compressibility chart, find Z for air at 5.65 MPa and 27°C.

▶ Solution:

From Table 11.1, for air, $p_c = 3.77$ MPa, $T_c = 132.5$ K.

$$\text{Therefore, } p_{\text{R}} = \frac{p}{p_c} = \frac{5.65}{3.77} = 1.5 \text{ and } T_{\text{R}} = \frac{T}{T_c} = \frac{300}{132.5} = 2.26$$

From the generalized compressibility chart, at $P_{\text{R}} = 1.5$ and $T_{\text{R}} = 2.26$, $Z = 0.97$

Example 11.5

Calculate the specific volume of gaseous propane at 7 MPa and 150°C using (i) ideal gas equation and (ii) generalized compressibility chart.

▶ Solution:

$$\text{(i) } v = \frac{RT}{P} = \left(\frac{8314.3}{44}\right) \times \frac{(273 + 150)}{7 \times 10^6} = 0.0114 \text{ m}^3\text{/kg}$$

(ii) From Table 11.1, for propane, $p_c = 4.26$ MPa and $T_c = 370$ K.

$$\text{Therefore, } p_{\text{R}} = \frac{p}{p_c} = \frac{7}{4.26} = 1.643, \text{ and } T_{\text{R}} = \frac{T}{T_c} = \frac{(273 + 150)}{370} = 1.143$$

From the generalized compressibility chart, corresponding to $p_{\text{R}} = 1.643$ and $T_{\text{R}} =$ 1.1.43, $Z = 0.56$

$$\text{Therefore, } v = \frac{ZRT}{p} = Z(v_{ideal}) = 0.56 \times 0.0114 = 0.00638 \text{ m}^3\text{/kg}$$

11.5 Other Equations of State

11.5.1 van der Waals Equation of State

In 1873, van der Waals, proposed an equation of state that supposedly compensated for the two assumptions made in deriving the ideal gas equation of state, namely, (i) intermolecular forces of attraction are negligible and (ii) the volumes of the molecules themselves are negligible in comparison to the volume occupied by the gas. He reasoned out that the actual volume of a gas at a certain pressure and temperature was less than the ideal volume. If the volume of molecules themselves was b (where b approximates to volume of all the molecules when they are packed together—which will be close to the specific volume of liquid) then the actual volume of the gas would be $(v-b)$, especially at high pressures, and not simply v as in the ideal gas equation. Similarly, he also reasoned out that the actual gas pressure is not the same as the ideal gas pressure because of attractive forces between the molecules and that these attractive forces are proportional to square of the density or to $\left(\frac{1}{v^2}\right)$. Thus the van der Waals equation of state for a real gas, in molar form, with the term $\left(\frac{a}{v^2}\right)$ accounting for the intermolecular forces of attraction and the term b accounting for the finite size of the molecules, is given by

$$\left(p + \frac{a}{\bar{v}^2}\right)(\bar{v} - b) = \mathcal{R}T \tag{11.5}$$

where a and b are the van der Waals constants which are derived from the critical properties of the gas under consideration. In (11.5), $\bar{v}$ is the molar specific volume in m^3/kmol and $\mathcal{R}$ is the universal gas constant, T is in K and the units of the constants a and b are compatible with the units of p and v. van der Waals equation of state can also be written on mass basis as

$$\left(p + \frac{a}{v^2}\right)(v - b) = \mathrm{R}T \tag{11.6}$$

where v is in m^3/kg and R, the particular gas constant is in J/kg K. The magnitudes of a and b in the above two equation would obviously be different. One must be careful to select the proper values of a and b from the property tables in conjunction with the units used for p, v, T and R.

The use of this equation is limited because of its inadequate accuracy over wide ranges as the constants a and b are evaluated with the knowledge of the properties at a single point—the critical point. This equation has a great historic importance as it was van der Waals who first attempted to model the behaviour of real gases. Also, the constants a and b are different for different gases and also that they depend to a certain extent on the temperature as well as pressure. Given in Table 11.2 are the van der Waals constants for a few gases.

Table 11.2 The van der Waals constants for various gases.

Gas	a MPa m^6 /$kmol^2$	b m^3/kmol
Acetylene	0.4520	0.0522
Air	0.1352	0.0365
Ammonia	0.4249	0.0373
Argon	0.135	0.0320
n-Butane	1.3858	0.1162
Carbon dioxide	0.3658	0.0428
Carbon monoxide	0.1470	0.0395

(Table 11.2 Contd.)

(Table 11.2 Continued)

Gas	a MPa m^6 /$kmol^2$	b m^3/kmol
Ethane	0.5573	0.06502
Ethylene	0.462	0.0583
Helium	0.00342	0.0235
Hydrogen	0.02481	0.0266
Methane	0.2281	0.0427
Nitrogen	0.1366	0.0386
Oxygen	0.1380	0.0317
Propane	0.9355	0.0901
Water	0.5528	0.0305

11.5.1.1 Van der Waals Constants in terms of Critical Properties

It can be seen that the van der Waals equation of state is a cubic equation in v. Thus, for a given pressure and temperature there are three possible values for specific volume. It can be observed from the general $p-V$ diagram for a pure substance that for each isotherm corresponding to temperatures less than the critical temperature, there exists a maxima (specific volume of saturated vapour) and minima (specific volume of saturated liquid). This is in contrast to the behaviour of an ideal gas. As temperature increases, the difference between the maximum and the minimum values of specific volume decrease. On the critical isotherm, the values of v coincide and it is observed that the critical isotherm on the $p-V$ plane has a horizontal inflection point at the critical state. Hence, the first and second derivatives of p with respect to v at the critical point are zero. This fact is used in evaluating the van der Waals constants in terms of the critical properties.

Shown in Figure.11.3 is the critical isotherm for a real gas.

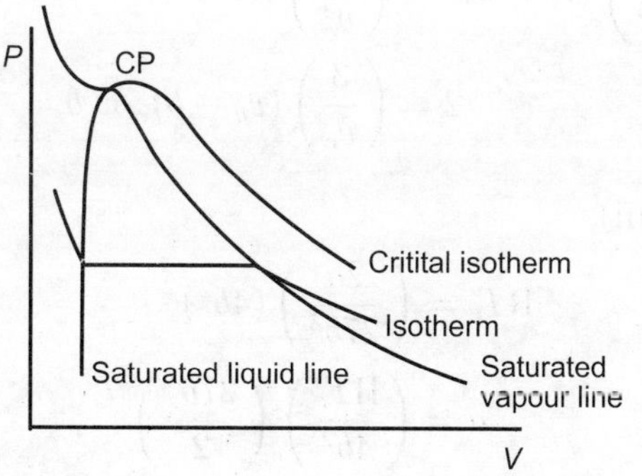

Fig. 11.3 The critical isotherm.

The van der Waals equation of state $\left(p + \frac{a}{v^2}\right)(v - b) = \mathrm{R}T$ can be rewritten as,

$$p = \frac{\mathrm{R}T}{(v - b)} - \frac{a}{v^2} \qquad \text{(i)}$$

At the critical state,

$$p_c = \frac{\mathrm{R}T_c}{(v_c - b)} - \frac{a}{v_c^2} \qquad \text{(ii)}$$

As the critical isothem has an inflection point at the critical state,

$$\left(\frac{\partial p}{\partial v}\right)_c = 0 \text{ and } \left(\frac{\partial^2 p}{\partial v^2}\right)_c = 0$$

and therefore,

$$\left(\frac{\partial p}{\partial v}\right)_c = \frac{-\mathrm{R}T_c}{(v_c - b)^2} + \frac{2a}{v_c^3} = 0$$

or,
$$\mathrm{R}T_c = \left(\frac{2a}{v_c^3}\right)(v_c - b)^2 \qquad \text{(iii)}$$

and,
$$\left(\frac{\partial^2 p}{\partial v^2}\right)_c = \frac{2\mathrm{R}T_c}{(v_c - b)^3} - \frac{6a}{v_c^4} = 0$$

That is,
$$2\mathrm{R}T_c = \left(\frac{6a}{v_c^4}\right)(v_c - b)^3$$

or,
$$\mathrm{R}T_c = \left(\frac{3a}{v_c^4}\right)(v_c - b)^3 \qquad \text{(iv)}$$

Eliminating $\mathrm{R}T_c$ between (iii) and (iv)

$$\left(\frac{2a}{v_c^3}\right)(v_c - b)^2 = \left(\frac{3a}{v_c^4}\right)(v_c - b)^3$$

Therefore,
$$2 = \left(\frac{3}{v_c}\right)(v_c - b), \text{ or } b = \frac{v_c}{3} \qquad \text{(v)}$$

Substituting for b in (iii)

$$\mathrm{R}T_c = \left(\frac{2a}{27b^3}\right)(4b^2)$$

Therefore,
$$a = \left(\frac{\mathrm{R}T_c}{4b^2}\right)\left(\frac{27b^3}{2}\right)$$

$$= \left(\frac{27}{8}\right)\mathrm{R}T_c b$$

$$= \left(\frac{27}{8}\right) RT_c \times \frac{v_c}{3}$$

$$= \left(\frac{9}{8}\right) T_c v_c R$$

that is,

$$a = \left(\frac{27}{8}\right) RT_c b = \left(\frac{9}{8}\right) T_c v_c R \tag{vi}$$

Substituting for a and b from (vi) and (v) in (ii)

$$p_c = \frac{RT_c}{(v_c - \frac{v_c}{3})} - \left(\frac{9}{8}\right) \frac{T_c v_c R}{v_c^2}$$

That is,

$$R = \left(\frac{8}{3}\right) \frac{p_c v_c}{T_c} \tag{vii}$$

Also, substituting for R from (vii) in (vi),

$$a = \left(\frac{9}{8}\right) T_c v_c \times \left(\frac{8}{3}\right) \frac{p_c v_c}{T_c} = 3 p_c v_c^2$$

11.5.1.2 Compressibility Factor at the Critical State

It has been shown earlier that for a van der Waal gas,

$$R = \left(\frac{8}{3}\right) \frac{p_c v_c}{T_c} \tag{viii}$$

From the definition, the compressibility factor at the critical state

$$Z_c = \frac{p_c v_c}{RT_c}$$

Substituting for R in this equation from (vii),

$$Z_c = \left(\frac{p_c v_c}{T_c}\right)\left(\frac{3T_c}{8 p_c v_c}\right)$$

$$= \frac{3}{8} = 0.375$$

Therefore the compressibility factor of a van der Waals gas at the critical state is $\frac{3}{8}$.

11.5.1.3 The van der Waals Equation of State in Reduced Properties

It has been shown above for a van der Waals gas that

$$a = 3 p_c v_c^2,\ b = \frac{v_c}{3} \text{ and } R = \frac{8 p_c v_c}{3 T_c}$$

Substitute for a, b and R in the van der Waals equation (11.6),

$$\left(p + \frac{3p_c v_c^2}{v^2}\right)\left(v - \frac{v_c}{3}\right) = \frac{8p_c v_c T}{3T_c}$$

By definition, $v_{\mathrm{R}} = \dfrac{v}{v_c}$ and $T_{\mathrm{R}} = \dfrac{T}{T_c}$

Therefore, $$\left(p + \frac{3p_c}{v_{\mathrm{R}}^2}\right)\left(v - \frac{v_c}{3}\right) = \frac{8p_c v_c T_{\mathrm{R}}}{3}$$

Dividing the left-hand side and the right-hand side by $p_c v_c$, we get the van der Waals equation of state on reduced properties as

$$\left(p_{\mathrm{R}} + \frac{3}{v_{\mathrm{R}}^2}\right)\left(v_{\mathrm{R}} - \frac{1}{3}\right) = \frac{8T_{\mathrm{R}}}{3}$$

Example 11.6

Determine the pressure exerted by 5 kg carbon dioxide at 27°C in a container of 1.5 m^3 capacity using (i) ideal gas equation of state and (ii) van der Waals equation.

► **Solution:**

(i) $p = \dfrac{m\mathrm{R}T}{V}$, where $m = 5$ kg, $\mathrm{R} = \dfrac{\mathcal{R}}{M} = \dfrac{8314.3}{44} = 188.96$ J/kg K, and

$$T = 273 + 27 = 300 \text{ K and } V = 1.5 \text{ m}^3$$

Therefore, $$p = 5 \times 188.96 \times \frac{300}{1.5} = 18896 \text{ N/m}^2 = 188.96 \text{ kPa}$$

(ii) From van der Waals equation, $p = \dfrac{\mathcal{R}T}{(\bar{v} - b)} - \dfrac{a}{\bar{v}^2}$

For CO_2, from Table 11.2,

$$a = 0.3658 \text{ MPa m}^6/\text{kmol}^2 = 365.8 \text{ kPa m}^6/\text{kmol}^2$$

and $$b = 0.0428 \text{ m}^3/\text{kmol}$$

$$\bar{v} = \left(\frac{V}{m}\right) M = \frac{1.5}{5} \times 44 = 13.2 \text{ m}^3/\text{kmol}$$

Substituting for a, b and $\bar{v}$ in van der Waals equation

$$p = 8.3143 \times \frac{300}{(13.2 - 0.0428)} - \frac{365.8}{13.2^2}$$
$$= 187.5 \text{ kPa}$$

11.5.2 Redlich and Kwong Equation of State

In 1949, Redlich and Kwong presented the following equation of state which has in it two constants a and b. These constants are not the same as the van der Waals constants, but are different functions of the critical state properties. This equation is more accurate than the van der Waals equation over a wide range.

$$p = \left\{ \frac{\mathcal{R}T}{(\bar{v} - b)} - \frac{a}{(\bar{v}^2 + \bar{v}b)T^{1/2}} \right\} \tag{11.7}$$

where $\bar{v}$ is the molar specific volume, and constants a and b are given by:

$$a = 0.427 \frac{\mathcal{R}^2 T_c^{2.5}}{p_c} \text{ and } b = 0.0866 \frac{\mathcal{R}T_c}{p_c}$$

11.5.3 Beattie--Bridgeman Equation

The Beattie–Bridgeman equation is another widely used real gas equation of state. There are five constants in this equation. This is used for formulating properties of vapour close to the liquid phase at pressures less than the critical pressure. The equation is more accurate when the specific volumes involved are greater than twice the critical volume.

$$p = \left\{ \frac{\mathcal{R}T(1 - \varepsilon)(\bar{v} + B)}{\bar{v}^2} \right\} - \frac{A}{\bar{v}^2} \tag{11.8}$$

where, $\varepsilon = \dfrac{c}{\bar{v}T^3}$, and the constants are given by

$$A = A_0 \left(1 - \frac{a}{\bar{v}}\right)$$
$$B = B_0 \left(1 - \frac{b}{\bar{v}}\right)$$

In the above relations, $\bar{v}$ is in m^3/kmol. The constants A_0, a, B_0, b and c for a few gases are given in Table 11.3. Their units are such that the constants can be directly incorporated into (11.8) when the units for pressure, volume, temperature and R are in kPa, m^3/kmol, K and 8.314 kJ/ kmol.K, respectively.

Table 11.3 Beattie–Bridgeman constants

Gas	A_0	a	B_0	b	c
Air	131.8441	0.01931	0.04611	– 0.001101	4.34×10^4
Argon	130.7802	0.02328	0.03931	0.0	5.99×10^4
Carbon dioxide	507.2836	0.07132	0.10476	0.07235	6.60×10^5
Carbon monoxide	134.5	0.0262	0.0504	– 0.00693	4.21×10^4
Helium	2.1886	0.05984	0.01400	0.0	40
Hydrogen	20.0117	– 0.00506	0.02096	– 0.04359	504
Nitrogen	136.2315	0.02617	0.05046	– 0.00691	4.20×10^4
Oxygen	151.0857	0.02562	0.04624	0.004208	4.80×10^4

11.5.4 Virial Equations of State

Virial equations of state that are being used extensively for computational purposes and for preparing extensive tables of property are two in number. They are:

$$\frac{pv}{RT} = A_0 + A_1 p + A_2 p^2 + A_3 p^3 + \cdots\cdots \tag{11.9}$$

and

$$\frac{pv}{RT} = B_0 + \frac{B_1}{v} + \frac{B_2}{v^2} + \frac{B_3}{v^3} + \cdots\cdots \tag{11.10}$$

where the As and Bs are functions of temperature only. The advantage of the virial equations is that the constants can be found out by the experimental p–v–T data.

Example 11.7

Determine the pressure in a rigid vessel of volume 0.015 m^3 which contains 3.5 kg nitrogen at 200°C using (i) ideal gas equation, (ii) generalized compressibility chart, (iii) van der Waals equation, (iv) Redlich–Kwong equation and (v) Beattie–Bridgeman equation.

▶ **Data:**

$M = 28$, $m = 3.5$ kg, $V = 0.015$ m^3 ($v = \dfrac{V}{m} = \dfrac{0.015}{3.5} = 0.004286$ m^3/kg). Molar specific volume $= v \times M = 0.004286 \times 28 = 0.12$ m^3/kmol), T = 273 + 200 = 473 K.

▶ **Solution:**

(i)

$$p = \frac{m\mathcal{R}T}{V} = m\left(\frac{\mathcal{R}}{M}\right)\frac{T}{V}$$

$$= 3.5 \times \frac{8314.3}{28} \times \frac{473}{0.015}$$

$$= 32.77 \times 10^6 \text{ N/m}^2 = 32.77 \text{ MPa}.$$

(ii) From Table 11.1, for N_2, $p_c = 3.39$ MPa, $T_c = 126.2\ K$ and $v_c = 0.0899\ \text{m}^3/\text{kmol}$.

$$\text{Therefore, } T_R = \frac{T}{T_c} = \frac{473}{126.2} = 3.75 \text{ and}$$

$$v_R = \frac{v}{\left(\frac{RT_c}{p_c}\right)} = \frac{0.004286}{\left[\left(\frac{8314.3}{28}\right) \times \frac{126.2}{3.39 \times 10^6}\right]} = 0.3877$$

From the generalized compressibility chart, $Z = 1.1$ for $T_R = 3.75$ and $v_R = 0.3877$

$$\text{Therefore,} \quad p = \frac{ZRT}{v}$$

$$= 1.1 \times \left(\frac{8314.3}{28}\right) \times \frac{473}{0.004286}$$

$$= 36{,}047{,}016 \text{ N/m}^2 = 36.047 \text{ MPa}.$$

(iii) From Table 11.2, for N_2, $a = 0.1366$ MPa m^6/kmol^2 and $b = 0.0386$ m^3/kmol.

$$p = \frac{\mathcal{R}T}{(\bar{v} - b)} - \frac{a}{\bar{v}^2}$$

$$= \left[8314.3 \times \frac{473}{(0.12 - 0.0386)}\right] - 0.1366 \times \frac{10^6}{0.12^2}$$

$$= 38{,}826{,}713 \text{ Pa} = 38.8 \text{ MPa}$$

(iv)
$$a = 0.427 \frac{\mathcal{R}^2 T_c^{2.5}}{p_c}$$

$$= 0.427 \times 8314.3^2 \times \frac{126.2^{2.5}}{3.39 \times 10^6} = 15{,}57{,}858$$

$$b = 0.0866 \frac{\mathcal{R}T_c}{p_c}$$

$$= 0.0866 \times 8314.3 \times \frac{126.2}{3.39 \times 10^6} = 0.0268$$

$$p = \left\{\left[\frac{\mathcal{R}T}{(\bar{v} - b)}\right] - \frac{a}{(\bar{v}^2 + \bar{v}b)T^{1/2}}\right\}$$

$$= \left\{8314.3 \times \frac{473}{(0.12 - 0.0268)}\right\} - \frac{1{,}557{,}858}{[(0.12^2 + 0.12 \times 0.0268) \times 473^{1/2}]}$$

$$= 42{,}19{,}5964 - 16{,}15{,}000 \text{ Pa} = 40{,}58{,}0964 \text{ Pa} = 40.58 \text{ MPa}$$

(v) From Table 11.3, For N_2,

$A_0 = 136.2315$, a $= 0.02617$, $B_0 = 0.05046$, $b = -\ 0.00691$ and $c = 4.2 \times 10^4$

$$\text{Therefore,} \qquad A = A_0(1 - a/\bar{v})$$

$$= 136.2315\left[1 - \frac{0.02617}{0.12}\right] = 106.52$$

$$B = B_0\left(1 - \frac{b}{\bar{v}}\right) = 0.05046\left(1 + \frac{0.00691}{0.12}\right)$$

$$= 1.148 \times 10^{-3} = 0.05336$$

$$\varepsilon = \frac{c}{\bar{v}T^3}$$

$$= 4.2 \times \frac{10^4}{0.12 \times 473^3} = 3.3 \times 10^{-3}$$

$$p = \left\{\frac{\mathcal{R}T(1-\varepsilon)(\bar{v}+B)}{\bar{v}^2}\right\} - \frac{A}{\bar{v}^2}$$

$$= 8.3143 \times 473 \times (1 - 1.148 \times 10^{-3}) \times \frac{(0.12 + 0.05336)}{0.12^2}$$

$$- \frac{106.52}{0.12^2}$$

$$= 47{,}290 - 7397 = 39{,}893 \text{ kPa} = 39.89 \text{ MPa}$$

Problems

11.1 Determine the specific volume of $R-134a$, at 0.5 MPa and 50°C using (i) the ideal gas equation of state and (ii) the generalized compressibility chart.

11.2 Invoking the law of corresponding states find the pressure and temperature of methane when it behaves similar to ethane at 10 bar, 27°C.

11.3 Determine the pressure of water vapour at 300°C and 0.02426 m^3/kg using (i) the steam tables, (ii) the ideal gas equation and (iii) the generalized compressibility chart.

11.4 Hydrogen at 200 bar and 250 K is stored in a cylinder of volume 2 m^3. Determine the mass of hydrogen using (i) generalized compressibility chart and (ii) ideal gas equation.

11.5 Compare the specific volume of steam at 30 MPa and 450°C as obtained from steam tables with that calculated using generalized compressibility chart.

11.6 If the compressibility factor and reduced pressure of ethane are 0.7 and 1.2, respectively, find the temperature.

11.7 Determine the specific volume of H_2 at 60 bar and 100 K using the (i) ideal gas equation, (ii) van der Waals equation and (iii) generalized compressibility chart.

11.8 Find the pressure of nitrogen gas at 200 K and 0.00375 m^3/kg using the (i) ideal gas equation, (ii) law of corresponding states, (iii) van der Waals equation, (iv) Redlich–Kwong equation and (v) Beattie–Bridgeman equation.

11.9 10 kg ammonia at 300 K is stored in a rigid tank of capacity 3 m^3. Determine the pressure of ammonia in the tank using the (i) ideal gas equation, (ii) van der Waals equation and (iii) Beattie–Bridgeman equation.

11.10 1 kmol of oxygen is compressed isothermally from 50°C, 0.15 m^3/kg to 0.06 m^3/kg. Assuming the gas to follow van der Waals equation of state, find the final pressure and the work interaction during the process described.

FUELS AND COMBUSTION 12

12.1 Introduction

The working substances dealt with in earlier chapters have been those which did not undergo any chemical change during a given process. In this chapter, the processes which involve change in chemical composition and aggregation of the working substance is examined. One such process of paramount importance in engineering is *combustion*. During combustion, a fuel and oxidizer are burnt to produce combustion products at high temperatures. There is a definite change in the chemical composition and aggregation of the working substance after the combustion process.

Thermal energy is required in abundance in several process industries. Also, majority of the world's power requirement is realized from power plants, the input for which is thermal energy at high temperatures in the combustion products. Motive power is obtained through internal combustion engines by burning a suitable fuel in the cylinders.

It is therefore essential for an engineer to have complete knowledge about all types of available fuel and the process of burning these to give high-temperature products.

This chapter, therefore, deals with (i) a brief description of fuels and their properties, (ii) calculation of minimum or theoretical quantity of air required for combustion, (iii) evaluation of actual air used in a combustion process, (iv) application of the first law of thermodynamics to combustion and (v) determination of energy released and maximum temperature obtainable by burning a fuel.

12.2 Fuels

Fuels are naturally available chemical species or synthetic ones which give high-temperature products when burnt with air or an oxidant. Generally, the major combustible elements (i.e., elements that can be burnt) in any fuel are carbon and hydrogen. There may also be negligible amounts of sulphur and phosphorous. These may be treated as impurities in the fuel. They may often have a profound significance on combustion but when they burn they have an insignificant effect on the thermodynamic analysis and energy released. Fuels are classified into three categories, namely (i) solid (ii) liquid and (iii) gaseous. Each of these may either be naturally available or synthetic. Most of these are available in nature whereas some are made synthetically.

While coal is a naturally available solid fuel, coke, bagasse, rice husk, saw dust etc. are examples of synthetic solid fuels. The major combustible element in coal is carbon. Associated with carbon are moisture, ash (non-combustible materials) and traces of other elements such as S, P, H_2, N_2 and O_2. The proportions of these elements vary with the geographical source of coal. The proportions of various elements, ash and moisture in coal are analysed on the basis of mass.

Generally, liquid fuels are mixtures of several hydrocarbons. Crude oil obtained from subsoil drilling is a natural liquid fuel. Most thermal energy requirements of the world are met by burning crude oil. Petrol, diesel, alcohol, kerosene, fuel oil, vegetable oil, organic liquids etc. are synthetic liquid fuels, most of which are obtained by fractional distillation of crude oil. The combustible elements in these fuels are generally C and H_2.

Natural gas available in the proximity of oil wells is a gaseous fuel. The approximate composition of the natural gas is, CH_4, 70%; H_2, 10%; N_2, 15%; O_2, 2% and CO_2, 3%. Other synthetic gases are (i) coal gas (approximate composition: 40 %H_2, 30 %CH_4, plus small percentages of CO, CO_2, C_2H_2, C_2H_4 and N_2) obtained by heating coal beyond 1300°C in the absence of oxygen, (ii) producer gas (approximate composition: 20–30% CO, 8–12% H_2, 50–55% N_2 and traces of CO_2) obtained by passing steam and air over hot coal or a bed of coke at about 1100°C, (iii) water gas, (approximate composition: H_2, 50%; CO, 40%; N_2, 5% and CO_2, 5%) obtained by passing steam and some air alternatively over a bed of hot coal or coke and (iv) biogas, (approximate composition: CH_4, 55%; H_2, 7–10%; CO_2, 35%; N_2, 2–6% and traces of H_2S) obtained by the anaerobic digestion or fermentation of vegetable and animal matter.

12.2.1 Fuel Analysis

The composition of a fuel is analysed based on either mass or volume (mol basis). When analysed on mass, the composition is given as mass fractions (y_i) of the constituents in the fuel. This is called the *gravimetric analysis.* In the other method of mass *analysis*, called the ultimate analysis, it gives the mass fractions of the chemical elements present in the fuel, irrespective of how they are chemically bonded in different species of the fuel, and the mass fractions of moisture and ash which are not chemical elements. Yet another type of analysis in vogue is the *proximate analysis* which is mostly used to analyse a sample of coal. It gives the mass fractions of arbitrarily designated constituents such as moisture, volatile matter, fixed carbon and ash.

The proportions of various chemical species present in a gaseous fuel are normally given by volume or mole fractions (x_i). The mass fractions and volume fractions of the

ith component in the fuel are related, as in (9.36), by: $y_i = x_i \dfrac{M_i}{M}$, where M_i and M are respectively the molar masses of the ith constituent and the fuel as such.

Example 12.1

The following is the volumetric analysis of a certain gaseous fuel: CO_2, 3%; CO, 19%; H_2, 41%; CH_4, 25%; C_2H_4, 9% and N_2, 3%. Find (a) the gravimetric and (b) ultimate analysis.

▶ Data:

$x_{CO_2} = 0.03, x_{CO} = 0.19, x_{H_2} = 0.41, x_{CH_4} = 0.25, x_{C_2H_4} = 0.09$ and $x_{N_2} = 0.03$.

▶ Solution:

(a) $$y_i = x_i \frac{M_i}{M} \text{ where } M = \sum x_i M_i$$

Here, $$M = 0.03 \times 44 + 0.19 \times 28 + 0.41 \times 2 + 0.25 \times 16$$
$$+0.09 \times 28 + 0.03 \times 28 = 14.82 \text{ kg/kmol.}$$

Therefore, $$y_{CO_2} = 0.03 \times \frac{44}{14.82} = 8.9\%$$
$$y_{CO} = 0.19 \times \frac{28}{14.82} = 35.9\%$$
$$y_{H_2} = 0.41 \times \frac{2}{14.82} = 5.5\%$$
$$y_{CH_4} = 0.25 \times \frac{16}{14.82} = 27\%$$
$$y_{C_2H_4} = 0.09 \times \frac{28}{14.82} = 17\%$$
$$y_{N_2} = 0.03 \times \frac{28}{14.82} = 5.7\%$$

(b) The fuel consists of C, H_2, O_2 and N_2. Grouping together the number of kmols of each of these elements in various species, we get the total kmol of each of these elements per kmol fuel:

Thus, $$n_C/\text{kmol fuel} = 0.03 + 0.19 + 0.25 + 0.09 \times 2 = 0.65$$
$$n_{H2}/\text{kmol fuel} = 0.41 + 0.5 + 0.18 = 1.09$$
$$n_{O_2}/\text{kmol fuel} = 0.03 + \frac{0.19}{2} = 0.125$$
and $$n_{N2}/\text{kmol fuel} = 0.03$$

Now, $$m_i = n_i M_i/\text{kmol fuel} = \left(\frac{n_i M_i}{M}\right) \text{ kg/kg fuel}$$

Therefore, $m_C = \frac{n_C M_C}{M} = \frac{0.65 \times 12}{14.82} = 0.526 \text{ kg/kg fuel} = 52.6\%$

Similarly, $m_{H_2} = \frac{1.09 \times 2}{14.82} = 14.7\%$

$$m_{O_2} = \frac{0.125 \times 32}{14.82} = 27\%$$

$$m_{N_2} = \frac{0.03 \times 28}{14.82} = 5.7\%.$$

12.3 Combustion

Combustion is the process of rapid oxidation of the combustible elements in a fuel with oxygen. During this rapid oxidation process the chemical energy of the fuel before combustion is converted to internal energy of the products after combustion. Due to this conversion of one mode of energy to another the temperature of the products after combustion will be very much higher than that of the fuel and the oxidant before combustion. In most cases, the oxygen required for combustion is obtained from atmospheric air.

In any combustion process, while the substances before combustion are called *reactants*, those resulting after combustion are called *products*. The combustible elements in any fuel are generally carbon and hydrogen. The other elements such as sulphur and phosphorous may exist in the fuel in negligible proportions. Though these two elements also burn during combustion their contribution to the increase in temperature of the products is insignificant.

The process of combustion is generally represented in the form of chemical equations. Two such equations for the combustion of simple fuels, for example, methane (CH_4) and carbon (C), are as follows:

$$CH_4 + 2O_2 \rightarrow CO_2 + 2H_2O \quad \text{(i)}$$

$$2C + 2O_2 \rightarrow 2CO + O_2 \quad \text{(ii)}$$

$$\text{(or,} \quad C + O_2 \rightarrow CO + \frac{1}{2}O_2)$$

- While CH_4 and $2O_2$ in (i) and 2C and $2O_2$ in (ii) are called the reactants, CO_2 and $2H_2O$ in (i) and 2CO and O_2 in (ii) are called the products.

- The combustion equation is always written on mol or kmol basis. In (i) it means, 1 kmol of CH_4 burns with 2 kmol of O_2 to give 1 kmol of CO_2 and 2 kmol of H_2O. In (ii) it means 2 kmol of C burns with 2 kmol of O_2 to give 2 kmol CO and 1 kmol O_2.

- It is not necessary that sum of the number of kmols of species in reactants is always equal to sum of the number of kmols of species in the products. In (i), while $\Sigma\,(n_i)_{reactants} = 1 + 2 = 3, \Sigma\,(n_i)_{products} = 1 + 2 = 3$. Here $\Sigma\,(n_i)_{reactants} = \Sigma\,(n_i)_{products}$. In (ii) $\Sigma\,(n_i)_{reactants} = 2 + 2 = 4, \Sigma\,(n_i)_{products} = 2 + 1 = 3$. Here $\Sigma\,(n_i)_{reactants} \neq \Sigma\,(n_i)_{products}$.

- Mass is always conserved during combustion. Hence, the mass of reactants is always equal to the mass of products. That is, $\Sigma\,(m_i)_{reactants} = \Sigma(m_i)_{products}$. We know that m, the mass, M, the molar mass (molecular weight) and n, the number of mols or kmols, are related by $m = nM$.

Therefore, the combustion equation (i) can be written on mass basis as,

$$(n_{CH_4}M_{CH_4})CH_4 + (n_{O_2}M_{O_2})O_2 \rightarrow (n_{CO_2}M_{CO_2})CO_2 + (n_{H_2O}M_{H_2O})H_2O$$

Here, $n_{CH_4} = 1, M_{CH_4} = 16, n_{O_2} = 2, M_{O_2} = 32, n_{CO_2} = 1, M_{CO_2} = 44$ and $n_{H_2O} = 2, M_{H_2O} = 18$. Substituting these values of n an M in the combustion equation above,

$$(1 \times 16)\text{ kg CH}_4 + (2 \times 32)\text{ kg O}_2 \rightarrow (1 \times 44)\text{ kg CO}_2 + (2 \times 18)\text{ kg H}_2\text{O}$$

Here, $\Sigma\,(m_i)_{reactants} = 16+64 = 80 = \Sigma\,(m_i)_{products} = 44+36 = 80$. Similarly, in (ii) $\Sigma\,(m_i)_{reactants} = (2 \times 12 + 2 \times 32) = 88$

$$= \Sigma\,(m_i)_{products} = (2 \times 28 + 32) = 88$$

12.3.1 Complete Combustion

Combustion is said to be complete when all the combustible elements in the fuel are converted to their final oxides. Equation (i) represents the case of a complete combustion.

12.3.2 Incomplete Combustion

Combustion is said to be incomplete if one or more combustible elements in the fuel are not completely oxidized. Thus equation (ii) is an example of incomplete combustion. If the combustion were to be complete, carbon should have been burnt to CO_2 according to $C + O_2 \rightarrow CO_2$.

12.3.3 Stoichiometric Air

Oxygen is essential for burning a fuel. Direct oxygen is rarely supplied for combustion except in special cases. Usually, air is supplied to burn the fuel. While oxygen in air is used for combustion, either partially or fully, the other components in air remain unaffected and unchanged. The minimum amount of dry air required to burn a fuel completely is called *stoichiometric air* or *theoretical air.* When a fuel is burnt under stoichiometric conditions, the combustion products will have only final oxides and nitrogen and no oxygen. The stoichiometric air is specified as kg air/kg fuel. Many times, it is also referred to as stoichiometric air–fuel ratio $(A/F)_{st}$. It is very essential to know how to evaluate the stoichiometric air.

12.3.4 Dry Air

Dry air is a mixture of O_2 and N_2 with small traces of other inert gases. From the point of view of combustion, it can be said that air is a mixture of O_2 and N_2. Traces of other gases are treated as a part of N_2. On volume basis (kmol basis) dry air consists of 21% O_2 and 79% N_2. Thus, 1 kmol air consists of 0.21 kmol O_2 and 0.79 kmol N_2. In other words, $\frac{1}{0.21}$ kmol air, that is, 4.76 kmol air consists of 1 kmol O_2 and 3.76 kmol N_2.

On mass basis, dry air consists of 23.2% O_2 and 76.8% N_2. Hence, one kilogram of air consists of 0.232 kg O_2 and 0.768 kg N_2.

If 'a' kmol of O_2 is required for combustion of a certain fuel, the amount of air required will be $4.76a$ kmol. As 4.76 kmol air consists of 1 kmol O_2 and 3.76 kmol N_2, in combustion equations, this amount is generally written as $a(O_2 + 3.76N_2)$ kmol.

12.3.5 Estimation of Stoichiometric Air

For example, (i) represents the combustion of CH_4 when it is burnt with O_2. If CH_4 is burnt with air, (i) takes the form:

$$CH_4 + 2(O_2 + 3.76N_2) \rightarrow 2CO_2 + 7.52N_2 + 2H_2O. \qquad \text{(iii)}$$

As all the elements are fully oxidized and that there is no oxygen in the combustion products, (iii) is the stoichiometric combustion equation.

It is not difficult to write equations, such as (iii), to describe the combustion of any fuel. To understand how it can be done, let us write (iii) in a general form as,

$$CH_4 + a(O_2 + 3.76N_2) \rightarrow bCO_2 + cN_2 + dH_2O. \qquad \text{(iv)}$$

After combustion, the elements are not destroyed but they are attached chemically in a different manner with other elements. Thus, each element is conserved. The numerical values of a, b, c and d can be obtained by balancing the coefficients of each element on both sides of the equation.

The coefficient of C on the left-hand side of (iv) is 1 and that on the right-hand side is b. Therefore, by C balance, $b = 1$

Similarly, by H_2 balance, $d = 2$

and by O_2 balance, $a = b + \frac{d}{2}$

Substituting for b and d, $a = 1 + \frac{2}{2} = 2$

By N_2 balance, $3.76a = c$, or $c = 3.76 \times 2 = 7.52$

When the values of a, b, c and d are substituted in (iv) we get the combustion equation, as in (iii).

$$\text{Now, stoichiometric air} = \left(\frac{A}{F}\right)_{st}$$

$$= (\text{mass of air/mass of fuel})$$

$$= a(n_{O_2}M_{O_2} + n_{N_2}M_{N_2})/(n_{CH_4} \times M_{CH_4}) \text{ from (iv)}$$

From (iii), $a = 2, n_{O_2} = 1, n_{N_2} = 3.76, n_{CH_4} = 1$ and $M_{O_2} = 32, M_{N_2} = 28, M_{CH_4} = 16$.

$$\text{Therefore,} \quad \left(\frac{A}{F}\right)_{\text{st}} = \frac{2(1 \times 32 + 3.76 \times 28)}{(1 \times 16)} = 17.16 \text{ kg/kg methane.}$$

Example 12.2

Find the stoichiometric air for the combustion of (*a*) carbon (*b*) hydrogen and (*c*) sulphur.

► Solution:

(a) Stoichiometric equation for the combustion of carbon is:

$$C + (O_2 + 3.76N_2) \rightarrow CO_2 + 3.76N_2$$

$$\text{Stoichiometric air} = \frac{\text{mass of air}}{\text{mass of carbon}}$$

$$= \frac{(32 + 3.76 \times 28)}{12} = 11.45 \text{ kg/kg carbon}$$

(b) Similarly, $H_2 + \frac{1}{2}(O_2 + 3.76N_2) \rightarrow H_2O + 1.88N_2$

$$\text{Stoichiometric air} = \frac{\frac{1}{2}(32 + 3.76 \times 28)}{2} = 34.3 \text{ kg/kg hydrogen}$$

(c) $S + (O_2 + 3.76N_2) \rightarrow SO_2 + 3.76N_2$

$$\text{Stoichiometric air} = \frac{\text{mass of air}}{\text{mass of carbon}}$$

$$= \frac{(32 + 3.76 \times 28)}{32} = 4.3 \text{ kg/kg sulphur}$$

12.3.6 Excess Air

In most combustion equipments, the combustion may not proceed to completion by just providing the stoichiometric air. This is because in the short period the fuel and air pass through the combustor all the combustible elements may not be able to find the required oxygen which is present in the stoichiometric air. This makes the combustion incomplete and the combustion products, besides the final oxides and nitrogen, also contain some intermediate oxides and unused oxygen. Incomplete combustion results in lower temperature of the products. It is for this reason that air is supplied in excess of the stoichiometric air to increase the possibility of every combustible element finding the required oxygen before the products are discharged from the combustor. One cannot be liberal in admitting excess air as it will increase the mass of the products which in turn reduces the temperature of the products. Thus, the actual air supplied in a combustion process is fractionally more than the stoichiometric air. Excess air is generally expressed as a percentage of the stoichiometric air. By definition,

$$\begin{aligned} \%\text{ excess air} &= \frac{100 \times (\text{actual air} - \text{stoichiometric air})}{\text{stoichiometric air}} \\ &= \frac{100 \times \left[\left(\frac{A}{F}\right)_{\text{act}} - \left(\frac{A}{F}\right)_{\text{st}}\right]}{\left(\frac{A}{F}\right)_{\text{st}}}. \end{aligned} \tag{12.1}$$

12.3.7 % Theoretical Air

Sometimes actual air is also expressed as % theoretical air which is defined as follows

$$\%\text{ theoretical air} = 100 + \%\text{ excess air} \tag{12.2}$$

For example, if actual air is 16.5 kg/kg fuel and stoichiometric air is 15 kg/kg fuel, then

$$\text{excess air} = \frac{(16.5 - 15)}{15} = 0.1 = 10\%$$

$$\text{and} \quad \%\text{ theoretical air} = 100 + \%\text{ excess air} = 100 + 10 = 110\%.$$

Example 12.3

A certain cooking gas is found to be a mixture of 20% propane and 80% butane on volume basis. Find the stoichiometric air.

► Solution:

1 kmol of the gas consists of 0.2 kmol propane and 0.8 kmol butane. The stoichiometric combustion equation is:

$$(0.2\ C_3H_8 + 0.8C_4H_{10}) + a(O_2 + 3.76N_2) \rightarrow bCO_2 + cN_2 + dH_2O \tag{i}$$

By C balance: $0.2 \times 3 + 0.8 \times 4 = b = 3.8$

By H_2 balance: $0.2 \times 4 + 0.8 \times 5 = d = 4.8$

By O_2 balance: $a = b + \frac{d}{2} = 3.8 + \frac{4.8}{2} = 6.2$

By N_2 balance: $3.76a = c = 3.76 \times 6.2 = 23.312$

Substituting for a, b, c and d in (i):

$$(0.2\ C_3H_8 + 0.8C_4H_{10}) + 6.2(O_2 + 3.76N_2) \rightarrow 3.8CO_2 + 23.312N_2 + 4.8H_2O \qquad \text{(ii)}$$

Therefore, stoichiometric air = mass of air/mass of fuel

$$= \frac{6.2(32 + 3.76 \times 28)}{(0.2 \times 44 + 0.8 \times 58)}$$

$$= 15.42 \text{ kg/kg cooking gas}$$

Example 12.4

The cooking gas mentioned in Example 12.3 is burnt steadily and completely in a burner. Given that the ratio of gas to dry air is 1: 40 by volume, find the excess air and % theoretical air.

► Solution:

As 1 kmol dry air contains 0.21 kmol O_2 and 0.79 kmol N_2, 1 kmol gas is being burnt in the burner with 40 (0.21 O_2 + 0.79 N_2) kmol air. Therefore, the combustion equation is:

$$(0.2C_3H_8 + 0.8C_4H_{10}) + 40(0.21O_2 + 0.79N_2) \rightarrow bCO_2 + cN_2 + dO_2 + eH_2O \qquad \text{(i)}$$

(There is no need to find the values of b, c, d and e to get the actual air, because the mass of air and fuel can readily be found from the left-hand side of the equation. However, the values of b, c, d and e are found for the sake of academic interest and possible future use elsewhere).

By C balance: $b = 0.2 \times 3 + 0.8 \times 4 = 3.8$

By H_2 balance: $0.2 \times 4 + 0.8 \times 5 = e = 4.8$

By O_2 balance: $40 \times 0.21 = b + d + \frac{e}{2} = 3.8 + d + \frac{4.8}{2}$, that is, $d = 2.2$

By N_2 balance: $40 \times 0.79 = c = 31.6$

Substituting for b, c, d and e in Eq.(i):

$$(0.2C_3H_8 + 0.8C_4H_{10}) + 40(0.21O_2 + 0.79N_2) \rightarrow$$
$$3.8CO_2 + 31.6N_2 + 2.2O_2 + 4.8H_2O \qquad \text{(ii)}$$

$$\text{actual air} = \frac{\text{mass of air}}{\text{mass of cooking gas}}$$

$$= \frac{40(0.21 \times 32 + 0.79 \times 28)}{(0.2 \times 44 + 0.8 \times 58)}$$

$$= 20.9 \text{ kg/kg gas}$$

As evaluated in Example 12.3,

$$\text{stoichiometric air} = 15.42 \text{ kg/kg cooking gas}$$

Therefore, $\quad \text{excess air} = \dfrac{(20.9 - 15.42)}{15.42} = 0.355 = 35.5\%.$

$$\% \text{ theoretical air} = 100 + \% \text{ excess air} = 100 + 35.5 = 135.5\%.$$

12.4 Analysis of Combustion Products

Fuel supplied to any combustion equipment can generally be metered using suitable measuring devices. But it is not easy to measure the quantity of air supplied. For example, the rate at which coal is fed into a furnace can be measured by weighing. But the rate of air supply cannot be metered as it is not only supplied through a conduit from blowers or fans but it can also diffuse into the furnace through several openings. Yet it is very necessary to know the rate of supply of air and control it as and when needed to have the fuel burnt completely with minimal excess air. In combustion technology, the rate of air supply is normally found by an inferential method using the results of experimental analysis of the combustion products and the composition of the fuel.

12.4.1 The Orsat Apparatus

A very simple and a common apparatus used to find the volume fractions of CO_2, O_2, CO and N_2 in combustion products, experimentally, is the Orsat apparatus. In its simplest form, an Orsat apparatus consists of a water-jacketed measuring burette (generally of 100 ml volume) with provision to suck a sample of combustion product into it and hold it at atmospheric pressure and temperature over a column of water. This sample is then forced into a glass cylinder containing a chemical solution (a strong KOH or NaOH solution) that has high affinity to absorb CO_2. When CO_2 in the sample is completely absorbed by the solution the remaining products are brought back to the measuring burette and held at the original pressure and temperature. Volume of the remaining products is measured under these conditions. The difference between the original volume and the volume after CO_2 is removed by absorption gives the volume of CO_2 in the sample. Hence, the ratio of this reduction in volume to the original volume gives the volume fraction of CO_2 in the sample. The remaining products in the burette are then forced to a cylinder containing a chemical (alkaline pyrogallic acid) that has affinity to absorb O_2. Similar methodology as the previous one is used to find the volume fraction of O_2 in the products. Volume fraction of CO is similarly measured by using a solution (ammoniacal cuprous chloride) that has affinity to absorb it. The remaining volume of products must correspond to nitrogen under the assumption that there is no unburnt hydrocarbon in the products. However, several refinements have been made in the simple apparatus to also find the unburnt hydrocarbons if they were to exist in the products of combustion. A few important points to note are:

- Orsat analysis is based on dry gas and no water vapour is considered. The sample in the burette of the Orsat apparatus does contain saturated water vapour. During measurement, as the temperature of the sample is maintained always at a fixed value the partial pressure of water vapour remains the same. Due to this, it can be shown that the results obtained would be the same as that obtained had there been no water vapour in the products. Even if some other equipment is used to do such an analysis the results obtained are called Orsat analysis results.
- The Orsat analysis results are always given on volume basis.

Example 12.5

Find the Orsat analysis of the combustion products in Example 12.4.

▶ Solution:

The products of combustion in Example 12.4 are:

$$3.8CO_2 + 31.6N_2 + 2.2O_2 + 4.8H_2O$$

Orsat analysis, as we know, gives the volume fractions of the constituents in the dry gas.

The dry gases in the products are

$$3.8CO_2 + 31.6N_2 + 2.2O_2$$

Volume fractions of the constituents are equal to their mole fractions.

that is, $\quad x_i = \dfrac{n_i}{\Sigma\, n_i}$

Here $\quad \Sigma\, n_i = 3.8 + 31.6 + 2.2 = 37.6$

Therefore,

$$x_{CO_2} = \frac{n_{CO_2}}{\Sigma\, n_i} = \frac{3.8}{37.6} = 10.1\%$$

$$x_{N_2} = \frac{n_{N_2}}{\Sigma\, n_i} = \frac{31.6}{37.6} = 84\%$$

$$x_{O_2} = \frac{n_{O_2}}{\Sigma\, n_i} = \frac{2.2}{37.6} = 5.9\%$$

Example 12.6

Find the volume fractions and mass fractions of the products of combustion in Example 12.4.

▶ Solution:

The products of combustion in Example 12.4 are:

$$3.8CO_2 + 31.6N_2 + 2.2O_2 + 4.8H_2O$$

Here, $\Sigma n_i = 3.8 + 31.6 + 2.2 + 4.8 = 42.4$

$$x_{CO_2} = \frac{n_{CO_2}}{\Sigma n_i} = \frac{3.8}{42.4} = 9\%$$

$$x_{N_2} = \frac{n_{N_2}}{\Sigma n_i} = \frac{31.6}{42.4} = 74.5\%$$

$$x_{O_2} = \frac{n_{O_2}}{\Sigma n_i} = \frac{2.2}{42.4} = 5.2\%$$

$$x_{H_2O} = \frac{n_{N_2}}{\Sigma n_i} = \frac{4.8}{42.4} = 11.3\%$$

The mass of combustion products is given by $\Sigma m_i = \Sigma n_i M_i$

$$\Sigma m_i = (3.8 \times 44 + 31.6 \times 28 + 2.2 \times 32 + 4.8 \times 18) = 1208.8$$

The mass fractions can now be calculated using $y_i = \frac{m_i}{\Sigma m_i} = \frac{m_i}{\Sigma n_i M_i}$

$$y_{CO_2} = \frac{n_{CO_2} \times M_{CO_2}}{\Sigma n_i M_i} = \frac{3.8 \times 44}{1208.8} = 13.8\%$$

$$y_{N_2} = \frac{n_{N_2} \times M_{N_2}}{\Sigma n_i M_i} = \frac{31.6 \times 28}{1208.8} = 73.2\%$$

$$y_{O_2} = \frac{n_{O_2} \times M_{O_2}}{\Sigma n_i M_i} = \frac{2.2 \times 32}{1208.8} = 5.8\%$$

$$y_{H_2O} = \frac{n_{H_2O} \times M_{H_2O}}{\Sigma n M_i} = \frac{4.8 \times 18}{1208.8} = 7.2\%$$

12.4.2 Finding Actual Air using Orsat Analysis

Mass balance in combustion is shown schematically in Figure 12.1 using the example of a steadily operating combustion chamber.

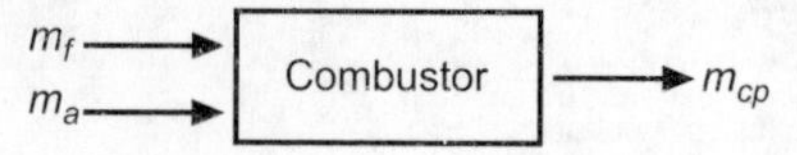

Fig. 12.1 Mass balance in combustion.

It is obvious from conservation of mass that,

$$m_f + m_a = m_{cp},$$

where, m_f, m_a and m_{cp} represent the mass-flow rates of fuel, air and combustion products respectively. Air supply consists of actual dry air (m_{da}) and atmospheric moisture ($m_{moisture}$) if any. The atmospheric moisture does not take part in the combustion process and simply exits along with the combustion products. Besides, dry gases (m_{dg}) such as CO_2, SO_2, CO, O_2, N_2 and unburnt hydrocarbons, the combustion products may consist of moisture (m_{H_2O}, cp) and solid refuse (m_R). The mass of moisture in the combustion products ($m_{H_2O,cp}$) is the sum of moisture in air and the moisture formed (m_{H_2O}) during combustion by burning H_2 in the fuel. This means,

$$m_{H_2O,cp} = m_{moisture} + m_{H_2O}$$

Therefore the mass balance equation can be written as:

$$m_f + (m_{da} + m_{moisture}) = m_{dg} + m_{moisture} + m_{H_2O} + m_R \tag{12.3}$$

In this relation, m_f and m_R are measurable and m_{H_2O} can be calculated by knowing the fraction of H_2 in the fuel. $m_{moisture}$ is calculated from the knowledge of the humidity in air. Thus, if m_{dg} is known, the mass of dry air used in combustion (m_{da}) can be calculated. The results of Orsat analysis helps in evaluating m_{dg}. Examples 12.7 and 12.8 are meant to explain how actual air is calculated using the Orsat analysis.

Example 12.7

A coal having an ultimate analysis of 65% carbon, 6% hydrogen, 10% oxygen, 4% nitrogen and 15% ash is burnt in a furnace with air. The solid refuse from the furnace is found to contain 25% carbon. Orsat analysis of the combustion products gave the following result: 14% CO_2, 1% CO, 3.5% O_2 and 81.5% N_2. Find (a) actual air (*b*) excess air (*c*) percentage theoretical air (*d*) mass fractions of the constituents of the combustion products and (*e*) dew point temperature, given that the barometric pressure is 1 bar.

▶ **Solution:**

(a) All carbon in the coal is not burnt as the solid refuse is found to contain unburnt carbon. Total carbon in the combustion products in the form of CO_2 and CO must be equal to burnt mass of carbon in the coal. Let this mass m be found first.

Per kg coal:

mass of C $= m_C = 0.65$ kg

mass of ash $= m_A = 0.15$ kg

mass of solid refuse $= m_R =$ (mass of ash + mass of carbon in refuse)

Therefore, $m_R = (m_A + C_R)$ where C_R is the mass of carbon in solid refuse.

From data, $C_R = 0.25 m_R = 0.25(m_A + C_R) = 0.25(0.15 + C_R)$

that is, $C_R = 0.05$ kg

Mass of carbon burnt $= m = (m_C - C_R) = 0.65 - 0.05 = 0.6$ kg.

Let x kilogram of coal when burnt with a kmol of O_2 in air give 100 kmol of dry combustion products and b kmol of H_2O. Noting that the number of kmol of each species in the coal is equal to mass of each species divided by its molar mass, the combustion equation can be written as:

$$x\left[\left(\frac{0.6}{12}\right)C + \left(\frac{0.06}{2}\right)H_2 + \left(\frac{0.1}{32}\right)O_2 + \left(\frac{0.04}{28}\right)N_2\right] + a(O_2 + 3.76N_2) \rightarrow$$

$$(14CO_2 + 1CO + 3.5O_2 + 81.5N_2) + bH_2O \qquad \text{(i)}$$

By C balance: $x\left(\frac{0.6}{12}\right) = 14 + 1$

Therefore, $x =$ mass of coal burnt $= 300$ kg.

By H_2 balance,

$$x\left(\frac{0.06}{2}\right) = b = 300\left(\frac{0.06}{2}\right), \text{ and therefore } b = 9$$

By O_2 balance, $x\left(\frac{0.1}{32}\right) + a = 14 + \frac{1}{2} + 3.5 + \frac{9}{2}$ and hence $a = 21.5625$.

By N_2 balance (as a check), $300\left(\frac{0.04}{28}\right) + 3.76a = 81.5$ and therefore $a = 21.5615$.

This value is in close agreement with that found through O_2 balance.

The mass of air used $= a(32 + 3.76 \times 28) = 21.5625(32 + 3.76 \times 28) = 2960$ kg

Therefore, actual air $= \frac{\text{mass of air}}{\text{mass of coal}} = \frac{2960}{300} = 9.867$ kg/kg coal

(b) Stoichiometric air can be easily calculated as follows:

Only combustible elements in the coal are C and H_2.

Soichiometric air for burning 1 kg coal, as in example 12.2(a) = 11.45 kg

Stoichiometric air for burning 1 kg H_2, as in exercise 12.2(b) = 34.3 kg

In one kg coal 0.65 kg C and 0.06 kg H_2 are burnt.

Stoichiometric air per kilogram of coal, therefore $= 0.65 \times 11.45 + 0.06 \times 34.3 = 9.5$ kg.

But, the coal already contains 0.1 kg O_2 per kilogram of coal.

As 0.232 kg O_2 is available in 1 kg air, the mass of air corresponding to 0.1 kg $O_2 = \frac{0.1}{0.232} = 0.4314$ kg.

This amount of air has to be deducted from the stoichiometric air calculated above to get the minimum quantity of air required for complete combustion.

Therefore, real stoichiometric air $= 9.5 - 0.4314 = 9.0686$ kg.

$$\text{Now, excess air} = \frac{(\text{actual air} - \text{stoichiometric air})}{\text{stoichiometric air}} = \frac{(9.867 - 9.0686)}{8.4976} = 8.8\%$$

(c) Percentage theoretical air = 100 + % excess air = 100 + 8.8 = 108.8%

(d) The combustion equation is obtained by substituting the values of x, a and b in Eq.(i)

$$300\left[\left(\frac{0.6}{12}\right)\text{C} + \left(\frac{0.06}{2}\right)\text{H}_2 + \left(\frac{0.1}{32}\right)\text{O}_2 + \left(\frac{0.04}{28}\right)\text{N}_2\right] + 21.5625(\text{O}_2 + 3.76\text{N}_2) \rightarrow$$

$$(14\text{CO}_2 + 1\text{CO} + 3.5\text{O}_2 + 81.5\text{N}_2) + 9\text{H}_2\text{O}$$

The mass of combustion products is obtained by summing the products of number of kmol and the molar mass of each species.

that is, $\Sigma m_i = \Sigma n_i M_i$

$$= 14 \times 44 + 1 \times 28 + 3.5 \times 32 + 81.5 \times 28 + 9 \times 18 = 3200 \text{ kg.}$$

Mass fraction of each species is given by,

$$y_i = \frac{m_i}{\Sigma m_i} = \frac{niMi}{\Sigma mi}$$

Hence,

$$y_{CO_2} = \frac{14 \times 44}{3200} = 19.25\%$$

$$y_{CO} = \frac{1 \times 28}{3200} = 0.875\%$$

$$y_{O_2} = \frac{3.5 \times 32}{3200} = 3.5\%$$

$$y_{N_2} = \frac{81.5 \times 28}{3200} = 71.3\%$$

$$y_{H_2O} = \frac{9 \times 18}{3200} = 5.075\%$$

(e) Molfraction of H_2O in the combustion products, x_{H_2O} is given by,

$$x_{H_2O} = \frac{n_{H_2O}}{\Sigma n_i} = \frac{9}{(14 + 1 + 3.5 + 81.5 + 9)} = 0.0826$$

Therefore, the partial pressure of $H_2O = x_{H_2O} \times p_{atm} = 0.0826 \times 1$ bar $= 8.26$ kPa.

Dew point temperature, t_{dp}, is defined as that temperature at which moisture in a mixture starts to condense when cooled at constant pressure. Thus, t_{dp} = saturation temperature of water corresponding to the partial pressure of water vapour in the mixture. From the steam tables, the saturation temperature corresponding to the pressure of 8.26 kPa is 42.5°C.

Hence, dew point temperature = 42.5°C.

Example 12.8

Ethylene (C_2H_4) is burnt in a steady flow combustor with dry air. Orsat analysis of the combustion products gave the following results: CO_2 9.7%, CO 1.1%, O_2 4% and N_2 85.2%. Find actual air and excess air.

▶ Solution:

Stoichiometric combustion equation is:

$$C_2H_4 + a(O_2 + 3.76N_2) \rightarrow (bCO_2 + cN_2) + dH_2O$$

By C balance, $b = 2$

By H_2 balance, $d = 2$

By O_2 balance, $a = b + \frac{d}{2} = 3$

By N_2 balance, $c = 3.76a = 3.76 \times 3 = 11.28$

Substituting the values of a, b, c and d in the combustion equation,

$$C_2H_4 + 3(O_2 + 3.76N_2) \rightarrow 2CO_2 + 11.28N_2 + 2H_2O$$

$$\text{Stoichiometric air} = \frac{3(32 + 3.76 \times 28)}{28} = 14.7 \text{ kg/kg ethylene.}$$

In the combustor, let x kmol of ethylene be burnt with p kmol of O_2 in air to get 100 kmol of dry combustion products of the given composition. The combustion equation would be:

$$xC_2H_4 + p(O_2 + 3.76N_2) \rightarrow (9.7CO_2 + 1.1CO + 4O_2 + 85.2N_2) + qH_2O$$

By C balance, $2x = 9.7 + 1.1$, that is, $x = 5.4$

By H_2 balance, $2x = q = 2 \times 5.4 = 10.8$

By O_2 balance, $p = 9.7 + \frac{1.1}{2} + 4 + \frac{q}{2} = 9.7 + \frac{1.1}{2} + 4 + \frac{10.8}{2} = 19.65$

Therefore, the actual combustion is:

$$5.4C_2H_4 + 19.65(O_2 + 3.76N_2) \rightarrow 9.7CO_2 + 1.1CO + 4O_2 + 85.2N_2 + 10.8H_2O$$

$$\text{Actual air} = \frac{19.65(32 + 3.76 \times 28)}{5.4 \times 28} = 17.8 \text{ kg/kg ethylene}$$

$$\text{Excess air} = \frac{(\text{Actual air} - \text{Stoichiometric air})}{\text{Stoichiometric air}} = \frac{(17.8 - 14.7)}{14.7} = 21\%.$$

12.5 First Law of Thermodynamics for a System Applied to Combustion

When a pure substance undergoes a change of state the first law of thermodynamics can be stated as

$$Q - W = (U_2 - U_1) = \Delta U.$$

Here ΔU the change in energy can be evaluated with the knowledge of the properties of the system initially and finally. For example, if the system is an ideal gas,

$$\Delta U = mc_v(t_2 - t_1).$$

Even if a system undergoes a chemical change, such as combustion, the form of the first law does not change and it can still be represented by the same relation. Perhaps suffix 2, indicating the final state, and suffix 1, indicating the initial state, may be replaced by P (products) and R (reactants), respectively. Thus,

$$Q - W = (U_P - U_R) = \Delta U'. \tag{12.4}$$

In the above equation, $\Delta U'$ is used in place of ΔU to distinguish between processes occurring with and without chemical reactions. Here, for the considered combustion reaction, it is not possible to evaluate the value of $\Delta U'$ from other properties such as p, v, T, c_v and c_p as was done earlier to evaluate ΔU in case of pure substances. This is understandable because, the products and reactants have different chemical compositions and hence their c_v and c_p are bound to be different and hence

$$\Delta U' \neq mc_v(t_P - t_R). \tag{12.5}$$

A pure substance like an ideal gas cannot undergo a state change without heat and/or work interactions. But it is not necessarily so in the case of combustion. Suppose a mixture of a fuel and air in an insulated rigid vessel is ignited by an infinitesimal spark the fuel burns spontaneously. While the reactants get converted to products, although there is no heat and work interactions, the products will be at a high temperature than the reactants. But, according to the first law of thermodynamics, as Q and W are each equal to zero,

$$Q - W = 0 = (U_P - U_R) = \Delta U'$$

Therefore, in a constant-volume adiabatic combustion process

$$U_P = U_R, \text{ and } t_p >> t_R \tag{12.6}$$

that is, the internal energy of the products at a high temperature is equal to the internal energy of the reactants at a lower temperature.

Suppose the combustion process is now carried out in a rigid but non-insulated vessel, and after the combustion the products are cooled to the initial temperature of the reactants, the first law for the entire process is still of the same form,

$$Q - W = (U_P - U_R)$$

Here W is zero and hence

$$Q_{v,t} = (U_P - U_R) \tag{12.7}$$

But $Q_{v,t}$($Q_{v,t}$ is used in place of Q as it indicates that the heat transfer is at constant volume and the initial temperature of the reactants and the final temperature of the products are the same) is negative as heat is transferred out of the vessel to cool the products. Therefore,

$$(U_P - U_R) < 0 \text{ and hence}$$

$$U_P < U_R, \text{ at } t_P = t_R. \tag{12.8}$$

It can hence be generalized that at any given temperature the internal energy of the products is less than that of the reactants.

12.5.1 *U–t* Diagram for a Combustion Process

For the sake of explanation and simplicity, let us consider the complete combustion of a kmol of CH_4 with 10% excess air in a closed rigid container. After the combustion, let the combustion products be cooled to the initial temperature of the reactants and the magnitude of the heat transfer out of the vessel is measured.

The stoichiometric combustion equation is

$$CH_4 + 2(O_2 + 3.76N_2) \rightarrow CO_2 + 7.52N_2 + 2H_2O$$

Therefore, the actual combustion equation is:

$$CH_4 + 2.2(O_2 + 3.76N_2) \rightarrow CO_2 + 0.2O_2 + 8.272N_2 + 2H_2O$$

One kmol of CO_2, 0.2 kmol O_2, 8.272 kmol N_2 and 2 kmol H_2O comprise the products. Assuming a common datum for internal energy of each component in the products and knowing the value of c_v for each it is possible to evaluate the internal energy of the products at a given temperature, t, using the relation,

$$\begin{aligned} U_P &= \Sigma n_i c_{vi}(t - t_0) \\ &= (1 \times c_{v,CO_2} + 0.2 \times c_{v,O_2} + 8.272 \times c_{v,N_2} + 2 \times c_{v,H_2O})(t - t_0), \end{aligned} \tag{12.9}$$

where t_0 is the datum temperature at which U of each species is assumed to be zero.

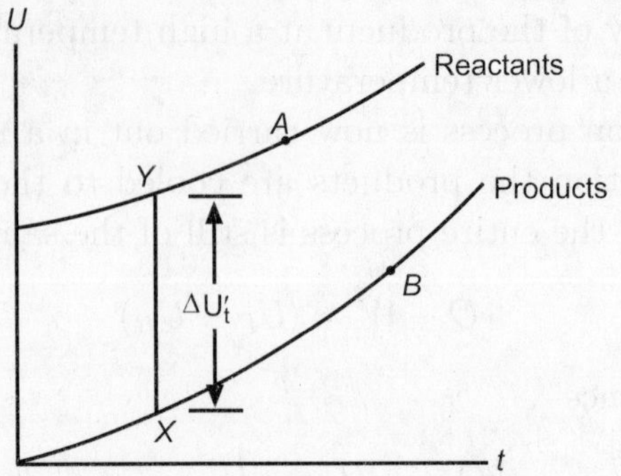

Fig. 12.2 U—t diagram for reactants and products.

If the specific heat values are in kJ/kmol K, then the unit of U_P is in kJ. As the products are obtained by burning 1 kmol of fuel (here it is 1 kmol of CH_4), more appropriately the unit of U_P is in kJ/kmol fuel.

Once the internal energies at different temperatures are evaluated, a plot of U versus t can be drawn for the products as in Figure 12.2.

Here, the reactants are 1 kmol CH_4, 2.2 kmol O_2 and 8.272 kmol N_2. A similar method can be adopted to evaluate the internal energy of the reactants at various temperatures using the relation,

$$\begin{aligned} U_R &= \Sigma n_i c_{vi}(t - t_o) \\ &= (1 \times c_{v,\mathrm{CH_4}} + 2.2 \times c_{v,\mathrm{O_2}} + 8.272 \times c_{v,\mathrm{N_2}})(t - t_o) \end{aligned} \tag{12.10}$$

If the specific heat values are in kJ/kmol K, then U_R is in kJ. As the reactants contain one kmol of fuel (here it is 1 kmol of CH_4), more appropriately, U_R is also in kJ/kmol fuel.

Variation of U_R with temperature can be plotted on a separate U–t diagram with an arbitrarily selected datum for the internal energies of the reactants. But, this variation cannot be plotted on the same U–t diagram for the products using the same datum used for the products. If done so, it implies that the internal energy of the products and the reactants are the same at the datum temperature. But we have already inferred that at any given temperature $U_R > U_P$. Thus, the following procedure is adopted to plot the variation of U_R on the same U–t diagram that gives the variation of U_P with temperature. Suppose the initial temperature of the reactants is t°C and the products are cooled to t°C after the constant-volume combustion process. Let point X on the U–t curve for the products indicate U_P at t. At the same temperature, according to (12.8) $U_P < U_R$ and according to (12.7),

$$Q_{v,t} = (U_P - U_R) = \Delta U'$$

where $Q_{v,t}$ is measurable experimentally. A point Y is marked vertically above X such that the distance between X and Y is equal to the magnitude of $Q_{v,t}$. This makes

$$(U_X - U_Y)_t = (U_P - U_R)_t = \Delta U' = Q_{v,t}.$$

Therefore, U_Y at point Y gives the value of U_R at t, with respect to U_P at the same t.

Now that the internal energy of the reactants is known at one temperature (t), its value at any other temperature can now be evaluated with the knowledge of specific heat and molfraction of each of the species in the reactants. Such a curve for the reactants is shown in Figure 12.2. This U–t diagram for the given reaction can now be used to find the change in internal energy in a constant volume combustion process, for any given values of initial temperature of the reactants and a final temperature of the products. The methodology adopted in doing so is explained below.

Let A be the initial state of the reactants, (U_A, t_A) and B the final state of the products, (U_B, t_B). The change in internal energy in this constant volume process is $= (U_B - U_A)$. And, this is also equal to the heat transfer (Q_{A-B}) during the said process between A and B. It can be seen from Figure 12.2 that,

$$\begin{aligned} Q_{A-B} &= (U_B - U_A) = (U_B - U_X) + (U_X - U_Y) + (U_Y - U_A). \\ &= (U_B - U_X) + \Delta U'_t + (U_Y - U_A) \end{aligned} \tag{12.11}$$

Here $\Delta U'_t = Q_{v,t}$ is an experimentally measured value at the temperature corresponding to state X. $(U_B - U_X)$ and $(U_Y - U_A)$ are respectively the changes in internal energy of products and reactants given by:

$$(U_B - U_X) = [\Sigma\, n_i \bar{c}_{vi}(t_B - t_X)]_{\text{products}} \quad \text{and} \tag{12.12}$$

$$(U_Y - U_A) = [\Sigma\, n_i \bar{c}_{vi}(t_Y - t_A)]_{\text{reactants}} \tag{12.13}$$

12.5.2 Internal Energy of Reaction

The change in internal energy in a constant-volume, isothermal (meaning the initial temperature of the reactants and the final temperature of the products are same) combustion reaction is called the internal energy of reaction. It is generally expressed in kJ/kmol fuel. It has a negative value for all exothermic reactions.

With reference to Figure 12.2, internal energy of reaction at a temperature t is given by,

$$\Delta U'_{v,t} = (U_X - U_Y)_t = (U_P - U_R)_t = Q_{v,t}$$

It is analysed already that $\Delta U'_{v,t}$ is negative. Generally $-\Delta U'_{v,t}$ is called the *heat of reaction at constant volume and temperature.*

12.5.2.1 Standard Internal Energy of Reaction

This is defined as $\Delta U'_{v,t}$ for a large volume (i.e. at a low pressure, generally taken as 100 kPa) and a temperature of 25°C. Also, the negative of this value is called the *standard heat of reaction at constant volume.*

12.6 First Law of Thermodynamics for a Control Volume Applied to Combustion

Suppose the reactants enter the combustion chamber at a certain pressure and temperature (state A) and the products leave the chamber steadily at a certain other or same temperature (state B), with negligible changes in kinetic and potential energies, the first law takes the form,

$$Q - W_x = H_B - H_A = \Delta H'$$

As during the reaction there is no shaft or electrical work, $W_x = 0$. Therefore,

$$Q = H_B - H_A = \Delta H'$$

12.6.1 *H–t* Diagram for a Combustion Process

All observations and arguments made in plotting a U–t diagram for reactants and products on a single diagram hold good here as well in drawing a H–t diagram. The variation of enthalpy with temperature for products, as shown in Figure 12.3, can be drawn with the knowledge of the values of c_p for all the species in the products and their molfractions. The same procedure is adopted to find and plot the variation of enthalpy of the reactants with temperature. The two curves will have to be relatively placed so that at any one temperature, t, corresponding to states X and Y, the vertical distance between the two is equal to $\Delta H'_{p,t}$. This is experimentally equal to the heat transfer in such a reaction at a low pressure and a temperature t.

H–t diagram in Figure 12.3 can be used to find $\Delta H'$ in any reaction with reactants at an arbitrary state A and the products at state B. The methodology generally adopted to do so is as explained in the following.

From the first law of thermodynamics,

$$Q_{A-B} = (H_B - H_A)$$

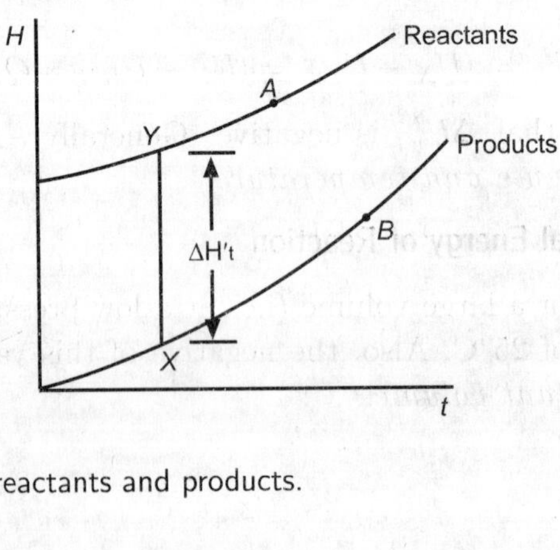

Fig. 12.3 H–t diagram for reactants and products.

Table 12.1 Ideal gas properties

Gas	Chemical Formula	M in kg/kmol (A)	(B)*	R kJ/kg K	C_p kJ/kg K
Air	–	28.97	28.97	0.287	1.005
Argon	Ar	39.948	40.0	0.208	0.520
Butane	C_4H_{10}	58.124	58.0	0.143	1.716
Carbon Dioxide	CO_2	44.010	44.0	0.189	0.842
Carbon Monoxide	CO	28.010	28.0	0.297	1.041
Ethane	C_2H_6	30.070	30.0	0.276	1.766
Ethylene	C_2H_4	28.054	28.0	0.296	1.548
Helium	He	4.003	4.0	2.077	5.193
Hydrogen	H_2	2.016	2.0	4.135	14.21
Methane	CH_4	16.040	16.0	0.518	2.254
Neon	Ne	20.183	20.0	0.421	1.039
Nitrogen	N_2	28.013	28.0	0.297	1.042
Oxygen	O_2	31.999	32.0	0.260	0.922
Propane	C_3H_8	44.097	44.0	0.188	1.679
Water	H_2O	18.02	18.0	0.461	1.86

(* Approximate values of M used in all problems in this book)

It can be seen from Figure 12.3 that

$$\begin{aligned}(H_B - H_A) &= (H_B - H_X) + (H_X - H_Y) + (H_Y - H_A) \\ &= (H_B - H_X) + \Delta H_t' + (H_Y - H_A) \qquad (12.14)\end{aligned}$$

Here $\Delta H_t' = Q_{p,t}$ is an experimentally measured value at the temperature corresponding to state X. Also,

$$(H_B - H_X) = [\Sigma n_i \bar{c}_{pi}(t_B - t_X)]_{\text{products}} \quad \text{and} \qquad (12.15)$$

$$(H_Y - H_A) = [\Sigma n_i \bar{c}_{pi}(t_Y - t_A)]_{\text{reactants}} \qquad (12.16)$$

Table 12.1 lists the values of c_p for a few gases.

12.6.2 Enthalpy of Reaction

The change in enthalpy in a constant-pressure, isothermal (meaning the initial temperature of the reactants and the final temperature of the products are same) combustion reaction is called the enthalpy reaction. It is generally expressed in kJ/kmol fuel. It has a negative value for all exothermic reactions.

With reference to Figure 12.3, enthalpy of reaction at a temperature t is given by,

$$\Delta H'_{p,t} = (H_X - H_Y)_t = (H_P - H_R)_t = Q_{p,t}$$

As in the case of $\Delta U'_{v,t,}, \Delta H'_{p,t}$ is negative. Generally $-\Delta H'_{p,t}$ is called the *heat of reaction at constant pressure and temperature.*

12.6.2.1 Standard Enthalpy of Reaction

This is defined as $\Delta H'_{p,t}$ for low pressure (generally considered to be 1 atm) and a temperature of 25°C. Also, the negative of this value is called the standard heat of reaction at constant pressure. The values of standard enthalpies of reaction for a few hydrocarbons are listed in Table 12.2.

Table 12.2 Standard enthalpy of reaction of some hydrocarbons in kJ/kg (at 25°C and 100 kpa)

		H_2O in Products in liquid phase		H_2O in Products in vapour phase	
Hydrocarbon	**Formula**	**Liquid HC**	**Gas. HC**	**Liquid HC**	**Gas. HC**
Methane	CH_4	–	–55496	–	–50010
Ethane	C_2H_6	–	–51875	–	–47484
Propane	C_3H_8	–49975	–50345	–45983	–46353
Butane	C_4H_{10}	–49130	–49500	–45344	–45714
Pentane	C_5H_{12}	–48643	–49011	–44983	–45351
Hexane	C_6H_{14}	–48308	–48676	–44733	–45101
Heptane	C_7H_{16}	–48071	–48436	–44557	–44922
Octane	C_8H_{18}	–47893	–48256	–44425	–44788
Decane	$C_{10}H_{22}$	–47641	–48000	–44239	–44598
Ethene	C_2H_4	–	–50296	–	–47158
Propene	C_3H_6	–	–48917	–	–45780
Butene	C_4H_8	–	–48453	–	–45316
Pentene	C_5H_{10}	–	–48134	–	–44996
Hexene	C_6H_{12}	–	–47937	–	–44800
Heptane	C_7H_{14}	–	–47800	–	–44662
Octane	C_8H_{16}	–	–47693	–	–44556
Decene	$C_{10}H_{20}$	–	–47547	–	–44410
Benzene	C_6H_6	–41831	–42266	–40141	–40576
Methylbenzene	C_7H_8	–42437	–42847	–40527	–40937
Ethylbenzene	C_8H_{10}	–42997	–43395	–40924	–41322
Propylbenzene	C_9H_{12}	–43416	–43800	–41219	–41603
Butylbenzene	$C_{10}H_{14}$	–43748	–44123	–41453	–41828

12.7 Relation between $\Delta H'_{p,t}$ and $\Delta U'_{v,t}$

If reactants and products are treated as ideal gases their internal energy and enthalpy happen to be functions of temperature only. Hence,

$$\Delta H'_{p,t} = \Delta H'_t$$

and

$$\Delta U'_{v,t} = \Delta U'_t$$

Now, as $H = U + pV$,

$$\begin{aligned}\Delta H'_t &= \Delta U'_t + \Delta(pV)_t \\ &= \Delta U'_t + (p_P V_P - p_R V_R)_t \\ &= \Delta U'_t + (n_P - n_R)RT,\end{aligned}$$

where n_P and n_R are, respectively, the number of kmols of products and reactants, R is the universal gas constant and T is the temperature in kelvin.

Example 12.9

Methane is burnt completely with 10% excess air in a combustor under steady flow conditions. The temperature of the reactants is 30°C and that of the products is 120°C. Find the heat transfer from the combustor per kg CH_4. Assume the following: standard enthalpy of reaction is −802.3 MJ/kmol CH_4 and the average values of c_p in kJ/kmol K, for various species are 36.0 for CH_4, 29.5 for O_2, 29.2 for N_2, 37.0 for CO_2 and 33.7 for H_2O.

► **Data:**

Fuel is CH_4; combustion is complete; 10% excess air, $t_R = 30°C$, $t_P = 120°C$, $\Delta H'_{25} = -802.3$ MJ/kmol CH_4, $c_{P,CH_4} = 36.0$, $c_{P,O_2} = 29.5$, $c_{P,N_2} = 29.2$, $c_{P,CO_2} = 37$ and, $c_{P,H_2O} = 33.7$, all in kJ/kmol K.

► **Solution:**

The combustion equation is,

$$CH_4 + 2.2(O_2 + 3.76N_2) \rightarrow CO_2 + 0.2O_2 + 8.272N_2 + 2H_2O$$

Referring to Figure 12.3,

$$\begin{aligned}Q_{A-B} &= (H_B - H_A) \\ &= (H_B - H_X) + (H_X - H_Y) + (H_Y - H_A) \\ &= (H_B - H_X) + \Delta H'_t + (H_Y - H_A)\end{aligned}$$

Here $t_A = 30°\text{C}$, $t_X = t_Y = 25°\text{C}$ and $t_B = 120°\text{C}$.

$$
\begin{aligned}
(H_B - H_X) &= [\Sigma\, n_i \bar{c}_{pi}(t_B - t_X)]_{\text{products}} \\
&= (1 \times \bar{c}_{p,\text{CO}_2} + 0.2 \times \bar{c}_{p,\text{O}_2} + 8.272 \times \bar{c}_{p,\text{N}_2} + 2 \times \bar{c}_{p,\text{H}_2\text{O}})(t_B - t_X) \\
&= (1 \times 37 + 0.2 \times 29.5 + 8.272 \times 29.2 + 2 \times 33.7) \times (120 - 25) \\
&= 33,425 \text{ kJ/kmol fuel.} \\
\Delta H_t' &= \text{Standard enthalpy of reaction} = -802.3 \times 10^3 \text{ kJ/kmol CH}_4
\end{aligned}
$$

$$
\begin{aligned}
(H_Y - H_A) &= [\Sigma\, n_i \bar{c}_{pi}(t_Y - t_A)]_{\text{reactants}} \\
&= (1 \times \bar{c}_{p,\text{CH}_4} + 2.2 \times \bar{c}_{p,\text{O}_2} + 8.272 \times \bar{c}_{p,\text{N}_2})(t_Y - t_A) \\
&= (1 \times 36 + 2.2 \times 29.5 + 8.272 \times 29.2) \times (25 - 30) \\
&= -1308.6 \text{ kJ/kmol of fuel.} \\
Q_{A-B} &= (H_B - H_X) + \Delta H_t' + (H_Y - H_A) \\
&= 33,425 - 802.3 \times 10^3 - 1308.6 \\
&= -770,183.6 \text{ kJ/kmol CH}_4 \\
&= \frac{-770\,183.6}{M_{\text{CH4}}} \text{ kJ/kg CH}_4 \\
&= \frac{-770\,183.6}{16} = -48,136.5 \text{ kJ/kg CH}_4
\end{aligned}
$$

12.7.1 Effect of Phase on $\Delta H'$ and $\Delta U'$

It has thus far been assumed that the reactants and products are in gaseous phases. However, a fuel can either be in liquid or gaseous phase and at least one component in the products, namely H_2O, can also exist in liquid or gaseous phase. At any one given temperature, the enthalpy of a substance in liquid phase is less than that of gaseous phase. Using a similar procedure explained earlier the variation of enthalpy of the reactants when in liquid phase can be drawn on the same H–t diagram that shows the enthalpy variation for the gaseous phase of reactants. The curve for liquid phase will be below that of the gaseous phase. Also, the enthalpy variation of products when H_2O is in liquid phase can also be drawn on the same diagram that gives the enthalpy variation of products when H_2O is in gaseous phase. This is shown in Figure 12.4. At any one temperature the distance between the curves for liquid and gaseous reactants is separated by $n\bar{h}_{fg}$, where n is the number of kmols of the reactant and h_{fg} is the molar enthalpy of vapourization of the reactant. The same logic is also applicable for the two curves for the products.

The following can be deduced from Figure 12.4: At any one temperature, for the combustion of liquid reactants to gaseous products

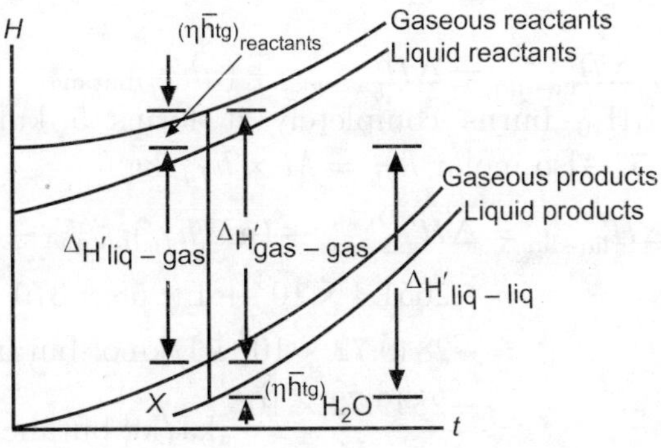

Fig. 12.4 H–t diagram for different phases of reactants and products.

$$\Delta H'_{\text{liq-gas}} = \Delta H'_{\text{gas-gas}} + (n\bar{h}_{fg})_{\text{reactants}} \tag{12.17}$$

Note: From Figure 12.4, $|\Delta H'_{\text{liq-gas}}| < |\Delta H'_{\text{gas-gas}}|$. We know both the enthalpies have negative values. Hence $n\bar{h}_{fg}$ has to be added to $\Delta H'$gas-gas to reduce its magnitude.

Similarly, for the combustion of liquid reactants to liquid products

$$\Delta H'_{\text{liq-liq}} = \Delta H'_{\text{gas-gas}} + (n\bar{h}_{fg})_{\text{reactants}} - (n\bar{h}_{fg})_{\text{products}} \tag{12.18}$$

The analysis made for $\Delta H'$ is valid for $\Delta U'$ as well.

Tabulated in Table 12.2 are values of $\Delta H'_{\text{std}}$ for a few hydrocarbons. There are four different values of $\Delta H'_{\text{std}}$ for each fuel depending upon the phase of the hydrocarbon and the phase of H_2O in the combustion products.

Example 12.10

Liquid butane (C_4H_{10}) at 25°C is burnt completely and the products are cooled to the initial temperature of the reactants and the H_2O in the products is found to be in liquid phase. Find the enthalpy of reaction in kJ/kg butane given that the standard enthalpy of reaction of gaseous butane is −2651.4 MJ/kmol when the H_2O in products is also in gaseous phase. Assume enthalpies of vapourization of butane and water, at 25°C, are, respectively, 370 and 2442 kJ/kg.

▶ **Data:**

Standard $\Delta H'_{\text{gas-gas}} = -2651 \times 10^3$ kJ/kmol butane, $(h_{fg})_{\text{butane}} = 370$ kJ/kg, $(h_{fg})_{\text{water}} = 2442$ kJ/kg. The reaction occurs at 25°C. Also, $M_{\text{butane}} = 58$ and $M_{H_2O} = 18$.

▶ Solution:

Referring to Figure 12.4, $\Delta H'_{\text{liq-liq}} = \Delta H'_{\text{gas-gas}} + (nh'_{fg})_{\text{butane}} - (nh'_{fg})_{H_2O}$

When 1 kmol of C_4H_{10} burns completely it forms 5 kmol of H_2O. Therefore, $n_{\text{butane}} = 1$ and $n_{H_2O} = 5$. Also molar $h_{fg} = M \times h_{fg}/\text{kg}$

$$\begin{aligned}\text{Therefore,}\qquad \Delta H'_{\text{liq-liq}} &= \Delta H'_{\text{gas-gas}} + (nMh_{fg})_{\text{butane}} - (nMh_{fg})_{H_2O}\\ &= -2651.4 \times 10^3 + 1 \times 58 \times 370 - 5 \times 18 \times 2442\\ &= -2849.72 \times 10^3 \text{ kJ/kmol butane}\\ &= \frac{-2849.72 \times 10^3}{M} \text{ kJ/kg butane}\\ &= \frac{-2849.72 \times 10^3}{58} = -49133 \text{ kJ/kg.}\end{aligned}$$

Note: Referring to Table 12.2, it is found that the standard enthalpy of reaction for butane in liquid phase and the water in products also in liquid phase is −49130 kJ/kg. This is almost the same value as found in the above example.

12.8 Enthalpy of Formation

The study made until now shows that the change in enthalpy in a reaction from any temperature of the reactants to any temperature of the products can be evaluated with the help of just the standard enthalpy of reaction and the properties of the constituents of both reactants and products. However, the only disadvantage in this method is that the values of standard enthalpies of reaction have to be exhaustively tabulated for all the multitudinous reactions. This is not an easy job.

Enthalpy of formation is a convenient alternative with the help of which standard enthalpy of reaction in any chemical reaction can be evaluated.

The enthalpy of formation of a chemical species is defined as the enthalpy of the product in a constant pressure (100 kPa) isothermal (25°C) reaction in which the species happen to be the only product and the reactants are chemical elements in their datum states. At the datum state of 100 kPa and 25°C the enthalpies of O_2, N_2, H_2 and Carbon (graphite) are assigned zero values.

Generally, the enthalpy of formation is denoted by h'_f (specific enthalpy in kJ/kg) or $\bar{h}'_f$ (molar enthalpy in kJ/kmol).

For example, the enthalpy of formation of CO, according to definition, is the enthalpy of CO in a constant pressure isothermal reaction occurring at 100 kPa and 25°C from carbon and oxygen.

$$C_{\text{solid}} + \frac{1}{2}O_2 \rightarrow CO$$

$$\begin{aligned}\Delta H'_{\text{standard}} &= \text{enthalpy of the product} - \text{enthalpy of the reactants}\\ &= H_P - H_R = \Delta H'_{\text{std}}\end{aligned}$$

Here, H_R is zero, as C and O_2 are at datum states, and therefore,

$$\Delta H'_{\text{std}} = H_P = (n\bar{h}'_f)_{CO} = (\bar{h}'_f)_{\text{CO}} \qquad (\text{as } n_{\text{CO}} = 1)$$

Enthalpy of formation of various chemical species is listed in Table.12.3.
The following example explains how the standard enthalpy of any reaction can be evaluated using various enthalpy of formation values:

Example 12.11

Find $\Delta H'_{\text{std}}$ for the reaction $C_3H_8 + 5O_2 \rightarrow 3CO_2 + 4H_2O$, assuming H_2O in the products is in liquid phase.

► **Solution:**

$$\begin{aligned}\Delta H'_{\text{std}} &= H_P - H_R \\ &= [(n\bar{h}'_f)_{\text{CO}_2} + (n\bar{h}'_f)_{\text{H}_2\text{O}}] - [(n\bar{h}'_f)_{\text{C}_3\text{H}_8} + (n\bar{h}'_f)_{\text{O}_2}] \\ &= [3(\bar{h}'_f)_{\text{CO}_2} + 4(\bar{h}'_f)_{\text{H2O}}] - [1(\bar{h}'_f)_{\text{C}_3\text{H}_8} + 5(\bar{h}'_f)_{\text{O}_2}]\end{aligned}$$

Table 12.3 Enthalpy of formation of various substances at 25°C and 100 kPa.

Substance	Formula	Phase	$\bar{h}'_f$(kJ/kmol)
Carbon monoxide	CO	Gas	−110,530
Carbon dioxide	CO_2	Gas	−393,510
Water	H_2O	Gas	−241,820
Water	H_2O	Liquid	−285,838
Methane	CH_4	Gas	− 74,850
Acetylene	C_2H_2	Gas	266,748
Ethene	C_2H_4	Gas	52,283
Ethane	C_2H_6	Gas	−83,850
Propane	C_3H_8	Gas	−104,680
Butane	C_4H_{10}	Gas	−126,650
Pentane	C_5H_{12}	Gas	−146,710
Hexane	C_6H_{14}	Gas	−166,940
Heptane	C_7H_{16}	Gas	−187,650
Octane	C_8H_{18}	Gas	−208,447
Octane	C_8H_{18}	Liquid	−249,952
Carbon(graphite)	C	Solid	0
Benzene	C_6H_6	Gas	82,927
Methanol	CH_3OH	Gas	−201,041
Ethanol	C_2H_5OH	Gas	−234,430
Hydrazine	N_2H_4	Liquid	50,626

Substituting the values of enthalpy of formation of all the species from Table 12.3,

$$\begin{aligned}\Delta H'_{\text{std}} &= [3(-393,510) + 4(-285,838)] - [1(-104,680) + 5(0)] \\ &= -2,219,202 \text{ kJ/kmol propane} \\ &= \frac{-2,219,202}{M_{C_3H_8}} = \frac{-2,219,202}{44} = -50,436 \text{ kJ/kg}\end{aligned}$$

Note: This value is almost the same as −50345 kJ/kg, which is the value listed in Table 12.2 for gaseous propane and liquid H_2O in products.

Example 12.12

Ethylene (C_2H_4) is burnt completely with 10% excess air in a steady-flow combustor. The reactants enter the combustion chamber at 50°C and the products leave the combustor at 120°C. Assume water in the products is in gaseous phase and find the heat transferred from the combustor per kg ethylene. Use enthalpy of formation and specific heat Tables.

▶ Solution:

The combustion equation is,

$$C_2H_4 + 3.3(O_2 + 3.76N_2) \rightarrow 2CO_2 + 2H_2O + 0.3O_2 + 12.408N_2$$

Figure 12.3 may be referred as the $H - t$ diagram for the reaction. Here, while A represents the initial state of the reactants, B represents the final state of the products.

From the figure, $H_B - H_A = (H_B - H_X) + (H_X - H_Y) + (H_Y - H_A)$

$$\begin{aligned}(H_B - H_X) &= [\Sigma\, n_i \bar{c}_{pi}(t_B - t_X)]_{\text{products}} \\ &= [(2 \times \bar{c}_p)_{CO_2} + (2 \times \bar{c}_p)_{H_2O} + (0.3 \times \bar{c}_p)_{O_2} + 12.408 \times c_p)_{N_2}] \\ &\quad \times (120 - 25)\end{aligned}$$

Picking up the values of c_p from Table 12.1, and noting that $\bar{c}_p = Mc_p$

$$\begin{aligned}(H_B - H_X) &= [(2 \times 44 \times 0.842 + (2 \times 18 \times 1.86) + (0.3 \times 32 \times 0.922) \\ &\quad +(12.408 \times 28 \times 1.042)] \times 95 \\ &= 48,635 \text{ kJ/kmol } C_2H_4\end{aligned}$$

$$\begin{aligned}\text{Similarly,}\quad (H_Y - H_A) &= [\Sigma\, n_i \bar{c}_{pi}(t_Y - t_A)]_{\text{reactants}} \\ &= [(1 \times \bar{c}_p)_{C_2H_4} + (3.3 \times \bar{c}_p)_{O_2} + (12.408 \times \bar{c}_p)_{N_2}] \times (25 - 50) \\ &= [(1 \times 28 \times 1.548) + (3.3 \times 32 \times 0.922) \\ &\quad +(12.408 \times 28 \times 1.042)] \times (-25) \\ &= -12,568 \text{ kJ/kmol } C_2H_4\end{aligned}$$

And, $(H_X - H_Y) = \Delta H'_{std} = (2 \times \bar{h}'_f)_{CO_2} + (2 \times \bar{h}'_f)_{H_2O} - (1 \times \bar{h}'_f)_{C_2H_4}$

$= 2 \times (-393,510) + 2 \times (-241,820) - (52,283)$

$= -1,322,943$ kJ/kmol

Therefore, $H_B - H_A = 48635 - 1,322,943 - 12,568 = -1,286,876$ kJ/kmol

$$= -\frac{1,286,876}{28} = -45,960 \text{ kJ/kg ethane.}$$

12.9 Adiabatic Combustion Temperature

Consider a fuel–air mixture burnt steadily at a constant pressure in an adiabatic combustor. Let the changes in kinetic and potential energies between the entry stream of reactants and the exit stream of products be negligible. Applying SS, SFEE to the combustor ($Q = 0, W = 0, \frac{\Delta V^2}{2} = 0$ and $\Delta gZ = 0$) it can be shown that

$$(H_P - H_R) = 0$$

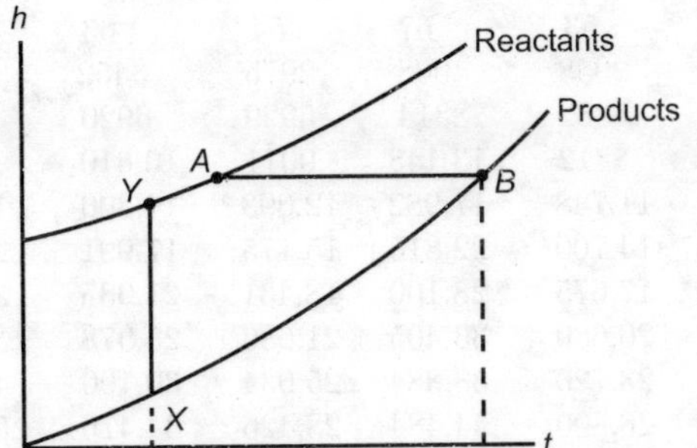

Fig. 12.5 H–t diagram for finding adiabatic combustion temperature.

That is, the combustion process is *isenthalpic*. Thus, the state of the reactants and the products is on a horizontal line as shown in H–t diagram of Figure 12.5. The temperature of the products being ejected is called the *adiabatic combustion temperature or adiabatic flame temperature* of the considered reaction. In other words, adiabatic combustion temperature is defined as the temperature of the products in a steady-flow constant pressure adiabatic combustion process with negligible changes in kinetic and potential energies. For this temperature to be maximum in any reaction, the combustion has to be complete under stoichiometric conditions.

For $(H_P - H_R)$ to be equal to zero, it is obvious from the figure that

$$(H_B - H_X) + (H_X - H_Y) + (H_Y - H_A) = 0$$

that is, $$[\Sigma n_i(\bar{h}_B - \bar{h}_X)]_{\text{products}} + \Delta H'_{std} + [\Sigma n_i(\bar{h}_Y - \bar{h}_A)]_{\text{reactants}} = 0 \qquad (12.19)$$

or, $$[\Sigma n_i \bar{c}_{pi}(t_B - t_X)]_{\text{products}} + \Delta H'_{std} + [\Sigma n_i \bar{c}_{pi}(t_Y - t_A)]_{\text{reactants}} = 0 \qquad (12.20)$$

Here t_B is the adiabatic combustion temperature and $t_X = t_Y = 25°C$.

The temperature t_B determined using (12.19) is most accurate. The data required to do so is the enthalpy of various species in both the reactants and the products at various temperatures with respect to enthalpies at X and Y. Listed in Table 12.4 are such enthalpies of a few chemical species. In this table, the enthalpy of each species at any given temperature is actually the difference between the enthalpies at the said temperature and 25°C. However, as explained in Example 12.13, the procedure involves a trial and error solution for the adiabatic combustion temperature.

It is easier to evaluate the adiabatic combustion temperature using (12.20). The procedure adopted requires the average values of c_p for various species in both the reactants and the products. These values can be taken from Table 12.1. However, the result obtained is not very accurate, as the procedure assumes a common value of c_p over a wide range of temperature which in reality is far from true.

Table 12.4 Ideal gas enthalpy ($\bar{h}_t - \bar{h}_{25C}$) (in kJ/kmol, Datum : 25°C = 298 K)

Temp. (K)	N_2	O_2	H_2	CO_2	CO	H_2O	CH_4	C_2H_2	C_2H_4
300	54	54	53	67	54	63	71	89	85
400	2971	3029	2958	4008	2975	3452	3870	4838	4895
500	5912	6088	5882	8314	5929	6920	8206	10,113	10,676
600	8957	9339	8812	13,148	9011	10,610	13,132	15,778	17,338
700	12,164	12,590	11,748	17,982	12,093	14,300	18,640	21,754	24,770
800	15,046	15,841	14,700	22,815	15,175	17,991	24,686	27,999	32,861
900	18,253	19,274	17,675	28,100	18,431	21,985	31,214	34,491	41,515
1000	21,460	22,707	20,680	33,405	21,686	25,978	38,182	41,221	50,668
1100	24,759	26,210	23,720	38,884	25,034	30,190	45,548	48,138	60,261
1200	28,108	29,765	26,800	44,484	28,426	34,476	53,271	55,256	70,249
1300	31,503	33,340	29,920	50,149	31,866	38,947	61,306	62,547	80,580
1400	34,936	36,966	33,083	55,907	35,338	43,447	69,616	69,994	91,209
1500	38,405	40,610	36,289	61,714	38,848	48,095	78,165	77,582	102,096
1600	41,903	44,279	39,540	67,580	42,384	52,844	88,925	85,298	113,209
1700	45,430	47,970	42,831	73,492	45,940	57,685	95,869	93,128	124,518
1800	48,982	51,689	46,164	79,442	49,522	62,609	104,975	101,063	135,998
1900	52,551	55,434	49,538	85,429	53,124	67,613	114,225	109,094	147,631
2000	56,141	59,199	52,950	91,450	56,739	72,689	123,603	117,211	159,398
2100	59,748	62,986	56,396	97,500	60,375	77,831	133,094	125,408	171,287
2200	63,371	66,802	59,877	103,575	64,019	83,036	142,686	133,679	183,285
2300	67,007	70,634	63,390	109,671	67,676	88,295	152,371	142,017	195,383
2400	70,651	74,492	66,932	115,788	71,346	93,604	161,239	150,418	207,571
2500	74,312	78,375	70,502	121,926	75,023	98,964	171,982	158,877	219,842
2600	77,973	82,274	74,100	128,085	78,714	104,370	181,894	167,390	232,189
2700	81,659	86,199	77,721	134,256	82,408	109,813	191,868	175,953	244,604
2800	85,345	90,144	81,368	140,444	86,115	115,294	201,897	184,563	257,177
2900	89,036	94,111	85,040	146,645	89,826	120,813	211,972	193,215	269,599
3000	92,738	98,098	88,738	152,862	93,542	126,361	222,088	201,905	282,157

Example 12.13

Find the adiabatic combustion temperature when methane is burnt completely with 300% excess air using (a) average c_p values and (b) the ideal gas enthalpy. Initially the reactants are at 25°C.

▶ Solution:

Figure 12.6 shows the combustion process on a H–t diagram.

For adiabatic combustion, in general,

$$(H_B - H_X) + (H_X - H_Y) + (H_Y - H_A) = 0$$

where A refers to the initial state of the reactants and B the final state of the products.

Because the reactants are at 25°C, the initial states of reactants coincide with Y.

Therefore,
$$(H_B - H_X) + (H_X - H_Y) = 0$$
$$\left[\Sigma n_i \left(\bar{h}_B - \bar{h}_X\right)\right]_{\text{products}} + \Delta H'_{\text{std}} = 0$$

The combustion equation is,

$$CH_4 + 3 \times 2(O_2 + 3.76N_2) \rightarrow CO_2 + 2H_2O + 3O_2 + 22.56N_2$$

Therefore,
$$[n(\bar{h}_B - \bar{h}_X)]_{CO_2} + [n(\bar{h}_B - \bar{h}_X)]_{H_2O} + [n(\bar{h}_B - \bar{h}_X)]_{O_2} + [n(\bar{h}_B - \bar{h}_X)]_{N_2} + \Delta H'_{\text{std}} = 0 \quad \text{(i)}$$

$$[n\bar{c}_p(t_B - t_X)]_{CO_2} + [n\bar{c}_p(t_B - t_X)]_{H_2O} + [n\bar{c}_p(t_B - t_X)]_{O_2} + [n\bar{c}_p(t_B - t_X)]_{N_2} + \Delta H'_{\text{std}} = 0$$

that is,
$$[(n\bar{c}_p)_{CO_2} + (n\bar{c}_p)_{H_2O} + (n\bar{c}_p)_{O_2} + (n\bar{c}_p)_{N_2}](t_B - t_X) + \Delta H'_{\text{std}} = 0 \quad \text{(ii)}$$

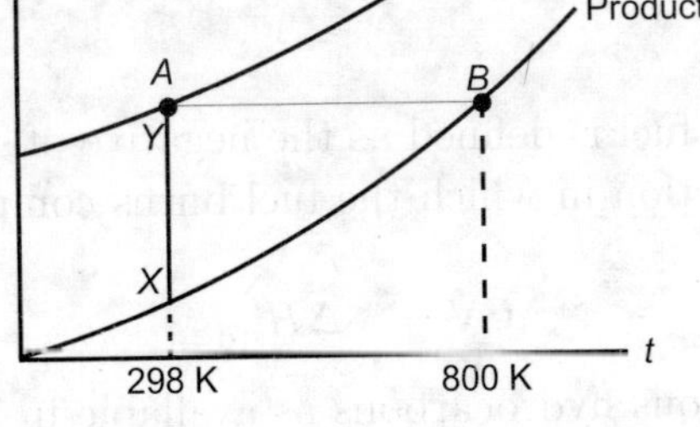

Fig. 12.6 H–t diagram for finding adiabatic combustion temperature.

(a) Picking up the various values of c_p from Table 12.1 and the value of $\Delta H'_{\text{std}}$ for CH_4 from Table 10.2, and substituting them in (ii),

$$[1\times 44\times 0.842+2\times 18\times 1.86+3\times 32\times 0.922+22.56\times 28\times 1.042](t_B-25)-16\times 50,010=0$$

Therefore, t_B the adiabatic combustion temperature = 965.5°C = 1238.5 K.

(b) In (i) $\Delta H'_{\text{std}} = -16 \times 50010 = -800160$ kJ/kmol

Therefore, (i) can be rewritten as,

$$[n(\bar{h}_B - \bar{h}_X)]_{CO_2} + [n(\bar{h}_B - \bar{h}_X)]_{H_2O} + [n(\bar{h}B - \bar{h}_X)]_{O_2} + [n(\bar{h}_B - h_X)]_{N_2} = 800160$$

The value of t_B found by trial and error method must be such that the left-hand-side terms add to 800160 kJ/kmol.

As a first trial let $t_B = 1000$ K

Taking the values of $(\bar{h}_B - \bar{h}_X)$ for each species, at 1000 K, from Table 10.4, and substituting them in the above relation,

$$1 \times 33,405 + 2 \times 25,978 + 3 \times 22,707 + 22.56 \times 21,460 = 637,617 \text{ kJ/kmol}$$

637,617 is less than 800,160 and hence the temperature must be more than 1000 K.

Now, let $t_B = 1200$ K.

Taking the values of $(h_B - h_X)$ for each species, at 1200 K, from Table 10.4, and substituting them in the above relation,

$$1 \times 44,484 + 2 \times 34,476 + 3 \times 29,765 + 22.56 \times 28,108 = 836,847$$

836,847 is greater than 800,160. Therefore, the real temperature must be between 1000 K and 1200 K.

The value of t_B that makes the sum equal to 800,160 can now be evaluated by linear interpolation and it happens to be 1163 K.

12.10 Calorific Value

The calorific value, CV, of a fuel is defined as the negative of standard heat of reaction at constant pressure of the reaction in which the fuel burns completely with oxygen.

$$\text{CV} = -\Delta H'_{\text{std}}$$

The values of $\Delta H'_{\text{std}}$ for various hydrocarbons as available in Table 12.2 when multiplied by −1 give the corresponding calorific values. It may be noted that CV is nothing but the heat transfer per kg or kmol fuel in a constant pressure isothermal reaction at 25°C.

It is known that H_2O in products may exist in liquid or gaseous phase in the products. The value of CV is different for these two situations.

$-\Delta H'_{\text{std}}$ when H_2O in the products is in liquid phase is called the *higher calorific value* (*HCV*) and $-\Delta H'$std when H_2O in the products is in gaseous phase is called the *lower calorific value* (*LCV*). The relation between the two, when the calorific values are given in kJ/kg, is

$$\text{HCV} = \text{LCV} + mh_{fg}$$

where m is the mass of H_2O formed during combustion and h_{fg} is the enthalpy of vapourization of water at 25°C which is equal to 2442 kJ/kg.

Example 12.14

It is given that the standard heat of reaction of gaseous hexane (C_6H_{14}) when H_2O in products is in gaseous phase is −45101 kJ/kg. Find the LCV and HCV of gaseous hexane. Assume that the enthalpy of vapourization of H_2O at 25°C is 2442 kJ/kg.

▶ **Solution:**

$$\begin{aligned} \text{LCV} &= -\Delta H'_{\text{std}} \text{ with } H_2O \text{ in products in gaseous phase} \\ &= -(-45101) = 45101 \text{ kJ/kg hexane.} \end{aligned}$$

Complete combustion of hexane with O_2 follows the equation:

$$C_6H_{14} + 9.5O_2 \rightarrow 6CO_2 + 7H_2O$$

Therefore, a kmol of C_6H_{14} gives 7 kmol H_2O.

That is, 86 kg C_6H_{14} gives 7×18 kg H_2O.

Hence, 1 kg C_6H_{14} gives $\dfrac{(7 \times 18)}{86} = 1.465$ kg H_2O

$$\begin{aligned} \text{HCV} &= \text{LCV} + (mh_{fg})_{H_2O} \\ &= 45101 + (1.465 \times 2442) = 48678 \text{kJ/kg hexane.} \end{aligned}$$

Illustrative Problems

12.1 A certain gaseous fuel has the following volumetric analysis: CO_2, 3%; CO, 19%; H_2, 41%; CH_4, 25%; C_2H_4, 9% and N_2, 3%. Find (a) gravimetric analysis, (b) ultimate analysis, (c) analysis of the combustion products when the fuel is burnt completely with 15% excess air, (d) Orsat analysis of the products in (c), (e) mass of combustion products in (c) per kilogram of fuel and (f) dew point temperature of the products in (c) given that the combustion occurs at a steady pressure of 1 bar.

▶ **Solution:**

(a) $y_i = \dfrac{x_i M_i}{M}$ where $M = \Sigma\, x_i M_i$.

$$\text{Here,}\quad M = 0.03 \times 44 + 0.19 \times 28 + 0.41 \times 2 + 0.25 \times 16 + 0.09 \times 28 + 0.03 \times 28$$

$$= 14.82 \text{ kg/kmol}$$

$$\text{Hence,}\quad y_{CO_2} = \frac{0.03 \times 44}{14.82} = 8.9\%$$

$$y_{CO} = \frac{0.19 \times 28}{14.82} = 35.9\%$$

$$y_{H_2} = \frac{0.41 \times 2}{14.82} = 5.5\%$$

$$y_{CH_4} = \frac{0.25 \times 16}{14.82} = 27\%$$

$$y_{C_2H_4} = \frac{0.09 \times 28}{14.82} = 17\%$$

$$y_{N_2} = \frac{0.03 \times 44}{14.82} = 5.7\%$$

(b) The fuel consists of C, H_2, O_2 and N_2.

Per kmol fuel,
$$n_C = 0.03 + 0.19 + 0.25 + 0.09 \times 2 = 0.65$$

$$n_{H_2} = 0.41 + 2 \times 0.25 + 2 \times 0.09 = 1.09$$

$$n_{O_2} = 0.03 + \frac{0.19}{2} = 0.125$$

$$n_{N_2} = 0.03$$

As
$$m_i \text{kg fuel} = \frac{n_i M_i}{M}$$

$$m_C = \frac{n_C M_C}{M} = \frac{0.65 \times 12}{14.82} = 0.526 \text{ kg/kg fuel}$$

similarly,
$$m_{H_2} = \frac{1.09 \times 2}{14.82} = 0.147 \text{ kg/kg fuel}$$

$$m_{O_2} = \frac{0.125 \times 32}{14.82} = 0.27 \text{ kg/kg fuel}$$

$$m_{N_2} = \frac{0.03 \times 28}{14.82} = 0.057 \text{ kg/kg fuel}$$

(c) Stoichiometric combustion equation for 100 kmol fuel is;

$$3CO_2 + 19CO + 41H_2 + 25CH_4 + 9C_2H_4 + 3N_2 + a(O_2 + 3.76N_2) \rightarrow bCO_2 + cH_2O + dN_2$$

By C balance: $b = 3 + 19 + 25 + 9 \times 2 = 65$

By H_2 balance: $c = 41 + 25 \times 2 + 9 \times 2 = 109$

By O_2 balance: $b + \frac{c}{2} = 3 + \frac{19}{2} + a = 65 + \frac{109}{2}$ and hence, $a = 107$
By N_2 balance: $d = 3 + 3.76a = 3 + 3.76 \times 107 = 405.32$
Substituting for the values of a, b, c and d in the stoichiometric equation,

$$3CO_2 + 19CO + 41H_2 + 25CH_4 + 9C_2H_4 + 3N_2 + 107(O_2 + 3.76N_2) \rightarrow$$

$$65CO_2 + 109H_2O + 405.32N_2$$

As the fuel is burnt completely with 15% excess air, the combustion equation is:
100 kmol fuel + $1.15 \times 107(O_2 + 3.76N_2) \rightarrow$
$65CO_2 + 109H_2O + 405.32N_2 + 0.15 \times 107(O_2 + 3.76N_2)$
Therefore, combustion products for 100 kmol fuel are
$65CO_2 + 109H_2O + 405.32N_2 + 16.05O_2 + 60.35N_2$

$$= 65CO_2 + 16.05O_2 + 465.57N_2 + 109H_2O$$

kmol of combustion products $= \Sigma n_i = 65 + 16.05 + 465.57 + 109 = 655.62$
The molfraction or volume fraction of each species is:

$$x_{CO_2} = \frac{n_{CO_2}}{\Sigma n_i} = \frac{65}{655.62} = 0.1$$
$$x_{O_2} = \frac{n_{O_2}}{\Sigma n_i} = \frac{16.05}{655.62} = 0.024$$
$$x_{N_2} = \frac{n_{N_2}}{\Sigma n_i} = \frac{465.7}{655.62} = 0.71$$
$$x_{H_2O} = \frac{n_{H_2O}}{\Sigma n_i} = \frac{109}{655.62} = 0.0166$$

(d) kmol of dry gases $= \Sigma n_i = 65 + 16.05 + 465.57 = 546.62$
Therefore, Orsat analysis is:

$$x_{CO_2} = \frac{n_{CO_2}}{\Sigma n_i} = \frac{65}{546.62} = 0.119 = 11.9\%$$
$$x_{O_2} = \frac{n_{O_2}}{\Sigma n_i} = \frac{16.05}{546.62} = 0.029 = 2.9\%$$
$$x_{N_2} = \frac{n_{N_2}}{\Sigma n_i} = \frac{465.7}{546.62} = 0.852 = 85.2\%$$

(e) mass of combustion products per kilogram of fuel =

$$\frac{\Sigma\,(n_i M_i)_{\text{comb.products}}}{\Sigma\,(n_i M_i)_{\text{fuel}}}$$

$$= \frac{(65 \times 44 + 16.05 \times 32 + 465.57 \times 28 + 109 \times 18)}{(100 \times 14.82)}$$

$$= 12.396 \text{ kg/kg fuel}$$

(f) mole fraction of H_2O in the combustion products = 0.166

Therefore, the partial pressure of water vapour,

$$p_{H_2O} = p \times 0.166 = 1 \times 0.166 = 0.166 \text{ bar}$$

$$t_{\text{dew point}} = t_{s|p=0.166} = 57°\text{C (from steam tables)}$$

12.2 (a) Find the stoichiometric air–fuel ratio for the combustion of ethane (C_2H_6).
(b) Find the Orsat analysis of the combustion products.
(c) Find the volumetric analysis of the total combustion products.

▶ Solution:

(a) The stoichiometric combustion equation is

$$C_2H_6 + a(O_2 + 3.76N_2) \rightarrow bCO_2 + cH_2O + dN_2$$

By C balance: $b = 2$

By H_2 balance $c = 3$

By O_2 balance $a = b + \frac{c}{2} = 2 + \frac{3}{2} = 3.5$

By N_2 balance, $d = 3.76a = 3.76 \times 3.5 = 13.16$

Substituting for a, b, c and d in the above equation,

$$C_2H_6 + 3.5(O_2 + 3.76N_2) \rightarrow 2CO_2 + 3H_2O + 13.16N_2 \tag{i}$$

$$\text{Stoichiometric air} = \left(\frac{A}{F}\right)_{\text{st}} = \frac{\text{mass of air}}{\text{mass of ethane}}$$

$$= \frac{3.5(32 + 3.76 \times 28)}{(24 + 6)} = 16 \text{ kg/kg ethane.}$$

(b) It can be seen from the right-hand side of (i) that the dry gases in the combustion products are 2 kmols of CO_2 and 13.16 kmols of N_2.

Thus, Σn_i, the total kmols of dry combustion products = 2 + 13.16 = 15.16.

As Orsat analysis gives the volume fractions (= molfractions), x_i, of dry gases,

$$x_{CO_2} = \frac{n_{CO_2}}{\Sigma n_i} = \frac{2}{15.16} = 13.2\% \text{ and}$$

$$x_{N_2} = \frac{n_{N_2}}{\Sigma n_i} = \frac{13.16}{15.16} = 86.8\%$$

(c) Again from the right-hand side of (i), the total in kmol of combustion products

$$\Sigma n_i = 2 + 3 + 13.16 = 18.16$$

$$\text{Therefore,} \quad x_{CO_2} = \frac{n_{CO_2}}{\Sigma n_i} = \frac{2}{18.16} = 11.0\%$$

$$x_{H_2O} = \frac{n_{H_2O}}{\Sigma n_i} = \frac{3}{18.16} = 16.5\%$$

$$x_{N_2} = \frac{n_{N_2}}{\Sigma n_i} = \frac{13.16}{18.16} = 72.5\%$$

12.3 A certain coal is found to have the following ultimate analysis: S, 0.5%; H_2, 5.5%; C, 75%; O_2, 10%; N_2, 1.5%; *ash*, 4% and moisture, 3.5%. This coal is burnt completely in a furnace with 40% excess air. Find the actual $\frac{A}{F}$ ratio.

► **Solution:**

100 kg coal contains $\frac{0.5}{32}$ kmol S, $\frac{5.5}{2}$ kmol H_2, $\frac{75}{12}$ kmol C, $\frac{10}{32}$ kmol O_2. The other species (N_2, ash and moisture) need not be considered as they do not take part in the combustion.

The stoichiometric combustion equation for 100 kg coal is:

$$\left(\frac{0.5}{32}\right) S + \left(\frac{5.5}{2}\right) H_2 + \left(\frac{75}{12}\right) C + \left(\frac{10}{32}\right) O_2 + a(O_2 + 3.76N_2) \rightarrow$$
$$bSO_2 + cH_2O + dCO_2 + eN_2$$

By *S* balance: $\quad b = \frac{0.5}{32} = 0.015625$

By C balance: $\quad d = \frac{75}{12} = 6.25$

By H_2 balance: $\quad c = \frac{5.5}{2} = 2.75$

By O_2 balance: $\quad \frac{10}{32} + a = b + d + \frac{c}{2} = 0.015625 + 6.25 + \frac{2.75}{2}$

therefore, $a = 7.3281$

$$\text{Now, stoichiometric } \frac{A}{F} = \frac{\text{mass of air}}{\text{mass of coal}} = \frac{a(32 + 3.76 \times 28)}{100}$$
$$= \frac{7.3281(32 + 3.76 \times 28)}{100}$$
$$= 10.06 \text{ kg/kg coal}$$

$$\text{By definition, excess air} = \left[\left(\frac{A}{F}\right)_{\text{act}} - \left(\frac{A}{F}\right)_{\text{st}}\right] \Big/ \left(\frac{A}{F}\right)_{\text{st}}$$

$$0.4 = \left[\left(\frac{A}{F}\right)_{\text{act}} - 10.06\right] \Big/ 10.06$$

$$\text{Therefore,} \quad \left(\frac{A}{F}\right)_{\text{act}} = 14.084 \text{ kg/kg coal.}$$

12.4 Orsat analysis of the combustion products obtained by burning a certain saturated hydrocarbon with air is given as follows: CO_2, 12.3%; CO, 0.65%; O_2, 1.95% and N_2, 85.1%. Identify the hydrocarbon and find the excess air.

▶ **Solution:**

A saturated hydrocarbon is of the form C_nH_{2n+2}. If $n = 1$, it is CH_4 (methane), if $n = 2$ it is C_2H_6 (ethane) and so on.

Let x kmols of the given C_nH_{2n+2} be burnt using a kmols of O_2 in air to give 100 kmols of dry combustion products and b kmols of H_2O.
The combustion equation would then be:

$$xC_nH_{2n+2} + a(O_2 + 3.76N_2) \rightarrow (12.3CO_2 + 0.65CO + 1.95O_2 + 85.1N_2) + bH_2O$$

By C balance : $\quad xn = 12.3 + 0.65 = 12.95 \qquad$ (i)

By H_2 balance: $\quad x(n+1) = b$

By O_2 balance: $\quad a = 12.3 + \dfrac{0.65}{2} + 1.95 + \dfrac{x(n+1)}{2}$

that is, $\quad xn + x = 2a - 24.6 - 0.65 - 3.9 = 2a - 29.15 \qquad$ (ii)

By N_2 balance : $\quad 3.76a = 85.1$ and therefore $a = 22.633 \qquad$ (iii)

Substituting for a in (ii) from (iii),

$$xn + x = 2 \times 22.633 - 29.15 = 16.116 \qquad \text{(iv)}$$

Substituting for xn in (iv) from (i),

$$12.95 + x = 16.116 \text{ and therefore, } x = 3.166$$

Substituting for x in (i),

$$3.166n = 12.95 \text{ and therefore, } n = 4.09 \approx 4$$

As $n = 4$, the hydrocarbon is $C_4H_{2\times4+2} = C_4H_{10}$, which is BUTANE.

Also, $b = x(n+1) = 3.166(4+1) = 15.83$

Substituting the values of x, a and n in the combustion equation,

$3.166C_4H_{10} + 22.633(O_2 + 3.76N_2) \rightarrow$

$$(12.3CO_2 + 0.65CO + 1.95O_2 + 85.1N_2) + 15.83H_2O$$

$$\text{Actual air} = \frac{22.633(32 + 3.76 \times 28)}{(3.166 \times 58)} = 16.92 \text{ kg/kg butane}$$

Stoichiometric combustion equation is

$$C_4H_{10} + a(O_2 + 3.76N_2) \rightarrow bCO_2 + cN_2 + dH_2O$$

By C balance: $b = 4$

By H_2 balance: $d = 5$

By O_2 balance: $a = b + \frac{d}{2} = 4 + \frac{5}{2} = 6.5$

By N_2 balance: $c = a \times 3.76 = 6.5 \times 3.76 = 24.44$

Substituting the values of a, b, c and d in the stoichiometric equation,

$$C_4H_{10} + 6.5(O_2 + 3.76N_2) \rightarrow 4CO_2 + 5H_2O + 24.44N_2$$

$$\text{Therefore, } \left(\frac{A}{F}\right)_{st} = \frac{6.5(32 + 3.76 \times 28)}{58} = 15.385 \text{ kg/kg butane}$$

$$\text{Excess air} = \frac{\left[\left(\frac{A}{F}\right)_{act} - \left(\frac{A}{F}\right)_{st}\right]}{\left(\frac{A}{F}\right)_{st}} = \frac{(16.92 - 15.385)}{15.385} = 10\%$$

12.5 Redo Example 12.12 by using the ideal gas enthalpy table instead of the specific heat table.

► **Solution:**

As in Example 12.12, the combustion equation is,

$$C_2H_4 + 3.3(O_2 + 3.76N_2) \rightarrow 2CO_2 + 2H_2O + 0.3O_2 + 12.408N_2$$

Figure 12.3 may be referred as the H–t diagram for the reaction. Here, while A represents the initial state of the reactants, B represents the final state of the products.

From the figure, $$H_B - H_A = (H_B - H_X) + (H_X - H_Y) + (H_Y - H_A) \quad \text{(i)}$$

Here, $(H_B - H_X) = \left[\Sigma n_i(\bar{h}_B - \bar{h}_X)\right]_{\text{products}}$ where $(\bar{h}_B - \bar{h}_X)$ at 120°C for each species can be picked up from the ideal gas enthalpy table.

As in Example 12.12 the products are: $2CO_2 + 2H_2O + 0.3O_2 + 12.408N_2$

Therefore, $(H_B - H_X) = \left[n_i(\bar{h}_B - \bar{h}_X)\right]_{CO_2} + [n_i(\bar{h}_B - \bar{h}_X)]_{H_2O} + [n_i(\bar{h}_B - \bar{h}_X)]_{N_2}$

Now 120°C = 273 + 120 = 393 K

In Table 12.4, enthalpies are not available at this temperature of 393 K. Therefore the required enthalpies have to be evaluated through linear interpolation between the tabulated values at 300 K and 400 K.

That is, $(\bar{h}_B - \bar{h}_X)_{393\text{K}} = (\bar{h}_B - \bar{h}_X)_{300\text{ K}} + [(\bar{h}_B - \bar{h}_X)_{400\text{ K}} - (\bar{h}_B - \bar{h}_X)_{300\text{K}}] \times \frac{93}{100}$

From Table 12.4, for CO_2, $(\bar{h}_B - \bar{h}_X)_{300\text{ K}} = 67$ kJ/kmol and $(\bar{h}_B - \bar{h}_X)_{400\text{ K}} =$ 4008 kJ/kmol

Therefore, for CO_2, $(\bar{h}_B - \bar{h}_X)_{393\text{ K}} = 67 + (4008 - 67) \times \frac{93}{100} = 3732$ kJ/kmol

Similarly, for H_2O, $(\bar{h}_B - \bar{h}_X)_{393\text{ K}} = 63 + (3452 - 63) \times \frac{93}{100} = 3215$ kJ/kmol

and for, O_2, $(\bar{h}_B - \bar{h}_X)_{393\text{ K}} = 54 + (3029 - 54) \times \frac{93}{100} = 2821$ kJ/kmol

for N_2, $(\bar{h}_B - \bar{h}_X)_{393\text{ K}} = 54 + (2971 - 54)\frac{93}{100} = 2767$kJ/kmol

Substituting the above values in Eq.(ii),

$$(H_B - H_X) = 2 \times 3732 + 2 \times 3215 + 0.3 \times 2821 + 12.408 \times 2767$$
$$= 49,073 \text{ kJ/kmol}$$

$(H_Y - H_A)$ at 50°C (= 323 K) can be found as above using the ideal gas enthalpy values for various species in the reactants.
The reactants, from the combustion equation are

$$C_2H_4 + 3.3O_2 + 12.408N_2$$

$$(H_Y - H_A) = \left[\Sigma n_i(\bar{h}_Y - \bar{h}_A)\right]_{\text{reactants}} = -\left[\Sigma n_i(\bar{h}_A - \bar{h}_Y)\right]_{\text{reactants}}$$

In general, $(h_A - h_Y)_{323\text{ K}} = (h_A - h_Y)_{300\text{ K}} + [(h_A - h_Y)_{400\text{ K}} - (h_A - h_Y)_{300\text{ K}}] \times \frac{23}{100}$

Picking up the required values of enthalpies from Table 12.4,

For C_2H_4, $(h_A - h_Y)_{323\text{ K}} = 85 + (4895 - 85) \times \frac{23}{100} = 1191$ kJ/kmol

For O_2, $(h_A - h_Y)_{323\text{ K}} = 54 + (3029 - 54)\frac{23}{100} = 738$ kJ/kmol

For N_2, $(h_A - h_Y)_{323\text{ K}} = 54 + (2971 - 54) \times \frac{23}{100} = 725$ kJ/kmol

Therefore, $(H_Y - H_A) = -[1 \times 1191 + 3.3 \times 738 + 12.408 \times 725] = -12622$ kJ/kmol

Again from example 12.12 , $(H_X - H_Y) = \Delta H'_{\text{std}} = -1,322,943$ kJ/kmol

Therefore, $H_B - H_A = 49,073 - 1,322,943 - 12,622 = -1,286,492$ kJ/kmol

$$= \frac{-1,286,492}{28} = -45,946 \text{ kJ/kg } C_2H_4.$$

12.6 A gaseous fuel consists of 50% octane and 50% hexane by volume. Find the HCV and LCV of this fuel given that the standard enthalpies of reaction of octane and hexane are, respectively, −44,788 and −45,101 kJ/kg with H_2O in products in gaseous phase and that the enthalpy of vapourization of water at 25°C is 2442 kJ/kg.

▶ **Solution:**

Complete combustion of this fuel with O_2 is given by:

$$0.5C_8H_{18} + 0.5C_6H_{14} + 11O_2 \rightarrow 7CO_2 + 8H_2O$$

LCV = $-\Delta H'_{\text{std}}$ when H_2O in products is in gaseous phase

1 kmol of fuel contains (0.5 × 114) = 57 kg octane and (0.5 × 86) = 43 kg hexane.

Therefore, 100 kg fuel contains 57 kg octane and 43 kg hexane. In other words, 1 kg fuel contains 0.57 kg octane and 0.43 kg hexane.

Now $\Delta H'_{std} = (m \times \Delta H'_{\text{std}})_{\text{octane}} + (m \times \Delta H'_{\text{std}})_{\text{hexane}}$

Using the given values of $\Delta H'_{\text{std}}$ for octane and hexane,

$$\begin{aligned}\Delta H'_{\text{std}} &= 0.57 \times (-44788) + 0.43 \times (-45101)\\ &= -44922 \text{ kJ/kgfuel}\end{aligned}$$

Therefore, $\quad \text{LCV} = -\Delta H'_{\text{std}} = 44922 \text{ kJ/kg fuel}$

According to the combustion equation 100 kg fuel gives 8 kmol H_2O

Therefore, 1 kg fuel gives 0.08 kmol H_2O = 0.08 × 18 = 1.44 kg H_2O.

$\text{HCV} = \text{LCV} + (m \times h_{fg})_{H_2O} = 44922 + 1.44 \times 2442 = 48438 \text{ kJ/kg}$.

12.7 A small gas turbine uses liquid hexane for fuel and 400% theoretical air. Air and fuel enter at 25°C and the products leave at 800 K. The specific fuel consumption is 0.9 kg/kWh. Determine the heat transfer from the engine per kg fuel. Assume complete combustion.

▶ **Solution:**

Applying SS, SFEE to the gas turbine,

$$\dot{Q} - \dot{W}_x = \dot{m}\Delta\left[h + \frac{V^2}{2} + gz\right] \tag{5.8}$$

Neglecting changes in kinetic and potential energies,

$$\dot{Q} - \dot{W}_x = m\Delta h \tag{i}$$

The stoichiometric combustion equation is

$$C_6H_{14} + 9.5(O_2 + 3.76N_2) \rightarrow 6CO_2 + 7H_2O + 9.5 \times 3.76N_2$$

400 % theoretical air is supplied to engine. As in (12.2), percentage of theoretical air = 100 + % excess air. Hence, % excess air = % theoretical air −100. Therefore, % excess air = 400 − 100 = 300%. The combustion equation would thus be,

$$C_6H_{14} + 3 \times 9.5(O_2 + 3.76N_2) \rightarrow 6CO_2 + 7H_2O + 2 \times 9.5O_2 + 3 \times 9.5 \times 3.76N_2$$

As shown in Figure 12.7, the fuel–air mixture enters the engine at state A and the products leave the engine at B.

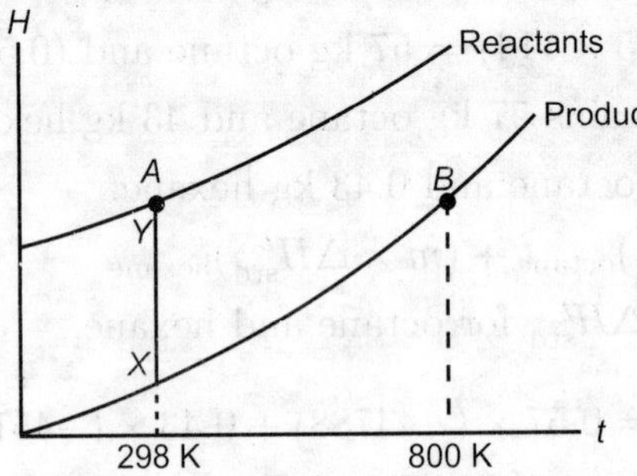

Fig. 12.7 H–t diagram discussed in Problem 12.7.

Thus, Δh in (i) will be equal to $(H_B - H_A)$ and this, from Figure (12.7) is given by

$$H_B - H_A = (H_B - H_X) + (H_X - H_Y) \text{ where}$$

$(H_B - H_X) = [\Sigma\, n_i(\bar{h}_B - \bar{h}_X)]_{\text{products}}$ where $(h_B - h_X)$ at 800 K for each species can be picked up from Table 12.4.

It is known from the combustion equation that the products comprise

$$6CO_2 + 7H_2O + 19O_2 + 107.16N_2$$

$$\begin{aligned} \text{Therefore, } (H_B - H_X) &= [n(\bar{h}_B - \bar{h}_X)]_{CO_2} + [n(\bar{h}_B - \bar{h}_X)]_{H_2O} + [n(\bar{h}_B - \bar{h}_X)]_{N2} \\ &= 6 \times 22,815 + 7 \times 17,991 + 19 \times 15,841 + 107.16 \times 15,046 \\ &= 21,173,728 \text{ kJ/kmol} \\ &= \frac{21,173,728}{M_{\text{hexane}}} = \frac{21,173,728}{86} = 25,276 \text{ kJ/kg.} \end{aligned}$$

From Table 12.2,

$(H_X - H_Y) = \Delta H'_{\text{std}} = -44,733$ kJ/kg (for liquid hexane and gaseous H_2O)

Therefore, $\Delta h = (H_B - H_A) = 25,276 - 44,733 = -19,457$ kJ/kg

Substituting this value in (i)

$$\dot{Q} - \dot{W} = -m \times 19,457 \tag{ii}$$

It is given that specific fuel consumption = 0.9 kg/kWh

That is, for $\dot{W} = 1$ kW, $\dot{m} = 0.9$ kg/h $= \dfrac{0.9}{3600}$ kg/s $= 2.5 \times 10^{-4}$ kg/s.

Substituting for $\dot{W}$ and $\dot{m}$ in (ii)

$$\dot{Q} - 1 = -2.5 \times 10^{-4} \times 19,457$$

Therefore, $\dot{Q} = -3.86$ kJ/s

That is, $Q = \dfrac{-3.86}{\dot{m}}$ kJ/kg $= \dfrac{-3.86}{(2.5 \times 10^{-4})} = 15{,}440$ kJ/kg.

12.8 Gaseous ethanol (C_2H_5OH) entering a steadily operating combustor at 25°C and 100 kPa burns completely with 25% excess air. Products of combustion leave the combustor at 127°C and 100 kPa. Given that the rate of heat transfer from the combustor is 900 kW, determine the mass flow rate of ethanol into the combustor in kg/s. Neglect the changes in kinetic and potential energies.

▶ **Solution:**

The stoichiometric combustion equation is,

$$C_2H_5OH + a(O_2 + 3.76N_2) \rightarrow bCO_2 + cH_2O + dN_2.$$

By C balance: $b = 2$

By H_2 balance: $c = 3$

By O_2 balance: $a + \dfrac{1}{2} = b + \dfrac{c}{2} = 2 + \dfrac{3}{2}$, and therefore $a = 3$

By N_2 balance: $d = a \times 3.76 = 3 \times 3.76 = 11.28$.

Substituting these values of a, b, c and d in the above stoichiometric equation,

$$C_2H_5OH + 3(O_2 + 3.76N_2) \rightarrow 2CO_2 + 3H_2O + 11.28N_2.$$

As 25% excess air is supplied, the actual combustion would be

$$C_2H_5OH + 1.25 \times 3(O_2 + 3.76N_2) \rightarrow 2CO_2 + 3H_2O + 11.28N_2 + 0.25 \times 3(O_2 + 3.76N_2)$$

or,
$$C_2H_5OH + 3.75O_2 + 14.1N_2 \rightarrow 2CO_2 + 3H_2O + 0.75O_2 + 14.1N_2 \quad \text{(i)}$$

B, the state of the products, and A, the state of the reactants, are shown in the H–t diagram of Figure 12.8.

Here, $H_B - H_A = (H_B - H_X) + (H_X - H_A)$

The products leave at 127°C = 400 K.

Therefore,
$$H_B - H_A = \Sigma\, n_i(\bar{h}_B - \bar{h}_X)_{i,400K} + \Delta H'_{\text{std}} \quad \text{(ii)}$$

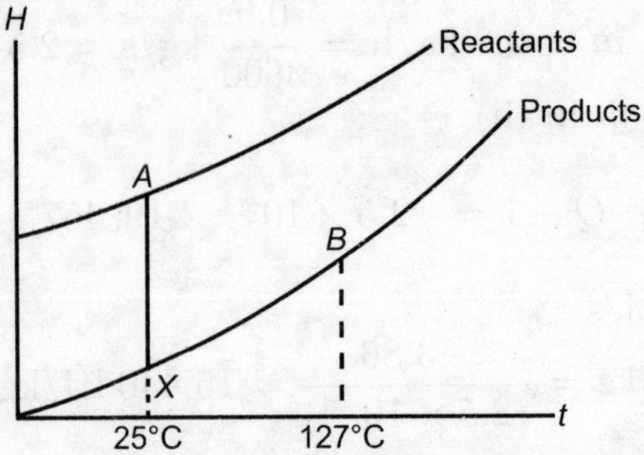

Fig. 12.8 H–t diagram.

The products, as in (i), comprise $2CO_2 + 3H_2O + 0.75O_2 + 14.1N_2$

Picking up the values of $(\bar{h}_B - \bar{h}_X)$, at 400 K, for each species, from Table 12.4,

$$\Sigma\, n_i(\bar{h}_B - \bar{h}_X)_{i,400\text{K}} = 2 \times 4008 + 3 \times 3452 + 0.75 \times 3029 + 14.1 \times 2971$$
$$= 62,535 \text{ kJ/kmol}$$

If available, the value of $\Delta H'_{\text{std}}$ for ethanol can be picked up from table 12.3 or it can be calculated using the values of enthalpies of formation of C_2H_5OH, CO_2 and H_2O. The latter method is used below.

Considering the reactants and products in (i), and noting that enthalpies of O_2 and N_2 at standard state are zero,

$$\Delta H'_{\text{std}} = (n\bar{h}'_f)_{CO_2} + (n\bar{h}'_f)_{H2O} - (n\bar{h}'_f)_{C_2H_5OH}$$
$$= 2 \times (-393,510) + 3 \times (-241,820) - 1 \times (-234,430)$$
$$= -1,278,050 \text{ kJ/kmol}$$

Substituting the values in (ii)

$$H_B - H_A = 62,535 - 1,278,050 = 1,215,515 \text{ kJ/kmol}$$
$$= \frac{-1,215,515}{M_{C_2H_5OH}} \text{ kJ/kg} = \frac{-1,215,515}{46} = -26,424 \text{ kJ/kg.}$$

Applying SS, SFEE to the combustor,

$$\dot{Q} - \dot{W} = \dot{m}\Delta(h + \frac{V^2}{2} + gz)$$

Here $\dot{W} = 0$ and changes in kinetic potential energies are given to be negligible.

Therefore, $\dot{Q} = \dot{m}\Delta h$ where $\dot{Q} = -900$ kW and $\Delta h = H_B - H_A = -26,424$ kJ/kg

Hence, $-900 = \dot{m} \times (-26,424)$ or $\dot{m} = 0.034$ kg/s.

Problems

12.1 A gaseous mixture at 1200 K contains 24%, CO; 11%, O_2; 14%, H_2; 6%, CH_4 and 45%, N_2 by volume. Find (a) the molar mass (molecular weight) of the mixture, (b) mass fractions of the constituents and (c) c_v for the mixture using the c_p values from the Table 12.1.

12.2 Bituminous coal is burnt completely with 60% excess air. Find the $\frac{A}{F}$ ratio and the volumetric analysis of the combustion products. The ultimate analysis of the coal is 72%, C; 4%, H_2; 12%, O_2; 6% moisture and 6% ash.

12.3 A gaseous fuel has the following composition by volume: 12%, CO; 25%, CH_4; 45%, H_2; 2%, O_2; 1.5%, CO and 14.5%, N_2. Find the volume fractions and mass fractions of the constituents of the combustion products given that the combustion is complete with 20% excess air.

12.4 A sample of coal having 10% ash when burnt in a furnace is found to have 40% carbon in solid refuse. It is given that the coal contains 70% carbon by mass. Find the mass of carbon in dry combustion products per kilogram of coal.

12.5 A gaseous fuel has the following volumetric composition: 30%, CO; 10%, H_2; 3%, CH_4; 5%, CO_2 and 52%, N_2. This fuel is burnt steadily in a combustion chamber at a pressure of 1 bar; the air – fuel ratio by volume being 2. Assume complete combustion and calculate: (i) actual $\frac{A}{F}$ ratio, (ii) excess air, (iii) Orsat analysis of the combustion products and (iv) dew point temperature.

12.6 A town gas has the following volumetric analysis: 10%, CO; 45%, H_2; 35%, CH_4; 4%, C_2H_4; 2%, O_2; 2%, N_2 and 2%, CO_2. For the complete combustion of this gas find (i) minimum air required in m^3/m^3 gas, (ii) minimum air required in kg/kg gas and (iii) Orsát analysis of the products of combustion.

12.7 A certain gas has the following volumetric analysis: 10%, CO; 45%, H_2; 35%, CH_4; 4%, N_2 and 6%, CO_2. Find the Orsat analysis of the combustion products when this fuel is burnt completely with 50% excess air.

12.8 Methyl alcohol (CH_3OH) is burnt with atmospheric air. Orsat analysis of the combustion products is given as follows: 10%, CO_2; 2.45%, O_2; 0.55%, CO and 87%, N_2. Find (i) $\frac{A}{F}$ ratio, (ii) percent theoretical air and (iii) volume fraction of moisture in the combustion products.

12.9 Ethyl alcohol (C_2H_5OH) is burnt with dry air. Orsat analysis of the combustion products is as follows: 11.5%, CO_2; 2.6%, O_2; 0.5%, CO and 85.4%, N_2. Calculate

(i) Stoichiometric $\frac{A}{F}$ratio, (ii) actual $\frac{A}{F}$ ratio, (iii) excess air and (iv) volume fraction of moisture in the combustion products.

12.10 3 kmol of butene (C_3H_8) when burnt with y kmol of dry air gave 100 kmol of dry combustion products comprising CO_2, O_2, CO and N_2. Given that volume fraction of N_2 in the dry gas is 83.14%, calculate (i) excess air and (ii) volume fractions of the other constituents in the dry gas.

12.11 For the reaction of carbon–oxygen system to carbon dioxide only, at a temperature of 800°C and a pressure of 1 atm, the enthalpy increase is -33.287×10^6 J/kg carbon. The mean specific heat of carbon between 25°C and 800°C is 1385 J/kg K and of oxygen is 1022 J/kgK. Determine the mean specific heat of CO_2 in this temperature range.

12.12 Calculate the calorific value of CO in kJ/kg using the enthalpy of formation tables.

12.13 Given that the standard enthalpy of reaction for gaseous ethane is $-47,590$ kJ/kg when H_2O in the products is in gaseous phase. Find the standard increase in internal energy in kJ/kg.

12.14 Using the tables of enthalpy of formation and standard enthalpy of reaction, calculate the enthalpy of formation of gaseous heptane (C_7H_{16}). Also calculate the internal energy of reaction in kJ/kg fuel for combustion of liquid heptane at 25°C with stoichiometric air assuming water formed is in liquid phase.

12.15 Using the table of enthalpy of formation calculate the following for the complete combustion of hydrogen with oxygen at 25°C and 100 kPa.: (i) The enthalpy of reaction in kJ/kmol and kJ/kg when H_2O formed is in (a) liquid phase and (b) gaseous phase , (ii) HCV and LCV.

12.16 Gaseous methane enters a combustion chamber of a gas turbine at 25°C and air from the compressor at 550 K. It is given that 99% of carbon burns to CO_2 and the rest to CO. What amount of excess air will be required if the temperature of the products is to be limited to 1200 K. Take the required specific heat values from the table.

12.17 Carbon monoxide is burnt adiabatically in steady flow at atmospheric pressure with 100% excess air. The reactants are supplied at 40°C. Sketch the enthalpy - temperature diagram and hence calculate the temperature of combustion products assuming the combustion to be complete. Take the c_p values from the tables.

12.18 Gaseous butane (C_4H_{10}) and 10% excess air are fed steadily at 25°C and 1 bar into a combustion chamber. The rate of heat transfer from the chamber is such that the products leave at 500 K and 1 bar. Assume complete combustion and that H_2O in the products is in the gaseous phase and calculate the heat transfer per kilogram of butane.

12.19 Indane the cooking gas is a mixture of 85% butane and 15% propane. A mixture of indane gas and 10% excess air enters a combustion chamber steadily at 35°C and 1 bar. The combustion products leave at 1 bar and a temperature of TK. Find T assuming complete combustion. Assume the following average values of c_p (in kJ/kg K) for the various species: 1.67 for C_3H_8; 1.71 for C_4H_{10}; 0.922 for O_2; 1.042 for N_2; 0.842 for CO_2 and 1.86 for H_2O.

12.20 Gaseous ethane is burnt completely in an adiabatic steady flow combustion chamber with 50% excess air. The reactants are at 25°C. (i) Using the enthalpy of formation tables determine the LCV of ethane. (ii) Find the temperature of the products picking up the required values of c_p from tables.

12.21 Gaseous methane is burnt completely in adiabatic steady flow with 40% excess air. The reactants are at 25°C. Determine (i) HCV and LCV of methane and (ii) the temperature of the combustion products assuming H_2O in the products to be in gaseous phase. Take the required data from the tables of enthalpy of formation and specific heats.

12.22 Redo Problem 12.21 using the standard enthalpy of reaction tables.

Objective Type Questions

I. State whether the following statements are TRUE or FALSE:

1. If p and v are the pressure and volume of a system, $(p + cv)$, where c is a constant, is a property of the system.
2. The control volume analysis is usually adopted to analyse flow processes.
3. The mass inside a control volume always remains the same.
4. A Bourdon pressure gauge cannot be used to measure vacuum pressures.
5. The thermodynamic definition of work associates the positive work with the raise of a weight as the sole effect external to a system.
6. The thermodynamic definition of work associates the negative work with the fall of a weight as the sole effect external to a system.
7. When a body falls freely in vacuum, the work interaction for the body is zero.
8. When a body is slowly raised in vacuum, the work interaction for the body is zero.
9. The mean effective pressure of an engine can be negative.
10. For an engine having an indicator diagram as shown in Figure OQ-1, Δp represents the mean effective pressure.

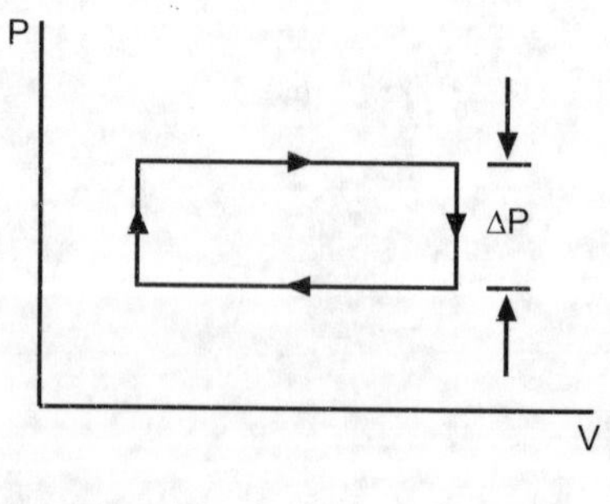

Fig. OQ-1 Indicator diagram for Question 10.

11. Figure OQ-2 shows a piston–cylinder device with a gas under pressure on one side of the piston of negligible mass. On the other side the pressure is zero. When the stops are removed the gas expands. The work for the gas is positive.

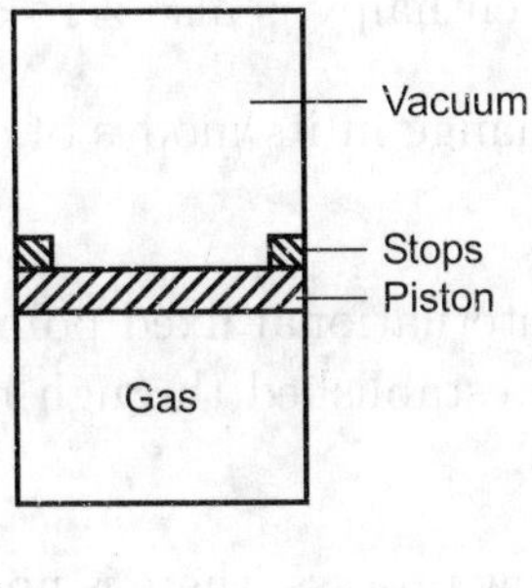

Fig. OQ-2 Schematic for Question 11.

12. In an automobile engine, the indicated work can be less than the work at the output shaft.

13. At a place where the atmospheric pressure is 1 bar, if the pressures of two systems A and B are $p_A = 1.5$ bar (abs) and $p_B = 0.5$ bar (gauge), $p_A - p_B = 1$ bar.

14. A gas contained in a rigid vessel is churned by a paddle wheel. Simultaneously, a current flows through a resistor inserted in the vessel and connected to an external battery. The displacement work for the gas is zero.

15. A gas contained in a rigid vessel is churned by a paddle wheel. Simultaneously, a current flows through a resistor inserted in the gas and connected to an external battery. The net work for the gas is negative.

16. The two-property rule states that the state of a given mass of a pure substance is fixed if any two properties of the system are given, provided the system is in equilibrium and that effects due to gravity, motion, electricity, magnetism etc. have no significant effects.

17. Hot coffee in a good thermos flask is in thermal equilibrium with the surroundings at the ambient temperature.

18. $(\delta Q - \delta W)$ for a system is an exact differential though δQ and δW are not exact differentials.

19. $\Delta E = Q - \int p dV$ is true for any process executed by a system, whether the process is quasistatic or not.

20. The energy of an adiabatic system can never increase.

21. The mechanical equivalent of heat is 4.187 Nm/J.

22. Heat transfer to a system need not necessarily increase the temperature of a system.

23. Suppose for a pure substance the value of specific internal energy is assumed to be zero at 1 bar, 15°C, its enthalpy is also zero at the same state.

24. A system can undergo a change in its modes of energy in the absence of heat and work interactions.

25. The temperatures of the international fixed points are universally assigned values and are not required to be established through measurements made using precise thermometers.

26. In a steady-state, steady-flow process, there is no accumulation of mass and energy inside a control volume

27. The primary objective of a heat pump is to cause heat transfer from a low-temperature region to a high-temperature region.

28. The ambient temperature is the high-temperature reservoir for a domestic refrigerator.

29. The Kelvin–Planck statement of the second law of thermodynamics implies that no heat engine can have a thermal efficiency of 100%.

30. The second law of thermodynamics implies that the COP of a heat pump can reach infinity.

31. The efficiency of a Carnot cycle can be increased by increasing the sink temperature while keeping the source temperature fixed.

32. A reversed Carnot cycle is also a reversible cycle.

33. Dissipative effects render a process reversible.

34. A reversible process is also a quasi-static process.

35. For a heat engine operating between two reservoirs at t_1°C and t_2°C (where $t_1 > t_2$), having heat transfers of Q_1 and Q_2, respectively, with these two temperatures, while the maximum efficiency is equal to $\frac{(t_1 - t_2)}{t_1}$, the actual efficiency is equal to $\frac{(Q_1 - Q_2)}{Q_1}$.

36. A reversed heat engine is an impossibility.

37. A heat engine, operating between two given temperatures, is reversible if its efficiency while operating directly is equal to the reciprocal of its COP while it operates as a refrigerator between the same two temperatures.

38. Consider two reversible heat engines X and Y operating between T_1 K and T_2 K. While X works using air, Y works using steam. Efficiency of X is equal to the efficiency of Y.

39. $(\text{COP})_{hp} = (\text{COP})_{ref} + 1$ is valid only for a reversible engine.

40. A reversible heat engine is operating between two reservoirs at t_1°C and t_2°C, where t_1 is greater than t_2. The efficiency of the engine is not equal to $\left(1 - \dfrac{t_2}{t_1}\right)$.

41. The COP of a heat pump is always greater than unity.

42. An internal combustion engine operates violating the second law of thermodynamics as it produces positive work while having heat interaction with only the surroundings at a fixed temperature.

43. All spontaneous processes are irreversible.

44. A Carnot cycle consists of two reversible isotherms and two reversible adiabatics.

45. The efficiency of a Carnot cycle can be increased by increasing the temperature of the source while keeping the sink temperature fixed.

46. A direct heat engine operates between T_1 K and T_2 K (where $T_1 > T_2$). The efficiency of the engine, reversible or irreversible, is given by $\left(1 - \dfrac{T_2}{T_1}\right)$.

47. A system undergoes a state change from 1 to 2. If t represents temperature in °C and T the corresponding temperature in K, then, $(t_2 - t_1) = (T_2 - T_1)$.

48. For a heat engine operating between two reservoirs at t_1°C($= T_1$ K) and t_2°C($= T_2$ K), $\left(1 - \dfrac{T_2}{T_1}\right) = \left(1 - \dfrac{t_2}{t_1}\right)$.

49. For an irreversible heat engine operating between two reservoirs at T_1 K and T_2 K, with heat transfers of Q_1 and Q_2, respectively, with these reservoirs, $\left(1 - \dfrac{Q_2}{Q_1}\right) > \left(1 - \dfrac{T_2}{T_1}\right)$.

50. In general, for a heat engine operating between two reservoirs at T_1 K and T_2 K, with heat transfers of Q_1 and Q_2, respectively, with these reservoirs, $\left(1 - \dfrac{Q_2}{Q_1}\right) \leq \left(1 - \dfrac{T_2}{T_1}\right)$.

51. On kelvin temperature scale, no system can have negative values for temperature.

52. In general, for a system undergoing a cyclic process, cyclic integral of $\dfrac{\delta Q}{T}$ is ≥ 0.

53. For all reversible paths between two given state points, $\displaystyle\int \frac{\delta Q}{T}$ has the same value.

54. Entropy is an extensive property.

55. Entropy 'S' is defined by, $\Delta S = \displaystyle\int \frac{\delta Q}{T}$.

56. For a system undergoing a state change, $\mathrm{d}S \leq \dfrac{\delta Q}{T}$.

57. Irrespective of the nature of the working substance a Carnot cycle is always a rectangle on a T–S diagram.

58. In general, during a process entropy of a system can never decrease.

59. Entropy of an adiabatic system can never decrease.

60. Two Carnot engines, A and B, operate separately between two reservoirs at T_1 K and T_2 K. While the working substance in A is steam, that in B is air. The two engines will work with the same efficiency.

61. The specific entropy of a pure substances, is related to other properties, with usual notations, by $T\mathrm{d}s = \mathrm{d}h + v\mathrm{d}p$.

62. The specific entropy of a pure substances, is related to other properties, with usual notations, by $T\mathrm{d}s = \mathrm{d}u + p\mathrm{d}v$.

63. An ideal gas is one which follows the relation, with usual notations, $pv = \mathrm{R}T$ at all ranges of pressure and temperature.

64. It is mandatory for a gas that follows the relation $pv = \mathrm{R}T$, to have its internal energy and enthalpy as functions of temperature only.

65. An ideal gas with constant specific heats is called a semi-perfect gas.

66. For an ideal gas $\dfrac{c_p}{c_v} < 1$.

67. $c_p - c_v = \text{R}$ is valid for both perfect and semi-perfect gases.

68. At any given state point on a T–S diagram, the constant pressure line is steeper than the constant volume line.

69. Only for an ideal gas, a reversible hyperbolic process is also an isothermal process.

70. An ideal gas undergoes a reversible polytropic expansion process according to $pV^n =$ constant. Heat is transferred to the gas if $n > \dfrac{c_p}{c_v}$.

71. An ideal gas undergoes a reversible polytropic expansion process according to $pV^n =$ constant. Heat is transferred from the gas if $n > \dfrac{c_p}{c_v}$.

72. An ideal gas undergoes a state change from 1 to 2. For any path between 1 and 2 the change in specific internal energy is given by $c_v(t_2 - t_1)$.

73. When an ideal gas expands according to $pV^\gamma =$ constant, where $\gamma = \dfrac{c_p}{c_v}$, the change in entropy of the gas during the process is greater than zero.

74. A mixture of ideal gases also behaves as an ideal gas.

75. The gist of Dalton's law of additive pressures is that each constituent gas behaves as though it existed alone at mixture temperature and volume.

76. The partial pressure of a constituent in an ideal gas mixture is the pressure it exerts if it were to exist alone at the mixture volume and temperature.

77. If y_i and p_i are, respectively, the mass fraction and partial pressure of the ith constituent of an ideal gas mixture, then $p_i = y_i p$, where p is the mixture pressure.

78. If x_i and R_i are the mole-fraction and particular gas constant of the ith constituent of an ideal gas mixture, then the gas constant of the mixture is given by $R = \Sigma\, x_i R_i$.

79. When an ideal gas A at a certain pressure and temperature is mixed adiabatically with another ideal gas B at the same pressure and temperature, the change in entropy during mixing depends only on the number of moles of the A and B and not on what gases they are.

80. When a certain mass of an ideal gas mixture is compressed adiabatically the mass fractions and mole-fractions of the constituents increase.

81. Compressibility factor of an ideal gas is equal to zero.

82. At critical state, the compressibility factor of a van der Waals gas is $\dfrac{3}{8}$.

83. For any real gas, at a temperature greater than 2.5 times the critical temperature, the compressibility factor is always greater than one for any pressure.

84. A real gas tends to behave as an ideal at high pressures and high temperatures.

85. At low temperatures and pressures the value of the compressibility factor of a real gas is always greater than one.

86. At high pressures, irrespective of the temperature, the compressibility factor of a real gas is always greater than one.

87. Two real gases, each at the same reduced pressure and reduced temperature have the same compressibility factor.

88. The three phases of a pure substance coexist in equilibrium at the critical state.

89. A rigid vessel contains a mixture of saturated steam and water at a certain pressure. The volume occupied by the steam is three times the volume occupied by the liquid. The dryness fraction of the mixture is 0.75.

90. The specific volume of steam at critical state is 0.00317 m^3/kg. A certain mass of wet steam with a specific volume of 0.00217 m^3/kg is heated in a rigid vessel. The steam never gets superheated.

91. Enthalpy of vapourization of water (latent heat) increases with an increase in pressure.

92. Steam with a dryness fraction of 0.9 is heated in a rigid vessel until it becomes saturated vapour. The steam pressure remains the same during the process.

93. When steam is throttled sufficiently it can become superheated.

94. On the Mollier diagram for water, the constant pressure lines in the wet region are straight lines.

95. The constant pressure lines in the wet region for water on Mollier diagram are parallel.

96. Enthalpy of sub-cooled water at a certain temperature can be approximated to enthalpy of saturated water at the same temperature.

97. With usual notations, the enthalpy of wet steam at a certain pressure is related to the enthalpies of the saturated states at the same pressure by $h = (1-x)h_f + xh_g$.

98. With usual notations, the entropy of wet steam at a certain pressure is related to the entropies of the saturated states at the same pressure by $s = xs_{fg} - s_f$.

99. At critical state enthalpy of vapourization of water is maximum.

100. In a turbine, saturated steam at 20 bar expands adiabatically to 0.1 bar. At the exit of the turbine, the steam shall be dryer in case of irreversible expansion compared to that in reversible expansion.

II. Choose the most appropriate answer(s) for the following questions. There may be more than one correct answer. Indicate your answer by marking the relevant letter(s) against each question.

1. In a thermodynamic system:

(A) there is flow of matter across the system boundary

(B) there is no flow of matter across the system boundary

(C) the shape, size and position of the system boundary may change

(D) the shape, size and position in space cannot change

2. The state of a system is identified by its:

(A) size (B) shape

(C) properties (D) surroundings

3. A tiny hole in the tube of a cycle tire filled with air causes the air to leak out slowly. If a boundary is drawn around the tube, it would constitute:

(A) a system

(B) a control volume

(C) neither a system nor a control volume

(D) a system as well as a control volume

4. The absolute pressure of a system, p, the gauge pressure of the same system, p_{gauge}, and the atmospheric pressure, p_{atm}, are related by,

(A) $p_{gauge} = p - p_{atm}$ (B) $p_{gauge} = p + p_{atm}$

(C) $p_{gauge} = p$ (D) $p_{gauge} = p_{atm} - p$

5. The absolute pressure of a system, p, the vacuum pressure of the same system, p_{vac}, and the atmospheric pressure, p_{atm}, are related by,

   (A) $p_{vac} = p - p_{atm}$ (B) $p_{vac} = p + p_{atm}$

   (C) $p_{vac} = p$ (D) $p_{vac} = p_{atm} - p$

6. A Bourdon gauge connected to a scooter tire reads 1.8 bar. If the atmospheric pressure is 1 bar, the absolute pressure of air in the tire is :

   (A) 1.8 bar (B) 2.8 bar

   (C) 0.8 bar (D) 1.4 bar

7. An extensive property P of a system, comprising a large number of sub-systems in thermodynamic equilibrium

   (A) is independent of the mass of the system

   (B) depends on the mass of the system

   (C) is given by $P = \Sigma P_i$

   (D) is given by $P = P_{All\ sub-systems} = P_{i,any\ sub-system}$

8. An intensive property p of a system, comprising a large number of subsystems in thermodynamic equilibrium

   (A) is independent of the mass of the system

   (B) depends on the mass of the system

   (C) is given by $p = \Sigma p_i$

   (D) is given by $p = p_{All\ sub-systems} = p_{i,any\ sub-system}$

9. A system is said to execute a quasi-static process when

   (A) the value of at least one of its properties change

   (B) the value of at least one of its properties is held constant

   (C) the final and initial values are known

   (D) it passes successively through equilibrium states

10. The magnitude of work in any interaction

    (A) does not depend on the shape of the system boundary

    (B) depends on the shape of the system boundary

    (C) depends only on the end states of the process

    (D) depends on the path of the process

11. Displacement work is the mode of work associated with

(A) the displacement of the system boundary
(B) either the normal or the shear stresses at the system boundary
(C) the normal stresses at the system boundary
(D) the shear stresses at the system

12. Displacement work done in a process, as the state of a system changes from 1 to 2 along the path represented by pV^n = Constant (where 'n' is positive, but not equal to 1), is given by

(A) $\dfrac{(p_1V_1 - p_2V_2)}{(1-n)}$ (B) $\dfrac{(p_1V_1 - p_2V_2)}{(n-1)}$
(C) $p_1V_1 \ln \dfrac{V_2}{V_1}$ (D) $p_1V_1 - p_2V_2$

13. When a system does positive work

(A) the surroundings do negative work
(B) the surroundings also do positive work
(C) the combined work for the system and surroundings is positive
(D) the combined work for the system and the surroundings is zero.

14. Engine indicator is a device to measure

(A) the work done on the piston during the expansion process
(B) the piston face pressure at each position of the piston
(C) the mean pressure only in one stroke of the piston
(D) the engine speed.

15. In the indicator diagram shown in Fig.OQ-3, the indicated work is given by the area

(A) cfeabc (B) bcdyxb
(C) abcda (D) bcfgb

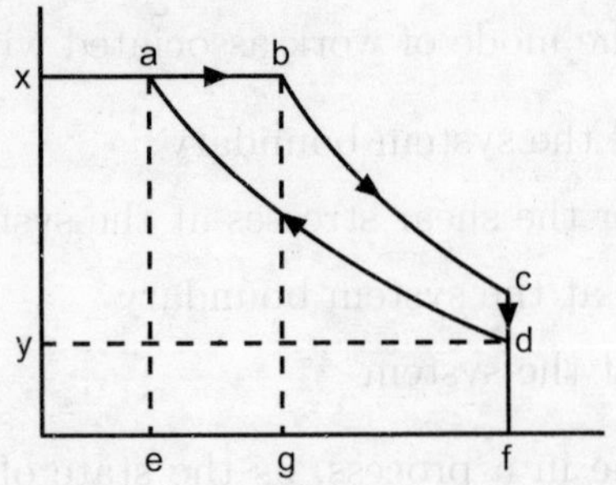

Fig. OQ-3 The Indicator diagram for Question 15.

16. If p_m is the mean effective pressure evaluated from an indicator diagram, W is the indicated work/cycle, V is the stroke volume, a and l are respectively, the area and length of the diagram, S is the spring constant, then

(A) $p_m = \frac{W}{V}$ (B) $W = p_m L$

(C) $W = V\frac{aS}{l}$ (D) $p_m = \frac{aS}{l}$

17. A single-cylinder, two-stroke reciprocating engine has a piston displacement volume of 750 cc and is run at 500 revolution per minute. The mean effective pressure is 2 bar. The indicated power in kW of the engine is

(A) 2.5 (B) 1.25

(C) 0.75 (D) 5.0

18. The torque in a shaft varies according to the equation $T = T_o \sin\theta$, where T is the torque, θ is the angle of rotation and T_o is a constant. The work done due to the shear force is

(A) $T_o \sin\theta$ (B) $T_o \cos\theta$

(C) T_o (D) $\int T_o \sin\theta d\theta$

19. A rotating fan-wheel in a closed room is allowed to coast to a full stop, while no power is being supplied to it. Consider the contents of the room as the system. The work interaction for the system

(A) is zero (B) is positive

(C) is negative (D) cannot be assessed

20. A pure substance undergoes a fully resisted adiabatic process. The work interaction for the substance is

(A) $= U_1 - U_2$ (B) $= U_2 - U_1$
(C) $< U_2 - U_1$ (D) $> U_2 - U_1$

21. Suppose a refrigerator having its door open is operated continuously in a kitchen and the kitchen is isolated from the rest of the house, the temperature of the kitchen

(A) decreases (B) does not change
(C) increases (D) increases first and then decreases

22. Directly or indirectly, the first law of thermodynamics tells us that

(A) $\int \delta Q = \int \delta W$
(B) $dE = \delta Q - \delta W$
(C) energy of a system is conserved
(D) $(\delta Q - \delta W)$ is an exact differential

23. When a system undergoes a change of state from 1 to 2 along path A, heat transferred to the system is 100 kJ and work done by it is 50 kJ. If the system is then restored to state 1 through an adiabatic path B, the work interaction during B is

(A) + 50 kJ (B) – 50 kJ
(C) + 100 kJ (D) – 100 kJ

24. Two kilograms of a pure substance undergoes a constant pressure process from 27°C to 37°C. It is given that internal energy and enthalpy of the pure substance are functions of temperature only and that c_v and c_p are respectively 0.7 kJ/kg deg and 1.0 kJ/kg deg. If the heat transfer to the system is 20 kJ, the work done by the system is

(A) 20 kJ (B) 34 kJ
(C) 0 (D) 6 kJ

25. The area enclosed by a thermodynamic cycle on a T–S diagram represents the

(A) net work
(B) net heat
(C) heat transferred to the system
(D) heat transferred from the system.

26. The relation $\oint T ds = \oint p dv$ is valid

(A) only for a reversible cycle
(B) only for an irreversible cycle
(C) for both reversible and irreversible cycles
(D) for no cycle.

27. In the adiabatic throttling of a gas flowing through a constant area pipe from cross–sections 1 to 2

(A) $h_1 = h_2$ (B) $u_1 = u_2$
(C) $p_1 > p_2$ (D) $v_1 = v_2$

28. When the valve on a rigid vessel is opened slowly, the fluid in it flows out gently and adiabatically. Flow work for the fluid

(A) is positive (B) is negative
(C) is zero (D) has no meaning

29. When the valve on a rigid vessel is opened slowly, the gas in it flows out gently and adiabatically. For the gas, u and h are functions temperature only. After a while the temperature of the remaining gas in the vessel would have

(A) increased (B) decreased
(C) remained the same (D) changed but difficult to predict how

30. A steam turbine receives steam at $p_1 = 10$ bar, $h_1 = 250$ kJ/kg and lets it out at $p_2 = 2$ bar and $h_2 = 100$ kJ/kg, with negligible changes in KE and PE and delivers a work of 100 kJ/kg. The heat loss from the turbine to the atmosphere is

(A) 50 kJ/kg (B) 0
(C) 200 kJ/kg (D) 150 kJ/kg

31. When an ideal gas is throttled adiabatically the change in its properties is given by

(A) $\Delta h = 0;\ \Delta s = 0$ (B) $\Delta h > 0;\ \Delta T = 0$
(C) $\Delta h > 0;\ \Delta s > 0$ (D) $\Delta h = 0;\ \Delta s > 0$

32. An example of a heat engine is

(A) a generator run by a steam turbine
(B) a closed cycle gas turbine power plant
(C) a car powered by a battery
(D) open cycle gas turbine power plant

33. Consider three heat engines X, Y and Z operating between 1000 and 300 K. It is claimed that, when heat transferred to each engine is 1000 kJ, engines X, Y and Z developed 800, 700 and 600 kJ of work, respectively. On which engine would you invest ?

(A) Z (B) X
(C) Y (D) Y and Z

34. Suppose a reversible refrigerator is operating between an evaporator temperature of 260 K and a condenser temperature of 310 K. The coefficient of performance of the refrigerator is

(A) 2.5 (B) 5.2
(C) 2.6 (D) 6.2

35. A reversible heat pump operates as shown in Fig.OQ-4. Its coefficient of performance and temperature T_2 are, respectively

(A) 2 and 200 K (B) 3 and 200 K
(C) 2.5 and 250 K (D) 1.5 and 200 K

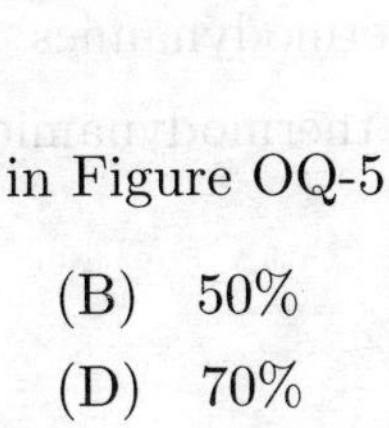

Fig. OQ-4 Schematic for Question 35.

36. A heat engine works as shown in Figure OQ-5. Its efficiency is

(A) 40% (B) 50%
(C) 60% (D) 70%

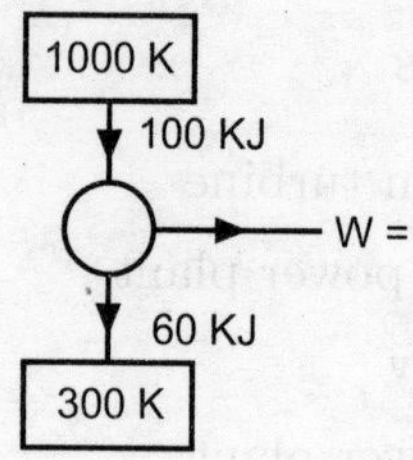

Fig. OQ-5 Schematic for Question 36.

37. A reversible heat engine (R) and an irreversible refrigerator (Ref) work between the same temperature limits as shown in Figure OQ-6

(A) $\eta_R = \dfrac{(T_1 - T_2)}{T_1}$ and $(\text{COP})_{Ref} = \dfrac{T_2}{(T_1 - T_2)}$

(B) $\eta_R = \dfrac{(T_1 - T_2)}{T_1}$ and $|Q_2| > |Q_4|$

(C) $(\text{COP})_{Ref} = \dfrac{Q_1}{W}$

(D) $(\text{COP})_{Ref} = \dfrac{Q_4}{W}$

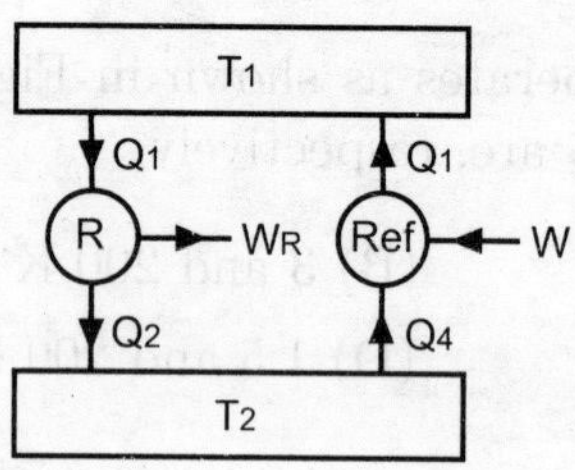

Fig. OQ-6 Schematic for Question 37.

38. A PMM II

(A) is a device operating in a thermodynamic cycle that produces work while having heat interaction with a single system at a fixed temperature

(B) violates the first law of thermodynamics

(C) violates the second law of thermodynamics

(D) is a possible device

39. Instead of $\dfrac{Q_1}{Q_2} = \dfrac{T_1}{T_2}$, if a new thermodynamic temperature scale was to be defined by $\dfrac{Q_1}{Q_2} = \dfrac{x_1^n}{x_2^n}$, where n is a positive integer and x_1 and x_2 are the temperatures on the new scale with $x_1 > x_2$, the efficiency of a reversible engine operating between x_1 and x_2 would be,

(A) $\dfrac{(x_1^n - x_2^n)}{x_1^n}$ (B) $\dfrac{(x_1 - x_2)}{x_1}$

(C) $\left[\dfrac{(x_1 - x_2)}{x_1}\right]^n$ (D) $\dfrac{(x_1 - x_2)}{nx_1}$

40. A reversible heat engine A operates between two reservoirs at T_1K and T_2K. Another reversible heat engine B operates between the reservoirs at T_2K and T_3K, where $T_3 < T_2 < T_1$. If heat transferred from A is equal to the heat transferred to B, and that A and B have same efficiencies T_1, T_2 and T_3 are related by

(A) $T_1 - T_3 = T_2$ (B) ${T_2}^2 = T_1T_3$

(C) $T_1 + T_2 = 2T_3$ (D) $T_2 + T_3 = T_1$

41. Two finite bodies with identical masses and specific heats but at different temperatures of T_1K and T_2K, where $T_1 > T_2$, are used as source and sink for a reversible engine, which operates until the temperatures of both the bodies attain a common value of T. The temperatures T_1, T_2 and T are then related by

(A) $2T = T_1 + T_2$ (B) $T^2 = T_1T_2$

(C) $T = T_1 - T_2$ (D) $2T = T_1 - T_2$

42. An irreversible process can be characterized as

(A) one that proceeds such that the system is always in equilibrium with its surroundings

(B) that which cannot be reversed without causing permanent changes in the surroundings

(C) a spontaneous process which proceeds in one direction only

(D) a process after the execution of which if it is reversed the system as well as its surroundings are restored to their initial states

43. A reversible process can be characterized as

(A) one that proceeds such that the system is always in thermodynamic equilibrium with its surroundings

(B) that which cannot be reversed without causing permanent changes in the surroundings

(C) a spontaneous process which proceeds in one direction only

(D) a process which when carried out in the reverse direction restores the system as well as its surroundings to their initial states

44. An engine is working in a cycle with the following three processes occurring sequentially: (i) a reversible adiabatic expansion (ii) reversible isothermal compression and (iii) irreversible adiabatic compression For the engine the

(A) net work is positive and the cycle is reversible

(B) net work is negative and the cycle is reversible

(C) net work is positive and the cycle is irreversible

(D) net work is negative and the cycle is irreversible

45. A frictionless piston–cylinder device contains a mixture of CH_4 and air, initially at pressure a of 2 bar (gauge) and a temperature of 50°C. The mixture expands very slowly and adiabatically until the pressure equals the atmospheric pressure and the temperature equals 30°C. A minute spark ignites the mixture and combustion takes place at constant pressure and heat is transferred out until the temperature reaches 30°C. Then the mixture is compressed to the original pressure and temperature. For the complete sequence of events,

(A) $Q = W$, and the sequence is reversible

(B) $Q = W + \Delta U$ with $\Delta U \neq 0$ and the entire sequence is reversible

(C) $Q = W + \Delta U$ with $\Delta U \neq 0$ and the entire sequence is irreversible

(D) $Q = W$, and the sequence is irreversible

46. A small sealed metallic vessel containing a mixture of fuel and air is immersed in a pool of water at 20°C. The mixture reacts due to a minute spark introduced into the mixture. After a while, the conditions are uniform. It can be assumed that the entire pool continues to be at the same temperature of 20°C. The process undergone by the contents of the vessel is

(A) internally and externally reversible

(B) internally reversible, externally irreversible

(C) internally irreversible, externally reversible

(D) internally and externally irreversible

47. The factors that make a process irreversible are:
    (A) resisted expansion
    (B) heat transfer with finite temperature difference
    (C) slow, frictionless, adiabatic compression
    (D) combustion

48. According to the Clausius inequality statement
    (A) $\oint \frac{\delta Q}{T} \leq 0$ (B) $\oint \frac{\delta Q}{T} \geq 0$
    (C) $\int \frac{\delta Q}{T} \leq 0$ (D) $\int \frac{\delta Q}{T} \geq 0$

49. In general, the entropy change of a pure substance as it undergoes a state change is equal to
    (A) $\int \frac{\delta Q_R}{T}$ (B) $\int \frac{\delta Q}{T}$
    (C) $\int \left[\left(\frac{dh}{T}\right) - \left(\frac{v}{T}\right) dp\right]$ (D) $-\Delta S_{surroundings}$

50. One kilogram of saturated liquid water is heated at a constant pressure of 1.013 bar until it becomes saturated vapour . The increase in entropy is equal to
    (A) $(s_g - s_f)$ at 100°C (B) $\frac{(h_g - h_f) \text{ at 1.013bar}}{373}$
    (C) $\frac{(u_g - u_f)_{\text{at 1.013bar}}}{373}$ (D) $\frac{(h_g - h_f)_{\text{at 100°C}}}{100}$
    where g and f represent saturated vapour and saturated liquid states, respectively.

51. For a system undergoing an irreversible process between two given states
    (A) $\Delta S = \int \frac{\delta Q}{T}$ (B) $\Delta S < \int \frac{\delta Q}{T}$
    (C) $\Delta S > \frac{\delta Q}{T}$ (D) $\Delta S \geq \frac{\delta Q}{T}$

52. For a system undergoing a reversible process, $\int T ds$ is
    (A) equal to heat transferred
    (B) greater than heat transferred
    (C) equal to zero
    (D) less than heat transferred

53. The entropy of a system

(A) can never increase

(B) can never decrease

(C) increases for an irreversible cycle

(D) can be changed by either heat or work interactions

54. When a system undergoes an irreversible process

(A) $\Delta S_{sys} = -\Delta S_{surroundings}$ (B) $\Delta S_{sys} + \Delta S_{surroundings} \leq 0$

(C) $\Delta S_{sys} + \Delta S_{surroundings} \geq 0$ (D) $\Delta S_{sys} + \Delta S_{surroundings} > 0$

55. For a practical heat engine operating between two reservoirs at T_1 K and T_2 K, where $(T_1 > T_2)$.

(A) $\dfrac{Q_1}{T_1} = \dfrac{Q_2}{T_2}$ and $\eta = 1 - \dfrac{T_2}{T_1}$ (B) $\dfrac{Q_1}{T_1} > \dfrac{Q_2}{T_2}$ and $\eta < 1 - \dfrac{T_2}{T_1}$

(C) $\dfrac{Q_1}{T_1} < \dfrac{Q_2}{T_2}$ and $\eta < 1 - \dfrac{T_2}{T_1}$ (D) $\dfrac{Q_1}{T_1} = \dfrac{Q_2}{T_2}$ and $\eta < 1 - \dfrac{T_2}{T_1}$

56. A reversible heat engine is supposed to be operating between two reservoirs at T_1 K and T_2 K at a place where surroundings are at T_0 K. If $(T_1 > T_2)$, and heat transferred to the engine is Q, the available energy of Q is given by

(A) $Q\left(1 - \dfrac{T_2}{T_1}\right)$ (B) $Q\left(1 - \dfrac{T_0}{T_1}\right)$

(C) $Q\dfrac{T_2}{T_1}$ (D) $Q\dfrac{T_0}{T_1}$

57. In the boiler of a steam power plant while the combustion product from the furnace cool from 1000 to 600 K, water is evaporated at 250°C from saturated liquid to saturated vapour. The enthalpy of vapourization is 1716 kJ/kg. During the time when 1 kg water evaporates it is found that the entropy of gas decreases by 2.1810 kJ/K. Given that the surroundings are at 300 K the decrease in available energy because of irreversible heat transfer, per kilogram of water, in kJ is

(A) 654.3 (B) 983.2

(C) 530.0 (D) 330.0

58. If heat is transferred isothermally to superheated steam at 10 bar, its

(A) temperature increases (B) pressure decreases

(C) volume decreases (D) entropy increases

59. On an $h-s$ diagram for a pure substance the constant pressure lines in the mixture region

(A) are slightly curved but parallel

(B) are straight lines and parallel

(C) are straight lines but not parallel

(D) are also the constant temperature lines

60. For a pure substance

(A) the enthalpy of vapourization decreases with increase in pressure

(B) the entropy of saturated vapour decreases with decrease in pressure

(C) the dryness fraction is one in the superheated region

(D) the entropy of saturated liquid decreases with a decrease in pressure

61. Which relation among the alternatives given below (where $p =$ pressure, $v =$ specific volume, $V =$ total volume, $m =$ mass, $M =$ molar mass, R = particular gas constant, $\mathcal{R} =$ universal gas constant and T = temperature in kelvin) does not represent the equation of state for an ideal gas.

(A) $pV = m\mathrm{R}T$, (B) $pv = \left(\frac{\mathcal{R}}{M}\right) T$

(C) $pV = m\mathcal{R}T$ (D) $p(Mv) = \mathcal{R}T$

62. For an ideal gas with constant specific heats identify the wrong statement from among the following alternatives

(A) Difference between the two specific heats is equal to the particular gas constant

(B) Internal energy and enthalpy are functions of temperature only

(C) Ratio of specific heats is equal to 1.4

(D) The temperature does not change when throttled adibatically

63. An ideal gas is throttled steadily from 10 bar, 300°C to 0.1 bar in an adiabatic duct. The temperature after throttling is

(A) = 300°C (B) > 300°C

(C) < 300°C (D) not calculable for want of sufficient data

64. Shown in Figure OQ-7, on T–S coordinates, are four hypothetical end states 2a, 2b, 2c and 2d, of an ideal gas when it is expanded adiabatically and irreversibly in a turbine from state 1 to pressure p_2. The plausible end state is

(A) 2a (B) 2b
(C) 2c (D) 2d

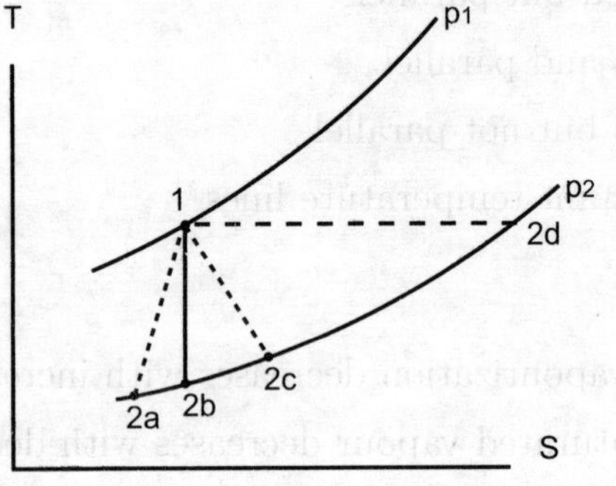

Fig. OQ-7 T–S diagram for Question 64.

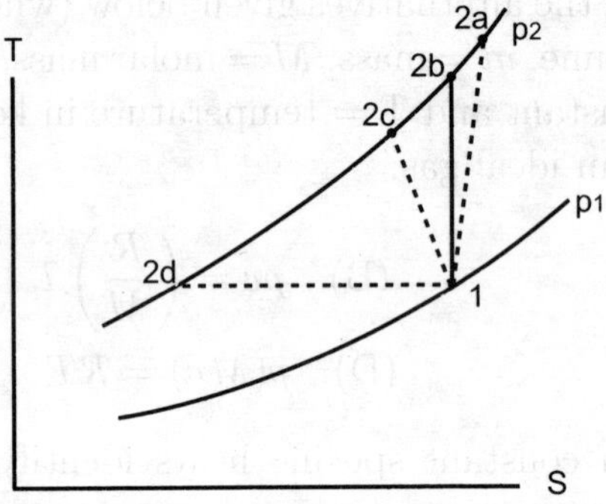

Fig. OQ-8 T–S diagram for Question 65.

65. Shown in Figure OQ-8, on T–S coordinates, are four hypothetical end states 2a, 2b, 2c and 2d, of an ideal gas when it is compressed adiabatically and irreversibly from state 1 to pressure p_2. The plausible end state is

(A) 2a (B) 2b
(C) 2c (D) 2d

66. An ideal gas is compressed irreversibly and adiabatically from state 1 to 2. The properties at the initial and end states are related by

(A) $\frac{T_2}{T_1} = \left[\frac{p_2}{p_1}\right]^{\frac{(\gamma-1)}{\gamma}}$ (B) $\frac{T_2}{T_1} = \left[\frac{p_2}{p_1}\right]^{\frac{\gamma}{(\gamma-1)}}$

(C) $\frac{T_2}{T_1} = \left[\frac{p_2}{p_1}\right]^{(1-\gamma)}$ (D) $\frac{T_2}{T_1} > \left[\frac{p_2}{p_1}\right]^{\frac{(\gamma-1)}{\gamma}}$

67. Consider two identical closed rigid containers A and B. While A contains 32 kg oxygen at 100°C, B contains 28 kg nitrogen at 100°C. If p_A and p_B are the corresponding pressures in the two containers,

(A) $p_A > p_B$ (B) $p_A = p_B$

(C) $p_A < p_B$ (D) p_A and p_B cannot be compared

68. Irrespective of its temperature, the value of the compressibility factor of a real gas at exceedingly low pressures tends to be

(A) zero (B) one

(C) greater than one (D) less than one

69. Irrespective of its temperature, the value of the compressibility factor of a real gas at exceedingly high pressures tends to be

(A) zero (B) one

(C) greater than one (D) less than one

70. The compressibility factor for a van der Waals gas at critical state is

(A) 0.375 (B) 0.5

(C) 0.75 (D) 1.0

Answers to Objective Type Questions I

Q.No.	Ans	Q.No.	Ans	Q.No.	Ans	Q.No.	Ans	Q.No.	Ans
1	T	21	F	41	T	61	F	81	F
2	T	22	T	42	F	62	T	82	T
3	F	23	F	43	T	63	T	83	T
4	F	24	T	44	T	64	T	84	F
5	T	25	T	45	T	65	F	85	F
6	F	26	T	46	F	66	F	86	T
7	T	27	T	47	T	67	T	87	T
8	F	28	T	48	F	68	F	88	F
9	F	29	T	49	F	69	T	89	F
10	T	30	F	50	T	70	F	90	T
11	F	31	F	51	T	71	T	91	F
12	F	32	T	52	F	72	T	92	F
13	F	33	F	53	T	73	F	93	T
14	T	34	T	54	T	74	T	94	T
15	T	35	F	55	F	75	T	95	F
16	F	36	F	56	F	76	T	96	T
17	F	37	F	57	T	77	F	97	T
18	T	38	T	58	F	78	F	98	F
19	F	39	F	59	T	79	T	99	F
20	F	40	T	60	T	80	F	100	T

Answers to Objective Type Questions II

Q.No.	Ans	Q.No.	Ans	Q.No.	Ans	Q.No.	Ans
1	B,C	21	C	41	B	61	C
2	C	22	B,C,D	42	B,C	62	C
3	C	23	A	43	A,D	63	A
4	A	24	D	44	D	64	C
5	D	25	A,B	45	C	65	A
6	B	26	C	46	C	66	D
7	B,C	27	A,C	47	B,D	67	B
8	A,D	28	A	48	A	68	B
9	D	29	B	49	A,C	69	C
10	D	30	A	50	A,B	70	A
11	A,C	31	D	51	C		
12	B	32	B	52	A		
13	A,D	33	A	53	D		
14	B	34	B	54	D		
15	C	35	B	55	C		
16	A,C,D	36	A	56	B		
17	B	37	B,D	57	D		
18	D	38	A,C	58	B,D		
19	A	39	A	59	C,D		
20	A	40	B	60	A,D		

BIBLIOGRAPHY

1. D B Spalding and E H Cole, *Engineering Thermodynamics*, Edward Arnold, London, 1967.

2. Gordon J Van Wylen and Richard E Sonntag, *Fundamentals of Classical Thermodynamics*, 2nd edition, SI Version, Wiley Eastern, 1989.

3. J B Jones and G A Hawkins, *Enginnering Thermodynamics*, 2nd edition, John Wiley and Sons, 1986.

4. P K Nag, *Basic and Applied Thermodynamics*, Tata McGraw-Hill., New Delhi, 2002.

5. Yunus A Cenegal and Michael A Boles, *Thermodynamics*, 2nd edition, Mc Graw-Hill, 1994.

6. Mark W Zemansky and Richard H Dittman, *Heat and Thermodynamics*, 7th edition, Mc Graw-Hill International, 1997.

7. Y V C Rao, *Theory and Problems in Thermodynamics*, Wiley Eastern, 1994.

8. K A Bhaskaran and A.Venkatesh, *Problems in Engineering Thermodynamics*, 2nd reprint, Tata McGraw-Hill, New Delhi, 1982.

9. J F Lee and F W Sears, *Thermodynamics*, Addison Wesley, Cambridge, 1955.

Index